Sabu Abdulhameed • N.S. Pradeep
Shiburaj Sugathan
Editors

Bioresources and Bioprocess in Biotechnology

Volume 1: Status and Strategies for Exploration

Springer

Editors
Sabu Abdulhameed
Inter University Centre for Bioscience, Department of Biotechnology and Microbiology
Kannur University
Kannur, Kerala, India

N.S. Pradeep
Division of Microbiology
Jawaharlal Nehru Tropical Botanic Garden and Research Institute
Thiruvananthapuram, Kerala, India

Shiburaj Sugathan
Division of Microbiology
Jawaharlal Nehru Tropical Botanic Garden and Research Institute
Thiruvananthapuram, Kerala, India

ISBN 978-981-10-3571-5 ISBN 978-981-10-3573-9 (eBook)
DOI 10.1007/978-981-10-3573-9

Library of Congress Control Number: 2016961510

Printed on acid-free paper

This Springer imprint is published by Springer Nature
The registered company is Springer Nature Singapore Pte Ltd.
The registered company address is: 152 Beach Road, #21-01/04 Gateway East, Singapore 189721, Singapore

Dedicated to Prof. M. S. Swaminathan for his pioneering scientific innovations to fill granaries and hunger-struck stomachs

Prof. M. S. Swaminathan

Foreword

It is a legal obligation for the parties to the Convention on Biological Diversity to evolve effective strategies and action plan for the conservation, enhancement and sustainable utilization of biodiversity, especially taking into consideration of benefitting such efforts to the livelihood means and services to local communities. Conservation and documentation of "high-value" wild plant diversity, particularly rare, endemic and threatened (RET) plant species, medicinal plants of ethnobotanical applications and agrobiodiversity and multiplicity of habitats and landscapes which provide are the significant areas of biodiversity action.

In addition to the CBD obligations, the national and state governments also have to work towards achieving the commitments under six other biodiversity-related conventions such as (i) CITES, (ii) CMS, (iii) the Treaty, (iv) the Ramsar Convention, (v) the Convention on World Cultural and Natural Heritage and (vi) the International Plant Protection Convention. In order to address the complete spectrum of biodiversity management, effective coordination and cooperation from all the concerned institutions and policy makers is required. Besides these conventions, the 2030 Sustainable Development Goals and the Paris Climate Action also have implications on the biodiversity management of any state.

So, it is important for any democratically elected government to formulate appropriate policies and practices that help the utilization of biodiversity and ecosystem services for developing in a sustainable and inclusive manner. The first and foremost action in this regard is to work towards achieving the Aichi Biodiversity Target Number 1, which aims to achieve the goal that "all people become aware of the values of biodiversity". Awareness level of the importance and implications of biodiversity and ecosystem services among the public as well as policy makers and the practitioners is very important for safeguarding and protecting in optimum and in a balanced way. The power of knowledge will help these actors to go for effective lobbying, advocacy and actions in biodiversity management.

The book entitled "Bioresources and Bioprocess in Biotechnology (Volume I: Status and Strategies for Exploration)" edited by Dr. Sabu A., Dr. N.S. Pradeep and Dr. Shiburaj S. is aimed towards this direction. The book chapters are contributed by eminent researchers in the broader area of sustainable biodiversity management. The volume comprises literature on the current status of biodiversity; biodiversity education; documentation, conservation and preservation; biodiversity law; and many other important topics related to biodiversity and its sustainable utilization.

Because of the importance of the topics covered, I think this volume will have wider acceptance and open up avenues for scientists to engage in consistent dialogue with the policy makers at different levels – state, local, national and global. It can also lead to a new alliance between scientific, governmental and commercial forces and find effective ways for lobbying, advocacy and innovative actions in sustainable biodiversity management.

In my opinion, the primary audience who will benefit from this book would be the research and teaching professionals, practitioners, policy makers, farmers/fishers and farmer/fisher representatives like NGO professionals, who work in the area of sustainable agricultural and rural development. This kind of a publication will be highly useful for those planners and practitioners in the area of biodiversity for climate adaptation.

Nadesapanicker Anil Kumar
Director, Biodiversity,
M. S. Swaminathan Research Foundation,
Kalpetta, 673577, Kerala, India

September 29, 2016

Preface

The evolution in diversity of life on earth is capable of transforming the world. Biodiversity and biological data thereof is increasingly being analyzed and integrated to enhance and revolutionize the researches and expand the knowledge of biodiversity. Biological diversity is the variety of the world's organisms, including their genetic diversity. All life on earth is part of one great, interdependent system. It interacts with and depends on the nonliving components of the planet such as atmosphere, oceans, freshwaters, rocks, and soils. Humanity depends totally on this community of life, the biosphere of which human populations are an integral part.

Bioresources are important components for progress and economic activities of any nation. But bioresources management and utilization for human welfare are very important for the optimum utilization of the bioresources. It includes genetic resources, organisms or parts thereof, populations, or any other biotic component of ecosystems with actual or potential use or value for humanity. Application of the knowledge of biodiversity and bioresources in an economic and sustainable manner will add value to human lives. Biotechnology is a technique that can be applied to study genetic diversity in crop plants, domesticated animals, industrially important microbes, bio-waste recycling, propagation and mass multiplication of threatened genetic resources, large-scale production of therapeutically important substances, genetic transformation, cryopreservation, DNA banking, etc. It involves manipulation of genetic information of a particular bioresource and the application, production, maintenance, conservation, security, and various other aspects of components of biodiversity.

Despite some conservation success especially at local scales, and increasing public and government interest, biodiversity continues to decline and is being lost as on today more rapidly than at any time in the past several million years. The current losses to biodiversity can be attributed to direct causes including habitat loss and fragmentation, invasion of introduced species, overexploitation of living resources, and modern agriculture and forestry practices.

Sustainable management of the ecosystems and the rich life within them remains one of the key natural resource management challenges. The conservation and sustainable use of biological diversity are of critical importance for meeting the need of food, fodder, fiber, health, water, and other needs of the growing world population for which purpose of, access to, and sharing of both genetic resources and technologies are essential. Here we are trying to document together the various aspects of biodiversity with a view to make it available for the judicious utilization by mankind.

Kerala, India

Sabu Abdulhameed
N.S. Pradeep
Shiburaj Sugathan

Contents

Contributors

Ananda Baskaran Venkatachalam Atlantic Centre for Transplantation Research, Dalhousie University, Halifax, Canada

Nadesapanicker Anil Kumar Community Agro Biodiversity Centre, M. S. Swaminathan Research Foundation, Puthoorvayal, Kalpetta, Wayanad, Kerala, India

R. Aswati Nair Department of Biochemistry and Molecular Biology, Central University of Kerala, Kasaragod, Kerala, India

School of Biotechnology, National Institute of Technology Calicut (NITC), Calicut, Kerala, India

V. Balakrishnan Community Agro Biodiversity Centre, M. S. Swaminathan Research Foundation, Puthoorvayal, Kalpetta, Wayanad, Kerala, India

Shalini Bhutani Kalpavriksh Environment Action Group, Pune, Maharashtra, India

A. Biju Kumar Department of Aquatic Biology and Fisheries, University of Kerala, Thiruvananthapuram, Kerala, India

Lekshmi K. Edison Division of Microbiology, Jawaharlal Nehru Tropical Botanic Garden and Research Institute, Thiruvananthapuram, Kerala, India

Madathilkovilakathu Haridas Inter University Centre for Bioscience, Department of Biotechnology and Microbiology, Kannur University, Kannur, Kerala, India

K.S. Jayakumar Biotechnology and Bioinformatics Division, Jawaharlal Nehru Tropical Botanic Garden and Research Institute, Thiruvananthapuram, Kerala, India

Kanchi Kohli Kalpavriksh Environment Action Group, Pune, Maharashtra, India

F. Nadiya Biotechnology and Bioinformatics Division, Jawaharlal Nehru Tropical Botanic Garden and Research Institute, Thiruvananthapuram, Kerala, India

Indu C. Nair Department of Biotechnology, SASSNDP Yogam College, Pathanamthitta, Kerala, India

Nandu Thrithamarassery Gangadharan Division of Microbiology, Jawaharlal Nehru Tropical Botanic Garden and Research Institute, Thiruvananthapuram, Kerala, India

Anoop Narayanan Department of Biological Sciences, Purdue University, West Lafayette, IN, USA

P. Padmesh Pillai Department of Genomic Science, Central University of Kerala, Kasaragod, Kerala, India

Shunmugiah Karutha Pandian Department of Biotechnology, Alagappa University, Karaikudi, Tamil Nadu, India

N.S. Pradeep Division of Microbiology, Jawaharlal Nehru Tropical Botanic Garden and Research Institute, Thiruvananthapuram, Kerala, India

S. Pradeep Kumar Kerala State Council for Science Technology and Environment, Thiruvananthapuram, Kerala, India

Parameswaran Prajeesh Community Agro Biodiversity Centre, M. S. Swaminathan Research Foundation, Puthoorvayal, Kalpetta, Wayanad, Kerala, India

P.E. Rajasekharan Division of Plant Genetic Resources, ICAR-Indian Institute of Horticultural Research, Bangalore, India

M.K. Ratheesh Narayanan Department of Botany, Payyanur College, Payyanur, Kerala, India

R. Ravinesh Department of Aquatic Biology and Fisheries, University of Kerala, Thiruvananthapuram, Kerala, India

Sabu Abdulhameed Inter University Centre for Bioscience, Department of Biotechnology and Microbiology, Kannur University, Kannur, Kerala, India

K.K. Sabu Biotechnology and Bioinformatics Division, Jawaharlal Nehru Tropical Botanic Garden and Research Institute, Thiruvananthapuram, Kerala, India

P.K. Satheeshkumar Division of Microbiology, Jawaharlal Nehru Tropical Botanic Garden and Research Institute, Thiruvananthapuram, Kerala, India

G. Seghal Kiran Department of Food Science and Technology, Pondicherry University, Puducherry, India

Shiburaj Sugathan Division of Microbiology, Jawaharlal Nehru Tropical Botanic Garden and Research Institute, Thiruvananthapuram, Kerala, India

A.R. Sivu Department of Botany, NSS College, Nilamel, Kollam, Kerala, India

K. Souravi Division of Plant Genetic Resources, ICAR-Indian Institute of Horticultural Research, Bangalore, India

T.S. Swapna Department of Botany, University of Kerala, Thiruvananthapuram Kerala, India

C.P. Unnikannan Weizmann Institute of Science, Rehovot, Israel

Gayathri Valsala Division of Microbiology, Jawaharlal Nehru Tropical Botanic Garden and Research Institute, Thiruvananthapuram, Kerala, India

Chethala N. Vishnuprasad School of Life Sciences, Institute of Transdisciplinary Health Science and Technology (TDU), Bangalore, India

Dharmaprakash Viszwapriya Department of Biotechnology, Alagappa University, Karaikudi, Tamil Nadu, India

Nimisha Vijayan Inter University Centre for Bioscience, Department of Biotechnology and Microbiology, Kannur University, Kannur, Kerala, India

About the Editors

Dr. Sabu Abdulhameed is teaching biotechnology at the Department of Biotechnology and Microbiology, School of Life Sciences, Kannur University, Kerala, India. He earned his PhD in biotechnology from Cochin University of Science and Technology (CUSAT), India. Prior to joining Kannur University, he was a scientist at the National Institute for Interdisciplinary Science and Technology (CSIR). Dr. Sabu has published several research papers in reputed Indian and international journals and owns three patents to his credit. He has already published three other books on various aspects of biotechnology. He was a visiting fellow at the University of Georgia, USA; Technical University of Budapest, Hungary; University of Debrecen, Hungary; University of Paul Cezanne, France; University of the Mediterranean, Marseille, France; and Autonomous University of Coahuila, Mexico. His current research interest includes production, purification, and characterization of industrial enzymes and characterization of bioactive molecules. Dr. Sabu also serves as a member in the editorial board of many research journals and is an expert member in many academic and scientific bodies (email: drsabu@gmail.com).

Dr. N.S. Pradeep is working as a senior scientist and head, Division of Microbiology, Jawaharlal Nehru Tropical Botanic Garden and Research Institute, Palode, Thiruvananthapuram, Kerala, India. He obtained his PhD from Gandhigram University, Madurai, India. He has more than 20 years of research experience in the area of microbial biotechnology and molecular taxonomy. Dr. Pradeep published several research papers in national and international journals and has a patent to his credit. He is the author/editor of several books and life member of the Association of Microbiologists of India, Mushroom Society of India, Indian Mycological Society, Indian Science Congress Association, Kerala Academy of Sciences, etc. Dr. Pradeep is actively involved in science extension and training, and he is a research committee member of the Integrated Rural Technology Centre, Kerala, India (email: drnspradeep@gmail.com).

Dr. Shiburaj Sugathan is currently working as senior scientist at the Division of Microbiology, Jawaharlal Nehru Tropical Botanic Garden and Research Institute, Palode, Thiruvananthapuram, Kerala, India. He has more than 20 years of experience in the area of microbial biotechnology and molecular microbiology and has published several papers in reputed national and international journals. He obtained his doctoral degree from the University of Kerala, India, and did his postdoctoral studies with the fellowship from Kerala Biotechnology Commission at Genetic Engineering Research Unit, School of Biotechnology, MKU, under the mentorship of Prof. K. Dharmalingam from 2004 to 2007. He was awarded the BOYCAST fellowship of the Department of Science and Technology, Government of India, and worked at Tom MacRae's lab, Dalhousie University, Halifax, Nova Scotia, Canada, from 2011 to 2012. He is a life member of the Association of Microbiologists of India, Proteomics Society of India, and Indian Mycological Society and member of the Asia-Pacific Chemical, Biological and Environmental Engineering Society (email: drshiburaj@gmail.com).

Part I

Bioresources

Bioresources: Current Status

1

N.S. Pradeep, Shiburaj Sugathan, and Sabu Abdulhameed

Abstract

The continued existence of mankind relies upon the accessibility of bioresources and their proper management. Overexploitation of bioresources leads to many environmental problems that the humanity is facing today. Destruction of vegetation has resulted in land degradation, denudation, soil erosion, landslides, floods, drought, and distorted ecosystem processes. Conventional resource management systems are considered as balanced system and often guaranteed reasonable sharing of benefits from natural resources. But people in most parts of the world have already forgotten the real values of bioresources and its conservation. It should be determined to conserve and sustainably utilize biological diversity for the benefit of present and future generations. This chapter briefly explains the knowledge of current status of bioresources associated with its management and utilization.

Keywords

Biodiversity • Ecosystem • Sustainable life • Bioprospecting

N.S. Pradeep (✉) • S. Sugathan
Division of Microbiology, Jawaharlal Nehru Tropical Botanic Garden and Research Institute, Palode, Thiruvananthapuram, Kerala 695562, India
e-mail: drnspradeep@gmail.com

S. Abdulhameed
Inter University Centre for Bioscience, Department of Biotechnology and Microbiology, Kannur University, Thalassery Campus, Kannur, Kerala 670661, India

S. Abdulhameed et al. (eds.), *Bioresources and Bioprocess in Biotechnology*,
DOI 10.1007/978-981-10-3573-9_1

1.1 Introduction

Bioresources or biological resources are biodiversity goods and include all products and services originating from natural environment that fulfill the requirements of humans, occurring at a frightening state due to the consequence of increasing population pressure, agricultural land degradation, and urbanization. According to the Biological Diversity Act 2002, "bioresources means plants, animals and microorganisms or parts thereof, their genetic material and by-products (excluding value added products) with actual or potential use or value, but does not include human genetic material." The availability of bioresources and their proper management are directly depended on the continued existence of mankind. Human race has been experimented with various sources of energy ranging from wood, coal, oil, and petroleum along with nuclear power through the different stages of development. Today humanity is facing various environmental problems due to the overexploitation of bioresources by the exploding population (Tynsong et al. 2012).

Energy considered as a fundamental part of a society plays a crucial role in the development of socioeconomic status by nurturing the quality and standard of living of life. Consumption pattern of energy is directly related with economic development of a particular region. As the economy grows, the energy demands also increase, creating a change in the consumption pattern and causing conversion loss and end-use efficiency. The countless energy sources like wood, coal, oil, petroleum, and nuclear power have trailed with the different developmental stages of mankind. However, unintentional developmental activities and promiscuous exploitation of resources cause serious environmental and ecological problems and create devastating changes in bioresources also (Ramachandra et al. 2004).

Drivers that cause to create changes in bioresources are *direct drivers* (local land use; species introduction or removal; technology adaptation and use; external inputs such as fertilizer use, pest control, and irrigation; harvest and resource consumption; climate change evolution; and volcanoes) and *indirect drivers* (demographic, economic, sociopolitical, cultural and religious, science and technology). Changes in indirect drivers such as population, technology, and lifestyle directly lead to changes in direct drivers. For example, the hunting of animals or the application of fertilizers for increasing food production causes changes in biodiversity and ecosystem services that in turn affect human well-being. In the past 50 years, changes in bioresources due to human activities were more fast, and the drivers of change that thrash biodiversity and ecosystem services are still steady, showing no data of declining over time.

1.2 Biodiversity and Its Link to Bioresources

Biodiversity takes part an important role in supporting and regulating bioresources. The Millennium Ecosystem Assessment of UNEP recognizes that biodiversity and bioresources, otherwise known as ecosystem services, provide a key active and constitutive part determining human well-being. These are essential for human well-being because biodiversity loss and deteriorating bioresources directly or indirectly cause worsening health, higher food insecurity, lower material wealth, increasing vulnerability, worsening social relations, and less liberty for choice and action. However, presently there are few studies that demonstrate the links between biodiversity and bioresources. Studies indicate that biodiversity that affects bioresources in turn changes human well-being. Advanced work that reveals the links between biodiversity and bioresources and also human well-being is vital but often difficult in illustrating the value of biodiversity.

Bioresources, at a given time, are strongly controlled by the ecological characteristics of the most abundant species, not by the number of species. The traits and its relative abundance of species determine its relative significance to ecosystem functioning. For example, the individuality of most abundant plant species such as their life, size, rate of assimilation of carbon and nutrients, mechanism of leaf decomposition, wood thickness, etc., usually drives ecosystem's processing of matter and energy. Accordingly protecting or renovating the composition of biological communities, rather than maximizing species numbers, is important to maintain bioresources. Loss of ecosystem functioning due to local extinctions by direct or indirect biodiversity alterations (species loss from a local area) can have remarkable impact on bioresources. However, local extinctions have received little attention compared with global extinctions.

Changes in biotic interactions, both direct (predation, parasitism, or facilitation) and indirect (predator preying on a dominant competitor such that the dominant is suppressed, which permits subordinate species to coexist), cause important consequences in bioresources. Exclusion or introduction of organisms in ecosystems that disrupt biotic interactions or ecosystem processes also leads to changes in bioresources. The impacts of either the removal of existing species or the inclusion of new species are difficult to anticipate because the network of interactions among species and the network of connections among bioresources are complex. Loss of individual species implicated in key interactions can also influence ecosystem processes and bioresources. For example, coral reefs, one of the most species-rich communities on earth, are directly dependent on a single key biotic interaction: symbiosis with algae. Coral reefs provided ecosystem services such as habitat construction, spawning places, and nurseries for fish, carbon and nitrogen fixing in nutrient-poor environments and nutrient cycling, and wave buffering and sediment stabilization. The disruptions of the symbiotic relation dramatically affect the climate change and variability on coral reefs.

Many rural communities used biological resources directly as an insurance and surviving mechanism. This is a biological "safety net" that has increased the security and flexibility of some local communities. The economic rights of poor are becoming dangerously threatened due to fluctuating product prices in the world. This can be overcome by the availability of an ecosystem-based food security, which provides an important insurance program. Survival of indigenous plants is most important for vulnerable people, who have little access to land, employment, or market opportunity. Maintaining agricultural biodiversity is an option for improving food security. Practices with wild relatives of domestic crops give genetic variability that is crucial for overcoming occurrences of pests and pathogens and also the environmental stresses. Increasing local diversity is a critical factor for the long-term productivity and viability of agricultural systems. Natural disasters increase human suffering, and economic losses can be prevented by the proper management of biodiversity. Rich sources of biodiversity like mangrove forests and coral reefs are excellent natural barriers against natural disasters like floods and storms. Their loss or reduction has increased the intensity of flooding on coastal communities.

Various sub-global assessments indicated that lots of people living in rural areas showed their interest in surviving ecosystem variability and bioresources as a part of risk management strategy. They maintain a diversity of bioresources as a solution to reduce their alternatives. According to these assessments, the diversity of species, food, and landscapes serves as "savings bank" for rural communities to survive with alteration and guarantee sustainable livelihoods.

Wild sources of food that provide a somewhat balanced diet are important for the poor and landless people. About 7000 species of plants and several hundred species of animals have been used for human food consumption. Some native and traditional communities currently consume 200 or more species. Worldwide overexploitation of marine fisheries has led to a decrease in the availability of wild fish protein, causing serious human health issues in many countries. Risk of exposure to many infectious diseases may depend on the preservation of biodiversity in natural ecosystems. So many evidences are accumulating to support that greater wildlife diversity may decrease the spread of many wildlife pathogens to humans.

In developing countries wood fuel provides more than half the energy used. In rural areas 95 % of energy is consumed in the form of firewood, while in urban areas 85 % is used in the form of charcoal. Areas with high population density may suffer with shortage of wood fuel without any access to alternative and inexpensive energy sources. People in that area are exposed to illness and malnutrition due to the lack of resources to warm homes, cook food, and boil water. The continued destruction of forests and watersheds diminishes the quality and availability of water delivered for agriculture and household use. For the clean drinking water to be available, it is necessary to protect the ecosystem than construct expensive buildings and operating water filtration plant. Also the loss or damage of ecosystem components may harm social relations because spiritual and religious values of many cultures are attached with ecosystems or their components like tree, hill, river, or groves.

Biodiversity loss in turn leads to the loss of bioresources and also means a loss of choices. For example, fishermen in local areas rely on mangroves as breeding grounds for local fish populations. So the destruction of mangroves causes loss of local fish stock and it affects their income for many generations. In some occasions the loss of biodiversity is irreversible, and the value placed on biodiversity for future generations can be significant. However, placing an economic figure on future values is very much difficult.

1.3 Bioresources for Sustainable Living

Biodiversity directly offers bioresources such as plants, animals, bacteria, and fungi that are needed by individuals in order to earn income and protect sustainable livelihoods. It also supplies a source of revenue through supporting the ecosystem services. For example, apple cultivation in the Himalayan region in India provided around 60–80 % of total household income. Honeybee diversity is also rich in this region and has an important role in pollinating crops and plants, and in turn this leads to the high productivity and ecosystem sustainability. In the early 1980s, market demand for particular types of apples led farmers to uproot pollinated varieties and plant new, sterile cultivars. These steps negatively affected pollinator populations by extreme use of pesticides. The consequence was a reduction in apple productivity and also the extinction of a lot of natural pollinator species.

Biodiversity also donates to a sort of other industries, including pharmaceuticals, cosmetics, and horticulture. Market movements widely fluctuate according to the industry and the country involved, but several bioprospecting activities and subsequent generation of revenue are expected to enhance over the next decades. The present economic trends suggest that pharmaceutical bioprospecting will increase, especially as new methods use evolutionary and ecological knowledge. Biodiversity loss can impose extensive costs at local and national scales. Current indications recommend that the preservation of the uprightness of local biological communities, both in expressions of the identity and the number of species, is vital for the protection of plant and animal productivity, soil fertility, and their steadiness in facing the altering environment.

The marine microbial community offers critical detoxification services. But very little information is available on how many species are participating in detoxification services; however, these services may significantly depend on one or a few species. Ecosystem services such as water filtration, reduction of eutrophication effects, etc., are provided by some marine organisms. For example, American oysters are used in Chesapeake Bay for filtering water as part of ecosystem services and proved that Chesapeake has much clearer water. A number of marine microbes can degrade toxic hydrocarbons into carbon and water due to oil spill, but this process requires oxygen. Consequently it is threatened by nutrient pollution, which produces oxygen scarcity.

Conservation of biodiversity is important as a supply of meticulous bioresources, for maintaining diverse ecosystem services, for sustaining the flexibility of ecosystems, and also for providing alternatives for next generations. The benefits that are provided by biodiversity to mankind have not been a sign of resource management; hence, the current rate of loss of biodiversity is higher.

The connection between biodiversity and bioresources depends on composition, comparative abundance, functional diversity, and taxonomic diversity. If various dimensions of biodiversity are driven to very low levels, particularly trophic or functional diversity within an ecosystem, both the level and stability of supportive services may decrease.

1.4 Value of Bioresources

Bioresources are represented as "natural capital assets" of a country. But the benefits that are attained from bioresources through better management are poorly discovered. Bioresources have great economic importance; habitually most of their values are not captured by the market. Therefore, the potential of bioresources are never estimated properly. Actually this misjudging is considered as an important factor because of rapid exhaustion of biodiversity and loss of territories and species. Reduction of these "natural capital assets" significantly declines the national wealth of countries with economies that are especially dependent on natural resources. The richness resulting from these ecosystem "surprises" can be extremely high. The majority of our biodiversity is on common land so its property rights are not obviously defined. Consequently, at the collection point, the bioresources derived from biodiversity do not expose its real "value."

Bioresources guarantee provision of energy to the poor and weaker groups. Bioresource utilization is still growing because in many areas, there is no practical alternative to biomass fuels for the poorest regions of the population. Dependence on biomass for energy is responsible for the unending deforestation and loss of vegetation. The production of biomass in all its forms for fuel, food, and fodder demands cautious planning of bioresources for ensuring environmentally sustainable land use (Ramachandran 2007). The costs and risks associated with biodiversity loss are expected to increase and it strangely affected the poor. The marginal value of biodiversity increases as the biodiversity and bioresources decrease. In rural areas, dependence on bioresource to meet the daily requirement is more than 85 %, due to the prevailing use of fuel wood for cooking, while about 35 % demand is in urban areas. The urban populations basically depend on commercial energy sources than the rural; they are primarily dependent on noncommercial energy sources such as fuel wood, cow dung, etc. High persisting use of biomass in rural areas also increases the dependence on fuel wood. The use of fuel wood in cooking has been indicated in rural systems increasing from 56 % in 1989/1990 to around 62 % in 1994/1995 (TERI 2001–2002). Different phases of bioresource reduction have an uneven impact on poor people. For example, the decline in fish populations mainly affects the communities that depend on fish as an important food. Likewise, the poor and vulnerable population suffers utmost due to the spoiling of dry land resources.

Now so many tools exist for computing the values of bioresources, despite only stipulating ones are regularly valued. Some are very hard to value, due to the lack of knowledge regarding the full costs, risks, and benefits. In general economists are looking forward to recognize the various explanations why bioresources are valuable to people. It supports the fact that bioresources directly or indirectly support people's own consumption. Various valuation methods are now available to estimate these different sources.

Many bioresources do not have markets or readily noticeable prices though lack of a price does not mean lack of value. A substantial body of research on nonmarket valuation is now available for some ecosystem services like clean drinking water, recreation, or commercially harvested species. The existence value of species and other "nonuse" values cause a greater challenge for measuring the complete value of conserving biodiversity and natural processes because ecosystems are dynamic and complex and also during the course of time the human preference may change which creates difficulties for trying to value the natural systems. It is clear in theory that getting realistic estimates of alternative value is difficult in practice. Better quantification of the benefits derived from ecosystems would provide greater movement for biodiversity protection and create a clearer picture of the equitability of the distribution of benefits.

The value related with private use of bioresource by individuals will typically ignore the "external" payback of conservation. For example, a farmer may benefit from rigorous use of the land but usually does not bear all the consequences caused by leaching of excess nutrients and pesticides into ground or surface water or the consequences of loss of habitat for native species.

The indirect value of bioresource conservation can be highly significant in contrast with the direct economic values derived from a particular area. The economic studies of changes to bioresources in specific locations such as mangrove forests, coral reefs, etc., and the costs of bioresource conversion are often found to be significant and sometimes exceed the benefits of the habitat conversion. The exhaustion and degradation of many bioresources represent the loss of a capital asset that is defectively reflected in conventional indicators of economic growth as well as growth in human well-being. These natural capital assets significantly change the balance sheet for economically developed countries which largely dependent on natural resources. Factors like edaphic, meteorological, geographic, agro-climatic conditions, socioeconomic status of the people, etc., are highly influencing the availability and accessibility of diversified bioresources.

1.5 Current Status of Bioresources

Freshwater Global freshwater utilization was estimated to expand 10% from 2000 to 2010; these rates reveal population growth, economic development, and changes in water use efficiency. Global water withdrawal is about 3600 cubic kilometers per year or 25% of the continental overflow to which the majority of the population has access during the year. If uses for navigation, waste processing, and habitat management are considered, humans use and regulate over 40% of renewable available supplies. Differential development pressures due to regional variations and efficiency changes during 1960–2000 increased water use of 15–32 % per decade. Areas for the largest amounts of renewable freshwater supply are forest and mountain ecosystems, having 57% and 28% of total runoff, respectively. These ecosystems offer renewable water supplies to two thirds of the global population. Sixteen percent and 0.2% of global runoff water is generated by cultivated and urban ecosystems, respectively, but it is associated with nutrient and industrial water pollutants because of its closeness to human settlements. From 5% to possibly 25% of global freshwater use exceeds long-term reachable supply. Much of this water is used for irrigation with severe losses in water-scarce regions. The water requirements for aquatic ecosystems and crop protection result in competition for the water resources. A number of the key consequences of this competition include transport of sediments, modification of habitat, chemical pollutants, and disruption of migration routes of aquatic biota. The struggle for freshwater in some areas in the world to completely expand the coastal zone results in oxygen depletion, coastal erosion, and harmful algal blooms.

The supply of freshwater can reduce severe pollutions due to anthropogenic activities in many parts of the world. More than the past half century, there has been a faster discharge of artificial chemicals into the environment. Inorganic nitrogen pollution of inland waterways, for example, has increased substantially. It is twofold in the preindustrial state and increases of more than tenfold occur in industrialized regions of the world. Most anthropogenic chemical substances are long-lived and changed into by-products whose characteristics are yet unknown. As a result of pollution, the capacity of the ecosystem to provide clean and consistent sources of freshwater is spoiled. Large numbers of dam building have created both positive and negative effects. Positive effects include stream stabilization for irrigation, control of flood, drinking water, and hydroelectricity. Negative effects include fragmentation and destruction of habitat, loss of species, stagnant water-related health issues, and loss of sediments and nutrients that support coastal ecosystems and fisheries.

Water shortage is a globally significant condition for 1–2 billion people worldwide, primarily causing problems in food production, human health, and economic development. The global increase in water insufficiency from 1960 to present measures to nearly 20 % per decade, with individual continents having values of 15 % to more than 30 % per decade. Unequal level of economic growth, education, and governance results in differences in managing the capacity for water scarcity. Diseases from inadequate water, sanitation, and hygiene cause a total of 1.7 million deaths and the loss of 50 million healthy lives annually. At present 1.1 billion people suffer from lack of safe drinking water and 2.6 billion lack basic sanitation. Human health and economic productivity can be improved by making some investments in drinking water supply and sanitation. Every day each person needs 20 to 50 liters of pure water for drinking and personal hygiene to survive. But the present state of freshwater resources is insufficiently monitored. New techniques are available for evaluation of impurities, but its universal application is lacking, and there are no regular epidemiological studies to recognize their impact on human well-being. The Convention on Biological Diversity has established ecosystem-based approaches; the Convention on Wetlands, the Food and Agriculture Organization, and others could significantly develop future condition of water-provisioning services for balancing economic development.

Food Over the past 42 years, global food production has increased by 168. The cereal production has increased by about 130%, but now it is growing more slowly. However, an estimated 852 million people were undernourished in 2000–2002; it was 37 million from the period 1997–1999, and of this around 96% live in developing countries. In most areas of the world, rising incomes, urbanization, and changing utilization patterns have increased per capita food consumption. Important drivers of food provision are food preferences which arise from cultural differences. As incomes increase, demands for high-value products such as livestock and fish also increased, but cereals remain as the major single component of global diets and to occupy the most areas of cultivated land. A diet with sufficient protein, fats, micronutrients, and other dietary factors is very important for well-being. Normal daily calorie intake has declined recently in the poorest countries. The world's poorest depend on starchy foods for energy, which leads to considerable protein, vitamin, and mineral deficiencies. Another health problem is overconsumption. Consumption of fruits and vegetables improves nutritional status and children's growth rates.

At present many countries face epidemic of diet-related obesity and noncommunicable disease, mainly among urbanized people where they have adopted diets with higher energy than traditional diets with diverse fruits and vegetables. Some other people suffer from diseases due to contaminated food. The risk of foodborne illnesses is increasing when the people eat more perishable foods such as meat, milk, fish, and eggs. Health threats from food differ by climate, income, diet, and public infrastructure. Improperly prepared or inadequately refrigerated animal foods cause health hazards. Microbial contamination is of particular concern in developing countries. Other health concerns related with food production are diseases transmitted from animals to humans, animal wastes with toxins, overuse of antibiotics in livestock production, etc. Poor people have less capacity to purchase food by spending money, so they completely depend on local food production for eliminating hunger and also promoting rural development. Food-insecure people in developing regions are increasing due to underdeveloped market infrastructures. In these areas, local food production has the main advantage for eradicating hunger and providing insurance against increasing food prices.

Maintaining a focus on increasing food production systems will be a main concern for both global food security and environmental sustainability. Government policies are also significant drivers of food production and consumption models. Many investments in rural roads, irrigation, and agricultural research serve to encourage food production. The increasing demand for livestock products is increasingly being met by industrial production systems, especially for chicken and swine mainly in Asia. This has contributed large increases in production. However, it creates serious waste problems and also high pressure on cultivated systems which provide feed inputs, which in turn cause demand for water and nitrogen fertilizer. Annual fish consumption is increasing, but total fish production has declined to some extent in industrial countries. Global fishing has been decreased since the late 1980s due to overexploitation. In 2002 the volume of fish consumption was 27 % of all fish harvested and 40 % of the total amount of fish products consumed as food.

Along with fish, wild plants and animals are also chief sources of nutrition in some diets, and some wild foods also have significant economic value. However, the importance of wild food is improperly understood and is excluded from economic analysis of natural resource systems as well as official statistics. Plants and animals are under pressure on unsustainable levels of harvesting, and there is a need for conservation of wild food resources to satisfy the nutritional needs of people who do not have access to agricultural land or resources.

Timber, Fiber, and Fuel Last four decades, global timber harvest has increased by 60%. The growth rate has slowed in recent years and is expected to grow more slowly in the near future. The term "timber" is used for denoting trees and their direct products. Pulp production from harvested wood was increased threefold since 1961. Timber provides common industrial wood products, but there will be scarcities of high-value species and best quality woods due to past overharvesting. In the future timber production from forest and agricultural areas will increase. Top quality woods from large and old trees of extremely valuable species are limited in most regions. They can be renovated through protection and sustainable forest management. In 2000, 5% of the global forest was covered by plantations, but they provided 35% of harvested round wood. It is predictable to increase 44% by 2020. Major changes in timber production resulted from a combination of economic stress, globalization, and changing national policies. The main issues associated with forest goods are the thinning and mining of plantations (Karnataka Forest Department Report 2001). International business in forest goods has increased much faster than the increase in production. Between 1961 and 2000, global exports of timbers were increased approximately 25-fold. International moves toward forest certification and sustainable forest management have prolonged rapidly in recent years. To date they have been used in industrially developed countries and developing countries. The global forestry sector annually offers survival and employment of 60 million people with 80 % residing in the developing world. Forest land in the industrial world is owned and managed privately or publicly. Most of the forest in the developing world is a public resource. These ownership factors create very different and diverse opportunities for people to interact with and benefit from forest products and services. Labor necessities per unit of production will continue to shrink in all regions due to technological change. Up to 15 % of global timber trade involves illegal activities, and this problem will require a major effort by both government and private industries. Fifty-five percent of global wood utilization is for fuel wood, and it is the primary source of energy for heating and cooking to 2.6 billion people. Global utilization of fuel wood emerged to have peaked in the 1990s and it is now supposed to be slowly declining. More precise data on fuel wood manufacture and utilization are hard to collect. The burning of fuel wood without suitable smoke venting generates significant health risks, and due to the shortage of woods, poor families may use dung or agricultural residues for heating and cooking. The consumption of cooked food may decline when adequate fuel is not available leading to unfavorable effects on nutrition and health. Through application of a variety of modern industrial processes, wood and forest biomass, manure, agricultural crops and residues, municipal and industrial wastes, and various other nonfossil organic resources can generate renewable energy and fuel. These technologies are being rapidly developed throughout the world. Biomass-based energy production, expected to expand slowly in the future, begins to affect energy economics because of shortages and supply disruptions.

Since 1961, global cotton production has doubled and silk production has tripled, and production of other agricultural fibers such as wool, flax, hemp, jute, and sisal has declined. Many species of wild mammals, reptiles, birds, and fishes are becoming endangered because they are frequently poached for their skins, furs, wools, and hairs that are traded in the international market to make products ranging from clothing and accessories such as footwear, shawls, and wallets to ornaments and furnishings.

Bioprospecting

Bioprospecting is the term used for the exploration of biodiversity for new biological resources of social and economic value, mainly carried out by industries that comprise pharmaceuticals, crop protection, botanical medicines, cosmetics, agricultural seeds, horticulture, environmental monitoring, and construction. Earth covers around 5 million to 30 million species, containing thousands of genes. But only less than 2 million species have been illustrated and the knowledge on the global distribution of species is limited. Less than 1% of species have provided the fundamental resources for the improvement of all civilizations. It is expected that the application of advanced technologies to the exploration of the unidentified species will yield more benefits for humanity. The fundamental resource for bioprospecting is biodiversity. A wide variety of microbial, plant, and animal species and their genes are sources of industrial products. Species-rich environments may be anticipated to supply many valuable products for long term. In order to provide more opportunities for bioprospecting in the future, conservation of biodiversity is necessary.

Well-planned bioprospecting contributes to ecosystem conservation and socioeconomic development. Bioprospecting can achieve various goals like building technological and scientific capacity to manage and study biodiversity; producing revenues for protected areas, conservation projects, and local communities; increasing knowledge of the commercial and noncommercial significance of biodiversity; enhancing biodiversity science; and creating business-dependent sustainable management of resources and profit generation for corporations and shareholders. New industries like bioremediation and biomimetics are well established and increase, while others have a less assured future. The current economic environment proposes that pharmaceutical bioprospecting is about to increase. By carefully implementing bioprospecting, biodiversity can be utilized properly for the sustainable management of natural resources, economic development, and poverty alleviation. Established biodiversity-based industries like farming, forestry, grazing, and fisheries, with its use such as food, medicines, and fiber production, can generate biodiversity knowledge.

Opportunities for bioprospecting industries in a country will depend on many aspects such as conservation status of its biodiversity and the trends in a diversity of markets. Threats like species losses can affect the improvement of new valuable products for humanity, such as medicines, new crop varieties, and industrial processes. Depletion of global biodiversity may change bioprospecting in many ways. Other threats include loss of traditional knowledge, the impact of modern agricultural technologies, and depletion of natural resources. Several international and national laws and regulation measures like ethics, high-quality contracts, and institutional policies have supported bioprospecting partnerships. The Convention on Biological Diversity (1992) and International Treaty on Plant Genetic Resources for Food and Agriculture (2001) are the two recent international agreements. Around 100 countries have introduced laws and policies for regulating biological resources and benefit-sharing. Further, some indigenous communities, researchers, and bioprospecting companies have developed documents containing ethical and legal framework for easy operation of bioprospecting.

Marine Fisheries All oceans are affected by humans with overfishing having direct impact on food provisioning services, which will affect future generations. Beyond the 50 meters of ocean depths are affected by fishing and pollution. Recent studies indicated that global fish catching peaked in the late 1980s and is now decreasing in spite of increasing fishing effort. Industrial fishing vessels are with greater efficiency in offshore and also in deeper waters to meet up the global demand for fish. Deep ocean fauna is also affected by fishing. Recent huge investments in fishing and advanced navigation aids cover almost whole regions of ocean, including polar, deep, and low-productivity areas. These catches easily deplete populations of long-lived species.

Overfishing causes negative impacts on marine biodiversity. Local extinctions of large, long-lived, slow-growing species resulted from the fragmented habitats and lowered biomasses due to unrestricted fishing. In addition, harsh fishing practices have long-term impacts on marine habitats. Trawling, dynamiting, and dredging are some of the destructive fishing practices that can change the structure of marine ecosystems with significant changes in their ability to provide services, such as food and income generation. Continuing losses in species and habitats due to destructive fishing eventually reduce the biodiversity of marine habitats. But in this case, some systems may recover the availability of some services and products somewhat quickly; more vulnerable structures, like cold-water corals and seamounts, may take hundreds of years to recover. The implementation of no-take marine assets combined with other intrusions, such as controls on fishing, would be more practical to fisheries management. Marine reserves can supply better fisheries management by rebuilding stocks through maintaining biodiversity, enhancing recruitment and spillover effects, buffering marine systems due to human disturbances, and maintaining the ecosystems.

Aquaculture is not a key to the crisis of dilapidated wild-capture fisheries. More successful approaches are good control and efficient management of wild fish capturing. One of the major problems related with aquaculture industry is the use of fishmeal and oil derived from wild-caught small pelagic fishes such as salmon and tuna. In countries like Chile, people are now largely using small pelagic fishes as a source of cheap protein for fishmeal. So the contribution of wild marine fish as an inexpensive source of protein for several countries is declining. In developing countries like China, per capita fish consumption has decreased from 9.4 kilograms per person in 1985 to 9.2 kilograms in 1997, and the current trend is also not at all hopeful. But in some areas, fish consumer prices have increased faster than the cost of living. All fish products are highly commercial and approximately 50% of fish exports are from developing countries.

Predictable uses of marine systems are farming of marine organisms, bioprospecting, oceanic mining, and carbon sequestration. However, the probable impacts of these activities are not well understood. In some cases, to test the theoretical basis for the activity, only restricted field studies have been accomplished. Policies will need to contract with the insecurity of probable impacts and the limited understanding of marine biodiversity. National and regional ocean policies having an integrated ecosystem management structure are likely to be implemented. Restoration of marine species and habitats might include such policies and thus form part of a preventive strategy for protecting against management errors.

Coastal Systems Coastal ecosystems or coastal lands are near-shore marine areas where freshwater and saltwater mix together. These are the most productive but highly threatened ecosystems. These ecosystems produce more bioresources than most other systems. On the other hand, these undergo some rapid environmental change. Globally there was loss or conversion of approximately 35% of mangrove area, and in the last few decades, approximately 20% of coral reefs have been destroyed and an annual loss of 20% coastal wetland also has been reported. Nearly 40% of the human populations in the world live within the 100 kilometers of the coast. Demographic trends imply that coastal populations are increasing rapidly through increased fertility, migration, and tourist visitation to these areas. Coastal population densities are nearly three times more than that of inland areas. Coastal communities and industries increasingly utilize fisheries, timber, fuel wood, construction materials, oil, natural gas, sand and strategic minerals, and genetic resources. Most of the ecosystem services are provided by coastal communities aggregating near the coastal systems. Around 71% of coastal population, living within 50 kilometers of estuaries, provides protein to a large percentage of the human coastal populations. However, many of these habitats are unprotected or slightly protected; as a result, bioresources in these areas are at high risk.

In all regions coastal fisheries have limited stocks of finfish, crustaceans, and mollusks. Illegal and destructive fisheries result in overexploitation and habitat damage. In response to declining capture of fisheries, demands for coastal aquaculture have been high. But in the last 10 years, the increase in aquaculture production has also caused habitat loss, pollution, and overexploitation of fisheries for fishmeal and fish oil. Overexploitation of other resources like mangroves for fuel wood, seaweeds for consumption, sand for construction material, etc., also destabilizes the ecological functioning of coastal systems. Development-related loss of habitats and services are the greatest threat to coastal systems. Humans are facing increasing coastal erosion and flooding, declining water quality, and increasing health risks as a result of coastal degradation and alteration. Urbanization, port development, aquaculture, resort development, and industrialization are regularly involved in the destruction of coastal forests, coral reefs, wetlands, etc. Worldwide sediment flows in rivers increased by about 20% as a result of enhanced human activities, but reservoirs and water diversions avoid around 30% of sediments reaching the oceans, which in turn reduced sediment delivery to coasts to 10 %.Within the last century, the average nitrogen loading has increased twofold that turns coastal areas as highly chemically altered ecosystems in the world. The resulting eutrophication causes coral reef changes and supplementary irreversible changes to coastal ecosystems. Nearly half the people living along coasts have no access to sanitation and thus face decreasing ecosystem services and increasing risks of disease. Mining and other industries near or within the coastal system cause toxic pollutions like heavy metal. The decreased water quality introduces harmful algal blooms and other pathogens which affect the health of both humans and marine organisms. Invasions of alien species also altered and threatened marine and coastal ecosystems.

The coastal systems' ability to provide highly valuable bioresources is closely linked to the adjacent marine, freshwater, and terrestrial systems. In some cases, dry land pollution in coastal region contaminates groundwater. Airborne dust causes desertification and pollution of coral reef ecosystems. Destruction of coastal wetlands has similarly been concerned in crop failures due to diminished coastal buffering leading to freezing in inland areas. The world's wealthiest populations occur mainly in coastal areas because the per capita income is four times higher in coastal areas than inland, and life expectancy is also higher in coastal regions, while infant death is thought to be lesser. Wealth inequality has deprived many coastal communities of access to resources. For example, access issues have in turn led to increased difference, such as in between small-scale artisanal fishers and large-scale commercial fishing enterprises. In the coastal system, alteration of coastal habitat for urbanization, industrial use, or other coastal developments often get short-term economic benefits, but the regulating and provisioning services are enduringly lost. Successful coastal area management necessitates the incorporation of management of other sectors. Watershed management is an essential element of efficient coastal management because coastal systems are strongly affected by inside and outside activities of coastal regions. In the Integrated Coastal Management Programme, marine protected area networks effectively protect the most ecologically vital habitats. Reestablishment of some coastal habitats such as marshlands and mangrove is being carried out now.

Inland Water Systems Inland water territories and species are in worse situations than those of forest, coastal systems, or grassland. During the twentieth century, it is anticipated that 50 % of inland water ecosystems will be lost. More than 50 % of inland waters exclusive of lakes and rivers have been vanished in parts of North America, Europe, and Australia. In addition to loss, degradation is also extensive. The species biodiversity of inland waters is among the most threatened of all ecosystems in many parts of the world. Global climate change is anticipated to intensify this loss and degradation. Loss and degradation of inland waters are well documented with respect to vegetation clearance, drainage, and infilling for the expansion of agriculture; the spread of infrastructures for urban and tourism; aquaculture, agriculture, or industrial purposes; the spread of invasive species; overharvesting through fishing and hunting; hydrologic modification; and pollution, salinization, and eutrophication. Global climate change is predicted to lead further degradation. Worldwide principal cause of the loss of inland water systems is the development of agricultural systems. In 1985, 56–65% of inland water systems had been exhausted for rigorous agriculture in Europe and North America, 27% in Asia, and 6% in South America.

The creation of dams and other structures has resulted in fragmentation and flow regulation of 60% of the large river systems in the world. In many countries water pollution and eutrophication are widespread and led to the degradation of many inland water systems. Toxic substances and artificial chemicals that are increasingly released into waterways have long-term effects on ecosystems and humans. The destruction caused by invasive species has been increasingly recognized worldwide. As a result of the decline of inland water systems, the bioresources derived from these systems and human well-being are at a high risk. Inland water fisheries are essential for human well-being. It is estimated that more than 50 million people are engaged directly in inland fisheries. Rice is another vital thing for human well-being. Currently around 50% of the world uses rice as a staple food crop, and this is likely to reach around 4 billion people depending on rice by the year 2020.

Supporting and regulating services are significant to support critical ecosystem functions. Links between surface and groundwater and flow regulation within and between inland waters are important ecosystem services that have been spoiled on a global scale. The natural flooding disruptions lead to decreased sediment transport and a loss of flood buffering and nutrient preservation. Flooding may cause severe adversity to humans, but sometimes it is essential for sustaining sediment-based fertility and supporting fish stocks in large rivers. In addition, inland water offers invaluable opportunities for recreation by many communities, mainly for tourism because they have significant aesthetic, artistic, educational, cultural, and spiritual values. River alterations provide improved transportation, better flood control and hydropower, and improved agricultural output. At the same time, water velocity in riverine systems has decreased because the rivers have been detached from their floodplains and other inland water habitats, and in some places, they have been converted to a chain of connected reservoirs. Groundwater revitalization has also been reduced. But in other places, infrastructure has improved the possibility of flooding by redirecting water and increasing flows.

Estimates of the level of inland water systems differ from 530 million to 1280 million hectares. Along with rivers, larger wetlands, lakes, and inland seas have been mapped; it is estimated that there are 10,000 lakes that cover 1 square kilometer, and peatlands cover more than 400 million hectares. Smaller habitats that are significant for many communities are not well mapped. The changes in inland water systems affected the migratory patterns of fish species and the composition of its habitats, opened up corridor for foreign species, distorted coastal ecosystems, and contributed general loss of freshwater biodiversity and inland fishery resources. The available information highlights on the global scales. Pioneering tools for successful evaluation of the status and development of inland water systems and the associated bioresources, especially in those parts of the world where data are lacking, are required.

Mountain Systems Half of the human population relies on mountains. Mountains take up about one fifth of the terrestrial surface. World's 20% of human population (1.2 billion) live in mountains or at their boundaries, and half of the population depend in one way or the other on mountain resources mainly for water. Rich biodiversity is the characterization of mountains. Commonly mountain regions are more diverse than lowlands because of the solidity of climatic life zones with altitude and different topoclimates causing small-scale habitat diversity. They maintain about one quarter of terrestrial biodiversity. Nearly half of the world's biodiversity hotspots are located in mountain regions. Geologically fragmented mountains maintain high ethnocultural diversity. Clean air in the mountains makes them target regions for recreation and tourism. On the other hand, mountains have spiritual significance and have scenic landscapes. It contains 32% of protected areas which come about 1.7 million square kilometers. Ecosystems in mountain are extremely fragile. Both natural and anthropogenic changes frequently occur in mountains. This may range from volcanic, seismic, and flooding events to global climate change and the thrashing of soils and vegetation due to inappropriate agricultural and forestry practices and innovation of industries.

Mountain biota is adapted to comparatively narrow ranges of temperature and precipitation. Human well-being depends on mountain wealth. These ecosystems are predominantly essential for the provision of clean water, and their ecological reliability is important for the protection of settlements and transport routes. They have rich biodiversity and supply considerable global plant and animal products. These services primarily depend on slope steadiness and erosion control supplied by a healthy vegetative wrap. The mountains provide water to nearly half the human population that may include many areas distant from mountains, and mountain agriculture provides survival for about half a billion people. Important mountain resources and services include water for hydroelectricity, mineral resources, flood control, medicinal plants, and timber. Mountain populations have a high diversity of cultures, including conventional agricultural knowledge and languages which commonly support sustainable production methods. In many mountain areas, tourism forms the backbone of regional as well as national economies.

In mountain regions both poverty and ethnic diversity are higher, and people are more vulnerable. About 90% of the global mountain population covers 1.2 billion total populations living in developing countries, and other 90 million mountain people are considered especially vulnerable to food insecurity. High risk to mountain ecosystem integrity is the land use pressure in many parts of the world. Forest destruction, industrial use, inappropriate cropping practices, and overgrazing cause permanent losses of soil and ecosystem function. It may increase environmental hazards in both mountains and neighboring lowland areas. Mountains habitually symbolize political borders, control transport to narrow corridors, and are refuges for minorities at political resistance. As such they are often crucial areas of armed clashes. Profits from mountain industries are not reinvested either in the management of highland resources or the stipulation of benefits to mountain neighborhood. In many mountain regions, both poverty and remoteness are responsible for poor medical care and education systems. Stronger highland-lowland linkages develop sustainability for both higher and lower populations. Environmental conservation and sustainable land exploitation in the world's mountains are crucial to human welfare for almost half the world's population who depend on mountain resources.

Polar Systems Alterations in polar community composition and biodiversity are disturbing human well-being. Main changes include the decrease of Antarctic marine food web top predators, altering food assets, increased shrub supremacy in Arctic wetlands, changes in insect loads that change food accessibility to wetland birds, increased abundance of snow birds which are degrading Arctic wetlands, overgrazing by domestic reindeer, and increase in the occurrence of invasive alien species. In polar regions climate change has considerably affected ecosystem services and human well-being. On an average, now the temperatures are warmer than at any time in the past 400 years. Warming-induced melt of permafrost is becoming more widespread, causing threshold changes in ecosystem services. The warmest regions on the planet are the Antarctic Peninsula and its neighboring oceanic sectors. In this area higher predator populations are determined as a result of elevated primary and secondary production, and the majority of other living resources have also been determined.

Trace-gas-induced global warming in polar regions diminishes the capability of polar regions to act as a cooling system for planet Earth. This climate is resulting from alterations in the physical system. In the Arctic regions, local warming interrelated with socioeconomic variations reduces survival of indigenous and other rural people having the greatest cultural and economic trust on these resources. Animals, such as moose in North America, have migrated northward in response to warming. Rising sea level in polar regions due to global warming will increase coastal erosion that currently threatens many coastal villages. Direct impacts of human actions on polar regions have been reserved, and nations with Arctic lands or Antarctic obligations have the economic resources to alleviate many current and predictable problems if suitable policies are applied. Consequently, in polar oceans and wetlands where the resource harvests are concentrated, there is high potential to provide important ecosystem services and biodiversity. However in the future, the sensitivity of polar ecosystems to disturbances related with resource withdrawal makes them susceptible to increases in resource demand.

Urban Systems The world's urban population was 200 million in 1900 and increased to 2.9 billion in 2000, and now it covers more than 50% of world's population. Urbanization is not itself intrinsically bad for ecosystems. Ecosystems in and around urban areas are more diverse than rural monocultures, and if they are properly managed, they also provide food, comfort, water services, amenities, cultural values, etc. Currently urban areas account for only 2.8% of the total land area, although it contains about half the world's population. Globally urban demographic and economic growth has increasing pressures on ecosystems. In respect to ecosystem services and human well-being, urban development has a tendency to create serious problems. The problems include rigorous environmental health problems within urban settlements due to insufficient access to ecosystem services like clean water. Urban improvements in some parts of the world have been taking place in a diverse technological and economic circumstance, with different environmental consequences. In low-income countries, interconnected problems involving local water, waste, sanitation, pests, etc., contribute diseases. This normally reveals a mixture of degraded or increasingly limited ecological services within the urban area. Problems relating to the degradation of ecosystems adjacent to urban settlements are undermining their capability to supply ecosystem services, particularly in large industrializing and motorizing cities. The global urban ecosystem changes are concentrated especially in upper- and middle-income countries. This usually reflects a mixture of environmental dislodgment and increasing rates of direct and indirect utilization of energy, materials, and ecosystem services.

When urban systems are handled more precisely and the loss of ecosystem services is firmly addressed, the repayment to human well-being can be significant. The motivated governmental and nongovernmental agencies can execute fundamental changes in local environmental management that decreases local burdens and supports vulnerable groups. A strong evidence for this is the experience of urban sanitary reform. The controlling of urban air and water pollution has also resulted in healthier air and water quality in some parts of the world. There is relatively little confirmation of significant steps to diminish the global ecosystem troubles of cities, although if it is done in a large scale, it would result in considerable decreases in those burdens.

Cultivated Systems More food has to be produced for increasing world population. As the requirement for food, feed, and fiber has increased, farmers are under pressure for expanding the cultivated area. This will intensify the burden on the carrying capacity of soil, freshwater, vegetation, and fauna. Over the past 40 years, strengthening of cultivated systems has been the primary source of increased global output. In some countries with high levels of productivity and low population expansion rates, the degree of distribution of land under cultivation is stabilizing or may be contracting. Globally, opportunities for additional extension of farming are getting reduced. Cultivated systems focus in the provision of food, fiber, and feed. Cultivation has affected other services in different ways such as alteration of biologically diverse natural grasslands, native forests, and wetlands into less diverse agro ecosystems; the choice of crop species and its cropping time and space; and the way in which the resources like crops, soil, and water are handled both at plot and landscape levels. Significant losses of ecosystem services occur as a direct result of conversion to agriculture. Subsequent impacts are due to the intensity of cultivation, the type and amount of water, nitrogen, pesticides, etc., applied.

Two important policies have emerged to raise economic benefits to farmers: first, increasing the efficiency of existing cropland through effective management of specialized cropping systems by using improved crop, soil, and water management practices and, second, manipulating more diverse crop systems that will give better living options as well as sustaining high levels of biological diversity and other environmental services. Area development for production decreases natural habitat and biodiversity due to land use conversion. Conversion of natural habitat at the global level for agricultural uses is possibly the single greatest threat to biodiversity. Therefore, sustaining yield increases unobtainable farmland to meet rising human food requirements will be vital for the protection of existing biodiversity.

Advances in ecological science united with field-based experimentation have yielded enhanced production so as to improve opportunities for wild biodiversity, habitat construction, wildlife passages, refuge, and buffers around susceptible areas. The economic advantages of pollinators, pests, soil bacteria, insects, birds, and other animals are increasingly being effective to farmers and the agricultural community. Rapid and broad spread of integrated pest management has an important role in sustainability which is mainly focused on platforms such as agroecology, eco-agriculture, and integrated natural resource management. Major global consumer of water is the cultivated systems. The area of irrigated cultivated systems is about 18%, which means around 40% of global food production is generated by such irrigation. In many areas water demand for food production is increasingly competing rising water demand for other uses. The irrigation efficiency can be increased by improving the water delivery systems and water application methods. Water harvesting practices like small tanks, runoff farming, etc., also proved as effective. Intensification of food production relating increased use of artificial nutrients and agricultural chemicals may lead to water pollution that contaminates freshwater, estuarine, and marine ecosystems and that also limits downstream water use. Water pollution can be greatly reduced by the application of new technologies or practices that enhance nutrient use efficiency and also minimize the need for pesticide. Inappropriate farming practices on sloping land can result in severe erosion that also contributes to pollution of water bodies, rivers, and marine ecosystems.

Cultivation has increased and modified the spatial outlines of nutrient cycling. The application of inorganic fertilizers primarily caused the disruption of the nitrogen cycle. Currently it included above 1 billion tons of nitrogen. Nitrogen is a major component of dietary protein and the most limiting plant nutrient. While naturally fixed nitrogen is an essential part of more productive cultivation, application of inorganic nitrogen enhances emission of nitrous oxide, a strong greenhouse gas, and also contributes acid rain, soil acidification, and eutrophication and finally to biodiversity loss. The best way to overcome these negative effects is to increase the efficiency in the application and handling of fertilizers and also increase use of organic nitrogen sources like mulching, legume crops, animal manure, etc., instead of inorganic fertilizers. In some parts of the world, overuse or inefficient use of nitrogen applied to crops leads to poor yields and often leads to cumulative depletion of soil nutrients. Carbon-based greenhouse gases are generally released when the natural ecosystems have been converted for cultivation, and carbon sequestration potential is also reduced. Now about 166 million tons of carbon per year is released into the atmosphere as a result of frequent cultivation, livestock production, and irrigated rice production. Greenhouse gases contributed by agriculture also give an opportunity for mitigation. The farming of befouls such as corn, sugarcane, and sorghum used for ethanol production has great potential. High demand for biofuels would result in expansion of cultivated areas or dislodgment of traditional crops. Better practices and new technologies may reduce the negative environmental cultivation impacts.

In some rich countries, public funds are increasingly used for the conduction of environmentally friendly practices, research, and development of new technologies and practices that reduce the trade-offs between food provision and other ecosystem services. To some extent farmer's choices about the scale of cultivation, the selection of the cultivation system, etc., can be determined by national policies, international agreements, and market forces. In many industrial countries, agricultural subsidies have encouraged overproduction although at the same time in poorer countries, it may reduce the economic capability of cropping systems. Some rich countries have introduced certain conservation programs to promote farmers to take environmentally susceptible production.

1.6 Conclusions

Bioresources are the significant constituents for development and economic undertakings of a nation. But its organization and exploitation for human welfare are most essential for the optimum consumption of the bioresources. Population explosion, encroachment, overexploitation, water scarcity, climate change, devastation of forests and inappropriate conservation of wetland ecosystems, industrial waste discharges into the aquatic environments, and inadequate and unhealthy management of sewage effluents certainly threaten all kinds of natural systems on earth. Both in global and regional levels, significant challenges will be faced to meet increasing food, feed, and fiber demand. Globally, in response to the rate of projected demand, the rate of cereal yields is falling below, which leads to a large expansion of cultivated area. It may reach low levels of soil quality, face irregular climatic patterns, and also introduce new pests and diseases, such as coffee and banana wilt and avian flu. Climate change associated with warmer global temperature is an emerging challenge supporting yield increases in currently favorable crop production and may decrease yield constancy in rainfall-dependent dry land cropping systems.

Bioresources management and conservation by the state government or central government are very rigorous. However, for the better living of ours and also of our future generations, every citizen of the country regardless of gender, caste, or religion has to be conscious of natural resources in our country and should take steps forward for the proper conservation and protection of our rich bioresources.

References

Biological Diversity Act (2002) http://www.nbaindia.org/act/act_ch3.htm

Karnataka Forest Department Report (2001) Government of Karnataka

Ramachandra TV (2007) Comparative assessment of techniques for bioresource monitoring using GIS and remote sensing. ICFAI J Environ Sci 1(2):7–47

Ramachandra TV, Joshi NV, Subramanian DK (2000) Present and prospective role of bioenergy in regional energy system. *Renew Sust Energy Rev* 4:375–430
Ramachandra TV, Kamakshi G, Shruthi BV (2004) Bioresource status in Karnataka. *Renew Sust Energy Rev* 8:1–47
Teri energy data directory and year book (2001–2002) TERI, New Delhi
Tynsong H, Dkhar M, Tiwari BK(2012) Traditional knowledge based management and utilization of bio- resources by war khasi tribe of Meghalaya, North-east India, 1(3), pp 162–174.

Conservation of Bioresources

2

P.E. Rajasekharan

Abstract

There are two well-known approaches for biodiversity conservation, viz., protecting the habitat by adopting in situ conservation and protecting individual species outside the habitat (ex situ). Biotechnology emerged as a universal tool for conserving biodiversity, management, and utilization. Through the application of biotechnological technique, the management of plant genetic resources for food and agriculture was made easy. It also helps to prevent the trend of genetic erosion in crops related to food and agriculture. The conservation of biodiversity is a complex process and could be achieved through protection of habitats, and it requires in depth analysis of available bio resources. The concept of conservation is changing over the past period and moved from static to dynamic phase with changing policies. The various approaches for conservation of bioresources with the examples are discussed within this chapter.

Keywords

Bio resources • Conservation • *In situ* • *Ex situ* • Biotechnology • Plant genetic resources

2.1 Introduction

Biological diversity exists at three levels in the farming system: at the ecosystem level, at the species level, and at the variety level. In practical terms this simply refers to the diversity of agricultural systems (agro ecosystems), the diversity in the

P.E. Rajasekharan (✉)
Division of Plant Genetic Resources, ICAR-Indian Institute of Horticultural Research, Hessaraghatta Lake post, Bangalore 560 089, India
e-mail: rajasekharan.pe@gmail.com

S. Abdulhameed et al. (eds.), *Bioresources and Bioprocess in Biotechnology*,
DOI 10.1007/978-981-10-3573-9_2

number of crop species grown in a particular agro ecosystem, and, finally, the diversity of different varieties of these crops. As far as crop diversity is concerned, genetic diversity within species is often more important than the diversity between species.

2.1.1 Need for Conservation

There is always selection in agriculture. Simmods (1962) have the opinion that when breeding achieves close adoption, it tends to reduce variability in long-term adaptation (fitness to agriculture environment) and adaptability (capacity for genetic change in adaptation). Through conservation we can address this antagonism by maintaining and rapidly deploying the variation that in the raw materials for the selection and continued evolution of crops.

Over the last decade, the need to develop practical strategies for supporting farming communities in conserving and using crop biological diversity in a sustainable manner has become widely recognized as important for ensuring food and livelihood security, especially in agriculturally marginal areas of the world. Approaches to conservation and utilization are changing rapidly, and feedback and flexibility are always needed. The increased worldwide awareness and recognition of the importance of conserving agro-biodiversity has not been met with increased funding to carry out all the necessary conservation efforts. Difficult policy choices have to be made, and efficiency becomes paramount as efforts are made to rationalize conservation. Crop networks play an important role in this by permitting elimination of duplicates and by setting stricter priorities on what to conserve and where.

2.2 *In Situ* Conservation

In situ technologies aim toward conserving the processes of evolution. The CBD (UNEP 1992) covering both wild and domesticated species uses a complex definition for in situ conservation:

> The conservation of ecosystems and natural habitats and the maintenance and recovery of viable populations of species in their natural surroundings, and in the case of domesticates or cultivated species, in the surrounding where they have developed their distinctive properties

This combination of ecosystem and population conservation is usual for wild species. There is limited experience for the *in situ* conservation of domesticated agrobiodiversity, little evaluation of attempts, and, as yet, no agreed protocols for its management *in situ* in compliance with CBD. *In situ* approaches may provide a linkage between conservation and development.

In situ conservation is becoming increasingly important in the conservation of plant genetic resources. A major GEF-funded program of work has been undertaken in Turkey to conserve crop wild relatives. This led to the development of a substantial national program to conserve crop wild relatives in situ and has provided much information on the financial and labor resources that might be needed and how such a program might be implemented. *In situ* conservation of crops (on-farm conservation) has also received increasing support and is the subject of extensive research in a number of countries around the world.

In situ conservation requires some significant changes in perspective for those used to ex situ conservation. In the first place, the concern is less with the maintenance of specific characteristics or specific genotypes than with the maintenance of evolutionary potential. The target populations have to exist within ecosystems that are themselves dynamic and subject to change over time and to succession. On-farm conservation will depend for its effectiveness on the extent to which local varieties continue to meet the needs of farmers and communities, and the approach taken needs to be one that is embedded in the community and reflects its values and concerns.

The research needed for in situ conservation of useful plants (both on farm and in natural habitats) is substantial. It is a fairly young endeavor compared to ex situ conservation, and the body of information that might help us in making choices about locations, populations, management procedures, and appropriate interventions has not yet been fully developed.

For crop wild relatives, a carefully selected set of detailed studies is required that complement the work done already in a few situations such as the wild wheats in Ammiad, Israel, and beans in Costa Rica. The exciting element of this research is that it will frequently have to be carried out alongside specific conservation activities and in full collaboration with communities living in the areas – a considerable challenge to research workers used to controlled experiments in defined environments. Since resources for the conservation actions needed will be extremely limited, methods also need to be found of identifying the key interventions required (which populations, of what species, in which areas) so that they, at least, can be implemented. Knowing what is most threatened and what is protected already (through being in protected areas or maintained *ex situ* or secured in other countries) will be an essential component of planning such interventions.

On-farm conservation requires a careful linking of research though multidisciplinary studies that will answer some important conservation questions (e.g., the amount of diversity in local varieties, its distribution, and variation from year to year) with an understanding of the social and economic dimension and a concern with development. In obtaining the information that will allow the scientific basis of on-farm conservation to be established, participatory procedures must be used which fully involve local communities and farmers. In order to make on-farm conservation effective, we need to understand the ways in which farmer management affects the extent and distribution of diversity and to determine how production environment and farmer decision making interact with crop biological properties to affect the diversity characteristics of local varieties.

2.3 Ex Situ Conservation

This process involves methods which are technology dependant, aimed at securing the genetic resources over prolonged durations. One of the most significant achievements of plant genetic resource work has been the establishment of procedures that provide for the safe, long-term maintenance of seeds. The procedures are generally well established, especially for our major seed-propagated crops. For this reason, perhaps, seed conservation research has received rather little attention in recent years, although some key questions remains unanswered.

How to manage different accessions of germplasm? To this effect different answers are there. Accession management has always been difficult, and apt procedures need to be in place for the same. The gene bank management is on a sticky wicket considering the fact that how to maintain all accessions without compromising efficacy and loss of diversity.

2.3.1 Cryobanks

In the omics era, cryopreservation is not only used as a means of conserving germplasm but as direct aid to genetic improvement facilitated by biotechnological research and development. The rationale for maintaining ex situ microorganisms in culture collections for the purpose of food, public health, poverty alleviation, industry, and international research networks is similarly justified. Developing rationales embrace the wider benefits of cryogenic storage, and they go beyond the traditional framework of the active and base collection concept for phytodiversity conservation. It is important to caution that this must not put at risk the basic need to preserve core germplasm collections. One of the greatest challenges in the development of the cryobank is to harmonize how different sectors offer their services; one of the ways of achieving this is through cooperatively engaging best practices (Benson 2008).

2.3.2 Seed Storage

Seed storage is the best form of ex situ conservation because of the convenience of use, and the protocols are well defined for majority of food crops. Viability and vigor monitoring of stored seeds are very important, and research in these area have made very good progress. Conservation in the form of seed is preferred because of the simplicity of technique, easiness to follow, practical feasibility, inexpensive, and better genetic stability (Rajasekharan 2015). If we compare different methods of conservation of plant genetic resources, long-term seed storage serves the best bet. Many factors affect the longevity in storage, viz., effect of environment during seed maturation, harvesting, drying, and subsequent period before seed is stored. According to the seed storage behavior, the seeds are classified into orthodox and recalcitrant. A third category called intermediate showing characters, between the two categories mentioned earlier, is also there. Orthodox seed could be dried to low levels of moisture contents, and longevity increases with decrease in moisture content and storage temperature. Orthodox seeds are tolerant to desiccation in all stages of their development and maturation. Cryopreservation of seeds extends viability compared to all other techniques (Rajasekharan 2015).

2.4 Seed Banks

What makes seed banks such an effective ex situ conservation technique is that the methodology can be applied to a wide range of species in a universal and straightforward way and that large amounts of intraspecific diversity can be conserved and for long periods of time without intervention. Additionally, germinating seeds to obtain fully grown plants is relatively simple compared with obtaining plants from *in vitro-stored* material. Just under half of the 6 million accessions are held in 12 national collections. To some degree, this is a function of the early establishment of their genetic resource collections. The collections include those in Russia and the United States (noted earlier), Japan (established in 1966), Germany (1970), Canada (1970), and Brazil (1974). Of the 1308 national or regional collections currently noted by FAO (1996a, b, c, d), only 397 within 75 countries are held in long- or medium-term seed banks. Medium-term storage might be assumed to be in the order of 10 or more years.

2.4.1 Pollen Storage

This is comparable to seed storage, since pollen can be dried (less than 5% moisture content on a dry weight basis) and stored below 0 °C. Their small size and desiccation tolerance render pollen grains particularly suitable for storage. However, relatively limited use has been made thus far of pollen for long-term germplasm conservation (Hoekstra 1995). However, it may be useful for storage of clonally propagated species. The international transfer of germplasm in the form of dry pollen is not generally restricted, so germplasm exchange through pollen can be viable technique. The life span of pollen is considerably shorter than that of seed, and it may vary according to species from just few hours to several months at room temperature. Taking the advantage and disadvantages together, pollen collection and storage are likely to be most useful and foreseeable further as a complementary conservation strategy for species that are normally conserved in field gene banks and have large and/or recalcitrant seeds. Some of the practical utilities are aimed at redesigning breeding strategies with cryopreserved viable pollen, facilitating hybrids between genera, species, and genotypes. Large-scale consolidation of potentially cryopreserved viable pollen from male parents facilitates uninterrupted supply of the male gametophyte for production of hybrids at a given location, and pollen can be transported to different locations where seed parents are grown for crossing. Cryogenic technology applied to pollen conservation facilitates integration of conventional breeding methods with modern biotechnological practices. The availability of pollen with good quality in a pollen cryobank will provide constant supply of the same for extended durations. Pollen in such a state can be termed value added by virtue of its potential extended life, for it has been able to be kept viable and fertile for extended duration to perform its natural function of fertilization, leading to the formation of fruit and seed. Cryopreserved pollen used in production of hybrids can be classified as value added. Hoekstra (1995) has assessed the merits and demerits of pollen as a genetic resource. Methods of pollen cryopreservation and procedures needed for developing pollen gene banks was reviewed by Towill and Walters (2000). Alexander and Ganeshan (1993) reviewed the work on pollen storage in fruit crops, and Ganeshan and Rajasekharan (1995) reviewed the work on ornamental species. Pollen collection is a critical component for a pollen bank (Hoekstra 1995). Pollen collection procedures depend on type of species, inflorescence, and peak anthesis period. Grout and Roberts (1995) detailed the methodology for pollen cryopreservation. Barnabas and Kovacs (1997) have stressed the importance and need for pollen conservation. Towill (1985) reviewed the work on cryopreservation of pollen in 30 species, and he reported that there were no significant differences in percentage of in vitro germination of pollen before and after cryogenic storage. Rajasekharan and Ganeshan (2003), Ganeshan et al. (2008), and Rajasekharan et al. (2013) reviewed pollen as a resource material for conservation of nuclear genetic diversity and the use on vegetable and ornamental crop breeding.

2.4.2 Pollen Cryobanks

In the last 30 years, many technologies have been developed for long-term cryopreservation of nuclear genetic diversity (NGD). Long-term cryopreserved pollen in the form of NGD of citrus, papaya, grape, mango, tomato, eggplant, onion, capsicum, rose, gladiolus, gerbera, carnation, and RET species of medicinal plants are maintained in liquid nitrogen. The pollen cryobank is managed by periodic replenishment of the cryogen, for maintaining a constant cryogenic temperature throughout the storage duration. This technology is globally acceptable and is useful for production of hybrid seeds in the abovementioned crops, besides its use in gene banks. For easy management of data on the pollen, a main database and three subdatabases are created. Genus, species, and cultivars information are stored in a separate database. Information pertaining to longevity, viability, and fertility of pollen after cryopreservation, media used for pollen germination, year-wise listing of collected samples, and year-wise list of samples cryopreserved in different years are available within the database, which is menu driven and user friendly. Although genetic conservation through pollen storage does not accomplish whole genome conservation, a breeder involved in genetic enhancement of a given crop can have access to a pollen cryobank facility, for nuclear genetic diversity (NGD) inputs in his amelioration program. Besides the already existing role of pollen cryobanks in breeding, there are many promising applications, which have come to focus with the recent advances in allied areas.

2.4.3 FGBs

These are used for the conservation of clonal crops, where seed is recalcitrant and where seed is rarely produced. The "rule of thumb" is to use the same propagation techniques as the farmers, for example, not disrupting adapted clones through genetic segregation in seed cycle. Management may be the same as used during routine farming, so the method can be adapted to local circumstances. Major constraints to field gene banks include cost and all the natural hazards of farming.

The conservation of vegetatively propagated crops such as potato, cassava, yams, sweet potato, sugarcane, coffee, cocoa, rubber, coconut, banana, mango, and many temperate fruit trees among others presents special problems for *ex situ* conservation. It will be seen that two categories are dealt with here. The first of these is relatively long-lived trees or bushes, while the second involves root or tuber-bearing species. Although some of the crops conserved in this way are sexually fertile, it is often not convenient to propagate them commercially from seed because of high levels of genetic heterozygosity, and breeders and horticulturalists commonly require uniform clones. Field gene bank conservation involves the collecting of materials from farmers' fields and gardens, or even from wild locations, and transferring them to a second site where they can be planted and monitored. It is also difficult to keep vegetatively propagated plants free from viruses, and this infection leads to degeneration of clonal stocks.

2.5 In Vitro Conservation

When one method of conservation is subject to unavoidable hazards, as with FGBs, an alternative method should also be used. The role of in vitro conservation techniques in the overall conservation strategies should be indicative of the fact that it should complement other conservation strategies within the total program of a given species or population. The methods chosen should be carefully considered taking into account the feasibility, practicality, economy, and security. In vitro collections of species could be maintained at the same or separate site but should have clear linkages with the field gene banks (Rajasekharan and Sahijram 2015).

The properties required for a successful storage system should satisfy the ability to:

1. Minimize growth and development in vitro.
2. Maintain viability of stored material at the highest possible level along with minimum risk of genetic stability.
3. Maintain full developmental and functional potential of stored plant material when it is returned to physiological temperatures.
4. Make significant savings in labor input, materials, and commitment of specialized facilities.

The distinct advantages and limitations of in vitro approaches:

Advantages

1. Collection may occur at any time, independent of flowering periods for each species.
2. There is potential for virus elimination from contaminated tissue through meristem culture.
3. Clonal material may be produced.
4. Rapid multiplication.
5. Germination of “difficult” immature seed/embryo rescue may be facilitated for breeding.
6. Distribution across borders may be safer.

Limitations

1. Initially expensive.
2. *In vitro* storage techniques, particularly cryopreservation procedures, are not yet well optimized for routine application across a broad range of species or genotypes within the species.
3. Somaclonal variation refers to all variability observed among tissue culture-regenerated plants; some of the methodological bases for variation are explant source, method of culture, hormone used, genotype, ploidy status, etc. (Larkin and Scowcroft 1981).

2.5.1 In Vitro Conservation: Practical Considerations

Issues of concern in potential application of in vitro techniques for germplasm conservation include:

1. Genetic Stability
2. Basic tissue culture competence

Genetic Stability Genetic instability created in tissue culture system is perceived as a major obstacle in using this technique for germplasm conservation. Variation may be observed at various levels, such as morphological, karyotypic, or biochemical. The use of auxiliary bud for multiplication reduces the probability of genetic variation. The genetic stability at all stages of *in vitro* conservation should be monitored. No well-defined techniques are available for conservation of endangered medicinal plants. Therefore, special attention is required in this regard since it is a question of species extinction, and it is essential to retain the quality and quantity of secondary metabolites.

Basic Tissue Culture Competence Difficulty in culture establishment, micropropagation, rooting and establishment of plants and extra vitrum – all these stages for any given plant must be optimized before going for in vitro conservation. Some species show recalcitrance in culture system for which special attention needs to be given.

2.5.2 DNA Storage

DNA storage is simple, easy, cheap, and widely applicable. The progress made in the research in genetic engineering resulted in breaking many barriers in transferring genes. Now several plants are available which are created by transferring genes across distantly related species and also from viruses, bacteria, fungi, and mice. For this purposes, many DNA repositories are established worldwide to store total genomic information of germplasm. Protocols to use the material stored in the form of DNA are yet to be developed, and hence the relevance of DNA banks for the conservation of plant genetic resources are yet to be brought to the mainstream conservation approaches. There is an increase in the importance of DNA banks. Preservation of DNA in the form of extracted DNA or genomic libraries has become a major thrust area, and translation of this information to a major conservation strategy demands further in depth study.

Practical application of DNA conservation:

1. Molecular phylogenetics and systematics of extant and extinct taxa
2. Production of previously characterized secondary compounds in transgenic cell cultures
3. Production of transgenic plants using genes from gene families
4. In vitro expression and study of enzyme structure and function using probes

2.5.3 In Vitro Gene Banks

Wherever tissue culture protocols are available for a plant species in vitro gene bank could be established. For medium storage, intervals between subcultures need to be extended by low light or low temperature or a combination of both. By this technique we will be in a position to reduce human resources required for handling the work. It can be readily applied for temperate species wherever the tissue culture protocols are available and difficult in the case of tropical species.

Advantages

- Stockpiling of cultures is useful for commercial tissue culture firms for proper scheduling of their operations.
- Disease-free plants could be produced.

- International exchange of germplasm in the case of vegetatively propagated species, viz., banana, potato, and cassava, is done only through tissue culture-raised plants.

Disadvantages

- Tissue culture protocols development for different species are expensive.
- Possibility of genetic instability due to somaclonal variation in tissue culture.
- Difficulty in developing protocols in the case of recalcitrant species (tropical species).

2.6 Complementarities of Conservation Methods

The objectives of any conservation strategy are to conserve a wide range of genetic diversity and to secure accessibility to be balanced between feasibility and cost-efficiency. Since any single method of conservation is not enough to conserve the whole range of genetic diversity, different and complimentary methods of conservation having different advantages and disadvantages are to be tried. Making such a choice will be difficult with a holistic view of proposed conservation effort, as a part of the process of development (Tables 2.1, 2.2, 2.3, 2.4, 2.5, and 2.6).

Table 2.1 Biological factors determining conservation methods

No	Biological factors	Preferred conservation methods	Remarks
1	Perennial species	In situ/field gene banks/seed and or pollen storage	If tree species be required for utilization purpose
2	Annual species	Seed and or pollen storage in vitro field gene bank	See also factors 3,4,6, and 7
3	Orthodox species	Seed storage	
4	Recalcitrant seeds	In vitro/in situ/field gene bank	
5	Synthetic seeds	As orthodox seeds	
6	Vegetatively propagated species with viable seeds	Field gene bank/pollen/in vitro/ cryopreservation	
7	Vegetatively propagated species with nonviable seeds	Field gene bank/pollen/in vitro/ cryopreservation	Field gene bank or genotype needs to be conserved
8	Long-living pollen	Pollen storage	
9	Tissue culturing feasibility	If low, look for alternative method	
10	Cryopreservation feasibility	If low, look for alternative method	
11	Genetic stability	If low for certain method, alternative method	

Table 2.2 Most common methods used for germplasm conservation and the corresponding PGR categories

No.	Methods	Predominantly conserved PGR categories by corresponding method
1	Biosphere reserve	Ecosystem/biodiversity by and large
2	Nature reserve	Specific habitat/wild and/or weedy species gene pool
3	Gene sanctuary	Ecosystem (specific)/wild species gene pool
4	On-farm conservation(mass reservoirs, bulk hybrid populations)	Agro-ecosystems/land races
5	Botanical garden/arboretum	Wild species, obsolete cultivars, tree crop germplasm
6	Field gene bank	Wild species, vegetatively propagated crops, tree crop germplasm
7	Plant organ storage	Vegetatively propagated crops, mainly in the form of roots, tubers, and bulbs
8	Seed storage	All plant species which produce fertile and orthodox seeds
9	Pollen storage	In principle all species which produce long-living pollen
10	*In vitro* storage	Wild and cultivated species which produce recalcitrant or no seeds, vegetatively propagated crops, disease-free germplasm as well as orthodox seeds
11	Cryopreservation	Germplasm mentioned above which permits cryopreservation
12	DNA and gene libraries	Special genetic stocks; in principle applicable for all germplasm

Table 2.3 Advantages and disadvantages of different ex situ conservation methods

Advantages	Disadvantages
In plantings and domestication programs	
Can conserve genetic resources in the habitats of expected use	Many areas required
Conservation of intrapopulation variation can be combined with conservation of interpopulation variation through a network of spatially separated areas	Spatial isolation to conserve population identity required
Can develop into multiple population conservation programs where new intrapopulation variation is developed as response to different conditions of growth or selection criteria	Lack of pollinators may cause problems
Can be combined with utilization	Relatively expensive if not combined with utilization
Can function as seed sources allowing rapid procurement of seed in commercial scale in early domestication	

(continued)

Table 2.3 (continued)

Advantages	Disadvantages
Seed banks	
Propagules ready for use (although the amount of seed typically is too limited to serve as input to commercial use)	Facilities required
Little space required (at least for species with small seeds)	Not applicable to species with recalcitrant seeds
Intra- and interpopulation can be easily conserved provided species range adequately sampled	Regular regeneration of seed lots pose severe practical problems even for many species
Seed can be conserved far away from the in situ environment if requested	With orthodox seed
	"Short-term storage rather than conservation" for the majority of species
Tissue culture banks	
Minimum space required	Expensive facilities required
Aseptic conservation (minimizes disease risk)	Sampling problems (representative individuals and within individual)
Time required to produce propagules for use is short	Difficult to conserve adequate number of genotypes
Germplasm can be conserved far away from the in situ environment if requested	Protocols are specific for species and often even for genotypes
	Problems of somaclonal variation and early maturation
Clonal archives	
Intra- and interpopulation variation can be conserved provided species range adequately sampled, but this will require a large number of genotypes	Suitable site(s) required
Useful method for unique phenotypes/ genotypes	Relatively expensive
Can readily provide scions for grafting, if requested	Risk of confusion with root stock (if grafted)
Clones can be conserved away from the in situ environment if requested	
Botanical gardens and arboreta	
Can be combined with demonstration and education	Suitable site(s) required
Botanical gardens are often part of very stable institutions and likely to be continuously maintained by trained staff	Difficult to collect seed due to hybridization
	In general not apt for conservation of inter and intrapopulation variation (requires a larger number of individuals than usually planted in botanical gardens/arboreta)

Table 2.4 Number of accessions of different crops in the various gene banks of the world

Crop	Accessions	Worldwide stored accessions
Wheat	850,145	14%
Rice	420,341	7%
Maize	261,584	4%
Total	1,532,070	25%
Other crops	4,615,802	75%
World total	6,147,872	

Source: Calculated according to FAO 1996a; WIEWS 1996

Table 2.5 *Ex situ* conserved PGRFA by storage type

Storage type	Number of accession	Distribution %
Seed banks	3,610,428	65%
Field banks	526,300	9%
In vitro	37,600	1%
Unknown storage facilities	1,380,177	25%
Total	5,554,505	

Source: Calculated according to FAO (1996a) and WIEWS (1996)

Table 2.6 DNA banks of the world

DNA bank	Web address
Australian Plant DNA Bank, Centre for Plant Conservation Genetics, Southern Cross University, Lismore, NSW, Australia	http://www.dnabank.com.au/index_files/index.php
BGBM DNA Bank, Botanic Garden and Botanical Museum, Berlin, Germany	http://www.bgbm.org/bgbm/research/dna/
DNA Bank Brazilian Flora Species, Rio de Janeiro Botanic Garden, Brazil	http://www.jbrj.gov.br/pesquisa/div_molecular/bancodna/index.htm
DNA Bank at Kirstenbosch, South African National Biodiversity Institute, Kirstenbosch, South Africa	http://www.nbi.ac.za/research/dnabank.htm
IRRI, DNA Bank, International Rice Research Institute, Philippines	http://www.irri.org/GRC/GRChome/Home.htm
Missouri Botanic Garden DNA Bank, St Louis, MO, USA	http://www.mobot.org/MOBOT/research/diversity/dna_banking.htm
National Herbarium Netherlands DNA Bank, The Netherlands	http://www.nationaalherbarium.nl/taskforcemolecular/dna_bank.htm
NIAS DNA Bank, National Institute of Agrobiological Sciences, Tsukuba, Ibaraki, Japan	http://www.dna.affrc.go.jp/
Plant DNA Bank Korea (PDBK), Graduate School of Biotechnology, Korea University, Seoul, Korea	http://www.pdbk.korea.ac.kr/

(continued)

Table 2.6 (continued)

DNA bank	Web address
Royal Botanic Garden Edinburgh DNA Bank, Edinburgh, Scotland	http://www.rbge.org.uk/rbge/web/science/research/
Royal Botanic Garden Kew DNA Bank, Richmond, England	http://www.rbgkew.org.uk/data/dnaBank/
TCD DNA Bank, Department of Botany, School of Natural Sciences, Trinity College, Ireland	http://www.dnabank.bot.tcd.ie/
Tropical Plant DNA Bank, Fairchild Tropical Botanic Garden and Florida International University, FL, USA	http://www.ftg.org/research

2.7 DNA Banks

Molecular techniques are becoming increasingly important in the study and management of genetic resources. DNA has been routinely extracted and stored from the nuclei, mitochondria, and chloroplasts of many plant species, together with derivatives such as RNA, cDNA, and genes. Technologies are available to allow all these to be stored quickly and at low cost in DNA banks as an insurance policy against loss of crop diversity. DNA storage has so far been undertaken with objectives other than conservation in mind; usually to allow genetic material to be made readily available for molecular applications, for distribution, or for training.

DNA banks can now be considered as a means of complimentary conservation. DNA storage is particularly useful for those species that cannot be conserved in traditional seed or field gene banks and nor conserved in situ due to high risk in that area. Although to date there are no cases where DNA banks have been assembled to specifically replace traditional methods of conserving genetic resources, the potential for DNA storage is promising due to the small sample size for storage of genetic information and the stable nature of DNA in cold storage. However, the use of DNA banks in conservation is limited as whole plants cannot be directly reconstituted from DNA nor are the original genotypes recovered. The genetic material must first be introduced artificially, through transformation or transduction using plasmids or liposomes, back into somatic cells that can then be grown into whole plants in vitro culture.

Advantages

DNA banking is an efficient, simple, and long-term method to conserve the genetic information.

Disadvantages

There are problems with gene isolation, cloning, and transfer of DNA back to a plant, and it currently does not allow the regeneration of the same genotype as the original sample.

2.8 PGRFA

The efforts to conserve PGRFA need to be undertaken with clear criteria. The expected benefits are uncertain. There is no data available to meet all relevant objectives. A balanced combination of *in situ* and *ex situ* conservation efforts is required. These are rarely managed with accuracy and efficiency. The data on how many and what kind of varieties is conserved *ex situ* is lacking. In general PGRFA encompasses material growing in farmer's field and their wild relatives as well as conserved, exchanged, utilized, and threatened ones. This is binding economically efficient approach to optimizing agro-biodiversity conservations.

Agro-biodiversity and PGRFA

Agro-biodiversity is defined broadly as "that part of biodiversity which nurtures people and which are nurtured by people" (FAO 1995a). For reason of functionality, agro-biodiversity is defined here as the diversity of existing domesticated plants and animals and is categorized here for agricultural crops as follows:

Genetic diversity – the diversity of genetic variation in one variety
Varietal diversity – the diversity of varieties in one crop species
Species diversity – the diversity of crop species in one region

Long-Term Storage Facilities

Many countries have long-term facilities for storage. Countries like Russia is with one of the world's largest gene bank which have no long-term facilities. Since the political transition, the Vavilov's Institute has had difficulties with its regeneration and collection activities, because the places of origin are no longer in the same country or in the same political sphere of influence, creating institutional and political problems (Virchow 1999). Consequently the quality of the Russian long-term storage activities can be assessed to be poor at present.

Storage Quality If viability falls below 85%, accession is threatened with extinction. Regeneration takes place every 10 years. A gene bank must regenerate 10% of its stored germplasm every year to meet the quality standard.

Accessibility Access should be there to the existing working collections for documentation and exchange of good accessions.

Documentation Documentation of the stored varieties is an essential component. *Many countries have poorly documented* ex situ *storage facilities, and only 25% of the countries show a good quality and standard in their documentation.*

2.8.1 Scope of Conservation and Documentation

The exploration, collection, conservation, and identification of potentially valuable plant genetic resources for food and sustainable agriculture, which are endangered by extinction, as well as other plant genetic resources, which have potential value for future development, are primary obligations for all countries and institutions adhering to the FAO International Undertaking on Plant Genetic Resources. The Global Plan of Action for the conservation and sustainable utilization of Plant Genetic Resources for Food and Agriculture (FAO 1996b) emphasizes the importance of ex situ conservation besides the in situ and on-farm conservation.

In concurrence with the global perspective, the focus in the developing countries' R&D is having increasing interest and awareness in the use of agro-biodiversity in enhancing sustainable use of available resources. This brings into focus the need for research on various conservation protocols and their relative significance and calls for a multidisciplinary approach to the issue of conservation (Soulé 1985). However, the rapid advances made elsewhere in the field of PGR (Hodgkin and Rao 2002) makes the selection criteria and decision on a specific conservation protocol difficult for developing countries. Limited and meager budget allocation for R&D, scarcity of capital for innovations, makes it difficult for planners to choose the direction of their research endeavors and also increases the chances of complete elimination of the option for innovation.

Speaking on the significance of economic imperative for the issue of PGR conservation, Hodgkin and Rao (2002) state:

> ……. "Some maintain that estimating the value of plant genetic diversity is futile as, it is invaluable. . However, economic aspects gain significance in view of the investment involved in the whole process of conservation. It is necessary that substantial economic analysis of costs and benefits from conservation work be undertaken, given the 'limited capital availability' situations……"

2.9 Ex Situ vs. In Situ

The costs of genetic resource conservation in gene banks are now borne largely by public investments and consumers benefiting indirectly from the genetic resource incorporated into improved crop varieties when output expands and prices declined (IPGRI 2001).

Ex situ methods suffer from some limitations (McNeely et al. 1990):

- Not feasible economically to keep more than limited sample of genetic diversity.
- Little directional habitat responsive evolution can take place ex situ so that the captive populations are not adapting to changing environmental conditions.
- Ex situ population is likely to have narrow genetic base and is unlikely to have been collected so as to ensure the representation of a wide range of genotypes.
- Ex situ conservation depends on continuity in policy and funds, which is far from assured, especially in tropics.

2.10 PGR Conservation and Need for Priority Setting

There is no accepted protocol existing for priority setting. So it is advised that a scheme of the same need must be achieved based on set of goals even though it is a complex task (McNeely et al. 1980). For different types of conservation actions, the established priorities will affect conservation of different entities.

Applicability

First and foremost is the suitability or applicability of a specific conservation protocol for a given objective. The in situ methods such as the biosphere or a gene sanctuary provide wider applicability and provide conservation for almost unlimited period of time, characterized by wider heterogeneity in its activities and institutions, while some of the *ex situ* methods are of value for more specific objectives, especially in conserving agro-biodiversity. As the main aim of conservation protocol revolves around obtaining maximum diversity with minimum number of accessions, within a minimum time and financial commitment, any *ex situ* program should aim for the selection of the host material from a genuine biodiversity "hot spot" (Ganeshaiah and Uma Shankar 1998).

The most effective and appropriate combination of conservation methods both in situ and ex situ should be used to meet the aims of security, efficiency, accessibility, and sustainability, balancing the advantages of one method against the disadvantages of others. Seeds are best suited for storage in gene banks but for species that do not set seeds or produce them sterile or recalcitrant, it is difficult to conserve. Vegetatively propagated crops will compulsorily go for methods of conservation that use material other than seed. Many vegetatively propagated species may be maintained as field collections with a complementary storage in vitro. Successful in vitro conservation of plantain, banana, cassava, yam, potato, strawberry, sweet potato, and *Allium* spp. has already been reported (Ashmore 1999). Slow growth storage is routinely used for the conservation of a number of species – notably Musa, potato, yam, and cassava – in regional and international germplasm conservation centers such as INIBAP, CIP, FAL, IITA, CIAT, and NBPGR.

Resource Requirement

The second most important factor that needs consideration in prioritizing conservation protocol is the resource requirement for different protocols. Resources other than the permanent structures like the land and infrastructure require special attention due to their specificity to the method in use. For techniques like cryopreservation and in vitro conservation, special skills for human resource, specific equipment, or chemical and regular supply of the same, especially along with others like controlled atmosphere, are of significance. The availability of these resources or their cost factors could alter the selection criteria based on the budgetary reasons. Resource requirement is often location specific, as it could change with various geographic regions.

Risk Involved in Conservation
Identifying and measuring risk is a complex phenomenon. Since the objective of a conservation protocol is the sustainable maintenance of the material, the main source of risk originates from the probability and time span taken for the change to occur. An in situ system is most likely to face this problem due to the high probability of segregation of gene expression and mutation with every successive generation of replication.

Regeneration
One of the most significant factors in PGR conservation is the time taken for regeneration of the conserved material and the retention capacity of the specific character. Success rate of reinstatement, change in the conserved germplasm, and the time taken for its regeneration are the most important features that weigh in favor of modern methods such as in vitro or cryopreservation as compared to the traditional methods. It is also of significance to mention here that the preference of a method needs to be assessed both from the local as well as global perspective in prioritizing its selection. About half of all gene bank accessions worldwide urgently require rejuvenation.

Conservation Costs Different approaches are required in identifying the specific costs of conservation. Method of estimating costs depends upon the approach taken. There are inherent difficulties and limitations in compiling data on current expenditure on PGRFA conservation. Costs of ex situ conservation depend on specific crop to be conserved and conservation method applied. There are difficulties and limitations in compiling data on costs only to know the order and magnitude of current efforts. Cost-effectiveness of conservation has not been discussed yet. Total costs of maintaining genetic resources US D 50 million/year and global conservation efforts amounts US D 337 million.

Conservation Benefits The value of genetically coded information can never be determined a priori but rather only from a posteriori observation that is as a result of their success on the market. To assign monetary value to the benefits generated by PGR conserved in ex situ collections is exceedingly difficult because they have multiple dimensions. The use of germplasm in plant breeding leads to changes in crop output, and breeder's requests represent the demand for germplasm that derived from demand for the crop in production and trade. Little information is available on the germplasm movement. Only some gene banks and national programs have drawn up some statistics. Over the past 3 years, for instance, the CGIAR centers have distributed an annual average of over 50,000 accessions to national program all over the world (SGRP 1996). Similarly between 1992 and 1994, the USA distributed over 100,000 samples each year (FAO 1996a, b, c, d). About 34% of all accessions are stored in public gene bank of developing countries, 49% in developed countries, 1% in private companies, and 0.2% in local conservatories (Iwanaga 1993). In India, Ethiopia, and China, the expected value for PGRFA conservation is assessed as very high (Virchow 1999).

2.10.1 Strategies for Fish Conservation

The conservation and management of aquatic resources is important for sustainable utilization of fisheries potential for the economic advancement of farmers and fishery workers. As the biotechnological and genomic revolutions are ready to take off, the sovereign rights of native inhabitants to intellectual property rights (IPR) related to aquatic genetic resources and associated traditional knowledge are also integral components of management strategies for genetic resources.

Genetic information has helped in the design and implementation of plans to protect and preserve threatened and endangered species. This information should be supported by comprehensive knowledge of the aquatic landscapes needed to preserve and restore habitats for native species (Lakra et al. 2007). To make the fisheries sector economically sustainable, the conservation management of aquatic resources is very important. The IPR and traditional knowledge related to aquatic resources are integral component for the management of these resources.

2.10.2 Conservation of Farm Animal Genetic Resources

In a local, national, regional, and world scale, conservation of farm animal genetic resources is a major concern. Farm genetic resources is mainly conserved in situ in farmers' field as a part of normal production system, which is also recommended by the national scientific community. As a complementary approach to preservation of genetic material, it is also practiced as an ex situ conservation strategy. In situ conservation is preferred because, if necessary, there is a possibility of reconstruction of population from conserved material in future; certain animal material like poultry embryo and semen of donkeys, sheep's, pigs, and chicken are impossible to cryopreserve, and keeping the live animals is the only way to adopt climate change.

2.10.3 Buffalo Conservation Strategies

In the case of buffalo, generally, sperm, oocytes, embryos DNA, and embryonic stem cell are conserved. Now there is a possibility of utilization of a variety of cells for conservation. Through this, conservation of endangered species is possible by:

1. Sperms and oocytes.
2. Embryos: In case of buffaloes, sheep, goats, and horse, cryopreservation of embryos have been successful and produced offsprings. This technique could be best tool in conservation of deployed zygote with all genetic information.
3. Storage of DNA: The genetic material of animals could be stored by cryopreservation of DNA.
4. Cloning of somatic cell: It offers the technique of production of exact replica of a given animal.

5. Embryonic stem cells: This potent embryonic cell has the capacity to develop viable embryos, and it is derived from culture of inner cell masses of young blastocyst.

In situ conservation is the best form of conservation of a population, and preserving their semen is ideal.

Data Bank Strategy For an effective conservation, the maintenance of database containing important breeds, population census, and ecological data is essential. NBAGR of Indian Council of Agricultural Research is engaged in creation and maintenance of this database.

Gene Bank Strategy Cryopreservation of semen from indigenous breeds is carried out to use the same in future. This needs to be replicated at least in two places in sufficient quantity. Periodical monitoring of the same is required to be done for effective utilization.

DNA Bank Strategy There is a provision of storing DNA fragment through cryopreservation. This has got several advantages over the storage of live cells since it is economical, occupies less space, and free from spread of diseases. Many countries are practicing this strategy.

Somatic Cell Strategy Dolly sheep heralded the advent of somatic cell technology in a big way. Future is bright to produce animals from a somatic cell through cloning.

2.11 Conservation of Avian Genetic Resources

IUCN (2012) categorizes 13.6% of non-avian species as endangered. So there is no need to emphasize the need of conservation of avian genetic resources. Since the conservation strategy for mammals cannot be applied directly to avian resources, there is a need to develop an integrated conservation system for these resources. This is mainly due to unique characteristics of avian egg which poses the question of developing of an alternative strategy of conservation for oviparous species of animals. There is a possibility of conserving embryos of chicken and retrieval of the same in full form.

2.12 Conservation of Microbial Resources

Exploration, identification, conservation, and utilization of microbial resources of the world need to get priority since it is important for mankind in particular and biota and environment in general. Many microbial collections are maintained by various countries for various purposes. These collections may be small or big catering the needs of scientific communities and bio-industries.

Microorganisms, their cells, or replicable parts (e.g., genomes, plasmids, viruses, cDNAs) are the tools for biotechnology and underpin the life sciences. The vast majority of microorganisms around the globe still remain hidden and need to be explored, identified, conserved, and utilized for the benefit of humankind in particular, and the biota and environment, in general. Microbial culture collections are established in many countries around the world having a variety of purposes. These range from small, specialized collections that support small groups of researchers to the large international public service collections that provide reference materials and services to the scientific community and bio-industries.

2.12.1 Biobanks

Biobanks are also known as Biological Resource Centers (BRCs) adding value to scientific research and industry and also helps in conservation. Without BRCs scientific research involving microbial cultures and cell lines are almost impossible. This can also be called depositories of biological standards holding taxonomic and other reference strains which are also cited in published research reports, with which commercial and ethical issues are also involved. Establishment and maintenance of BRCs involved knowledge and skill which are abundantly required. Even though it is an apt preservation technique, appropriate quality assurance needs to be in place for the recovery of cultures which ensures materials to perform like the original ones. There are many important issues involved in the operation of BRCs in long term.

2.13 Conclusions

Conserving biological resources is an absolute necessity for existence of the universe. Whether it is life in the form of a single-celled microorganism or in the form of a multicellular complex tree or animal, it has got its own relevance in the ecosystem. Deterioration of biological resources takes place in an alarming pace, and it is the responsibility of everybody to conserve the remaining resources to maintain a biologically rich ecosystem for the future generations. Technology has developed in all spheres of life, and one must make sure that the breakthroughs in technology are properly utilized in the conservation strategies.

References

Alexander MP, Ganeshan S (1993) Pollen storage. In: Chadha KL, Pareek OP (eds) Advances in horticulture, Fruit crops Part I, vol I. Malhotra Publishing House, New Delhi, pp 481–486

Ashmore SE (1999) Status report on the development and application of in vitro techniques for the conservation and use of plant genetic resources. IPGRI, Rome

Barnabas B, Kovacs G (1997) Storage of pollen. In: Shivanna KR, Sawney VK (eds) Pollen biotechnology for crop production and improvement. Cambridge University Press, Cambridge, pp 293–314
Benson EE (2008) Cryopreservation of phytodiversity: a critical appraisal of theory & practice. Crit Rev Plant Sci 27:141–219
FAO (1995) Report of the sixth session of the commission on plant genetic resources, 19–30 June 1995. Document CPGR-6/95 REP. FAO, Rome.
FAO (1996a) Report on the state of the world's plant genetic resources for food and agriculture. FAO, Rome, p 75
FAO (1996b) Global plan of action for the conservation and sustainable utilization of plant genetic resources for food and agriculture. FAO, Rome, p 63
FAO (1996c) Report of the international technical conference on plant genetic resources, Leipzig, Germany, 17–23 June 1996. Document: ITCPGR/96/Rep. FAO, Rome
FAO (1996d) Respective country report. Submitted to FAO in the preparatory process for the international technical conference on plant genetic resources, 199
Ganeshaiah KN, Uma Shaanker R (1998) Contours of conservation – a national agenda for mapping biodiversity, special issue. Curr Sci 75(3):292–298
Ganeshan S, Rajasekharan PE (1995) Genetic conservation through pollen storage in ornamental crops. In: Chadha KL, Bhattacharjee SK (eds) Advances in horticulture 1995, vol.12-Part-1. Ornamental Crops Malhotra Publishing House, New Delhi, pp 87–108
Ganeshan SPE, Rajasekharan S, Shashikumar WD (2008) Cryopreservation of pollen. In: Reed BM (ed) Plant cryopreservation: a practical guide. Springer, New York
Grout BWW, Roberts AV (1995) Storage of free pollen, pollen embryos and the zygotic embryos of seed by cryopreservation and freeze drying. In: Grout B (ed) Genetic preservation of plant cells *in vitro*. Springer, Berlin, pp 63–74
Hodgkin T., Ramanatha Rao V (2002) People, plants and DNA: perspectives on the scientific and technical aspects of conserving and using plant genetic resources. In Engels JMM, Ramanatha Rao V, Brown AHD, Jackson MT (eds) Managing plant genetic diversity. CAB International Wallingford and IPGRI, Wallingford/Rome, pp 469–480
Hoekstra FA (1995) Collecting pollen for genetic resources conservation. In: Guarino L, Rao VR, Reid R (eds.) Collecting plant genetic diversity Technical guidelines. CAB International, Wallingford pp 527–550.
IPGRI (2001) Annual report IPGRI. Rome
IUCN Red List (2012) URL: http://www.iucnredlist.org/documents/summary statistics/2012 2 RL Stats Table1.pdf. Last Accessed 3 Dec 2012
Iwanaga M (1993) Enhancing the links between germplasm conservation and use in a changing world. Int Crop Sci 1:407–413
Lakra WS, Mohindra V, Lal KK (2007) Fish genetics and conservation research in India: status and perspectives. Fish Physiol Biochem 33:475–487
Larkin PJ, Scowcroft WR (1981) Somaclonal variation – a novel source of variability from cell cultures for plant improvement. Theor Appl Genet 60(4):197–214
McNeely, Written JA, Written AJ, Home A (1980) Saving Siberia: a conservation master plan WWF. Bogor
McNeely JA, Miller KR, Reid WV. Mittermeier RA Werner TB (1990) Conserving the world's biological diversity. The World Bank/WRI/IUCN/Conservation International/WWF, Washington, DC
Rajasekharan PE (2015) Gene banking for e*x situ* conservation of plant genetic resources. In: Bir Bahadur et al. (eds) Plant biology and biotechnology: volume II: plant genomics and biotechnology, doi: 10.1007/978-81-322-2283-5_23. Springer, New Delhi
Rajasekharan PE, Ganeshan S (2003) Feasibility of pollen cryopreservation in Capsicum sp Capsicum eggplant. News Lett 22:87–90

Rajasekharan PE, Sahijram L (2015) In vitro conservation of plant germplasm. In: BirBahadur et al (eds) Plant biology and biotechnology: Volume II: Plant genomics and biotechnology. doi: 10.1007/978–81–322-2283-5_22. Springer, New Delhi

Rajasekharan PE, Ravish BS, Vasantha Kumar T, Ganeshan S (2013) Pollen cryobanking for tropical plant species in conservation of tropical plant species, Normah MN,·Chin Barbara HF, Reed M (eds). Springer, New York, pp 65–75

SGRP (1996) Report of internally commissioned external survey of CGIAR gene banks operations. IPGRI, Rome

Simmonds NW (1962) Variability in crop plants, its use and conservation. Biol Rev 37:442–465

Soulé ME (1985) What is conservation biology?: a new synthetic discipline addresses the dynamics and problems of perturbed species, communities, and ecosystems. Bio Sci 35(11):727–734

Towill LE (1985) Low temperature and freeze/vaccum-drying preservation of pollen. In: Kartha KK (ed) Cryopreservation of plant cells and organs. CRC Press, Boca Raton, pp 171–198

Towill LE, Walters C (2000) Cryopreservation of pollen. In: Engelmann F, Takagi H (eds) Cryopreservation of tropical plant germplasm—current research progress and applications. JIRCAS/IPGRI, Tsukuba/Rome, pp 115–129

UNEP (1992) Conserving the worlds biological diversity. IUCN, Gland

Virchow D (1999) Conservation of genetic resources: costs and implications for sustainable utilisation of PGRFA. Springer, Berlin

WIEWS (1996) Data information from world information at early warning system. FAO

Biodiversity and the Processes of Law

3

Shalini Bhutani and Kanchi Kohli

Abstract

National laws on biological resources have emerged in response to the international legal framework on biological diversity – the United Nations Convention on Biological Diversity (CBD), 1993. The Convention is the international law for conserving biodiversity, ensuring sustainable use of its components and sharing benefits arising out of the use of genetic resources. This body of law comprises principles, guidelines and protocols for national practice. Two key protocols have developed under the Convention through intergovernmental processes; one is the *Cartagena Protocol on Biosafety* in effect from 2003. The other is the *Nagoya Protocol on Access to Genetic Resources and the Fair and Equitable Sharing of Benefits Arising from their Utilization to the Convention on Biological Diversity*, which entered into force in October 2014. Both these represent different dimensions of the relationship between bioresources and modern biotechnology. Distinct legal and regulatory regimes are developing for each of them at the country level.

In India the national law in compliance with the Convention, namely, the Biological Diversity Act, 2002, began to be implemented after executive rules were issued by the central government in 2004. Rules for biosafety predate this Act and the Convention. This chapter traces the broad trends that have emerged in the decade (2004–2014) of implementation of the Act, with specific focus on

Both co-coordinate the pan-India *Campaign for Conservation and Community Control over Biodiversity.*

S. Bhutani (✉) • K. Kohli
Kalpavriksh Environment Action Group, Pune 411004, Maharashtra, India
e-mail: emailsbhutani@gmail.com; kanchikohli@gmail.com

S. Abdulhameed et al. (eds.), *Bioresources and Bioprocess in Biotechnology*,
DOI 10.1007/978-981-10-3573-9_3

the two aspects in the CBD protocols. Firstly, it focuses on the emphasis on access, which is of interest to the plant biotechnology industry. Secondly, it highlights how the issue of biosafety has been handled under the legal regime. This chapter elaborates how the due processes of law for biodiversity and the regime for biotechnology interface with each other in a megadiverse country.

Keywords
Access • Biodiversity • Bioresources • Biotechnology • Biosafety • Convention • Law • Protocol • Regulatory

3.1 Introduction

Process is understood to be a continuous ongoing action, as it is with lawmaking and implementing the law once it is made. The term 'due process of law' implies that an individual cannot be deprived of her life, liberties or property without there being appropriate law to do so. This puts an obligation on the state to put in place laws that guard ordinary peoples against arbitrary and abusive actions. The laws may also need to be constantly updated or amended to meet the intended objectives. The processes of law and their purposes are influenced by the political economy. So it is with biological resources, hereinafter called bioresources.

The needs of the life science industry have to a large extent determined the body of law on bioresources. Nonetheless, the objectives of a law on bioresources, particularly in a country like India, can only but be multiple. For on the one hand, it is rich in bioresources and attractive to bioprospectors, while on the other hand, it is equally keen to spur bioenterprises in its territory. The law has to effectively regulate both aspects in the domestic space. At the same time, as India is a party to the CBD and subscribes to the international law, its own law has to be in line with the Convention that has developed at the international level.

Therefore, it becomes even more critical to understand the role of law and how it regulates biodiversity conservation, sustainable use and modern biotechnology and in doing so how it is able to balance seemingly conflicting ends. Firstly, as with every other law, its role is of regulating access to and use of biological material and knowledge based on which research is to be carried out. Second is to impose penalties if certain legally prescribed rules for biosafety or procedures laid down for approval of genetically modified (GM) products are violated. In doing so the aim is to either prevent from harm or provide redress for damage that might have been caused by living modified organisms (LMOs). Third is to ensure that the benefits that accrue from the access to otherwise commonly or privately held biological materials are shared equitably amongst local communities of the provider country.

Yet mere existence of international and national legal regimes does not always translate into compliance. And between governments, political and economic interests do come into play in treaty negotiations and subsequently in their application. There are practical challenges too around interpretation and implementation;

nonconformity or selective application also shapes the legal narrative on bioresources. This narrative in concept and as much in practice is dynamically evolving in response to developments at the global level and demands at the national and local level.

3.2 International Law

The United Nations (UN) Convention on Biological Diversity (CBD) is one of the most significant outcomes of the UN Conference on Environment and Development at Rio de Janeiro, Brazil, in 1992 (United Nations 1992). At Rio, the CBD emerged from the worldwide concern to protect biodiversity loss and check 'biopiracy' in the global south. Even though the process to formulate such an international law had started in 1988 (CBD 2015 undated), it was at that Rio Earth Summit at 1992 that the CBD was opened for signature. The Convention finally entered into force on 29 December 1993.

Bioresources acquire a specific definition under international law. According to Article 2 of the CBD text, 'biological resources' are:

> *genetic resources, organisms or parts thereof, populations, or any other biotic component of ecosystems with actual or potential use or value for humanity.*

But this may not offer enough guidance when applied in the real time locally. For the Indian experience has shown that there could be varying perceptions about what constitutes a bioresource (*more on this in the section on India's Legislation*). In the same article, the CBD defines 'biotechnology' broadly as:

> *any technology that uses biological systems, living organisms or derivatives thereof, to make or modify products or processes for specific use.*

The CBD's international protocols[1] – kinds of thematic sub-treaties – take forward and elaborate two very specific aspects of the use of bioresources. The Cartagena Protocol deals with the specific issue of biosafety when products of modern biotechnology are used. And the Nagoya Protocol deals with the issue of access and benefit sharing (ABS) with respect to bioresources.

3.3 Cartagena Protocol

The *Cartagena Protocol on Biosafety* (CPB) entered into force on 11 September 2003 as a supplementary agreement to the CBD. Amongst the articles in the CBD text, Article 19 specifically deals with the *Handling of Biotechnology and*

[1] In international law, a protocol is a legal instrument that is subordinate to a convention and is meant to take forward the convention's objectives, while also either amending the convention or further detailing an aspect of it (as in the case of the Biosafety Protocol of the CBD).

Distribution of its Benefits. Paragraph 3 of the said Article expressly requires that countries who are members of the CBD:

> *shall consider the need for and modalities of a protocol setting out appropriate procedures, including, in particular, advance informed agreement, in the field of the safe transfer, handling and use of any living modified organism resulting from biotechnology that may have adverse effect on the conservation and sustainable use of biological diversity.*

The Protocol comprises the set of rules to be followed by member countries of the Protocol in case of movement of living modified organisms (LMOs) across borders. It prescribes safety measures for the transboundary movement of LMOs. LMOs as defined by the Protocol 'biotechnology' is the application of:

> *a. In vitro nucleic acid techniques, including recombinant deoxyribonucleic acid (DNA) and direct injection of nucleic acid into cells or organelles, or*
> *b. Fusion of cells beyond the taxonomic family, that overcome natural physiological reproductive or recombination barriers and that are not techniques used in traditional breeding and selection;* – Paragraph 3(i)

The Protocol is premised on the precautionary approach. This is derived from Principle 15 of the Rio Declaration on Environment and Development (UNEP 1992).[2] Accordingly, advocating precaution its 40 articles elaborate the international regime on biosafety. It is based on the idea that a country cannot exercise caution and regulate LMOs unless it is aware of them being transported into its area. Therefore, it requires for Advance Informed Agreement (AIA) to be signed before LMOs are shipped to another country. This means both the exporting and the importing country regulations on biosafety must provide for an AIA procedure. The Protocol in a way accepts that there will be trade in LMOs between countries. For most products of modern biotechnology have commercial applications. However, as per the Protocol, GM products to be exported as food and feed and for processing do not require an AIA (CBD 2000).

The implications of the precautionary approach in environmental decision-making are that:

- It requires preventive action in the face of uncertainty about a technology.
- It shifts the burden of proof on the technology provider and scientific community to persuade users about the safety of the technology.
- It puts the liability on the proponents and developers of the activity, for example, in the context of modern biotechnology for any false or misleading information, the responsibility will be that of all the people and institutions associated with the technology.

[2]Principle 15: In order to protect the environment, the precautionary approach shall be widely applied by States according to their capabilities. Where there are threats of serious or irreversible damage, lack of full scientific certainty shall not be used as a reason for postponing cost-effective measures to prevent environmental degradation. (UNEP 1992)

- It requires a defining liability of a user of a technology done so with the knowledge of the risks involved.
- It requires an assessment of a wide range of alternatives to the possibly harmful actions be explored before deploying the risky technology.
- It insists that public participation be increased in decision-making.

It follows that a law premised on the precautionary principle (PP)[3] must incorporate all these elements.

Meanwhile, there is no agreement worldwide on the degree of risk from LMOs. Nonetheless, what makes biosafety a risk-prone endeavour is that LMOs do not follow the laws of the land; they follow the laws of nature. The risk factor is due to reasons internal to the LMO and its new genome; it is also external vulnerabilities that it can generate for human and ecological health. Risk in this context is the probability or chance of danger or harm to human and ecological health from the use of LMOs and their application in the open environment. The Cartagena Protocol prescribes all three elements of what must form a risk analysis framework:

- Risk assessment
- Risk management
- Risk communication

Negotiations for this Protocol began under the CBD in 1994, and it took nearly a decade-long process for countries to agree on a text and for the text to enter into force. This was due to the opposing viewpoints on the issue of safety of LMOs. For there are countries that consider the risk factor and are opposed to LMOs, while those pro-LMOs are opposed to any strong legal restrictions on their use.

As of 2014, 194 countries world over are members of the CBD. But the success of this international law depends on the capacities of individual countries and their government's commitment to the very idea of biosafety. Amongst the many challenges in the implementation of CBD is the fact that one of the key proponents of biotechnology products – USA – is not a party to the CBD. Moreover, in its own biosafety framework, the USA regards GM products, such as novel foods 'substantially equivalent' to those that are made without the use of genetic engineering (GE). As per the USA, products of GM do not trigger any special regulatory consideration. The Coordinated Framework for Regulation of Biotechnology (1986) in the USA focuses on the nature of the product and not on the process by which they are produced. In fact, the reason that the USA has to date kept out of the CBD is because of the Convention's treatment of biotechnology. Given that it is not a party to the CBD, it is also not a member of either of the protocols that have developed under it.

The European Union (EU) as a whole has rejected the idea of substantial equivalence with respect to novel foods, containing or consisting LMOs (Schuazu 2000).

---

[3] In broad terms, the precautionary principle works on a premise that an action should not be taken if the consequences are uncertain and could be potentially dangerous.

This points to the competing regulatory approaches between the USA and EU on the issue of biosafety. The differences are not simply in the domestic regulatory framework, but also in the systemic principles that make the foundation of the regulation. EU member countries by and large are for the precautionary approach towards modern biotechnology. As per this approach, biosafety laws can be made less stringent only once there is adequate proof that LMOs/GM products are safe. This is seen more in line with the idea of sustainable use, an idea taken forward in the CBD. The value of EU collectively holding to its position on biotechnology is better understood by the fact of how critical Europe's role is for global environmental governance and the international law in this area (Vogler 2005).

In Asia, there are few, if any, coordinated regional-level regulatory approaches to biosafety despite the fact that most countries in the region are members of the CPB. Those that are there, such as the South Asia Biosafety Program, are activities funded by US donor agencies to engage the governments in India, Bangladesh and Pakistan in the making of national regulatory frameworks for GM products.[4] Individually these countries are at different stages of updating their biosafety regulatory framework.

3.4 Nagoya Protocol

Before a plant, genetic material or any other bioresource can be altered at the genetic level, it has to be accessed by and be physically available to the potential user for such alteration. The raw (genetic) material as it exists in nature may be from a particular region in the world, while the laboratory or enterprise with the technology to either undertake research on it or use it for developing a commercial product might be in the location of another political territory. This creates another legal challenge – to evolve international rules that are globally respected and locally applicable, for lawful access to bioresources.

These are the rules contained in the other significant Protocol under the CBD, which is the *Nagoya Protocol on Access to Genetic Resources and the Fair and Equitable Sharing of Benefits Arising from their Utilization to the Convention on Biological Diversity*. It entered into force in October 2014 after 6 years of intergovernmental negotiations (Secretariat of the Convention on Biological Diversity 2011). As of October 2015, 68 countries including India are members of the Protocol. As the title suggests, it deals specifically with the issue of access and benefit sharing (ABS). The Nagoya Protocol is premised on the principle that states have *sovereign rights* over their bioresources. This principle has been enunciated by its parent convention, namely, the CBD. Post-CBD, bioresources are no longer the 'common heritage of humankind' for anyone to access freely.

[4]The South Asia Biosafety Program (SABP) is an international development programme conducted with the support of the US Agency for International Development (USAID).

The assertion of sovereign rights over bioresources means that national governments can legally provide for the recognition of their people's rights over bioresources in their local areas. The idea of sovereignty does not give the government itself the power to sell. In practical terms it means that local communities have a legally protected position from which to negotiate the terms of access, when the life sciences industry seeks bioresources from their areas. It follows that the Nagoya Protocol does not apply to areas beyond national jurisdiction (ABNJ), where countries do not have sovereign rights, for instance, the ocean bed or marine areas outside territorial waters of a country.

The Convention and the Nagoya Protocol seek to ensure that access to local bioresources does not happen without due process of law. It is meant to effect returns for the provider country or community in return for access. One of the three key objectives of the CBD, as stated in Article 1, is:

> *fair and equitable sharing of the benefits arising out of the utilisation of genetic resources, including by appropriate access to genetic resources and by appropriate transfer of relevant technologies, taking into account all rights over those resources and to technologies, and by appropriate funding.*

'Appropriate access', after due consideration of rights of indigenous peoples and local communities over the bioresources present in their areas, has to be provided for by the law implementing the CBD and its Nagoya Protocol.

The Convention insists that access to genetic resources and reciprocal transfer of technologies must be relevant to the purposes of conservation and sustainable use of biodiversity. This is clearly laid out in Articles 15 and 16 of the CBD. Biodiversity-rich countries like India are required by CBD to facilitate access to their genetic resources by non-Indians. But technology-rich countries are too required to provide access to technology. The purpose of the Nagoya Protocol is to elaborate the mechanism of ABS to be applied by both provider and user countries. But the provider countries are under relatively more pressure to set up appropriate processes of law for the access of bioresources by outsiders and non-national persons, natural or legal.

Few other (user) countries have policies or guidelines to regulate their domestic players seeking access from (provider) communities/countries. For example, Australia has a national genetic resources policy: *Nationally Consistent Approach for Access to and the Utilisation of Australia's Native Genetic and Biochemical Resources* (NCA), 2002 (Natural Resource Management Ministerial Council 2002). The government of Japan has too framed Guidelines on Access to Genetic Resources for Users in Japan, 2005 (METI and JBA 2005). Given the nature of the issue, it is not sufficient to simply have a domestic-level ABS law. The user countries must be willing to co-operate and comply with the ABS regime of the country providing bioresources. Further, user countries must also provide for legal and administrative mechanisms for benefit sharing in their own jurisdictions when using India's bioresources/people's knowledge.

EU Regulation No. 511/2014 on compliance measures for users from the Nagoya Protocol adopted by the European Commission on 16 April 2014 entered into force on 9 November 2015. The EU Regulation defines 'user' as *a natural or legal person that utilises genetic resources or traditional knowledge associated with genetic resources*. It lays down a set of obligations for users (in Chapter II, Article 4), for them to be compliant with the NP, requiring access to be only on mutually agreed terms. The Regulation makes it mandatory for (first) accessors and subsequent users to maintain an internationally recognised certificate of compliance. The first such certificate, issued to a researcher from the University of Kent in the UK, was deposited by India at the CBD's ABS Clearing House on 7 October 2015 (Secretariat of Convention on Biological Diversity 2015).

3.5 India's Legislation

In India there are a range of legislation and regulations, which are relevant to the conservation and use of bioresources and associated people's knowledge. But those key for bioresources and biotechnology are:

- The Biological Diversity Act, 2002 (and corresponding rules, notifications, guidelines, etc.)
- Rules for Manufacture, Use, Import, Export and Storage of Hazardous Microorganisms/Genetically Engineered Organisms or Cells, 1989 (issued under the Environment Protection Act, 1986)

This chapter discusses the processes of these two frameworks, which are most critical from the regulatory point of view for access and biosafety. Other than these, several intellectual property laws [such as the Indian Patent Act, 1970 (and its three amendments), as well as the Protection of Plant Varieties and Farmers' Rights Act, 2001, and its implementing rules and notifications] and guidelines of the central government also have a bearing on bioresources. But these are not the intended subject matter of this chapter and therefore not elaborated.

The legal regime discussed in this chapter was drafted when the policymakers saw India primarily as a provider country. But in the last two decades, given the macroeconomic 'reforms', there has been a discernible shift in governmental perception in also considering India's interests as a user country. The life sciences industry is being actively encouraged in the country. Thus, India's policymakers do not wish to impose too onerous access conditions through the regulatory regime, as it also sees itself as a user country that in reciprocity wants to be able to access technology with equal ease from developed countries. This in part explains the processes of law on bioresources and particularly how its ABS and biosafety provisions are evolving.

3.5.1 The Biological Diversity Act

The Biological Diversity Bill was first placed before the lower house of the Parliament of India in 2000. From there it was referred to a Department-related Parliamentary Standing Committee on Science and Technology, Environment and Forests. The Committee gave its report in December 2001(Rajya Sabha Secretariat 2001). Both houses of the Indian Parliament passed the Biological Diversity (BD) Act in 2002. After this lawmaking process, the Bill became an Act coming into force in 2003, precisely a decade after the CBD entered into force.

The BD Act draws its objectives from the CBD and thereby conservation and sustainable use of biodiversity are the first two of its aims. The third objective is to ensure equitable sharing of benefits arising out of use. For the purposes of this chapter, it would also be important to understand the manner in which the BD Act defines 'bioresources':

> *"biological resources" means plants, animals and micro-organisms or parts thereof, their genetic material and by-products (excluding value added products) with actual or potential use or value, but does not include human genetic material;*

According to Section 2(c) of the national BD Act, biological resources are:

> *plants, animals and micro-organisms or parts thereof, their genetic material and by-products (excluding value-added products) with actual or potential use or value, but does not include human genetic material.*

The 'use' envisaged in the biodiversity framework is primarily that for the needs of the life sciences industry.

The extent of coverage of the definition for bioresources in the national law was contested before the National Green Tribunal (NGT) in 2013[5] (Kohli and Bhutani 2013), a decade after the Indian BD Act came into force. The biodiversity board of Madhya Pradesh (MP) chose to take a substantially expanded meaning of the terms 'biological resources' and 'commercial utilisation' used in the BD Act. That by expectation of the MP SBB would widen the net to bring in activities under the ABS regime, which could effect more benefit sharing.

The Act also lays out an institutional framework for the implementation of the provisions of the law. The National Biodiversity Authority (NBA) is the apex institution set up to implement the BD Act and was established on 1 October 2003, as a functionary of India's Environment Ministry. The NBA is at the head of the institutional structure, with state-level biodiversity boards (SBBs) to be established in all 29 states of India.[6] The All India Biotech Association (AIBA) that gave its suggestions to the Parliamentary Committee considering the BD Bill, 2000, had

[5] Cases against the MP SBB by companies such as Agro Solvent and Lilason Breweries fought before the Bhopal Bench of the NGT.

[6] List of State Biodiversity Boards as on 18 November 2015. http://nbaindia.org/link/241/34/1/SBBs.html

recommended that the biotech industry be given representation in the NBA and the SBBs. Out of the 29 SBBs, 21 have also either drafted or notified their state-level Biological Diversity Rules. The law also requires local-level biodiversity management committees (BMCs) to be set up in every village or municipality. Each of these institutions has a role to play in the process of regulating access, determining benefits and monitoring violations.

Sections 3, 4 and 6 of BD Act together with Rules 14–19 of the BD Rules lay down the procedure to be followed for access to Indian bioresources and/or associated traditional knowledge. The law, in line with the CBD, makes clear that its main focus is to regulate the use of bioresources and related people's knowledge by non-Indian persons. Thereby, the Act focuses on regulating access by non-Indian persons, both natural and legal. The procedure for access by Indian persons is less regulated, with both local people and traditional healers not under its purview, and neither there being as strict rules for Indian companies as compared to non-Indian. While foreign persons have to take permissions prior to any sort of access including research, biosurvey and commercial utilisation, the law requires Indian enterprises to merely inform the relevant SBBs in the state in which they are based. When it comes to seeking intellectual property (except in the case of plant variety protection), both Indians and foreign entities have to mandatorily take permission from the NBA. In all these instances, the NBA and the SBBs are required to consult the local-level BMC, in both rural and urban areas.

A schematic diagram depicting the access procedure designed by the NBA explains the process.

The NBA sets up several thematic expert committees to develop guidelines and oversee the implementation of specific aspects of the law. One of these is the Expert Committee on Access and Benefit Sharing (EC on ABS) to carry out its functions in this area.[7] This is a standing committee that is periodically reconstituted with members drawn from diverse backgrounds. The EC on ABS processes access applications. After the 18-step access procedure is followed, the NBA enters into an ABS agreement with the access applicant (Fig. 3.1).

The access procedure involves consulting not only SBBs but also the local-level BMCs, which are to be set up in every local body (see Step 5 in the diagram). This is how the idea of sovereignty principle that CBD lays down is sought to be realised through the India's BD Act. Yet the Act only insists on consultation and not prior informed consent of BMCs. In practice this consultation if at all is taking place wherever they have been established in the country. The local-level procedure is also meant to identify the legitimate 'benefit claimers', to whom benefits can be channelised to once an ABS agreement is in place.

Growers and cultivators from within the local communities in an area do not need to intimate any government body, i.e. the SBBs, to access resources, as domestic companies (whether small firms or large corporations) need to do before such

[7]NBA Office order dated 6 January 2015 on Reconstitution of Expert Committee on Access and Benefit Sharing for processing the applications received by NBA. http://nbaindia.org/uploaded/committee/OO_Re__EC_on_ABS.pdf

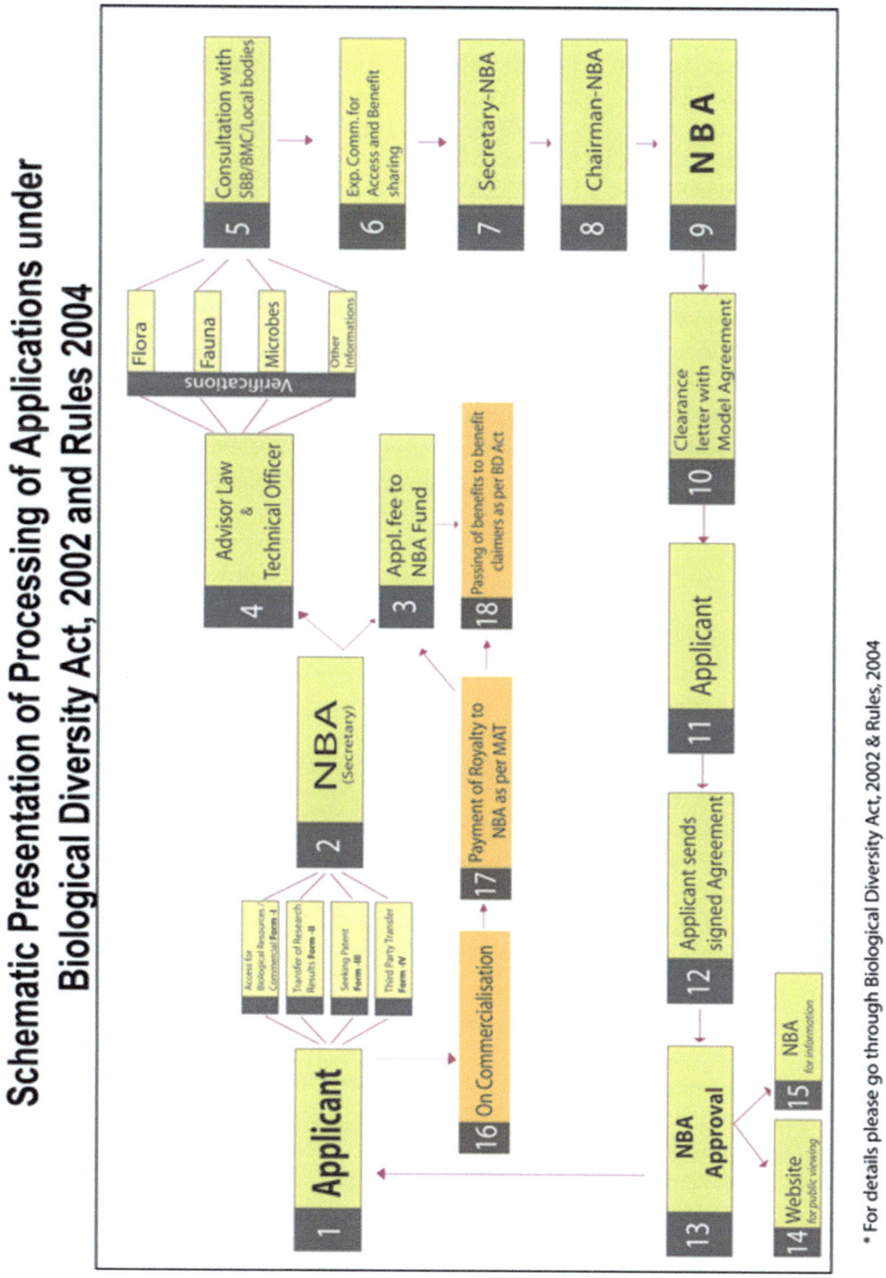

Fig. 3.1 Access application process (Source: NBA website http://nbaindia.org/content/684/62/1/applicationprocess.html)

access [Section 7]. There are three other exemptions to the ABS requirements, which are important to understand the regulatory framework around access and determining benefits. These are:

1. Exemption [under Section 6(3) of the BD Act] to any person making an application for any intellectual property right (IPR) under the Protection of Plant Varieties and Farmers' Rights Act of 2001 (as these are dealt by the Ministry of Agriculture, rather than Environment Ministry).
2. Exemption (under Section 40 of the BD Act) for 190 bioresources, categorised into three, medicinal plants, spices and horticultural crops, and listed as normally traded commodities (NTCs) pursuant to the provisions of the law[8]; however, this only applies when the species is traded as a commodity and not when used as a raw material in R&D.
3. Collaborative research projects (under Section 5 of the BD Act) that involve the transfer or exchange of bioresources or information between institutions, including government-sponsored institutions of India; these projects have to be approved by the central government and conform to their relevant policy guidelines, details of which have to be given in the proforma designed by the NBA.[9]

Access vis-à-vis Indian bioresources/people's knowledge is usually granted for four kinds of uses, when users approach the NBA through the prescribed forms:

(i) Form I – Research/commercial purposes/biosurvey/bioutilisation
(ii) Form II – Transfer of results of research
(iii) Form III – Approval for obtaining IPR on inventions based on any research or information on a bioresource obtained from India
(iv) Form IV – Third-party transfer of already accessed bioresource/knowledge

With the 'new' ABS guidelines issued by the NBA in 2014 (discussed in Section 3.2), another category for grant of access has been added. This is through a prescribed Form B, which is for processing access for bioresources to Indian researchers/government institutes for conducting non-commercial research or research for emergency purposes. As of 11 November 2015, the NBA had granted a total of five such approvals for access.[10] Applicants seeking access to any Indian bioresource or related people's knowledge must approach the NBA with the prescribed form along with payment of the requisite fees. Any of these access types could be relevant for the biotech industry.

The publicly available information on the access approvals shows how the maximum number of approvals granted from when NBA started processing applications in 2006 up to 11 November 2015 is 209 (Fig. 3.2). Under the Form III category for

[8] Vide notification S.O. No. 2726(E) dated 26 October 2009 issued by MoEFCC.

[9] Prescribed proforma. http://nbaindia.org/uploaded/docs/Proforma-act2002.doc

[10] As per NBA data available on http://nbaindia.org/content/683/61/1/approvals.html as on 18 November 2015.

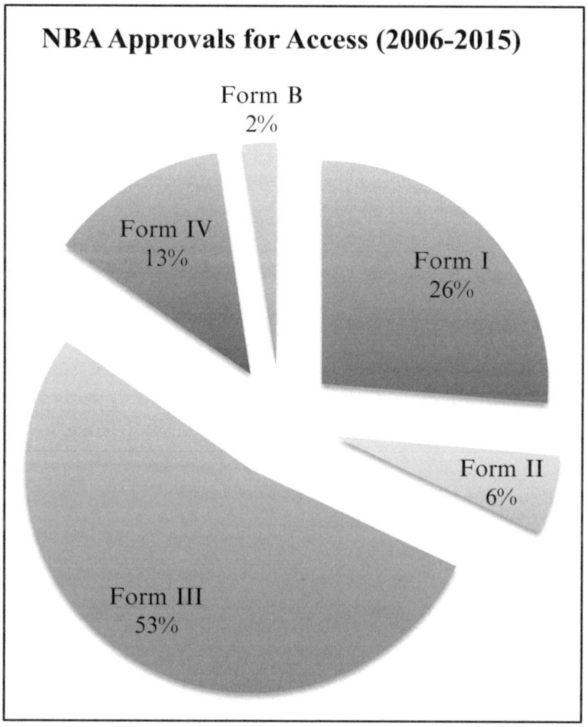

Fig. 3.2 Approvals for access granted by the NBA (2006–2015) (Compiled by the authors from data on the NBA website, as accessed on 18 November 2015)

seeking IPR, more than half – 110 out of a total of 209 applications approved – are for IPR on new methods, processes, herbal or medicinal compositions, etc. There are only about five applicants for IPR, which are either from departments of biotechnology from different universities in India or are from the Department of Biotechnology of the Government of India's Ministry of Science and Technology itself.[11] Only after the approval by NBA can patent examiners proceed to allow patents on biotechnology-related inventions, with due reference to the guidelines by the Indian Patent Office in this regard.

The interesting thing to note respect to the IPR approvals is that all except one (in which the co-applicant is from the USA) are for Indians.[12] But the main objective of the BD Act was to arrest 'biopiracy' by foreign persons. Nonetheless, the provision in the law [Section 18(4)] that empowers the NBA to *take any measures necessary to oppose the grant of IPR in any country outside India on any bioresource obtained from India or knowledge associated with such bioresource which is derived from India* has never been invoked.

[11] The full list as of 18 November 2015 can be accessed from the NBA website: http://nbaindia.org/uploaded/Approvals/FormIII_11NOV15.pdf

[12] The full list as of 18 November 2015 can be accessed here: http://nbaindia.org/uploaded/Approvals/FormIII_11NOV15.pdf

With reference to the context of biotechnology, the information on access approvals also reveals that very few of the access approvals explicitly mention if and when the access is for biotechnological purposes. For example, out of the 55 cases of Form I access approvals, two were granted to a professor, in the Department of Biochemistry and Biotechnology, University of Munster, Germany, one for leaves from certain trees in the biodiverse Mudumalai Wildlife Sanctuary in South India to be able to isolate microorganisms in the leaf tissues and the other for access to soil samples from a chitin-/chitosan-producing plant.[13] ABS agreements were signed between the applicant and NBA in 2007 and 2008, respectively.

Perhaps the most talked of access approvals for the biotech industry are the three granted in 2007 wherein Form II approval for transfer of research results (in this case brinjal varieties) was granted to M/s Maharashtra Hybrid Seeds Company Ltd. (Mahyco).[14] This cross-border transfer of Indian bioresources entailed shipping out parental seeds gathered from the crossing of Mahyco's transgenic Bt brinjal/eggplant event (EE-1) containing cry1Ac gene in the Mahyco Research Centre, Maharashtra, with brinjal/eggplants that were imported from:

(i) East West Seeds, Bangladesh
(ii) Bangladesh Agriculture Research Institute, Bangladesh
(iii) University of the Philippines, Philippines

In January 2013 it was reported by the media that due to public pressure, the NBA had filed a legal case against the US transnational Monsanto and its Indian counterpart Mahyco, for failing to seek the approval of NBA before accessing six local brinjal varieties, which it was using to develop its GE brinjal products. However, the issue of possible genetic contamination of biodiversity in the receiving countries, namely, Bangladesh and the Philippines, never really became an issue, despite all three countries being members of the Cartagena Protocol.

Form IV access is also sought by several private seed companies and agricultural research institutes for the export of seed and planting material from India. Notable in this context is the access approval granted by the NBA to the international agricultural research centre based in India – ICRISAT[15] – for the export of seeds of transgenic groundnut to South Africa for testing.[16] Likewise, M/s Bayer Bioscience Pvt. Ltd. was granted approval by the NBA to export Bt cotton hybrids to Pakistan for research and trial purposes, for which an ABS Agreement was signed on 9 July

[13] Application nos. 92 and 151: http://nbaindia.org/uploaded/Approvals/FormI_11NOV15.pdf

[14] Access application nos. 68, 69 and 70 for which agreements were signed between the company and NBA on 24 April 2007. http://nbaindia.org/uploaded/Approvals/Form-II_30.09.2015.pdf

[15] The International Crops Research Institute for the Semi-Arid Tropics (ICRISAT) in Patancheru, in the Southern Indian state of Telangana, functions under the aegis of the CGIAR; it had developed transgenic groundnut events by introducing the coat protein gene of groundnut rosette assistor virus (GRAV cp) into groundnut varieties JL 24 and ICGS 44.

[16] Access application no. 97 by Dr. William Dar, Director General, ICRISAT, followed by an ABS Agreement signed with NBA on 8 October 2007 for a one-year period.

2010.[17] Apart from pure research, seed companies too seek NBA approval for the export of GE seeds to other countries for field trials, as with Bt cotton hybrids to Pakistan by Rasi Seed Pvt. Ltd., Nuziveedu Seeds Pvt. Ltd., Nath Biogenes (I) Ltd. and Prabhat Agri Biotech Ltd. all in the year 2012.[18] Transfer of bioresources between different country offices of transnational seed corporations or between them and their sub-licensees also requires seeking NBA approvals if Indian bioresources are being sent out. This has been the case with Rasi Seed Pvt. Ltd. exporting Bt cotton hybrids to Monsanto Pakistan for field trials in 2012, as well as Bayer sending foundation seeds of commercial hybrid rice cultivars to Bayer Crop Science Inc. Philippines in 2014.

The industry experience gives it a basis to seek for changes in the ABS regime. While it continues to seek access, it would rather not be under strict legal obligations for either biosafety or benefit sharing. The resistance from industry can be anticipated; it will always seek favourable conditions for its own functioning. This is also the motivation of the Ayurvedic Drug Manufacturers' Association (ADMA), which is an important stakeholder that relies on access for an assured supply of bioresources. Yet it does not want any benefit-sharing obligations. The NBA needs to bring this largest single industry sector that benefits from access to bioresources on board for the success of its own benefit-sharing regime.[19]

3.5.2 Guidelines for ABS

India's BD Act has a clear definition for those the law regards as legitimate 'benefit claimers', once access to bioresources or people's knowledge takes place. They are *conservers of biological resources, their by-products, creators and holders of knowledge and information relating to the use of such biological resources, innovations and practices associated with such use and application.*[20] Meanwhile, SBBs have been struggling to harness back benefits from the bioindustry, whether for generating funds for *in situ* conservation or for sharing with local communities.

This triggered off much discussion in India on the need for guidance on access and benefit sharing. In the absence of clear instructions from the Centre, some state governments, like MP and Kerala, began to issue their own access forms and guidelines. The MP SBB issued a notice requiring all those using bioresources for commercial purpose to apply in the prescribed Form I to the Member Secretary of the SSB.[21] Kerala directed not only industries registered in India and commercially utilising bioresources but also local self-government institutions to regulate access to the people's biodiversity registers (PBR), where local knowledge is documented

[17] Form IV application no. 376 by M/s Bayer Bioscience Pvt. Ltd., Hyderabad, India.

[18] The complete list of these Form IV approvals is available on the NBA website: http://nbaindia.org/uploaded/Approvals/FormIV_11NOV15.pdf

[19] http://nbaindia.org/blog/646/47//RepresentativesofA.html

[20] Article 2(a) of the BD Act.

[21] MP SBB Notice. http://www.mpsbb.info/ImportantNotice.aspx

by BMCs under the processes of the BD Act. The Principal Secretary, Environment of the State Government of Kerala in a circular issued in early 2013, gave specific instructions to local bodies against permitting an external agency to access the PBR without the knowledge of the government or the Kerala SBB.[22]

The need for guidelines clearly came from the practical difficulties encountered by the implementing SBBs, as well as the lack of benefits accruing to local communities through BMCs. In a letter by the Member Secretary of the MP SBB to the NBA dated 3 April 2013, there is an emphatic mention that *in the absence of any guideline by the NBA for access and benefit sharing to the State Biodiversity Board, we are not able to implement third and most important objective of the Biological Diversity Act, 2002 and* i.e. *access and benefit sharing.* The then environment minister Jairam Ramesh had also reiterated that by virtue of the powers vested by the BD Act, SBBs can regulate the use of bioresources by the domestic industry. Local-level BMCs in each state also have the power [under Section 41(3) of the BD Act] to levy charges for the access of bioresources for commercial purposes.[23] Realising that the BMCs cannot actively pursue ABS unless they themselves are aware of their rights and responsibilities, the NBA issued a set of guidelines for the operationalisation of BMCs (National Biodiversity Authority 2013).

Finally, in April 2013, the NBA drafted and made public two sets of guidelines, one on access and the other on benefit sharing, seeking comments on the same. These two separate documents were subsequently merged. However, the document was not publicly accessible until May 2014, after the NBA was asked by the NGT in March 2014 to issue guidelines in the light of increasing confusions on the issue, which led to more than a score of cases being filed by industry before the NGT and several benches of the High Court of MP. Eventually, on 21 November 2014, the NBA and the Ministry of Environment, Forests and Climate Change (MoEFCC) issued a consolidated set of *Guidelines on Access to Biological Resources and Associated Knowledge and Benefit Sharing Regulations, 2014* as mandatory for ABS (MoEFCC 2014). These guidelines not only seek to streamline the access procedure but also expand on the concept of benefit sharing, adding to what is already provided for on the issue in the BD Act.

These guidelines for ABS are pursuant to India's commitment to the Nagoya – the international regime (IR) on ABS, which entered into force on 12 October 2014.[24] The MoEFCC in August 2014 had designated the NBA as the 'competent national authority' for the purposes of the Nagoya Protocol. The IR forms the essential backdrop for the ABS guidelines in India. To operationalise the guidelines, the

[22]The text of the state government circular in Malayalam is available here: http://keralabiodiversity.org/images/news/circular_pbr.pdf

[23]Section 41(3): The biodiversity management committees may levy charges by way of collection fees from any person for accessing or collecting biological resource for commercial purposes from areas falling within its territorial jurisdiction.

[24]The full text of the *Nagoya Protocol on Access to Genetic Resources and the Fair and Equitable Sharing of Benefits Arising from their Utilization* can be accessed at https://www.cbd.int/abs/about/

NBA constituted another Expert Committee to revise the existing ABS Agreement formats.[25]

The BD Act lists six broad types of benefit sharing that can be realised when either access takes place or approval for IPR is granted. This includes joint ownership of IPR to either NBA or an identified benefit claimer, transfer of technology, involvement in research and development endeavours, setting up venture capital funds or payment of compensation. The 2014 Guidelines broadly convey that sharing of benefits may be done either through a monetary or nonmonetary mode. A list of options, both monetary and nonmonetary, is contained in an Annexure I to the notified guidelines. They prescribe that when India's bioresources are accessed and commercially utilised, the applicant shall have the option to pay the benefit sharing ranging from 0.1 to 0.5 per cent at graded percentages of the annual gross ex-factory sale of the product, which shall be worked out on the basis of annual gross ex-factory sale minus government taxes. The experience of the 10-year practice of implementation of the BD Act in India prior to the issuance of the guidelines has shown a preference for monetary benefits in ABS arrangements, though the collection in the National Biodiversity Fund has been far less than anticipated. And the said fund is meant to finance conservation and also facilitate benefit sharing with local communities (Table 3.1).

The new guidelines only marginally add to what the existing BD Act and Rules lay down on the ABS issue. The only area where the guidelines introduce something new is a category of access that allows an Indian researcher or government institution to carry or send Indian bioresources out for basic research to avert emergencies like epidemics, pandemics, etc. That apart the ABS guidelines issued do not fully explain the reasoning or the process through which the percentages have been determined or how the various figures for payment have been arrived at. They also don't explain why in some instances there are direct payments to local-level committees and in others it is not envisaged. Thereby, the intended users of the guidelines do not get any guidance on that aspect of the thought process of the policymakers. A supplementary note to the notification issuing the guidelines, indicating the reasoning, would have helped to better understand the real objectives (Kohli and Bhutani 2015).

The BD Rules, 2004, also give administrators the power to restrict or prohibit access to biological resources on account of overriding public interest or for protection of environment and conservation of biological diversity. But since information on the rejected applications is not publicly available from the NBA, it is not easy to assess how often this power is used in thc name of conservation, particularly when access applications relate to biotechnology. The NBA only makes known that a total of 1131 applications were received by it since 2003 up to 16 November 2015. The section on 'closed' applications on the NBA website does not give the reasons for

[25] Expert Committee to revise the existing ABS Agreement formats as on 18 November 2015 is here: http://nbaindia.org/uploaded/pdf/Oo_extension_Tenure_EC%20_revising_existing_Agreement_formats.pdf

Table 3.1 Understanding the 2014 ABS guidelines[a]

1. Access of a bioresource for commercial utilisation/biosurvey/bioutilisation for commercial utilisation		
Condition	**Payment by trader**	**Payment by manufacturer**
Where no prior benefit-sharing (BS) agreement with Joint Forest Management Committee/Gram Sabha/ forest dweller/cultivator	1–3% of the purchase price	3–5% of purchase price
Further sale of biological resource by a trader to another trader/manufacturer	1–3% of the purchase price (*in case there is proof of supply chain, then BS only on amount for which BS has not been paid earlier*)	3–5% of purchase price
Where there is prior BS agreement with JFMC/Gram Sabha/forest dweller/cultivator	Not less than 3% of purchase price	Not less than 5% of purchase price
High economic value bioresource like Red Sanders	Upfront payment of not less than 5% of 5.0%, on the proceeds of the auction or sale amount, as decided by the NBA or SBB into a designated fund	Upfront payment of not less than 5% of 5.0%, on the proceeds of the auction or sale amount, as decided by the NBA or SBB into a designated fund
Where access leads to commercial utilisation, optional benefit sharing on ex-factory sale price	Rs.1 lakh (0.1%); Rs.1–3 lakhs (0.2%); above 3 lakhs (0.5%)	Rs.1 lakh (0.1%); Rs.1–3 lakhs (0.2%); above 3 lakhs (0.5%)
2. Access for transfer of research results		
Condition	**Payment to NBA**	**Payment to SBB/BMC**
With complete details disclosed of potential commercial value	3–5% of the monetary consideration	
3. Access for intellectual property rights		
Condition	**Payment to NBA**	**Payment to SBB/BMC**
In case of commercialisation	Monetary and/or nonmonetary benefit as agreed with NBA	
Applicant assigns licences the process/ product/innovation to a third party for commercialisation	3–5% of the fee received (in any form including the licence I assignee fee) and 2–5% of the royalty amount received annually from the assignee/licensee, based on sectoral approach	

(continued)

Table 3.1 (continued)

4. Transfer of research results for research/commercial utilisation		
Condition	**Payment to NBA**	**Payment to SBB/BMC**
When the resource is not of high value	Monetary and/or nonmonetary benefit as mutually agreed	
	2–5% (following a sectoral approach) of any amount and/or royalty received from the transferee, throughout the term of the agreement	
Where resource is of high value	In addition to the above, also an upfront payment, as mutually agreed between the applicant and the NBA	

[a]This table first appeared in a paper by Kohli and Bhutani in the *Economic and Political Weekly* Kohli and Bhutani (2015)

such closure in 173 cases,[26] though as per the procedure applications are usually terminated due to the applicant giving incomplete information (National Biodiversity Authority 2014).

3.5.3 Rules for Biosafety

The BD Act has very specific provisions in its text on biotechnology. It lays down legal obligations on biosafety for the government of India to follow. According to Section 36(4)(ii) of BD Act, it is mandatory for the government of India to undertake measures:

> *to regulate, manage or control the risks associated with the use and release of LMOs resulting from biotechnology likely to have adverse impact on the conservation and sustainable use of biological diversity and human health*

No other legislation in India imposes such legal obligations on the state with respect to biosafety. Yet despite the fact that the first LMO – transgenic Bt cotton crop – was approved for commercial release in India in 2002, this provision of the law has never been invoked by the central government in the history of the BD Act, 2002. But there are few instances of state governments attempting to use the BD Act

[26]The list of closed applications is available on NBA website at http://nbaindia.org/text/22/Closed.html, as accessed on 18 November 2015.

in the interest of biosafety. One such instance is that of the government of Karnataka, which through its SBB directed that for R&D and biosafety trials for any (GE) Bt crop, the prior permission of the SBB would have to be taken.[27]

The only other environmental legislation – the Environment (Protection) [EP] Act, 1986 – was made before LMOs had become an issue. Under the EP Act, the key rules on biosafety are the *Manufacture, Use, Import, Export and Storage of Hazardous Micro-organisms/Genetically engineered organisms or cells, 1989.*[28] These Rules were issued by the Ministry of Environment, Forests and Climate Change (MoEFCC) under India's key environmental law – the EP Act, 1986. The Rules (1989) establish the Genetic Engineering Appraisal Committee (GEAC) as the main body under the Environment Ministry to both process applications for GE use and oversee biosafety. As an expert body, it was set up to scientifically appraise and recommend approvals on the commercial application of genetic engineering (GE) in agriculture, pharmaceutical and other related sectors. In the current biosafety regulatory framework in the country, GEAC is one of the three approval-granting authorities with respect to GE. The GEAC follows a case-by-case system of screening applications. The MoEFCC then takes the final call (Bhutani et al. 2014).

The other relevant bodies at the central level are the Recombinant DNA Advisory Committee (RDAC) and the Review Committee on Genetic Manipulation (RCGM). The former is meant to review developments in biotechnology at national and international levels and recommend suitable and appropriate safety regulations for India in recombinant research, use and applications.[29] The latter is to monitor the safety-related aspects in respect to ongoing research projects and activities involving GE organisms/hazardous microorganisms.[30] Both these committees are in the Department of Biotechnology (DBT), which functions under the Ministry of Science and Technology. Within each institute undertaking R&D in modern biotechnology, the Institutional Biosafety Committee (ISBC) is required to be set up. While at the state and district levels, the SBCC and the DLCCs, respectively, are required to be functioning.

Apart from the Rules of 1989 enforced by the Environment Ministry, the biosafety regime also comprises a set of executive rules, which are the responsibility of other governmental ministries and departments under them.[31] For instance, the DBT regulates biosafety through the following executive rules:

---

[27] Letter No. FEE 77 ENV 2011 dated 21 January 2012 from Secretary, Government of Karnataka, Forest Ecology and Environment Department to the Member Secretary, RCGM, DBT, Government of India.

[28] The full text can be downloaded from the Ministry website as accessed on 18 November 2015: http://envfor.nic.in/sites/default/files/geac/notification.html

[29] Rule 4(1) of the GE Rules, 1989.

[30] Rule 4(2).

[31] The full list of rules and their texts can be downloaded from the DBT website: http://dbtbiosafety.nic.in/

1. Recombinant DNA Safety Guidelines, 1990
2. Revised Guidelines for Research in Transgenic Plants, 1998
3. Protocols for Food and Feed Safety Assessment of GE Crops, 2008

Meanwhile, in May 2003, the Union Ministry of Agriculture in India had set up a task force, chaired by the agricultural scientist Prof. M S Swaminathan for the formulation of a policy on application of genetic engineering in agriculture. The task force recommended that a national law be legislated and an independent biotechnology regulatory authority, which would oversee biosafety concerns, be established.

The government of the day also moved to revamp the biosafety regime of the country through first drafting a National Biotechnology Development Strategy, 2007 (Department of Biotechnology 2007). This was followed by the Union Ministry of Science and Technology, through the DBT drafting a new biosafety law: National Biotechnology Regulatory Authority Bill, 2008.[32] The public opposition to it led to the attempt being shelved. Thereafter, discussions on a stand-alone law on biosafety – the Biotechnology Regulatory Authority of India (BRAI) Bill – have (re)surfaced. The National Biotechnology Development Strategy, drafted back then in 2005 as a 10-year vision document, is also up for revision. All these processes of law and policymaking will have to factor in the demands of the Cartagena Protocol.

Anyhow, law is not only made by legislatures or contained simply by executive authorities. The judiciary is also a source of law. Judgments and judicial orders passed by courts form another body of law on a subject. Over the years, ordinary citizens and non-governmental organisations have filed cases in public interest for a proper biosafety system to be set up in the country. The key ongoing litigation in the highest court of the land – the Supreme Court of India – is that of *Aruna Rodrigues and Others* versus *Union of India and Others* [Special Leave Petition (Civil) No. 260 of 2005].[33] In this case, the petitioners approached the Court under its writ jurisdiction[34] to ask for necessary directions so that biosafety regulation is undertaken by independent scientific agencies.

3.6 Conclusions

The legal and administrative frameworks to govern access to bioresources and enforce biosafety may seem to be developing quite independently of each other both at the international and the national level. And their respective processes, as the

[32] The text of the proposed law can be downloaded as on 18 November 2015 at: http://dbtindia.nic.in/Draft%20NBR%20Act_%2028may2008.pdf

[33] The full text of the original petition filed in 2005 can be accessed here: http://ddsindia.com/www/PDF/PIL_October27.pdf

[34] The Supreme Court of India has original jurisdiction to issue writs – a formal written order – for the protection of fundamental rights under Article 32 of the Constitution of India and under Article 139 to enforce other than fundamental rights.

Indian regulatory regime clearly exhibits, take place under different central ministries and government departments. Yet the two aspects of governance with respect to bioresources, namely, ABS regulations and biosafety mechanisms, have obvious interlinkages. The access framework is a means for the bioindustry to acquire through due process of law the bioresources it needs for its operations.

The experience of the law on bioresources is an example of how the dominant economic system can come to bear on implementation. Both ABS procedures and biosafety rules are premised on the fundamental principle of privatisation of bioresources and the development of proprietary technologies. In such a scenario, the prevalent thinking is that to sell more is to have more. Given the orientation, how the implementation of the law will proceed will depend not only how the bioindustry responds to the regulatory regime but also how much the BMCs at the local level act as spaces for community sovereignty, rather than merely environmental watchdogs or benefit claimers when bioresources are traded. Though evidence points to the fact, very few benefit-sharing cases have resulted despite the due process of law followed for the access procedures.

Nevertheless, ABS has dominated both the CBD landscape globally and the implementation of the BD Act domestically. The operating principle is that access to biological material for bioenterprises is inevitable and not to be restrained. This is seen as a means to integrate with the global trading system. Therefore, the regulatory system is constantly under pressure to safeguard bioresources while it tries to better itself in selling those very bioresources through an ABS system. The challenge for the state functionaries is to be able to use the BD Act for both marketing and conserving bioresources.

Conservation is also an aspect that brings the objectives of the two CBD protocols – Cartagena Protocol and the Nagoya Protocol – in convergence. In fact, it can be said that the ABS regulatory regime requires conservation to be considered at the point of both access and when imposing the terms and conditions for benefit sharing, and biosafety regimes require that conservation be undertaken whenever the release of an LMO is being considered or is already approved for released. After all, both processes of law have to meet the objectives of their parent treaty CBD. Bioresources and people's knowledge systems around them are the actual and potential source of products and processes that can serve the needs of society. Yet at the ground level in areas where both bioresources are accessed from and where outputs of modern biotechnology are either field tested or released in the open environment, there remain genuine concerns about conservation despite the law for biodiversity.

References

Bhutani S, Menon M, Kohli K (2014) Fielding trials for genetic engineering. Economic & Political Weekly Vol XLIX, No. 25, June 21, 2014

CBD (Convention on Biological Diversity) (2000) Cartagena protocol on biosafety to the convention on biological diversity. Accessed from https://www.cbd.int/ on 18 Nov 2015

CBD (Convention on Biological Diversity) (2015 Undated) History of the convention. Accessed from https://www.cbd.int/ on 18 Nov 2015

Department of Biotechnology (2007) National biotechnology development strategy 2007, Ministry of Science and Technology, Government of India, New Delhi

Kohli K, Bhutani S (2013) The legal meaning of biodiversity. Economic & Political Weekly Vol XLVIII, No. 33, August 17, 2013

Kohli K, Bhutani S (2015) Access to India's biodiversity and sharing its benefits. Economic & Political Weekly Vol L, No. 31, August 01, 2015

METI (Ministry of Economy, Trade and Industry) and JBA (Japan Bioindustry Association) (2005) Guidelines on access to genetic resources for users in Japan, Ministry of Economy, Trade and Industry, Tokyo

MoEFCC (Ministry of Environment, Forests and Climate Change) (2014) Guidelines on access to biological resources and associated knowledge and benefits sharing regulations, 2014, Government of India, New Delhi

National Biodiversity Authority (2013) Guidelines for operationalisation of biodiversity management committees, National Biodiversity Authority, India

National Biodiversity Authority (2014) Procedure for closing of applications received under the biological diversity act 2002. Accessed on 18 Nov 2015 on http://nbaindia.org/uploaded/pdf/Clsoing_procedure_for_application. pdf

Natural Resource Management Ministerial Council (2002) Nationally consistent approach for access and utilisation of Australia's native genetic and biochemical resources, Department of Environment and Heritage, Australia

Rajya Sabha Secretariat (2001) Ninety-fifth report dated December 2001 on The Biological Diversity Bill, 2000 Department-Related Parliamentary Standing Committee on Science & Technology, Environment & Forests, Rajya Sabha, New Delhi.

Schuazu M (2000) The concept of substantial equivalence in safety assessment of foods derived from genetically modified organisms AgBioTechNet 2000, Vol 2, April, ABN 044 http://www.bfr.bund.de/cm/349/schauzu.pdf

Secretariat of the Convention on Biological Diversity (2011) Nagoya Protocol on access to genetic resources and the fair and equitable sharing of benefits arising from their utilization to the convention on biological diversity: text and annex. Accessed from https://www.cbd.int/ on 18 Nov 2015

Secretariat of the Convention on Biological Diversity (2015) Press release: the first internationally recognised certificate of compliance is issued under the Nagoya Protocol on Access and Benefit-sharing. Accessed, CBD, 7 Oct, 2015

UNEP (United Nations Environment Program) (1992) Rio declaration on environment and development, UNEP

United Nations (1992) Convention on biological diversity. Accessed from https://www.cbd.int/ on 18 Nov 2015

Vogler J (2005) The European contribution to global environmental governance. Int Aff 81(5):835–850 http://www.cesruc.org/uploads/soft/130311/1-130311152133.pdf

Shalini Bhutani is a legal researcher and policy analyst, who works independently on law and policy issues in the agriculture and biodiversity sector in the Asian region. She also lectures in different universities on legal and regulatory affairs.

Kanchi Kohli is a member of Kalpavriksh Environment Action Group. Her research focuses on the interface of law and governance in areas of environment, forests and biodiversity.

Bioresources for Food and Nutrition Security: The Case of Wild Edibles of Western Ghats

4

M.K. Ratheesh Narayanan, Nadesapanicker Anil Kumar, and Parameswaran Prajeesh

Abstract

Wild edibles (foods) are plant and animal resources outside the agricultural areas that are harvested for the purpose of human consumption. This chapter provides the diversity and trends in the use and management of wild edible species with reference to different sociocultural groups from the Wayanad District, in Kerala, part of the Western Ghats. Three hundred fifty-three species of seven major groups of wild foods that include leafy greens, tubers and roots, fruits and seeds, mushrooms, honey, crabs and fishes are discussed. The places where the wild edibles proliferate, the communities that come in search of these edibles, the colloquial terms that are commonly used to describe them and the gender roles that come into play during harvesting and processing of the wild foods have also been traced out. Tribal and rural families of the region continue to collect and conserve a wide range of plants to meet their diverse food needs, and women are more skilful in managing the surrounding landscape and are the chief knowledge holders and conservationists. It is also noted that the wild edibles have a critical role to play in dealing with the issue of undernutrition, and hence dynamic conservation of agrobiodiversity needs to be placed high in the national development agenda for leveraging nutrition in agriculture and alleviating poverty and malnutrition.

Keywords

Wild edibles • Wayanad • Western Ghats • Gender • Agrobiodiversity

M.K. Ratheesh Narayanan
Department of Botany, Payyanur College, Payyanur 670327, Kerala, India

N. Anil Kumar (✉) • P. Prajeesh
Community Agro Biodiversity Centre, M. S. Swaminathan Research Foundation, Puthoorvayal, Kalpetta, Wayanad 673577, Kerala, India
e-mail: anilmaruthur@gmail.com

S. Abdulhameed et al. (eds.), *Bioresources and Bioprocess in Biotechnology*, DOI 10.1007/978-981-10-3573-9_4

4.1 Introduction

There are at least 3000 edible plant species known to mankind, but just about 30 crops alone contribute to more than 90% of the world's calorie intake, and only 120 crops are economically important at the national scale (FAO 1993). This shows that several hundreds of species remain discarded or unnoticed at the hands of various human societies. A wide variety of cultivated plants used by mankind today have been derived from such wild related crop plants (Swaminathan and Jana 1992). Among these, many are nutritionally or otherwise important. Reid and Miller (1989) have given the example of quinoa (*Chenopodium quinoa*), a staple grain of Incas. It is a little-known crop to the modern world, yet is one of the world's most productive sources of protein. There are plant and animal resources outside of agricultural areas that are harvested or collected for the purpose of human consumption. These are incorporated into the normal livelihood strategies of many rural people, be they pastoralists, shifting cultivators, continuous croppers or hunter-gatherers. They are usually considered as an additional diet to rural and forest dwelling people's daily food consumption pattern and are usually described as 'wild foods' (Bell 1995). Wild plants and animals have provided an important source of food since time immemorial (Gammie 1902; Fernold and Kinsey 1958; Medsager 1957; Bell 1995). Various wild species of *Dioscorea*, taros and *Amaranthus*, which are the source of vitamins and nutrients, supplement the food needs of a multitude of families who live near forests (Roy et al. 1998). Many studies have revealed the role of wild food – the 'forest food' – particularly during stress seasons in the household food security and importantly the nutritional security of poor tribal and rural families (Singh and Singh 1981, FAO 1982, 1984, 1989; Falconer 1990). Animal food from forest and wild trees contributes a significant portion to the wild food consumed by the forest-dependent communities (Falconer and Arnold 1991). There are a number of studies from Northeast Thailand analysing the wild food plant diversity of rice fields which serves the utilities of food, medicines, timber, latex, oil and fodder. Most of these wild food plants are found in paddy fields compared to secondary woody areas and home gardens (Grandstaff et al. 1986; Watanabe et al. 1990; Price 1997; Prachaiyo 2000; Maneechote 2007, Cruz-Garcia and Price 2011).

Authors have differently classified and grouped the wild edibles (wild foods). The grouping followed was based on either (1) the parts of the plant consumed or (2) the type of resources. Those groupings ran as tubers and roots, leafy vegetables and greens, buds and flowers, fruits, seeds and nuts, typical famine food plants, wild food plants with famine food components, wild food plants attracting additional consumer categories during food shortage periods, on-farm food crops with famine food components, food eaten only in an emergency, wild plants gathered by people living in and around forests, food gathered from the forest by the local people and also available for sale in rural and semiurban market, cultivated edible forest species available for sale, etc. (Negi 1994; Arora and Pandey 1996; Guinand and Dechassa 2001; Kanvinde et al. 2001).

Arora and Pandey (1996) have reported 1532 wild food species as edible from India, mostly from the Western Ghats and Himalayan regions. A wild food also contributes to the household income security of millions of forest-dependent communities. In India, those who collect species such as gooseberry, *Garcinia*, *Parmelia* and honey for market are mostly dependent on it as their major source of income (Muralidharan et al. 1997). The wild foods used by various tribes in India also have received considerable attention (Jain 1964; Maheshwari and Singh 1986; Narayanan et al. 2004). But it was studied that the integrity of wild genetic resources is threatened by a number of human activities (McNeely et al. 1990). These include deforestation and changes in land use, inappropriate forest harvesting techniques, pollution and climate change and the often uncontrolled and undocumented movement of forest germplasm for plantation establishment, leading to the loss of locally adapted populations (WRI et al. 1992). Still the striking wild biodiversity that survives despite the calamitous human interventions lends unparalleled uniqueness to the Indian subcontinent and especially in the Western Ghats. This chapter gives an account of the wild foods recorded from Wayanad District in Kerala, part of the Western Ghats.

Wayanad: An Agrobiodiversity and Sociocultural Hotspot in the Western Ghats Wayanad is a hilly terrain in southern Western Ghats and lies at an average altitude of 750 m above sea level (Fig. 4.1). The district of 2136 km^2 is unique for its rich wealth of flora and fauna and for the diverse cultures that inhabit the land. Wayanad is a high-range agro-ecological zone having moderately distributed monsoons (Kerala Agricultural University 2011). Narrow valleys surrounded by low-range undulating hills and steep slopes characterise typical paddy fields in Wayanad. The total geographic area is 212,966 ha with a total cropped area of 174,190 ha (Department of Economics and Statistics 2015). The contribution to the state's foreign exchange earnings through cash crops (pepper, cardamom, coffee, tea, ginger, turmeric, rubber and areca nut) is significant (Kumar et al. 2010). Floristic exploration of the district has recorded nearly 49% of the flora of the Kerala state and more than 10% of the flora of India. This study has reported a total of 596 endemic taxa in which 15 are exclusive to the district (Narayanan 2009). According to Nair (1911), *Wayanad* means upper land or from *Vayalnadu* meaning land (*nadu*) of paddy fields (*vayal*) or from *Vananadu* meaning land of forests (*Vanam*).

Wayanad is notable for its large tribal population, which accounts for 18.53% and is the largest among the districts in the state (Office of the Registrar General and Census Commissioner 2011). They can be broadly classified into farming communities (Kurichya, Mullukuruma), agricultural labourers (Paniya, Adiya), artisan communities (Uralikuruma) and hunter-gatherer communities (Kattunaikka). Others are Thachanadan Moopan, Karimbalar, Pathiya and Wayanadan Kadar. These are the communities who still hold knowledge on biodiversity and follow traditional conservation practices. The Paniya constitutes the single largest Scheduled Tribe in Kerala and mainly lives in the Wayanad District and the neighbouring areas of Karnataka State. They have a distinct language of their own, closely related to Malayalam. The Paniya was brought to Wayanad by the Gounders who trained them to be agricultural labourers in their fields. The Kattunaikka is referred to variously as 'Jenu Kuruman', 'Tenu Kurumban' and 'Naickan'. The term

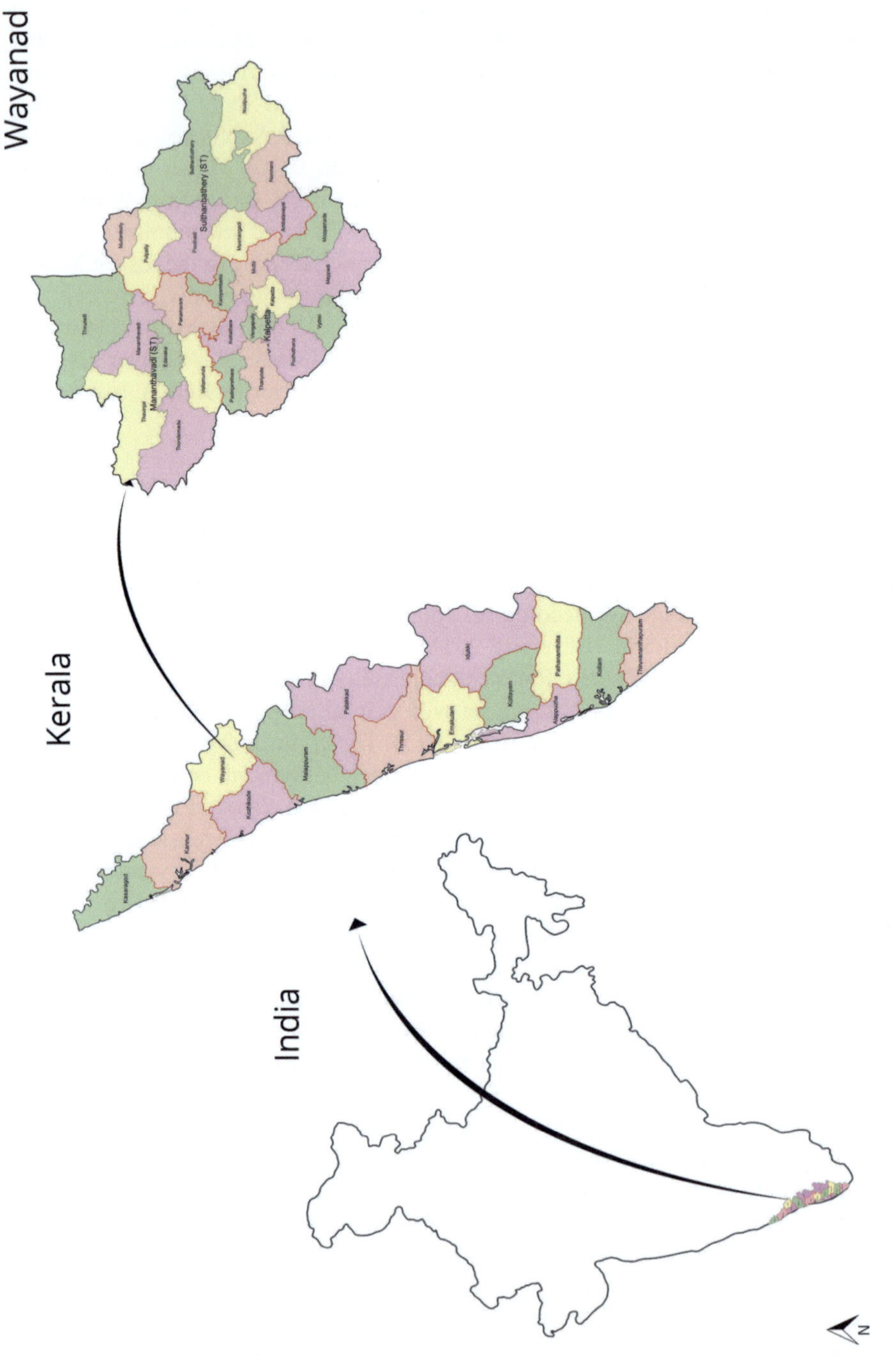

Fig. 4.1 Location of Wayanad (From MSSRF archive)

Kattunaikkan has been derived from the word 'Kadu', meaning forest, and 'Naikkan', meaning leader or headman. The community is predominantly distributed in Wayanad District. The Mullukuruma, often referred to as just Kuruma, is a Scheduled Tribe found in Wayanad and also has the largest settler population in Kerala. No religion is predominant in the district; the major religious groups of the state are all more or less equally represented. Christians who are the single largest religious group constitute one fourth of the population. They are primarily settlers who emigrated from Travancore area of south Kerala. Muslims who constitute another near one fourth of the population are the 'Mapilas' who have come from Malappuram and Kozhikode districts. Hindus of different castes, like Nairs, Thiyyas, Wayanadan Chetty, etc. who settled here from different parts of Kerala, form the rest of the population. A small representation by way of the Gounders who came from Karnataka constitutes the Jain community of Wayanad, many of whom are planters with sizeable holdings (Luiz 1962; Thurston 1909; Nair 1911; Janah 1994; Indian Institute of Management 2006).

4.2 Wild Edibles of Wayanad

Leafy Greens Wild leaves are among the most widely consumed wild foods. Most of the leafy wild food plants are locally referred to and classified as 'weeds', sprouting and flourishing after rains. Women use them in soups, stews and relishes that add flavour to staples. While some leaves are high in fats, others are high in protein, and most are good source of vitamins and minerals. Ninety-four wild edible greens have been reported, but only a few species are widely used (Tables 4.1 and 4.2). The household survey conducted has revealed that the Paniya families consume about 83 species followed by the Kattunaikka families who consume 43 species, the Kuruma

Table 4.1 Most frequently used leafy greens in Wayanad

No.	Local name(s)	Botanical name
1	Ponnamkanni/Minugalasoppu	*Alternanthera sessilis* R. Br.
2	Mudungachapu/Kattuthakkali/Chukkootti	*Solanum nigrum* L.
3	Churuli	*Diplazium esculentum* (Retz.) Sw.
4	Mullancheera/Mullukeera/Mullancheera/ Chuvappu Cheera	*Amaranthus spinosus* L.
5	Kalicheera/Kuppakeera/Vazhacheera	*Amaranthus viridis* var.
6	Karinthal/ Kollithal	*Colocasia esculenta* (L.) Schott
7	Marakeera	*Embelia tsjeriam-cottam* A.DC.
8	Vasalacheera	*Basella alba* L.

Narayanan et al. (2004)

Table 4.2 Leafy greens of Wayanad

Sl no.	Local name	Botanical name	Knowledgeable sociocultural group[a]
1	Alanchappu	*Bidens biternata* (Lour.) Merr.	P, Ku, K, O
2	Aliyanchappu	*Zehneria maysorensis* (Wt. and Arn.) Am.	P, Ku, K
3	Ambal	*Nymphaea nouchali* Burm. f.	P, Ku, K
4	Attanga	*Cucumis prophetarum* L.	P, Ku, K
5	Ayanichakka	*Artocarpus hirsutus* Lamk.	P, Ku, K, O
6	Brahmichappu	*Bacopa monnieri* (L.) Pennell	P, Ku, K
7	Chakka	*Artocarpus heterophyllus* Lamk.	P, Ku, K, O
8	Cheenaparangi	*Capsicum annuum* L.	P, Ku, K, O
9	Cherukadaladi	*Cyathula prostrata* (L.) Bl.	P, Ku, K, O
10	Cherucheera	*Alternanthera bettzickiana* Br.	P, Ku, K, O
11	Cherukadaladi	*Cyathula prostrata* L.	P, Ku, K, O
12	Chooral	*Dendrocalamus strictus* (Roxb.) Nees	P, Ku, K
13	Choracheera	*Alternanthera dentate* Br.	P, K, O
14	Chorakam	*Polygonum chinense* L.	P, Ku, K
15	Choriyanam	*Laportea interrupta* (L.) Chew.	P, Ku, K, O
16	Chorkam	*Polygonum glabrum* L.	P, Ku, K
17	Churuli	*Diplazium esculentum* (Retz.) Sw.	P, Ku, K, O
18	Eenthukumpu	*Cycas circinalis* L.	P, Ku, K, O
19	Hattakkeerai	*Justicia nilgherrensis* (Nees) T. Anders	P, Ku, K
20	Hinnisan kaya	*Tamilnadia uliginosa* (Retz.) Tirveng	P, Ku, K
21	Kadambu	*Barringtonia racemosa* Bl.	P, Ku, K
22	Kadukucheera	*Blumea barbata* DC.	P, Ku, K, O
23	Kaida	*Pandanus fascicularis* Lamk.	P, Ku, K
24	Kalicheera	*Amaranthus viridis* var.	P, Ku, O
25	Kannisoppu	*Commelina benghalensis* L.	P, Ku, K
26	Kara	*Catunaregam spinosa* (Thunb.) Tir.	P, Ku, K
27	Karimthalu	*Colocasia esculenta* (L.) Schott	P, Ku, K, O
28	Karinkoovalam	*Monochoria vaginalis* Presl.	P, Ku, K
29	Kattucheera	*Amaranthus caudatus* L.	P, Ku, K, O
30	Kattueenthu	*Phoenix sylvestris* Roxb.	P, Ku, K
31	Kattukaipaka	*Momordica dioica* Roxb.	P, Ku, K, O
32	Kattumandaram	*Bauhinia purpurea* L.	P, Ku, K
33	Kattumudunga	*Lycianthes laevis* (Dunal) Bitter	P, Ku, K
34	Kattupaval	*Momordica subangulata* Bl.	P, Ku, K
35	Kattupayar	*Canavalia cathartica* Thouars.	P, Ku, K
36	Kattupayar	*Mucuna monosperma* DC.	P, Ku, K
37	Kattuthakkali	*Passiflora calcarata* Mast.	P, Ku, K, O
38	Kattuvenda	*Abelmoschus angulosus* Wall.	P, Ku, K, O
39	Kayalkkalli	*Bambusa arundinacea* Willd.	P, Ku, K, O

(continued)

Table 4.2 (continued)

Sl no.	Local name	Botanical name	Knowledgeable sociocultural group[a]
40	Keezharnelli	*Phyllanthus niruri* L.	P, Ku, K, O
41	Kollithalu	*Colocasia esculenta* (L.) Schott	P, Ku, K, O
42	Koombichapu	*Adenia hondala* (Gaertn.) de Wilde	P, Ku, K
43	Koovilisoppu	*Crotalaria laevigata* Lam.	P, Ku, K
44	Kozhimullan	*Hygrophila schulli* Ham.	P, Ku, K, O
45	Kozhivalan	*Achyranthes bidentata* Bl.	P, Ku, K
46	Kozhuppa cheera	*Portulaca oleracea* L.	P, Ku, K
47	Kumbil	*Gmelina arborea* Roxb.	P, Ku, K
48	Kundimaruma	*Sonerila rheedii* Wt.	P, Ku, K
49	Kuniyanchappu	*Diplocyclos palmatus* (L.) Jeffrey	P, Ku, K
50	Kunni	*Abrus precatorius* L.	P, Ku, K
51	Kuppacheera	*Amaranthus viridis* L.	P, Ku, K, O
52	Malampuli	*Begonia malabarica* Lamk.	P, Ku, K, O
53	Malampuli	*Begonia integrifolia* Dalz.	P, Ku, K
54	Malampunna	*Dillenia indica* L.	P, Ku, K
55	Malanchuruli	*Dryopteris cochleata*	P, Ku, K
56	Maracheera	*Waltheria indica* L.	P, Ku, K
57	Marachembu	*Remusatia vivipara* Schott.	P, Ku, K
58	Maradasoppu	*Capparis* sp.	P, Ku, K
59	Marakkeera	*Embelia tsjeriam-cottam* A.DC.	P, Ku, K
60	Minnamkkanni	*Alternanthera pungens* Kunth.	P, Ku, K
61	Motampuli	*Physalis minima* L.	P, Ku, K
62	Mudungachappu	*Solanum nigrum* L.	P, Ku, K, O
63	Mukkapeera	*Mukia maderaspatana* (L.) M. Roem.	P, Ku, K
64	Mullancheera	*Amaranthus spinosus* L.	P, Ku, K, O
65	Mullancheera	*Chuvappu Amaranthus spinosus* L.	P, Ku, K, O
66	Murikkinchappu	*Erythrina stricta* Roxb.	P, Ku, K, O
67	Muthilila	*Centella asiatica* (L.) Urban	P, Ku, K, O
68	Muyalcheviyan	*Emilia sonchifolia* (L.) DC.	P, Ku, K
69	Naikkadugu	*Cleome viscosa* L.	P, Ku, K
70	Nakkuneeti	*Ophioglossum reticulatum* L.	P, Ku, K
71	Njetipanakumpu	*Arenga wightii* Griff.	P, Ku, K
72	Noolithali	*Antidesma acidum* Retz.	P, Ku, K
73	Palancheera	*Ceropegia stocksii* Hook.	P, Ku, K, O
74	Palankeera	*Ceropegia metziana* Miq.	P, Ku, K
75	Palcheera	*Euphorbia hirta* L.	P, Ku, K
76	Panamchapu	*Caryota urens* L.	P, Ku, K
77	Panchithalu	*Cryptocoryne spiralis* Fisch.	P, Ku, K
78	Parachava	*Dryopteris cochleata* J. Sm.	P, Ku, K

(continued)

Table 4.2 (continued)

Sl no.	Local name	Botanical name	Knowledgeable sociocultural group[a]
79	Parippukkeera	*Chenopodium album* L.	P, Ku, K
80	Paruthiyila	*Hibiscus hispidissimus* Griff.	P, Ku, K, O
81	Poninthavara	*Cassia occidentalis* L.	P, Ku, K
82	Ponnamkanni	*Alternanthera sessilis* R. Br.	P, Ku, K, O
83	Poola	*Bombax ceiba* L.	P, Ku, K, O
84	Poovarasu	*Thespesia populnea* Soland.	P, Ku, K
85	Puliyarila	*Oxalis corniculata* L.	P, Ku, K, O
86	Sambar cheera	*Talinum cuneifolium* Willd.	P, Ku, K, O
87	Thaivasoppu	*Pteridium aquilinum*	P, Ku, K
88	Thavara	*Cassia tora* L.	P, Ku, K, O
89	Thazhuthama	*Boerhaavia diffusa* L.	P, Ku, K, O
90	Unnithandu	*Costus speciosus* (Koen.) Smith	P, Ku, K
91	Valiyakadaladi	*Achyranthes aspera* L.	P, Ku, K, O
92	Vallimaruma	*Cissus discolor* Bl.	P, Ku, K
93	Vasalachapu	*Basella alba* L.	P, Ku, K, O
94	Vattachappu	*Marsilea quadrifolia*	P, Ku, K
95	Vayalthalu	*Colocasia esculenta* (L.) Schott.	P, Ku, K, O

[a]*P* Paniya, *K* Kattunaikka, *Ku* Kuruma, *O* others
Narayanan et al. (2004)

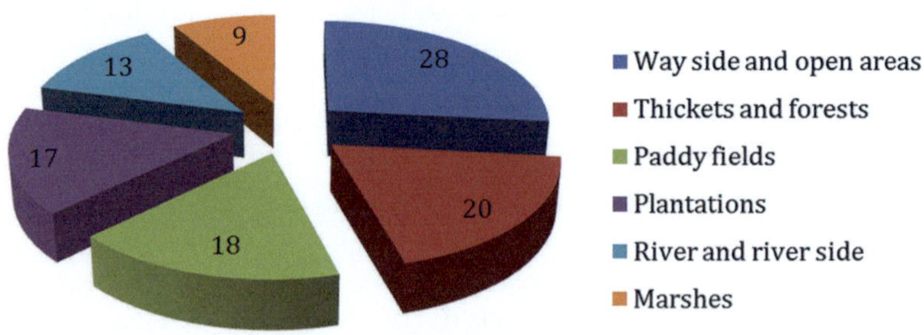

Fig. 4.2 Leafy green and its abundance in relation with different landscapes (Narayanan et al. 2004)

families who consume about 21 types of wild edible leaves and the settlers who restrict themselves to between 8 and 14 types of leafy greens. Most of these species are herbs (90%), and very few are trees. An analysis of dependency on various landscapes for collecting these plants (Fig. 4.2) shows that wayside and open areas provide the maximum species (28) followed by thickets and forest (20 species), paddy fields and associated ecosystems (18), plantations (17), river and riversides (13) and finally the marshy areas (9).

Tubers and Roots More than 25 wild plant species/types in Wayanad are known for edible roots, tubers and rhizomes and are eaten by the Adivasi and non-Adivasi communities of the district. Of these, 19 are species/varieties of *Dioscorea*, which is the main tuber plant known and used in this region. Wild *Dioscorea* species are still a major source of food for forest-based communities like Kattunaikka, and these serve as a 'life-saving' plant group during periods of food scarcity. The communities who are dependent on wild *Dioscorea* for their food classify each member of this genus, based on characteristics like edibility, taste, colour, size, direction of growth, fibre content, cooking properties and occasionally the proliferation underground. The Kattunaikka calls these tubers as 'Galasu', and Narayanan et al. (2004) documented that they know about 21 different Galasu. Among the varieties known to them, Vennigalasu (*D. hamiltonii*), Hehkkugalasu (*D. belophylla*) and Kavalagalasu (*D. oppositifolia*) are seen in interior evergreen and moist deciduous forests and Eragalasu (*D. wightii*) in rocky grasslands. Nooragalasu (*D. pentaphylla*), Naragalasu (*D. wallichii*) and Hendiridaegalasu (*D. bulbifera*) are found in wayside bushes and Boojikavalagalasu (*D. pubera*) in marshy areas. The Kattunaikka collect *Dioscorea* from almost all these places but more frequently from the forests and other such unmanaged habitats (Table 4.3).

Among the different species of *Dioscorea*, Nallanoora (*D. pentaphylla* var. *pentaphylla*) is the most commonly consumed tuber. As the name indicates, 'nalla' means safe or good to eat. The tuber is single, cylindrical, up to 1 m in length, less fibrous and powdery when cooked and tastes good. This variety is common in the fringes of deciduous forests. Korana (*D. pentaphylla var. rheedii*) is commonly used for various culinary preparations, occasionally as stewed cake, because of its high fibre content. Unless thoroughly washed before cooking, it can leave an itching sensation in the throat. Chenakorana (*D. pentaphylla* var. *communis*) has got the shape of a 'Chena' (elephant foot yam) and is fibrous in nature. Hendikorana (*D. pentaphylla var. linnaei*) tuber has got the shape of 'Hendi' (wild boar in the Kattunaikka dialect), with thick black-coloured root hairs all over the tuber. Kavala (*D. oppositifolia*) is another very popular tuber among all the tribes of Wayanad. It is excellent in taste and is commonly found in moist forests in which the Kattunaikka community depends more. Salugalasu, which is identified as *D. tomentosa*, is not consumed regularly due to its high mucilaginous content and is eaten only during times of acute famine. It has peculiar kind of fibres, which leave an itching sensation when consumed, particularly on children, making communities other than the Kattunaikka shy away from this tuber. The Paniya community, the study recorded, use roots and tubers of 19 plant species as their food. As in the case of Kattunaikka, *Dioscorea* (Kattukachil or Kattukizhangu) forms an important source of their food. They consume nine kinds of *Dioscorea* tubers, in which the most preferred ones are Kavalakizhangu (*D. oppositifolia*) and Noorakizhangu (*D. pentaphylla* var. *pentaphylla*). They consider the Noorakizhangu and Kavalakizhangu to be rich in 'Podi'

Table 4.3 Wild *Dioscorea* species consumed by Kattunaikka of Wayanad

Sl no.	Scientific name	Local name	Available landscapes
1	*Dioscorea pentaphylla* var. *pentaphylla*	Noora	Moist deciduous forest
2	*Dioscorea pentaphylla* var. *rheedii*	Korana	Moist deciduous forest
3	*Dioscorea pentaphylla* var. *communis*	Chenakorana	Moist deciduous forest
4	*Dioscorea pentaphylla* var. *linnaei*	Hendhikorana	Moist deciduous forest
5	*Dioscorea hispida*	Kottunoora	Moist deciduous forest
6	*Dioscorea* sp.	Moodavenni	Moist deciduous forest
7	*Dioscorea hamiltonii*	Kaluvenni	Evergreen and moist deciduous forest
8	*Dioscorea belophylla*	Hekku	Moist deciduous forest
9	*Dioscorea* sp.	Hekkuheruman	Moist deciduous forest
10	*Dioscorea* sp.	Heruman	Moist deciduous forest
11	*Dioscorea wallichii*	Narra	Wayside bushes
12	*Dioscorea* sp.	Narramooyan	Wayside bushes
13	*Dioscorea oppositifolia*	Kavalakalasu	Evergreen and moist deciduous forest
14	*Dioscorea wightii*	Erekalasu	Rocky grass lands
15	*Dioscorea intermedia*	Shoddikalasu	Dry deciduous forest
16	*Dioscorea pubera*	Boojikavala	Marshy areas
17	*Dioscorea pentaphylla*	Noora korana	Wayside bushes
18	*Dioscorea tomentosa*	Salu	Moist deciduous forest

Narayanan et al. (2004) and Balakrishnan (2009)

(starch) and 'Kozhuppu' (fat) and the Narakizhangu (*D. wallichii*) to be rich in 'Naru' (fibre). Noora and Kavala do not need any detoxification before cooking. Kuruma, Wayanadan Chetty and other settler communities know only three species of *Dioscorea*. Fifteen to 20 years ago, men of these communities used to collect Kavala and Noora, but today wild tubers do not flavour their diets. They consider it too tedious, a job to search and dig out the tuber, being otherwise engaged. They grow several tubers in their home gardens, and these are none too costly in the markets either. All the different sociocultural groups have got *Dioscorea alata* as a cultivated species in their home garden. An interesting side is that in Chooralmala area of the district, Muslim, Hindu and Christian women buy Kavala and Noorakizhangu from the Paniya in exchange for money or rice. Many of the youngsters of these communities are but totally ignorant about these tubers or their importance mainly because wild *Dioscorea* is no more a part of their diet. Nannari (*Hemidesmus indicus*), Muthanga (*Cyperus rotundus*), Shatavari (*Asparagus racemosus*), Unnithandu (*Costus speciosus*), various species of wild curcuma and wild ginger are some of the other wild plants used for roots, rhizomes and tubers by various sociocultural groups of the study area and are often used as important ingredients in certain traditional medicines.

Fruits and Seeds Fruits and seeds are an important group of edibles, which contribute to the Adivasi communities' nutritional requirements. Much of their vitamin and mineral needs are met by this category of food. Information on 60 such fruits and seeds (fruits 54, seeds 6) was collected by Narayanan and Anil 2007 (Table 4.4). Among the fruit-yielding plants, 33 are trees, usually found in the forests and hills. Fruit trees like Plavu (*Artocarpus heterophyllus*), Mavu (*Mangifera indica*), Athi (*Ficus racemosa*) and Njaval (*Syzygium cumini*) are protected on waysides and in the agricultural landscapes in the study sites. The fruits of these trees (except *Ficus*) are widely used by people across communities on a regular basis. Various *Ficus* varieties are protected, either for their sanctity or because the birds feed on them or because they host nocturnal animals like bats which are beneficial to the crops. Paniya communities are the largest consumers of various wild fruits. There are about 50 species that are consumed by this community alone, largely collected from forests, wooded hills or such unmanaged areas. The non-Adivasi communities restrict themselves to the fruits of jack, mango, gooseberry and njaval trees and generally avoid the lesser-known fruits from the forests.

Mushrooms About 2500 species of mushrooms are reported from across the world, but only a few of the wild mushrooms are eaten by the rural population. Besides their diverse and interesting culinary uses, mushrooms are much endowed with nutritional and medicinal value. Some mushrooms are reported to contain cancer-fighting properties, and several aid the body's immune system. Many of these are good sources of protein, vitamins and minerals. The carbohydrate content in mushrooms is very low; therefore, these are specially recommended to diabetic and anaemic persons, owing to their high folic acid content. Thirty-five different wild mushrooms are consumed by the people of Wayanad (Table 4.5), especially the Kattunaikka who call mushroom as 'Anavae'. They classify the mushrooms as Maranavae mushrooms seen on the bark of different trees, Huthaanavae those seen on termite mounts and Mannanavae which are on the forest floor and associated habitats. The availability is seasonal and specific to their habitats and host plants. This delicacy is usually available after the onset of the monsoons. The most commonly consumed mushrooms are Arikkoon, Puttukoon and Perumkali.

Some mushrooms are named after the host trees; for example, the Anavae seen in Njeral (*Syzygium cumini*) is called Njeralanavae, on Jal (*Dalbergia latifolia*) Jalanavae, on Kaval (*Erythrina indica*) Kavalanavae and so on. There are three species of Huthaanavae. The one, which is milky white in colour, is called Vellanavae, the pale white in colour is Ummanavae and the large-sized off-white coloured is typical Huthaanavae. The community considers the mushrooms seen associated with 'Huthu' (termite mount) to be non-toxic and do not feel the need to process it in any manner. 'Kumman' is the Paniya word for mushrooms, which is considered a delicacy. This community uses about 25 species of Kumman, many of which are collected from plantations. Mushrooms are classified into two groups by them:

Table 4.4 Wild edible fruits and seeds of Wayanad

Sl no.	Local name	Botanical name	Knowledgeable sociocultural group[a]
1	Arinjaval	*Syzygium densiflorum* Wall.	P, K
2	Athapala	*Chrysophyllum lanceolatum* Bl. (DC.)	P
3	Athipazham	*Ficus racemosa* L.	K, Ku, P, O
4	Ayanichakka	*Artocarpus hirsutus* Lam.	K, Ku, P, O
5	Chadachikkaya	*Grewia tiliaefolia* Vahl.	P, K, O
6	Chakadahannu	*Schefflera oleracea*	L, K
7	Chalir	*Flacourtia montana* Graham	P, Ku, K,O
8	Chammikkaya	*Aponogeton appendiculatus* van Bruggen	P
9	Chekkipazham	*Ixora coccinea* L.	K, P, K
10	Deprahannu	*Diospyros* sp.	K
11	Eachil	*Aporosa lindleyana* Baill.	P, O
12	Edavahannu	*Leea indica* L.	K
13	Eenthukaya	*Cycas circinalis* L.	P, K, Ku, O
14	Elanchipazham	*Mimusops elengi* L.	P, O
15	Geru hannu	*Buchanania axillaries* (Desr.) Ram	K
16	Hallaekaya	*Grewia sp.*	K
17	Kalanthatta	*Sterculia foetida* L.	P, O
18	Kandakarichunda	*Solanum xanthocarpum* Schrad.	P, K, Ku, O
19	Karinjavel	*Syzygium gardneri* Thw.	P, K, Ku
20	Karuvachakka	*Solena amplexicaulis* (Lam.) Gandhi	P, K, O
21	Kattambazham	*Spondias indica* Wt.	P, Ku, O
22	Kattuchakka	*Artocarpus heterophyllus* Lam.	K, Ku, P, O
23	Kattujadikka	*Myristica malabarica* Lamk.	P, O
24	Kattukariveppu	*Clausena heptaphylla* (Roxb.) Wight	P, K
25	Kattukodampuli	*Garcinia gummi-gutta* (L.) Robs.	P, K, Ku, O
26	Kattumanga	*Mangifera indica* L.	P, K, Ku, O
27	Kattu-munthiri	*Rubus fulvus* Focke	P, K, Ku, O
28	Kattunjaval	*Syzygium laetum* Ham.	P, K
29	Kattuthakkali	*Passiflora calcarata* Mast.	P, O
30	Kirinda	*Scleropyrum pentandrum* (Dennst.) Mabb.	P
31	Kongini	*Lantana camara* L.	P, K
32	Koovalam	*Aegle marmelos* Corr.	P, O
33	Kottamullu	*Ziziphus jujuba* Lamk.	P, Ku
34	Kottapazham	*Ziziphus oenoplia* (L.) Mill.	P, Ku, K, O
35	Kottilampazham	*Elaeocarpus tuberculatus* Roxb.	P, K, O

(continued)

Table 4.4 (continued)

Sl no.	Local name	Botanical name	Knowledgeable sociocultural group[a]
36	Kulayari	*Bambusa arundinaceous* Willd.	P, K, Ku
37	Kurukkanchunda	*Solanum ferox* L.	P, Ku, K, O
38	Motampuli	*Physalis minima* L.	P, K, Ku, O
39	Mottilthoory	*Baccaurea courtallensis* Wt.	P
40	Mudungakaya	*Solanum nigrum* L.	P, Ku, K, O
41	Mukayani	*Bridelia retusa* Spreng	P
42	Mulluvalli	*Toddalia asiatica* (L.) Lamk.	P
43	Neeli pazham	*Bischofia javanica* Bl.	P, K
44	Nelli	*Emblica officinalis* Gaertn.	K, Ku, P, O
45	Nendravally	*Bridelia scandens* (Roxb.) Willd.	P
46	Njara pazham	*Syzygium cumini* (L.) Skeels	P, K, Ku, O
47	Njara pazham	*Syzygium caryophyllatum* (L.) Alston	P
48	Njenumkaya	*Gnetum ula* Brogn.	P, O
49	Njotanjodian	*Physalis mouritiana* L.	P, O
50	Palakkai	*Palaquium ellipticum* (Dalz.) Engl.	P, K, O
51	Palapazham	*Glycosmis pentaphylla* (Retz.) DC.	P, Ku
52	Pillandi	*Melastoma malabathricum* L.	P, O
53	Pindichakka	*Randia uliginosa* DC.	K, Ku, P
54	Poochapazham	*Syzygium zeylanicum* (L.) DC.	P, O
55	Poodapazham	*Passiflora foetida* L.	P, K, O
56	Putharichunda	*Solanum torvum* Sw.	P, K, Ku, O
57	Thanikkuru	*Terminalia bellerica* Roxb.	P, O
58	Thodali	*Ziziphus rugosa* Lamk.	P, Ku, K,O
59	Tholnjaval	*Syzygium hemisphericum* Bedd.	P, K, Ku
60	Undanjaval	*Syzygium mundagam* Bourd.	P, K, Ku

[a]*P* Paniya, *K* Kattunaikka, *Ku* Kuruma, *O* others
Narayanan et al. (2004)

Marakkumman (those seen on trees) and Mannukkumman (those seen in soil). These are further classified based on substrate, shape, size, etc. For example, 'Valakkumma' refers to mushrooms that sprout in valam (cow dung), 'Vaikkolkkumma' sprouts on vaikkol (paddy straw), 'Kathukkumman' the one that has the shape of 'Kathu' (human earlobes) and 'Ambukkumman' the one that has the shape of 'Ambu' (arrow). The Kuruma refers to mushrooms as 'Koonu', the same popular Malayalam word for it, and it finds pride of place in the family menu when available. Though around 14 species are consumed by Kuruma, they only prefer the 'Koonu' seen on soil, especially, those associated with termite mounts. The most commonly consumed mushrooms are Perumkoonu, Arikkoon, Nedumthali and Puttukoon. Compared to the Adivasi communities, others use only very few

Table 4.5 Wild edible mushrooms of Wayanad

Sl no.	Local name	Landscape
1	Nettanavae	Forest floor
2	Mavanavae	Dried mango trees
3	Huthaanavae	Termite pit soil
4	Monjanavae	Soil
5	Kattanavae	On dried bamboo
6	Chulliyanavae	Grasslands
7	Maranavae	Dried trees
8	Kozikalanavae	Bamboo forest
9	Kachikalanave	Soil
10	Chorakalanavae	Riverside, toil
11	Thorathalanavae	Tree
12	Vendageenkananavae	On *Lagerstroemia* trees
13	Jalanave	On *Terminalia* trees
14	Komananavae	On *Mangifera* trees
15	Uppuhuriyananavae	Forest floor
16	Karanavae	Plantations
17	Kaykananavae	Plantations
18	Therikanavae	Plantations
19	Sunkgeenkan	Trees
20	Kavananavae	Soil
21	Pellikuthananavae	Forest floor
22	Njerananavae	On *Syzygium* trees
23	Ummananavae	Soil
24	Marageenkananavae	Tree
25	Penankivi	Soil
26	Vellanave	Termite pit soil
27	Vendanavae	Forest floor
28	Mukkanavae	Plantations
29	Kolanavae	Forest floor
30	Naymulanavae	Paddy, grassland
31	Kodankimianavae	On dried *Erythrina* bark
32	Mayilpeelikumman	Near bamboo forest
33	Mottananavae (karadi)	Forest floor
34	Vennanave	Forest floor
35	Valanavae	Cow dung

Narayanan et al. (2004)

mushrooms, largely the Arikkoon and Puttukoon, which are considered highly delicious. Among the non-Adivasi communities, Muslims refer to different varieties of mushrooms by distinct names like Mothirakkanikoon, Pavakoon, etc.

Honey Honey is one of the most important minor forest produce for most Dravidian Adivasi communities. It features as a major constituent of their diet as well. The Adivasi communities classify different types of honey according to the type of bees and the nature of hives from which they are collected, and they vary greatly in taste and therapeutic properties. Adivasi communities are among the most skilled in locating, extracting, processing and preserving honey and among the most knowledgeable about the dietary and curative properties of honey. Kattunaikka are also known as Jenukkurumba because of their exceptionally high skill in 'Jen' (the Kattunaikka word for honey) collection. They recognise five types of 'Jen': Kombujen/Daddanjen, Thudajen, Gandajen/Dojjan, Nasarjen/Kothukujen and Kusumbujen/Cherujen. The bulk of the honey comes from hives seen in the branches of tall trees or rock crevices (Kombujen, more than 25 kg/hive). The bees that produce this honey are larger in size and yellowish in colour. Honey is also differentiated based on the season of availability, as Kannijen which is available in bulk during Kanni Masam (Sept–Oct) and Karthijen which is available in Karkkidakam (July). The availability and quality of honey are based on the flowering of forest trees in the locality. Flowering of Venthekku (*Lagerstroemia lanceolata*) signals abundant availability of Kombujen. The hives of Thudajen/Pothujen are mainly seen in tree crevices or in inactive white ants' pits. The colour of the honey is reddish brown, and its availability is usually based on the flowering of Mattimaram (*Terminalia tomentosa*). The Jenukkurumbas are experts in inducing the small bees of the *Melipona* species to hive in bamboo stems, to get an almost 'captive supply' of the delicious cherujen. Cherujen is the most precious of the various varieties of honey, sparsely produced, difficult to extract and believed to possess excellent nutritional and medicinal attributes. It is available round the year. The Kattunaikkas are expert honey tasters and will link the distinct tastes to specific forest flowers. Kedujen/Kothukujen is mainly seen on dry branches of bushy plants, and the bees are comparatively smaller.

Crabs Crabs are yet another wild food habitually consumed by the dominant Adivasi communities of Wayanad (Table 4.6). The Paniya collect five kinds of crabs regularly. These are Vellanjendu, Palnjendu, Kundunjendu, Kottinjendu and Karinjendu. All of these varieties are abundantly available during the rainy season in habitats like paddy field and associated habitats. The Kuruma consume three of these, Kolathinjendu (Vellanjendu), Karinjendu and Kundunjendu. Kattunaikka refer to the crabs as 'Nelli', and their names for various varieties of crabs are Kottinelli, Gundranelli, Valanelli, Halnelli and Kallunelli. Except the Halnelli, all the other varieties are consumed by the Kattunaikka community. Palnjendu is a small crab, found inside the root clusters of paddy seedlings. These are caught while transplanting the paddy seedlings, and along with fish, they make a delicious supper. This crab, white in colour, occasionally sheds its outer shell and is then very

Table 4.6 Crabs in the dietary system of different sociocultural groups

Sl no.	Local name	Knowledgeable sociocultural group[a]	Available landscape
1	(Vellanjendu) Kolathinjendu Valanelli/ Vayalnjendu	P, Ku, K, O	Paddy fields
2	Palnjendu/Muthachinjendu	P	Paddy fields and associated marshy areas
3	Kundunjendu, Gundranelli	Ku, K	Marshy areas and plantations
4	Kottinjendu, Kottinelli	K	Marshy areas
5	Karinjendu, Kallunelli, Puzhanjendu	P, Ku, K, O	River and streams

[a]*P* Paniya, *K* Kattunaikka, *Ku* Kuruma, *O* others
Narayanan et al. (2004)

slimy. During this time, it has little flesh, and if the legs are broken, a milky juice oozes out (thus earning it the name Palnjendu). It is not eaten when its shells are shed. Kundunjendu/Gundranelli is the smallest crab and the tastiest one as well. Their tentacles are longer, compared to the other crabs. Normally the shell of this crab is brown in colour and changes to white on ageing. The shell of this variety has got numerous lines, and it is available in plenty after the monsoon rains. Vellanjendu/Valanelli found in the paddy fields is pale yellow in colour with lines on the shell. This is the most frequently consumed crab by the Paniya and the Kattunaikka. Other communities (Chetty, Hindu, Muslim and Christian) distinguish mainly two kinds of crabs and have named them based on the habitat. Vayalnjendu are crabs found in paddy fields, and Puzhanjendu are those found in rivers. Only the crabs that are found in the river are preferred by these communities.

Fish The Paniya community has knowledge of 36 edible fishes, and almost all these species are available throughout the year (Table 4.7). Some of these species are also known to Kuruma community. Out of these, about 20 are strictly inhabitants of rivulets and streams but were seen to enter the paddy fields during heavy rain. The most frequently harvested fish by Paniya are Kalluppatti, Koyma, Konjai, Kannappae, Parel, Kaduvae, Muzhu and Kommai, which are commonly available in streams irrespective of the season. A big-sized fish called Malanjil found in plenty once has now become very rare. They attribute the disappearance of this species to the chemical pollution of streams from the pesticides sprayed in the plantations.

4.3 Gender Dimensions of Wild Edible Collection and Utilisation

Gender studies show that a majority of plant species and varieties used for food and medicine are conserved and managed at household level by women (Gurung and Jeannette 1997; Balakrishnan 2000). The role of gender in enhancing food security

Table 4.7 Wild edible fish species of Wayanad

Sl no.	Local name	Landscape	Knowledgeable sociocultural group[a]
1	Mushu	Stream, river	P, K, Ku, O
2	Kaduve	Stream, river	P, K, Ku, O
3	Parel	Paddy fields	P, K, Ku, O
4	Njenu	Stream	P
5	Kalleppatti	River	P, K, Ku, O
6	Koyma	Stream, river	P, Ku
7	Kanneppe	Stream, river	P, Ku
8	Aaral	Big river	P, K, Ku, O
9	Thodameen	Stream, river	P, K, Ku, O
10	Kaichalu	Big river	P, Ku
11	Kalancheppi	River	P
12	Chakkamullan	River	P, K, Ku, O
13	Chempally	Big river	P, K, Ku, O
14	Chethil	Big river	P, K, Ku, O
15	Philoppy	Paddy fields, stream	P, Ku, O
16	Kottavala	River, stream	P, K, O
17	Malanjil	River	P, K, O
18	Thalammakkannan	Stream	P, K, Ku, O
19	Thuppal kothi	Streams	P
20	Pullipparal	Streams	P, Ku, O
21	Vattapparal	Streams	P, Ku
22	Paralam paral	Streams	P
23	Kakkaparal	Streams	P, Ku
24	Thalavannan paral	Streams	P, O
25	Kammai	Streams	P
26	Kooriparal	Streams	P, O
27	Attuvala	River	P, K, Ku, O
28	Pulvala	River	P, Ku, O
29	Pullumeen	Streams	P, K, Ku
30	Chillumullam	Streams	P
31	Choorikoyma	Streams	P
32	Cheriyakoyma	Streams	P
33	Konachan red	Streams	P, K, Ku, O
34	Konje	Streams	P, K, Ku, O
35	Kumbalappotti	Stream	P, Ku
36	Vannal	Paddy	Ku

[a]*P* Paniya, *K* Kattunaikka, *Ku* Kuruma, *O* others
Narayanan et al. (2004)

has been also a major topic of discussion (Price 1993; Kanvinde 1999). Many authors have attempted to understand the roles, responsibilities and relations of women and men in collection, processing, cooking, consumption and management of various food species available to them in the wilderness. Kanvinde et al. (2001) show that women are responsible for all the leafy greens collected, processed and cooked. Borjas (Borjas 2001) argues that women's ability to conserve biodiversity and influence the way that others conserve is eroding rapidly. The author cites several reasons for this: the dwindling of forests and other land resources that women depend on, introduction of cash crops and modern varieties that displace the traditional crops, out-migration of males from rural areas that leaves women to manage agriculture without access to labour and other such critical resources, erosion of women's rights to private land and the disparagement of their native diets.

Gender roles and responsibilities assigned by the society give women the predominant role in collection and processing of wild greens for consumption. As in the case of other socially assigned female roles, this requires patience and is time-consuming. Women consider it as their responsibility, and this role does not vary on account of religion, ethnicity or class. Women perform cent percent of all labour inputs required, from collection to processing and serving. They have knowledge about each and every plant, such as its location, availability, factors influencing palatability, nutritional value and so on. For collection of the greens, women of Paniya community walk considerably long distances compared to other categories of women. The traditional dressing style of Paniya women is attuned to storing and carrying comfortably the collected foodstuffs from field. At times they are seen using bamboo baskets or areca palm sheath to carry the collected materials. Compared to the other women, Paniya and Kattunaikka women are more experienced and knowledgeable regarding collection and storage. Men, especially the Paniya and Kuruma youth and certainly the menfolk of settled communities, look upon wild green collection as beneath their dignity, but there are exceptions also. Generally though, it is considered that the man's role is to get the staple food like rice or tubers, and it is the women's duty to add diversity and flavour by getting the leaves. Since many of the leaves are seasonal, they ensure year-round supply to supplement their diet by zeroing in on commonly available greens specific to the seasons. Women thus use the resource in a variety of beneficial ways, not restricting themselves to just the commonly used leaves. They also adopt various processing methods to make the edibles consumable and palatable. According to the usefulness of each species, the women adopt various management mechanisms for its conservation and sustainable usage.

In the collection, processing and management of *Dioscorea*, we can clearly see differences in gender roles in each sociocultural group. In the case of Kattunaikka, the entire task is shared by both men and women except for tool making for collection. In the case of Paniya, there is a clear division of tasks except for the responsibility of ensuring the long-term availability of tuber. Both men and women make it a point to retain a piece of tuber in the pit for regeneration. In the case of Kuruma, there is no involvement of women in the collection and management, and men play a predominant role in the identification and collection of other tubers. Though women are also involved in collection to support men, processing is left completely to women, and they hold the related knowledge. Men and women of Kattunaikka community are well versed in the identification of *Dioscorea* in terms of its availability, habitat and associated plants. They are also adept at identifying the matured and sweet tuber ideal for consumption. Commonly, men and women go together for collection. Kattunaikka women used to introduce several of the species of *Dioscorea* into their home gardens. For the Paniya community, tuber collection is a highly seasonal activity because of the distance from the forest and is undertaken only two or three times in a year. Before venturing out for collection, both the men and women discuss and plan the locations, timing, tools and materials to be taken, etc. Women avoid going alone to the interior forests for fear of wild elephants. The digging out of the tubers is a collective effort involving both men and women. Men are experts in making the tool to dig out the tuber. While men dig out the tubers spotted, women go around locating fresh *Dioscorea* plants and spotting the exact location of the tuber underground, which needs a lot of patience. Digging out the tuber is a collective effort, and while men are digging out the tubers, women help them by removing the soil, stones, etc. from the pit. Women also dig out the tubers but only those that are shallowly rooted. It was observed that Paniya men and women above 40 years of age have a clear idea about the tuber's growing pattern. However, men are better informed in identifying different roots and tubers based on the shape and texture of their leaves and wines.

It is observed that while men and children seek after fruits and seeds based on their individual preferences, women consider the requirements of the family as a whole. Women go out specifically to collect such fruits accompanied by children or sometimes men. When women go alone, they use long poles to harvest fruits like jack and gooseberry. It is not an unusual practice for them to leave the upper branches of fruit trees unharvested or selectively harvested for the birds and other animals to savour. Gooseberry (*Emblica officinalis*) is one of the most widely collected non-wood forest produce, and it is an important source of income for Kattunaikka families. Both women and men are engaged in its collection and sale. Kuruma women are rather selective in their choice of fruits. They accord greater preference to mango and jack. Women have specific knowledge about recipes and storage techniques. An important seed, which the Paniya and occasionally the Kattunaikka largely use, is that of bamboo (Mulayari). Bamboo flowers very rarely, and at the onset of the flowering season, the Paniya women start preparations for the collection of seeds. Women go to the bamboo brakes before fruit setting and clear the undergrowth and prepare a clean bed for the seeds to fall. The collection may continue for days, and together men and children join in gathering the seeds.

Mostly women and girls are involved in the collection, processing, preservation and preparation of mushrooms. Women, by and large, hold the knowledge about its quality, use and conservation. Women also take interest in protecting the wild habitats and areas in the home gardens where mushrooms naturally sprout. It was noticed that women protect the termite mounts and the tree species, which host certain mushrooms. While the collection and processing are in the women's domain, men help in some specific activities like collecting mushrooms that are seen on tall tree trunks or in the interior of forest. The women use different criteria, like culinary qualities, palatability, and nutrition, while selecting and conserving mushrooms. Kattunaikka women, among the others, are more adept at distinguishing between poisonous and non-poisonous varieties. While collecting mushrooms from tree trunks, maximum care is given to harvest only the fruiting body and spare the basal portion to sprout again. Among the Paniya community, men sometimes help women in collection. But women of the community consider it as their responsibility to collect, process and cook such delicacies. Normally men bother to collect only those mushrooms, which are either very tasty or rare in distribution, which they come across while out on other errands. Men also avoid those mushrooms which require patient tedious labour to collect. Paniya women are also experts in identifying edible and nonedible mushrooms. The Paniya women in general distinctly remember the time of the year and the specific locality where mushrooms are available. Among the Kuruma community, collecting mushrooms is considered exclusively the woman's job. Despite the growing trend in this community of women being increasingly confined to their households and men assuming the role of sole breadwinners, the Kuruma women retain enough knowledge about the nature and kind of mushrooms, which can be stored or eaten instantly. The Paniya women have to walk long distances these days for mushrooms as they are seen only in less polluted habitats, which are often far off from their habitation. The Kuruma women search for mushrooms only in their surroundings, mostly in the hill slopes or in coffee or teak plantations. But for a little support extended in collection by the men of Paniya and Kattunaikka communities, women shoulder the entire responsibility of mushroom harvesting, processing and dish preparation.

The Kattunaikka have developed over time a host of techniques for honey collection. A honey expedition into the forest requires much preparation and proactive thinking. If the honey is located far away from the dwellings, a whole group of men and women join the foray and stay in the forests through the period of collection, which might stretch to 2 weeks or more. Both men and women are good at locating the hives and easily pick up the sound of different bees in flight. These bees are watched and followed to the hives. Sometimes both men and women together set out to the forests, about a month before the season, to locate the beehives. Usually a group of three to five members is involved in honey collection. While in the specific act of extraction, especially when the beehives are located on tall trees, women may not be involved, they are crucial in providing the logistics. These include procuring the necessary tools, organising the required provisions for the period of extended stay in the woods and cooking the food for the whole group during the days, in other words, most of the tedious work. The bees are driven away by driving smoke into the hives using fresh green grass, which is collected by women. Collecting Pothujen/Thudajen from tree crevices and termite mounts is considered comparatively easy, and women participation in it is greater. Kattunaikka women take special interest in collecting Cheruthen, as it is believed to be highly nutritious and medicinal and is best suited for infants. The collection of this type of honey involves much labour and a great deal of patience, and men normally avoid the task. Other communities like Paniya and Kuruma also collect honey occasionally. The women of these communities hardly involve themselves in honey collection or processing, the exception being the hill Paniyas, who are experts in honey collection, and both men and women are involved in its collection.

Normally among Paniya and Kuruma communities, the women and children are involved in catching crabs. The men have no dearth of skills in the job either; some of the older male members told us of a variety of techniques in unearthing the crabs from their holes and seizing it without hurting your fingers. Among the Kattunaikka, as a practice both men and women are together involved in the collection. Normally in the Paniya and the Kuruma tribes, women take independent decisions to go hunting crabs and take the children along. Among the Kattunaikka, both men and women set out together for collection, except to the habitats in the close vicinity of their homes, where the women alone go, along with children. Invariably women do the processing. Women are good at identifying the presence of the crabs judging from the nature of the small soil mount deposited outside the crab's pit. In all the communities, women and girl children do the processing and cooking. Paniya women have profound knowledge about the breeding time of crabs.

Fishing is an important activity for men, women and children among the Paniya community studied. Though fish is available irrespective of seasons, the fishing activity is more during summer as the reduced water level in the streams makes fishing easier then. There is a flurry of fishing at the onset of monsoon as well, but this is mainly by women of the Kuruma community. Kuruma and Paniya women are capable of approximating quantity and kind of fish available, judging from the intensity and pattern of rain. Kuruma men use Vala (fish net), Ambu (arrow) and Choonda (hook) for catching fish. Kuruma women use a special trap called Chada, made of bamboo splices. Making the Chada calls for good skills, and placing it aptly needs expertise as well, both of which are the forte of Kuruma women. Fishing is of ritualistic significance as well for the Kuruma. A practice that has all but died out is that on the third day after marriage, the bride is escorted by the women of the groom's settlement to the nearby stream, where she is expected to demonstrate her fishing skills. After the breaking of pollution on the third day after a death (Pulakuli), men of the household go for hunting and the women for fishing. The fish and meat so obtained are cooked and first ritualistically offered to the ancestral spirits and then consumed by the members of the clan. Among the Paniya and the Kattunaikka, fishing is a collective effort involving all the family members. Still, the decisions regarding the choice of location, the time of fishing and the mode of catching fish are all taken by the men. The women of non-Adivasi communities are not involved in fishing. Irrespective of the communities studied, women do the processing and cooking of fish. Analysis of the gender roles in collection of fish by Paniya shows that 10, out of 34 species, are collected only by men and three species only by women, while all the other species are collected jointly by men, women and children. The species collected only by men are mostly seen in the big rivers and streams inside thick evergreen forests. Women, evidently, find these habitats inaccessible.

4.4 The Changing Scenario

Habitat Loss for Wild Edibles: The Conversion of Paddy Fields The paddy fields of Wayanad had been a veritable treasure trove of a variety of leafy greens and a host of other wild food, regularly accessed by the Adivasi communities, especially the Paniya and the Kuruma. Paddy fields, as they existed nearly two decades ago in Wayanad, provided food, employment and ecological security to these people. Apart from greens, a number of other species of high food and health value such as fish, crustaceans like crab and snails and medicinal plants were associated with this agroecosystem. The Kurichya and Kuruma completely relied on paddy cultivation and this ecosystem for their food security. The Paniya community depended on paddy fields for employment. Women of this community were among the most adept at all tasks related to paddy cultivation, and they depended on the wage earnings from it as their principal source of income. The paddy fields were also among their principal sources for a variety of wild food. Even today, the Paniya women know and use 19 plant species from the paddy fields and its mud bunds. Besides this, a number of rituals and traditions of the Paniya, Kuruma and Kurichya communities are strongly intertwined with this ecosystem. From an ecological view point, the paddy fields situated in the low-lying areas of the undulating Wayanad terrain act as a trough collecting and retaining a large quantity of water that is used by a number of plants and animals (most of them of direct use to the dependent communities). Conversion of this land for cultivation of perennial crops (or, as is the common practice, for banana cultivation) limits the storage capacity of this ‘sponge’ leading to water shortages in nearby wells during dry seasons and floods during rainy seasons.

Paddy fields in the district, which once occupied a major portion of the cultivable area, have marked a drop by 66% (from 30,000 ha to 10,230 ha) during the period 1980–1981 to 2012–2013 (DES 1983, 2013). This agroecosystem is under increasing threat in the form of habitat loss, land reclamations and other alternate unsustainable land use. The shift in land use from paddy cultivation to the banana crop, with its attendant reduction in the water content of the soil and the high infusion of chemical fertilisers and insecticides, has taken a heavy toll. Another very important social and economic repercussion is the loss of employment opportunities of Paniya women, which has forced them to go in search of jobs even to remote places. There is a clear need for a more rational and sustainable management of remaining paddy fields in the district, not merely because the production of the staple food of populace is affected but also because its preservation is inextricably linked to the food supply chain. The availability and benefits of greens, fish and crabs and a host of other locally important products depend on the paddy fields remaining intact.

Overapplication of Chemicals Studies have shown that the unscientific application of chemical pesticides, weedicides, fertilisers, etc. in the coffee, tea, cardamom and banana plantations has considerably reduced the population of common edible greens and mushrooms. Paniya women of Mutharikkunnu cited an incident where five members of a family had to be hospitalised after consuming greens collected from a banana plantation sprayed with toxic pesticides. The feeling that wild greens may not be safe anymore has also reduced its consumption according to these women.

Invasion of Alien Species All open clearings like waysides, grazing lands, new plantations and the fallow paddy fields are the usual sites for green leaf collection. A variety of alien species that have appeared suddenly and are getting naturalised rapidly throng these locations. The climate of the district is highly suitable for the fast growth of many of these exotic species. Some of these have replaced the edible greens; for example, species such as *Cassia tora*, *Alternanthera sessilis*, *Amaranthus viridis*, *Amaranthus spinosus*, *Colocasia esculenta* (Karathalu), etc. are edged out by exotics like *Lantana camara*, *Parthenium hysterophorus*, *Drymaria cordata*, etc. Sizeable areas of Muthanga wildlife sanctuary in Wayanad are highly infested with *Lantana camera* and *Chromolaena odorata*– two noxious exotics. They thrive in the area, which has been clear-felled of eucalyptus plantations. Interestingly, the eucalyptus plantation itself came into being after pristine natural forests were cleared to plant them as part of the social forestry scheme. *Mikania micrantha* is another troublesome weed found in almost all the forest fringes in the district. *Mucuna pruriens*, *Parthenium hysterophorus*, *Bidens biternata* and several species of *Blumea* are some of the quick-growing alien species, which have proliferated in different habitats in the area. Interestingly, some of these alien species, as mentioned earlier, are included in their collection of greens by the Paniya women. *Bidens biternata* (Alanchappu as the Paniya has named it) has turned out to be a delicious supplement in their food. This species, found as a weed in plantations, is sought to be controlled by the planters using strong weedicides. The *Bidens biternata* flourishes in all the open landscapes, particularly in the human-managed ones like the coffee plantations. Only the Paniya women go for this species, but it underscores the ingenuity of the Adivasi women in discovering new edible plants to replenish their food basket. It is also an assertion that "traditional knowledge" is dynamic, evolving and ever changing, with both additions and deletions over time.

4.5 Conclusions

Three hundred fifty-three species of wild edibles have been studied from Wayanad District of Kerala, part of Western Ghats. This account has also traced out the places where the wild edibles proliferate, the communities that come in search of the edible forest yield during different seasons, the colloquial terms that are commonly used to describe them and the gender roles that come into play during harvesting and processing of the wild foods. An examination of the pattern of accessing wild food from various landscapes by the different communities has brought out the following observations. The Paniya community successfully explores all the landscapes ranging from forests to grazing lands but more from the habitats outside forests – thickets, plantations, paddy fields, swamps, waysides and grazing lands/mountains. The Kattunaikka, on the other hand, forage mostly the forests, and the

Kuruma access the plantations and paddy fields. The non-Adivasi communities access largely three landscapes – the paddy fields, thickets and plantations – to collect only the most valuable species or varieties. Adivasi and rural families of Wayanad continue to conserve a wide range of plants to meet their diverse food needs. Women are more skilful in managing the surrounding landscape and are the chief knowledge holders and conservationists.

India is one of the agrobiodiversity-rich countries of the world with over 160 crop species with hundreds of varieties, 325 crop wild relatives and around 1500 wild edible plant species and diverse domesticated diversity of animals and birds (National Academy of Agricultural Sciences 1998). In stark contrast, the country has also the second-highest estimated number of undernourished people in the world (FAO et al. 2015). Kumar et al. (2015) has noted that Agrobiodiversity (including the wild edibles discussed so far) has a critical role to play in dealing with the issue of undernutrition, and hence dynamic conservation of agrobiodiversity needs to be placed high in the national development agenda for leveraging nutrition in agriculture and alleviating poverty and malnutrition.

Acknowledgement The authors are very grateful to all those men, women and children of different sociocultural groups of Wayanad District who generously contributed their knowledge and materials used to produce this account. We also acknowledge Ms. M. P. Swapna, our former colleague at MSSRF, for her unconditional assistance without which this account would not have materialised. Authors gratefully acknowledge Ms. Mina Swaminathan for her wholehearted encouragement and guidance and Drs. Sumi Krishna, Hemal Kanvinde and Meera Devi for their kind supports. The inspiration given by Prof. M. S. Swaminathan is deeply acknowledged. The financial support provided by Uttara Devi Resource Centre for Gender and Development (MSSRF) is also acknowledged. The executive directors and other colleagues, especially of the biodiversity programme of MSSRF, are gratefully acknowledged for helping in bringing out this research study.

References

Arora RK, Pandey A (1996) Wild edible plants of India: diversity, conservation and use. National Bureau of Plant Genetic Resources, New Delhi, p 294

Balakrishnan R (2000) Gender-defined strategies for biodiversity management for household food security. Gender dimensions in biodiversity management and food security: policy and programme strategies for Asia. M. S. Swaminathan Research Foundation, Chennai, MSSRF/PR/00/37

Balakrishnan V (2009) Ethnobotany, diversity and conservation of wild yams (Dioscorea) of Southern Western Ghats, India, Ph.D. thesis, University of Madras, Chennai

Bell J (1995) The hidden harvest. Seedling, Genetic Resources Action International, Barcelona. Last viewed 26 September 2016, https://www.grain.org/article/entries/318-the-hidden-harvest

Borjas PH (2001) Women in the plant world: the significance of women and gender bias for botany and biological diversity. Wageningen Universiteit. Last viewed 26 September 2016, http://edepot.wur.nl/234682

Cruz-Garcia G, Price LL (2011) Ethnobotanical investigation of 'wild' food plants used by rice farmers in Kalasin, Northeast Thailand. J Ethnobiol Ethnomed 7(33):1–20

Department of Economics and Statistics (1983) Statistics for planning. Department of Economics and Statistics, Government of Kerala, Thiruvananthapuram, Kerala

Department of Economics and Statistics (2013) Agricultural statistics, Department of Economics and Statistics, Government of Kerala, Thiruvananthapuram, Kerala. Viewed 26 September 2016, http://www.ecostat.kerala.gov.in/index. php/reports/154.html
Department of Economics and Statistics (2015) Agricultural statistics 2013–14. Department of Economics and Statistics, Government of Kerala, Thiruvananthapuram, Kerala. Viewed 26 September 2016, http://www.ecostat.kerala.gov.in/docs/pdf/reports/agristat/1314/agristat1314.pdf
Falconer J (1990) Hungry season food from forests. Unasylva 40:160
Falconer J, Arnold JEM (1991) Household food security and forestry- an analysis of socio-economic issues. Food and Agriculture Organisation, Rome
FAO, IFAD and WFP (2015) The state of food insecurity in the World 2015: meeting the 2015 international hunger targets: taking stock of uneven progress. FAO, Rome. Viewed 26 September 2016, http://www.fao.org/3/a-i4646e.pdf
Fernold ML, Kinsey AC (1958) Edible wild plants of Eastern North America. Harper and Brothers, New York, p 452
Food and Agriculture Organization (1982) Fruit bearing forest trees. Technical Notes, FAO Forestry paper 34. FAO, Rome
Food and Agriculture Organization (1984) Food and fruit bearing forest species 2: examples from Southeastern Asia. Forestry paper 44/2. FAO, Rome
Food and Agriculture Organization (1989) Forestry and nutrition – a reference manual. FAO, Bangkok
Food and Agriculture Organisation (1993) The sixth world food survey. FAO, United Nations, Rome
Gammie GA (1902) A note on plants used during famine and seasons of scarcity in Bombay presidency. Rec Bot Survey India 2:2
Grandstaff S, Grandstaff TB, Rathakette P, Thomas DE, Thomas JK (1986) Trees in paddy fields in Northeast Thailand. In: Marten GE (ed) Traditional agriculture in Southeast Asia a human ecology perspective. Westview Press, London, pp 273–292
Guinand Y, Dechassa L (2001) Wild food plants in Southern Ethiopia: reflections on the role of 'famine-foods' at the time of drought. UN-EUE Survey, Addis Ababa
Gurung P, Jeannette D (1997) Gender dimensions in bio-diversity management: Bhutan and Nepal. Report Submitted to FAO, RAP, Bangkok
Indian Institute of Management (2006) Wayanad initiative: a situational study and feasibility report for a comprehensive development of Adivasi communities in Wayanad. Viewed 26 September 2016, http://www.scribd.com/doc/4074255/Wayanad-Initiative
Jain SK (1964) Wild plant foods of the Tribals of Bastar (Madhya Predesh). Proc Nat Inst Sci India. Part B, Biol Sci 30B
Janah S (1994) The Tribals of India. Oxford University Press, Oxford
Kanvinde HS (1999) Maldivian gender roles in bio-resources management. RAP Pub. 1999/15. FAO, RAP, Bangkok
Kanvinde HS, Kumar AN, Rasheed PA (2001) Wild food management in Wayanad, Kerala; an exploratory study. M. S. Swaminathan Research Foundation, Chennai, MSSRF/RR/01/07
Kerala Agricultural University (2011) Package of practices recommendations: crops, 14th edn. Kerala Agricultural University, Thrissur, p 360
Kumar AN, Gopi G, Parameswaran P (2010) Genetic erosion and degradation of ecosystem services of wetland rice fields: a case study from Western Ghats, India'. In: Lockie S, Carpenter D (eds) Agriculture, biodiversity and markets. Livelihoods and agro-ecology in comparative perspective. Earthscan, London, Washington DC, pp 137–153
Kumar AN, Nambi VA, Geetha Rani M, King EDIO, Chaudhury SS, Mishra S (2015) Community agro biodiversity conservation continuum: an integrated approach to achieve food and nutrition security. Curr Sci 109(3):474–487
Luiz AAD (1962) Tribes of Kerala. Bharatiya Adimjati Sevak Sangh, New Delhi

Maheshwari JK, Singh JP (1986) Contribution to the ethnobotany of Bhoxa Ttribe of Bijnor and Pauri Garhwal Districts, U.P. J Econ Tax Bot 5(3):253–259

Maneechote C (2007) Utilization of weeds and their relatives as resources in Thailand. In: Kim KU, Shin DH, Lee IJ, Daegu (eds) Utility of weeds and their relatives as resources. Kyungpook National University, Daegu

McNeely JA, Miller KR, Reid WV, Mittermeier RA, Werner TB (1990) Conserving the world's biological diversity. IUCN, World Resources Institute, Conservation International, WWF-US and the World Bank, Washington, DC

Medsager OP (1957) Edible wild plants. Macmillan and Co., New York

Muralidharan PK, Sasidharan N, Seethalakshmi KK (1997) Biodiversity in tropical moist forest: a study of sustainable use of non-wood forest products in the Western Ghats, Kerala. KFRI Research Report No. 133. ISSN 0970–8103

Nair GC (1911) Wynad: its people and traditions. Higginbotham & Co, Madras. Viewed 26 September 2016, https://archive.org/stream/wynaditspeoplest00goparich/wynaditspeoplest00goparich_djvu.txt

Narayanan RMK (2009) Floristic study of Wayanad District giving special emphasis on conservation of Rare and Threatened plants'. Ph.D. thesis, University of Calicut, Kerala

Narayanan RMK, Swapna MP, Kumar AN (2004) Gender dimensions of wild food management in Wayanad, Kerala. M. S. Swaminathan Research Foundation, Chennai, MSSRF/RR/04/12

National Academy of Agricultural Sciences (1998) India policy paper for conservation: management and use of agrobiodiversity, Policy paper 4. National Academy of Agricultural Sciences, New Delhi, p 8. Viewed 26 September 2016, http://naasindia.org/Policy%20Papers/pp4.pdf

Narayanan RMK, Anil Kumar N (2007) Gendered knowledge and changing trends in utilization of wild edible greens in Western Ghats, India. Indian J Tradit Knowl 6(1):204–216

Negi SS (1994) India's forest, forestry and wildlife. Indus Publishing Company, New Delhi

Office of the Registrar General and Census Commissioner (2011) Census of India 2011. Office of the Registrar General and Census Commissioner, Government of India, New Delhi. Viewed 26 September 2016. http://www.censusindia.gov.in/2011census/PCA/PCA_Highlights/pca_highlights_file/kerala/Exeutive_Summary.pdf

Prachaiyo B (2000) Farmers and forests: a changing phase in Northeast Thailand. Southeast Asian Studies 38:3–156

Price LL (1993) Women's wild plant food entitlements in Thailand's Agricultural Northeast. UMI Press, Ann Arbor (dissertations)

Price LL (1997) Wild plant food in agricultural environments: a study of occurrence, management, and gathering rights in Northeast Thailand. Human Org 56:209–221

Reid WV, Miller KR (1989) Keeping options alive: the scientific basis for conserving biological diversity. World Resources Institute, Washington DC

Roy B, Halder AC, Pal DC (1998) Plants for human consumption in India. Flora of India Series 4. Botanical Survey of India, Calcutta

Singh RS, Singh NI (1981) A preliminary ethnobotanical study on wild edible plants in the markets of manipur-1. J Econ Taxon Bot 6(3):699–703

Swaminathan MS, Jana S (1992) Biodiversity; implications for global food security. Macmillan Publications, Madras, p 326

Thurston E (1909) Castes and tribes of Southern India. Madras Government Press

Watanabe H, Abe KI, Hoshikawa T, Prachaiyo B, Sahunalu P, Khemnark C (1990) On trees in paddy fields in Northeast Thailand. Southeast Asian Studies 28:45–54

WRI, IUCN, UNEP (1992) Global biodiversity strategy, guidelines for action to save, study, and use Earth's biotic wealth sustainably and equitably. World Resource Institute, World Conservation Union, and United Nations Environment Programme, p 244

Climate Change and Biodiversity

5

A. Biju Kumar and R. Ravinesh

Abstract

Climate change is a vital environmental issue affecting nature and distribution of biodiversity. As the economy of developing countries depends primarily upon climate-sensitive sectors such as agriculture, fisheries and forestry, a holistic analysis of climate change and its implications on biodiversity is warranted in order to achieve sustainable development. The impacts of climate change on critical ecosystems and livelihood activities and on insert biological resources, especially those that are vulnerable due to other prevailing anthropogenic interventions, are discussed in the chapter. The strategies to reduce climate change impacts, including adaptation and mitigation measures, which integrate ecosystem approach into broader cross-sectoral policies as a compliment to structural and technological measures, would ultimately help achieving the targets set for biodiversity conservation. The ecosystem approach of the Convention on Biological Diversity provides a flexible management framework to address climate change mitigation and adaptation activities in a broad perspective and can help to balance ecological, economic and social considerations in projects, programmes and policies related to climate change mitigation and adaptation. The research options on the appropriate mitigation and adaptation strategies so as to manage the rich biodiversity and to specifically address the impacts of climate change on species and ecosystems coupled with the need to integrate biodiversity into developmental plans and policies are discussed.

Keywords

Biodiversity • Climate change • Mitigation • Adaptation • Sustainable development • Aichi target • IPCC • Ocean acidification

A. Biju Kumar (✉) • R. Ravinesh
Department of Aquatic Biology and Fisheries, University of Kerala, Thiruvananthapuram 695581, Kerala, India
e-mail: bijupuzhayoram@gmail.com

S. Abdulhameed et al. (eds.), *Bioresources and Bioprocess in Biotechnology*,
DOI 10.1007/978-981-10-3573-9_5

5.1 Introduction

The United Nations Framework Convention on Climate Change (UNFCCC), in its Article 1, defines climate change as 'a change of climate which is attributed directly or indirectly to human activity that alters the composition of the global atmosphere and which is in addition to natural climate variability observed over comparable time periods'. The Intergovernmental Panel on Climate Change (IPCC) reports unequivocally established the warming of climate system, and since the 1950s, many of the observed changes are unprecedented over decades to millennia. The atmospheric concentrations of the greenhouse gases (GHG) such as carbon dioxide, methane and nitrous oxides responsible for climate change have increased alarmingly in the last five decades compared to any other era in planetary history (IPCC 2014a, b).

Some changes in extreme weather and climate events observed since 1950s are linked to human activity, resulting in impacts such as decrease and increase in temperature extremes, increase in extreme high sea levels due to melting of glaciers and increase in the number of heavy precipitation events. Projections of IPCC suggest an increase in global mean temperature in 2100 of 3.7–4.8 °C, relative to pre-industrial levels, in the absence of new policies to mitigate climate change. While the per capita contribution of greenhouse gas emission of industrialised countries is on average 2.5 times of those from developing countries, global GHG emissions have continued to grow and reached 49.5 billion tons (gigatons or Gt) of carbon dioxide equivalents (CO_2eq) in the year 2010, higher than any level prior to that date, with an uncertainty estimate at ± 10% for the 90% confidence interval (Victor et al. 2014).

The first report published by Indian Network for Climate Change Assessment (INCCA) of the Ministry of Environment and Forests and Climate Change projected a net increase in annual temperatures in the 2030s ranging between 1.7 °C and 2.2 °C, with extreme temperatures increasing by 1–4 °C, with maximum increase in coastal regions; seasons may be warmer by around 2 °C towards the 2030s (INCCA 2010). Trend analysis of past precipitation data in India has not shown any major widespread change in the patterns, though general circulation models (GCMs) show that in the future, these patterns are likely to change. Potential climate change impacts affecting water availability include changes in precipitation amount, intensity, timing and form (rain or snow), changes in snowmelt timing and changes to evapotranspiration. In India, winter precipitation is projected to decline, and this is likely to lead to higher need for rabi irrigation, lesser storage and increased water stress during the pre-monsoon months (INCCA 2010).

The current trajectory of global greenhouse gas emissions is not consistent with limiting global warming to below 1.5 relative to pre-industrial levels, as suggested in the United Nations Paris Climate Change Conference in December 2015, with a global effort to significantly reduce the risks and impacts of climate change. With increasing evidences of human interferences in the climate system, there are mounting numbers of reports highlighting the risks of climate change on both human and natural systems (Field et al. 2014). The social impacts of climate change include food and water shortages, increased poverty, increased displacement of people (often referred to as climate refugees) and coastal flooding. Globally, habitat loss and degradation, unsustainable exploitation, invasive species and climate change are the main threats facing the biodiversity, the natural capital of earth, and they have contributed to a decline of over half of the Living Planet Index since 1970, including the species of mammals, birds, reptiles, amphibians and fish (WWF 2016).

5.2 Climate-Biodiversity Linkages

Biodiversity, the variety of genes, species and ecosystems that constitute life on earth and the intricate interrelations of living creatures and the environment that evolved along with time, is the basis of human development and sustenance, as it ensures the sustainability of life support systems—air, water and food—besides the scope for future values, which remains unknown for many species. The biodiversity loss or the 'sixth extinction' is currently at its peak, with everyday species extinctions continuing at up to 1000 times or more the natural rate, highlighting the need for effective conservation programmes (Barnosky et al. 2011).

The United Nations Convention on Biological Diversity (CBD), agreed at the 1992 UN Conference on Environment and Development (Earth Summit), is one of the most widely ratified treaties in the world, with the fundamental objectives to conserve biodiversity, sustainable utilisation of biodiversity and sharing of benefits arising out of biodiversity. Since the World Summit on Sustainable Development in 2002, 193 parties to the CBD have committed themselves to substantially reducing rates of biodiversity loss by 2010 ('Biodiversity Target'), and this goal was incorporated into the UN Millennium Development Goals (MDG) in 2005. Following this, a wide array of international, national and regional strategies and action plans have been forwarded to achieve the goals of biodiversity conservation. There are also a series of conservation initiatives by civil society organisations across the globe. India also prepared the National Biodiversity Strategies and Action Plan (NBSAP) and enacted the Biological Diversity Act (2002) and the Biological Diversity Rules (2004). The United Nations has declared 2010 as the International Year of Biodiversity (IYB) to create awareness about the crucial importance of biodiversity to society, to communicate the human costs of biodiversity loss and to enthuse people, particularly youth, throughout the world in the fight to protect biodiversity.

In October 2010, in Japan, governments agreed to the 'Strategic Plan for Biodiversity 2011–2020 and the Aichi Targets' as the basis for halting and eventually reversing the loss of biodiversity of the planet. To build support and momentum for this urgent task, the United Nations General Assembly declared the period 2011–2020 to be 'the United Nations Decade on Biodiversity, with a view to contributing to the implementation of the Strategic Plan for Biodiversity for the period 2011–2020'. The United Nations Decade on Biodiversity will serve to support the implementation of the Strategic Plan for Biodiversity and promote its overall vision of living in harmony with nature, with a goal to mainstream biodiversity at different

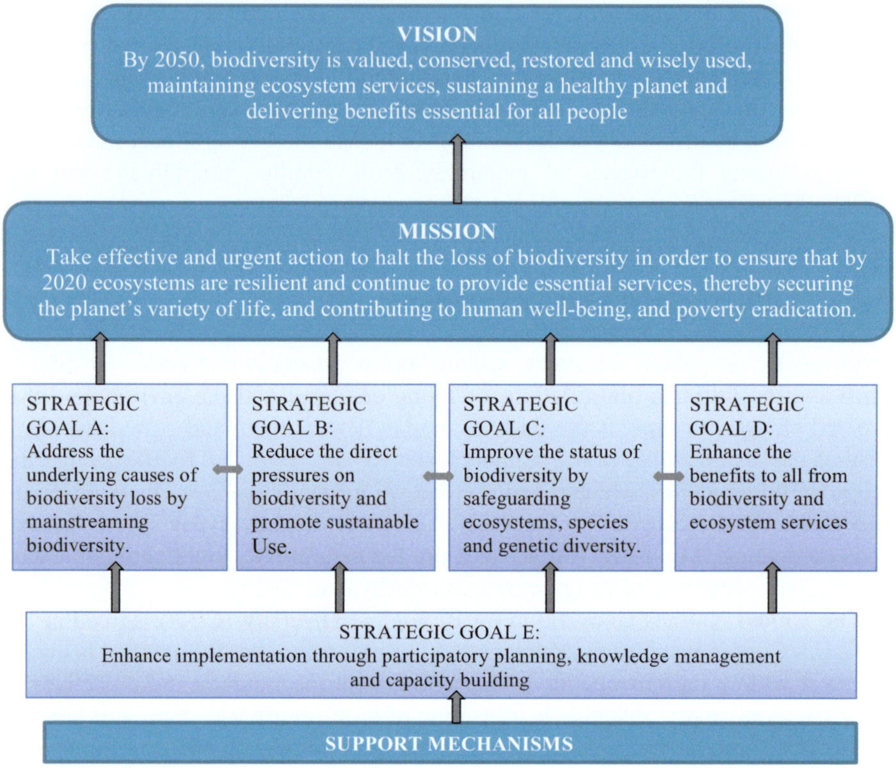

Fig. 5.1 Strategic Plan for Biodiversity 2011–2020: vision, mission and strategic goals

levels (Fig. 5.1). At the national and regional levels, this period should serve as a time to develop, implement and communicate the results of national strategies for implementation of the Strategic Plan for Biodiversity.

For achieving the Aichi Targets by 2020, especially the five strategic goals such as addressing the underlying causes of biodiversity loss by mainstreaming biodiversity across government and society, reducing the direct pressures such as climate change on biodiversity and promote sustainable use, improving the status of biodiversity by safeguarding ecosystems, species and genetic diversity, enhancing the benefits to all from biodiversity and ecosystem services and enhancing implementation through participatory planning, knowledge management and capacity building, better planning and concerted action are required. Aichi Target 14 requires that 'by 2020, ecosystems that provide essential services, including services related to water, and contribute to health, livelihoods and well-being, are restored and safeguarded, taking into account the needs of women, indigenous and local communities and the poor and vulnerable' (SCBD 2010).

The origin and expansion of life on earth can be linked to various climatic conditions prevailed on earth. Climate is one of the most important factors controlling the growth, abundance, survival and distribution of species as well as regulating natural ecosystems in a variety of ways. In recent decades, changes in climate have caused impacts on natural and human systems globally. Impacts due to climate change are observed in almost every part of the natural world and its interdependent social and economic systems (Field et al. 2014). Many terrestrial, freshwater and marine species have shifted their geographic ranges and activities in response to climate change. Past climate changes were slower than those anticipated for the twenty-first century, but even these drove significant ecosystem shifts and extinctions (Williams et al. 2011). Climate change will compound the impacts of other drivers of biodiversity loss such as habitat modification, over-exploitation, pollution and invasive species (Field et al. 2014). The economy of developing countries like India depends primarily upon climate-sensitive sectors such as agriculture, fisheries and forestry (Chaturvedi et al. 2014), and therefore climate change and its impacts on biodiversity need to be monitored closely in order to achieve sustainable development.

5.3 Impacts of Climate Change

Although many anthropogenic interventions (changing land-use patterns, pollution, habitat loss, modification and fragmentation, over-exploitation of species and introduction of non-native species) impact biodiversity, the effects of climate change are becoming much more effervescent on natural and human systems on all continents and across the oceans. The climate changes, especially global warming, have affected the timing of reproduction of flora and fauna, migration of animals, the length of the growing season, species distributions and population sizes, spread of invasive species and the frequency of pest and disease outbreaks (IPCC 2002). Climate change is projected to affect individual organisms, populations, species distributions and ecosystem composition and function both directly (e.g., through vagaries in precipitation, temperature, sea level changes, ocean acidification, etc.) and indirectly (e.g., through climate changing the intensity and frequency of events such as forest fires, storms, etc.). Other environmental issues such as habitat loss and fragmentation, pollution and invasive alien species would further exacerbate the implications of climate change on biodiversity.

The IPCC reports, which analysed the impacts of climate change on biodiversity, reveal that there has been a discernible impact of regional climate change on species and ecosystems, and the impacts are pronounced in high-altitude and high-latitude ecosystems (IPCC 2002, 2014a, b). The risk of extinction of species will further increase for many species that are already vulnerable due to various reasons referred above. Species with limited climate ranges and smaller populations enjoying narrow distributional ranges such as highly endemic species in the Western Ghats and Himalayas, biota restricted to islands and climate-sensitive species found in the coastal and marine ecosystems such as mangroves, coastal wetlands and coral reefs are more vulnerable to extinction (IPCC 2002).

Evidence of climate change impacts is strongest and most comprehensive for natural systems, and it impacts lives, livelihoods, health, ecosystems, economies, societies, cultures, services and infrastructure due to the interaction of climate changes or hazardous climate events occurring within a specific time period and the vulnerability of an exposed society or system (Field et al. 2014). The discussion is limited to impact analysis of climate change on key ecosystems and sectors where the impacts will be primarily on biodiversity and various livelihood activities.

5.3.1 Water Resources

Evidence of regional climate change impacts on elements of the hydrological cycle suggests that warmer temperatures in some regions lead to intensification of the hydrological cycle. Changes in stream flow, floods, droughts, water temperature and water quality have been observed, and they have affected biodiversity and the goods and services ecosystems provide (IPCC 2002). Gosain et al. (2011) projected the impact of climate change on the 17 most important river basins in India up to mid-century and towards the end of the century. They estimated a decline in rainfall in 14 out of the 17 river basins towards the 2030s and the 2080s. In almost all river basins, rainfall declines from 4 to 23%, following changes in precipitation, resulting in decline in water yield in river basins. Rajendran et al. (2013) project spatially heterogeneous increase in warm days and extreme hot events (highest decile) over India; projected changes in extreme rainfall events (above 95 percentile) show intensification of extreme rainfall over most parts of India by the end of the century with opposite change over the west coast. The reduction in surface water availability may influence hydropower generation and environmental flow in river basins (INCCA 2010).

Altered water temperature due to global warming also impacts water quality, and, in reservoirs and lakes, it influences the potential for algal blooms, which can further reduce oxygen levels (IPCC 2014a, b). Climatic variables such as air temperature and precipitation affect environmental flows in rivers and the mobility and dilution of contaminants, hence affecting water quantity as well as water quality parameters (Dhanya and Arun Kumar 2015). Further, changes in water availability may affect concentrations of suspended sediment, nutrients and chemical contaminants in rivers and lakes. Changes in precipitation intensity and frequency will also influence non-point source pollution, and overall changes in water quality may lead to increased incidence of water-borne diseases. Increase in river water temperatures is likely to affect fish breeding, migration and harvests, as many endemic species have restricted distribution, especially in biodiversity hotspots such as the Western Ghats.

5.3.2 Forest Ecosystems

India is one of the 12 mega biodiversity countries of the world, which represents 11% of world's flora in about 2.4% of global land mass (MoEF 2014). The role of forest in addressing the issues related to climate change is better realised due to the dynamic functions and ecosystem services offered by the rich biodiversity in these areas. O' Briena et al. (O'Briena et al. 2004) estimated the vulnerability of India under multiple stressors such as climate change, and Chaturvedi et al. (2011) project that 39% and 35% of the forest grids in India will likely undergo change due to climate impacts; forests of northeast and southern Western Ghats and eastern parts of India are projected to be least vulnerable on account of their high biodiversity, low fragmentation, high tree density as well as low rates of vegetation change.

Chitale et al. (2014) studied the future distribution of 637 endemic plant species from three biodiversity hotspots in India, including Himalaya, Western Ghats and Indo-Burma, based on modelling studies and suggested that the endemic flora will be adversely impacted, even under moderate climate scenario, with predicted shifts in northern and north-eastern direction in Himalaya and Indo-Burma, while in southern and south-western direction in Western Ghats, due to cooler climatic conditions in these regions. The model predicts a 23.99% range reduction and a 7.70% range expansion in future distribution by 2050, while a 41.34% range reduction and a 24.10% range expansion by 2080.

The Western Ghats region, part of the Western Ghats-Sri Lanka biodiversity hotspot, harbours species with extremely high levels of endemism, and climate change variations may severely impact the rich biodiversity and their distribution pattern. Eighty five percent of threatened vertebrates are inadequately represented in the existing global network of protected areas (PAs), with only a marginal progress in coverage since the last decade (Venter et al. 2014). Endemic species with restricted distribution and/or small populations are in particular vulnerable to extinctions (Hawkins et al. 2000) and of priority in conservation policy-making (Rosenfield 2002). Yet, many endemic species, particularly invertebrates, and those from the lower vertebrate groups are often ignored in both policy-making and conservation action, with freshwater-dependent fauna being the most affected (Darwall et al. 2011). A stronger baseline data on endemic biodiversity is one of the prerequisites for conservation as climate impacts will be much pronounced in these species.

5.3.3 Agriculture and Food Security

In an agrarian society like India, climate plays a vital role in deciding the success of agriculture, and even minimal warming will lead to loss in crop yields (Parry et al. 2007). IPCC has projected that by the end of the twenty-first century, rainfall over India will increase by 10–12%, and the mean annual temperature will rise by 3–5 °C (IPCC 2014a, b). Climate change and global warming may precariously affect agricultural production, reduction in crop area, pest and diseases and labour and livelihood of farmers by unpredictably changing the abundance of seasonal rainfall and extreme events (Chaturvedi et al. 2014). Farmers can no longer rely on the timing of seasons and the availability of rainfall, and this scenario is rather disturbing for the farmers in India as they rely on monsoon for agriculture. Studies conducted by the Indian Agricultural Research Institute (IARI) indicate the possibility of loss of 4–5 million tons in wheat production with every rise of 1 °C temperature throughout the growing period, and the losses for other crops are expected to be smaller, especially for kharif crops (Aggarwal et al. 2009). Erratic monsoons will have serious effects on rain-fed agriculture with projected decreases in the productivity of crops including rice, maize and sorghum (especially in the Western Ghats, coastal region and north-eastern regions) and apples (in the Himalayan region) (Kumar et al. 2011).

Studies indicate that increased droughts and floods are likely to increase production variability and lead to considerable effects on microbes, pathogens and insects needed for the maintenance of healthy agricultural systems (Gautam et al. 2013). Increasing glacier melt in Himalayas could affect availability of irrigation especially in the Indo-Gangetic plains, which, in turn, would have consequences on food production. Aggarwal et al. (2011) estimated the impact of climate change on livestock and concluded that animal distress could lead to effects on reproduction and subsequently loss of 1.5 million tons of milk by 2020. The resilience of many ecosystems (their ability to adapt naturally) is likely to be exceeded by an unprecedented combination of change in climate; associated disturbances such as flooding, drought, wildfire, pests, ocean acidification; and other global change drivers such as land-use change, pollution and over-exploitation of resources (IPCC 2014a, b). Smallholder and subsistence farmers and fisher folk are likely to suffer multiple and localised impacts of climate change. Frequent exposure to flooding may lead to ecological changes in the agricultural fields.

5.3.4 Coastal and Marine Ecosystems

Sea level along the Indian coast has been rising at the rate of 1.3 mm/year and is likely to rise in consonance with the global sea level rise (SLR). Significant coastal inundation may occur with a 1 m sea level rise, especially in low-lying areas, with increase in environment-related health risks (INCCA 2010). In the coastal and marine ecosystems, pattern of species richness, especially that of fish and invertebrates, is strongly related to climatic factors. Species in the marine ecosystems respond to ocean warming by shifting their latitudinal range (Perry et al. 2005) and depth range (Dulvy et al. 2008), which may lead to local extinctions and invasions, resulting in changes in pattern of marine species richness and disrupting marine biodiversity and ecosystems, and impact commercial fisheries (Cheung et al. 2009).

Another emerging issue in oceans in recent times is the frequent incidences of jellyfish blooms worldwide in marine ecosystems, indicating a state shift in pelagic ecosystems (Purcell et al. 2007). The coastal waters around India are suspected to be the areas prone for jellyfish blooms (Brotz et al. 2012). Jellyfish exhibits wide fluctuation in biomass in response to anthropogenic perturbations or changing oceanographic condition such as eutrophication, hypoxia and climate-induced regional regime shift, and researchers opine that global warming might lead to increasing populations of jellyfish because it could affect the distribution, growth and larval production (Lee et al. 2013). Because climate changes have complex ecosystem-level effects, the proximate causes of jellyfish increases are difficult to deduce. There are, however, reports that global warming might lead to increasing populations of jellyfish because it could affect the distribution, growth and larval production (Richardson et al. 2009). In the clear absence of predatory fish species and turtles capable of feeding specifically on jellyfish, the population of jellyfish in Indian waters also would increase in the coming years (Biju Kumar 2012).

In the past 200 years, it is estimated that the ocean has absorbed more than a quarter of the carbon dioxide released by human activity, increasing ocean acidity, a phenomenon often referred to as ocean acidification. Marine organisms such as corals, molluscs, crustaceans and echinoderms which play an important roles in ecosystem functioning may exhibit growth retardation and low survival rates with ocean acidification, though these responses are variable, and some species can live at low pH conditions (CBD 2014). Non-calcifying phytoplankton can show increased photosynthesis and growth under high carbon dioxide conditions that rely on carbonate ions to form their calcareous shells or skeletons in a process known as calcification. Ocean acidification may result in destabilisation of skeletal structures of marine organisms with calcareous exoskeleton.

Changes in precipitation frequency and intensity coupled with variations in pH, water temperature, wind, dissolved oxygen, dissolved CO_2, salinity variations and pollution may lead to poor water quality of marine ecosystems. Human activities, especially increased nutrient loads that set in motion a cascading chain of events related to eutrophication, accelerate development of hypoxia (lower oxygen concentration) in many areas of the world's coastal ocean. The changed environmental conditions could favour the growth of harmful algal blooms (HABs). Increase in HABs may negatively impact the environment, human health and economy of the communities (Wells et al. 2015). Occurrence, increase in frequency, intensity and spatial coverage of harmful algal blooms in the EEZ of India indicated a sharp increase, with the frequent contribution of toxic species such as *Alexandrium* spp., *Gymnodinium* spp., *Dinophysis* spp., *Coolia monotis*, *Prorocentrum lima* and *Pseudonitzschia* spp. (Padmakumar et al. 2012).

It was generally held that there is an increase in the biomass of phytoplankton during the recent decades in western Indian Ocean as a result of climate change and global warming. However, the current study points out an alarming decrease of up to 20% in phytoplankton in this region over the past six decades, and these trends in chlorophyll are driven by enhanced ocean stratification due to rapid warming in the Indian Ocean, which suppresses nutrient mixing from subsurface layers. Future climate projections suggest that the Indian Ocean will continue to warm, driving this productive region into an ecological desert (Roxy et al. 2016).

Coral reefs are the most diverse, biologically complex and economically important marine ecosystems on earth. The reef-building scleractinian corals are currently facing multiple stresses caused by shifts in the marine environment associated with global warming, ocean acidification, sedimentation, pollution and eutrophication. In the Indian Ocean, for example, 45% of living coral was killed across the 1998 warm temperature anomaly, and the recovery of corals from the impact varied amongst various regions (Ateweberhan and McClanahan 2010). Many of the corals have undergone coral bleaching events due to the loss of intracellular endosymbionts (*Symbiodinium* or zooxanthellae) through either expulsion or loss of algal

Fig. 5.2 Coral bleaching observed at Lakshadweep Islands, India

pigmentation under elevated temperature regime (Fig. 5.2). Though coral bleaching may be reversible under normal temperature regime, if that sea temperature stays too high for too long, the corals may eventually die. Widespread coral bleaching events and mass mortalities occurred in the Pacific Ocean due to El Niño events which began in the early 1980s due to strong temperature anomalies, and the 1997–1998 El Niño-Southern Oscillation (ENSO) event, which elevated sea surface temperatures (SSTs) of tropical oceans by more than 3 °C, triggered massive coral bleaching in the east and west coasts of India and in Lakshadweep (Arthur 2000). Climate change impacts on coral bleaching have also been reported from Pirotan Island, Gulf of Kachchh Marine National Park, Gujarat (Adhavan et al. 2014). By using the relationship between past temperatures and bleaching events and the predicted SST for another 100 years, Vivekanandan et al. (2009) predicted that reefs should soon start to decline in terms of coral cover and appearance.

Mangroves, which show an amazing example of natural adaptation to their unique location at the interface of the sea, are vulnerable to climatic change and resultant sea level rise (Table 5.1). By altering ecobiological processes, the intertidal and supratidal zones may extend further inland, and limitation of the landward margin results in a vertical rise, leading to waterlogging and ultimately killing the mangroves and dependent biodiversity (Jagtap and Nagle 2007). Changes in growth pattern of dominant species of mangroves such as *Sonneratia apetala*, *Avicennia alba* and *Excoecaria agallocha* have been recorded from Indian Sundarbans due to climate change (Raha et al. 2012).

Coastal freshwater wetlands may be vulnerable to saltwater intrusion with rising sea levels, but in most river deltas, local subsidence for non-climatic reasons will be more important (Syvitski et al. 2009). Humans have been the primary drivers of changes in coastal aquifers, lagoons, estuaries, deltas and wetlands and are expected to further exacerbate human pressures on coastal ecosystems resulting from excess nutrient input, changes in run-off and reduced sediment delivery (IPCC 2014a, b).

Table 5.1 Predicted effects of climate change factors on mangrove

Factors	Processes affected	Impacts
Rising sea level	Forest health	Forest mortality, dieback from the seaward edge
	Forest productivity	Migration landward, but dependent on sediment inputs, topography and human modifications
	Recruitment	
	Inundation period	
	Sedimentation rates	
Extreme storms	Forest productivity	Forests damaged or destroyed
	Recruitment	Ground elevation change
	Sedimentation rates	Erosion or sediment smothering
Increased waves and wind	Sedimentation rates	Changes in forest coverage, depending on whether coasts are accreting or eroding
	Recruitment	
Increased air and sea temperature	Respiration	Reduced productivity at low latitudes and increased winter productivity at high latitudes
	Photosynthesis	
	Forest productivity	
Enhanced CO_2	Photosynthesis	Increased productivity, subject to limiting factors of salinity, humidity and nutrients
	Respiration	Soil elevation gain
	Biomass allocation	
	Forest productivity	
Increased rainfall	Sediment inputs	Increased sediments and maintenance of surface elevation
	Ground water	Increased ground water
	Salinity	Increased diversity
	Productivity	Increased productivity
		Increased recruitment
Reduced rainfall	Sediment inputs	Reduced sediments and relative subsidence
	Ground water	Migration landward
	Salinity	Reduced ground water
		Reduced photosynthesis
		Reduced productivity
		Species turnover
		Reduced diversity
Reduced humidity	Photosynthesis	Reduced productivity
	Forest productivity	Species turnover
		Reduced diversity

Ellison (2012)

The sea level rise recorded in Indian seas during 1970–2010 was more than 8 cm (Unnikrishnan and Shankar 2007), and the projection for 2050 and 2100 is more than 20–40 and 50–70 cm, respectively. Under the influence of climate change, beaches, sand dunes and cliffs currently eroding will continue to do so under increasing sea level. The human settlements (especially those of the fisher folks), transportation and tourism infrastructure at or near the coast are vulnerable to more frequent flooding, rising sea levels and possible increase in the magnitude and frequency of tropical storms and other natural calamities due to climate change.

5.3.5 Vulnerable Species

Many species in terrestrial, coastal and marine ecosystems are on imminent threat of extinction due to various ecosystem changes including climate change. Endemic species restricted to narrow environmental niches, especially those in mountainous ecosystems and islands, may face impacts of climate change more seriously than other species in similar ecosystems. Further, climate-sensitive species with specific phenological or physiological requirements (e.g., species with temperature-dependent sex determination such as turtles and crocodiles and amphibians with permeable skin and eggs) are more susceptible. Impacts of climate change are likely due to direct physiological stress, habitat changes and change in disturbance regime (IPCC 2002). The risk of extinction is always high for species with restricted distribution and limited climatic ranges, with lesser population densities. This demands extension of protected areas and measures to ensure connectivity between critical habitats.

Marine turtles, which are otherwise facing lot of struggles towards survival, are seriously impacted by changing climatic conditions, and the issues include (i) erosion of turtle nesting beaches, (ii) skewed sex ratios of hatchlings, (iii) destruction of feeding niches, (iv) modification of migration pathways and (v) flooding of nesting sites (Hawkes et al. 2009). They occupy a wide range of marine habitats, and many aspects of their life history have been demonstrated to be closely tied to climatic variables such as ambient temperature and storminess and therefore could be used as indicators of climate change. As a species with temperature-dependent sex determination, one of the serious challenges will be that warmer incubation temperatures may produce female hatchlings. Laloë et al. (2014) estimated that light-coloured beaches produce 70.10% females, whereas dark-coloured beaches produce 93.46% females in Cape Verde Islands, Atlantic, one of the largest sea turtles rookeries in the world.

Gahirmatha Marine Sanctuary in Orissa, India, is one of the biggest mass nesting grounds of olive ridley turtles in the world. Delaying of nesting of the olive ridley turtle in this coast could be attributed to the accretion of sand on the nesting beach due to the northerly winds; turtles always prefer to nest after the accretion of new sand that occurs around this time. If they nest during the accretion, the eggs will get buried and hatchling success will be reduced. The changes in beach morphology in Kerala coast due to natural and anthropogenic reasons have reduced turtle nesting sites.

Fisheries play an important role in food supply, food security and livelihood security of thousands of fishermen and associated fish supply chains living in coastal areas. Temperature is known to affect fish distribution and migration; the combined

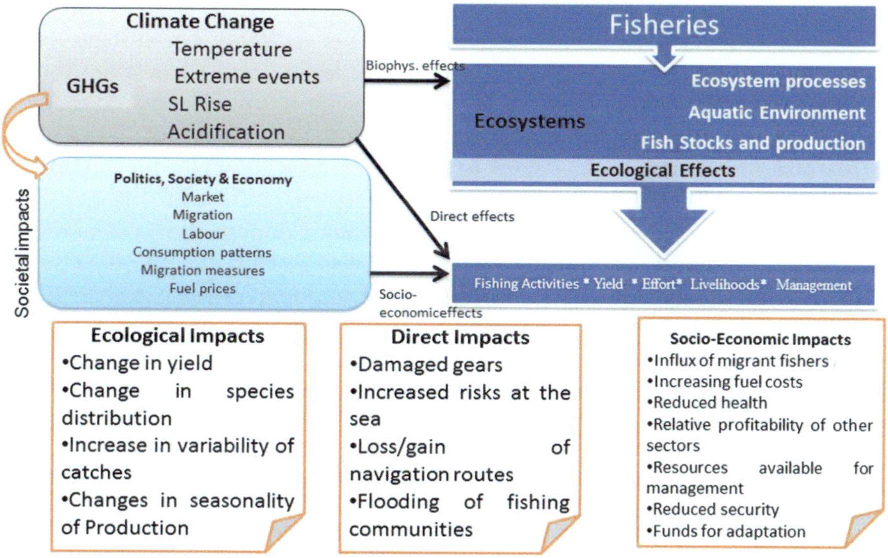

Fig. 5.3 Impacts of climate change on fisheries

effects of changes in distribution, abundance and physiology may reduce the body size of marine fishes, particularly in the tropics and intermediate latitudes (Cheung et al. 2013). Growth rate of fish increases with increasing temperature within the optimal temperature window. It is likely that the food utilisation parameters may be operating at an elevated level in fishes at higher temperatures, demanding higher food supply to attain faster growth rate. Fishes may change their phenology of reproductive activity to adapt to elevated temperatures for spawning and larval survival. At population level, temperature and other factors related to climate change may strongly influence distribution and abundance, evidences for which are accumulating in Indian seas; extension of distributional boundary of small pelagics, extension of depth of occurrence and phenological changes were recorded from Indian waters (Vivekanandan 2013). All these climate change implications would ultimately affect the fishery sector and coastal livelihood activities (Fig. 5.3).

Studies on the impact of climate change on fisheries carried out by the Central Marine Fisheries Research Institute (CMFRI) show that different Indian marine species respond to climate change as follows: (i) small pelagic fishes that may extend their boundaries, (ii) some species that may be migrated to deeper waters as well and (iii) phenological changes. The major pelagic species that represent major portion of marine fish landings in Indian coast are the oil sardine (*Sardinella longiceps*) and the Indian mackerel (*Rastrelliger kanagurta*), contributing up to 40% marine fisheries. The oil sardine is restricted in distribution between latitude 8°N and 14°N and longitude 75°E and 77°E (Malabar upwelling zone along the south-west coast of India) where the annual average sea surface temperature ranges from 27 to 29 °C. While almost the entire catch of oil sardine was from the Malabar upwelling zone till 1985, their landings from latitude 14°N to 20°N are consistently increasing in the last few decades (Vivekanandan et al. 2009). The surface waters of the Indian seas are warming by 0.04 °C per decade, and the warmer tongue (27–28.5 °C) of the surface waters is expanding to latitudes north of 14°N, enabling the oil sardine to extend their distributional range to northern latitudes (Anon. 2010). The studies done by the Central Marine Fisheries Research Institute (CMFRI) showed that elevated SST, favourable wind (and perhaps current) and increasing CUI have induced higher chlorophyll—a concentration during southwest monsoon, which has resulted in increasing the recruitment and catches of oil sardine during post-southwest monsoon season along the Kerala coast (Vivekanandan et al. 2009).

The Indian mackerel is commonly found in the Indian and West Pacific Oceans and their surrounding seas; in India, they are abundantly found along the southwest coast. Recent reports indicate that this fish, in addition to extension of northern boundaries, is found to descend to deeper waters in the last few decades. The mackerel populations normally occupying surface and subsurface waters have now started moving down the water column and are often caught in large numbers in bottom trawl nets operated by large mechanised boats at about 50–100 m depth (Vivekanandan et al. 2009).

The increase in catch of puffer fish *Lagocephalus inermis* biomass along Kerala coast in the recent decades may be related to the decline of predators (Mohamed et al. 2013). The examples recorded from Kerala coast show that differential physiological effects of temperature on individual species are key to understanding and projecting climate-induced changes in species interactions and in community composition (Portner and Farrell 2008). Further, habitat destruction, pollution, energy production, mining and aquaculture are all affecting marine ecosystems and may exacerbate the effects of climate change.

Amphibians are particularly vulnerable to climate change. As they occupy a vital position in the food web, susceptibility to water-soluble toxins through permeable skin and a life history that requires both terrestrial and aquatic habitats, frogs are good indicators of environmental conditions. Frogs in high mountainous areas are most affected by global warming. Of the more than 7000 amphibian species known globally, about 32% are threatened with extinction, according to the International Union for Conservation of Nature (IUCN). Climate change may also exacerbate the spread of chytridiomycosis, caused by the parasitic fungus *Batrachochytrium dendrobatidis* (Bd), and has led to the recent decline or extinction of 200 frog species worldwide (Catenazzi et al. 2014).

5.4 Strategies to Combat Climate Change

Climate change involves complex interactions between climatic, environmental, economic, political, institutional, social and technological processes. It cannot be addressed or comprehended in isolation of broader societal goals (such as equity or sustainable development), or other existing or probable future sources of stress. Strategies to reduce climate change impacts on biodiversity therefore include a mixture of adaptation and mitigation measures such as containing human population growth, reducing greenhouse gas emissions as per agreed targets, promotion of biodiversity to ensure carbon capture, addressing pre-existing stressors on biodiversity and taking adequate steps to adapt to climate change and to improve conservation efforts by expanding protected areas for conservation of biodiversity.

5.4.1 Adaptation

The IPCC defined adaptation as adjustment in natural or human systems in response to actual or expected climatic stimuli or their effects, which moderate harm or exploit beneficial opportunities. Adaptation is also defined as an understanding of how individuals, groups and natural systems can prepare for and respond to changes in climate or their environment and is crucial to reducing vulnerability to climate change. Biodiversity and healthy well-functioning ecosystems provide natural solutions that build resilience and help society adapt to the adverse impacts of climate change. They also support poverty alleviation by providing safer and more secure livelihoods, especially for the poor and vulnerable. With climate change already well underway and further change unavoidable, adaptation is gaining significantly more focus, especially in countries like India where the economy is dependent on climate-sensitive sectors.

Biodiversity is linked to climate change adaptation through three ways:

1. It can play a role in societal adaptation, and ecosystem-based adaptation can provide cost-effective strategies across the major sectors involved in adaptation (e.g., coastal defence, water sector, agriculture, etc.).
2. Societal adaptation strategies can have significant impacts on biodiversity, and these impacts are negative, but where appropriate natural resource management is used (e.g., in improved agricultural practices), adaptation strategies may prove beneficial for biodiversity (CBD 2009).
3. Biodiversity conservation itself is a sector that requires its own adaptation strategies. Such strategies, which involve improved protected area design, maintaining habitat connectivity in the wider landscape and reducing other anthropogenic pressures, are likely to increase the resilience of biodiversity to climate change (CBD 2009).

In short, the benefits of adaptation activities can be achieved through maintaining and restoring native ecosystems; protecting and enhancing ecosystem services; actively preventing and controlling invasive alien species; managing habitats for rare, threatened and endangered species; developing agroforestry systems at transition zones; paying attention to traditional knowledge and monitoring results and changing management regimes accordingly. Adaptation activities that can be beneficial to biodiversity include the establishment of a mosaic of interconnected terrestrial, freshwater and marine multiple-use reserve protected areas designed to take into account projected changes in climate and integrated land and water management activities that reduce non-climate pressure on the biodiversity and hence make the system less vulnerable to changes in climate. Adaptation activities can also threaten biodiversity either directly through the destruction of habitats (e.g., building sea walls), thus affecting coastal ecosystems, or indirectly through the introduction of new species or changed management practices such as aquaculture.

Agriculture can be used as an example to explain the adaptation strategies to climate change. The sustainable use of genetic resources for food and agriculture will be the foundation for many of the adaptation strategies required in food and agriculture. Agricultural biodiversity has never been properly integrated in agricultural adaptation strategies to climate change, which creates a challenge for the future. Enhancing ecosystem services through use of agricultural biodiversity will be crucial, given that it contributes to adaptation, mitigation and resilience. For example, farmers in Kuttanad area of Kerala have been practising below sea-level farming for several decades using salt-resistant strains of paddy ('Pokkali'), and popularisation and genetic improvement of this variety could facilitate farming in coastal flood plains.

Good management of agricultural biodiversity allows production systems to adapt to changing conditions while maintaining productivity. Enabling the sustainable use of agricultural biodiversity has a huge potential for developing win-win strategies with multiple benefits such as coping with climate change, conserving biodiversity and improving human well-being. Huge gains can be made for a greener future by simply reducing agricultural waste and inefficiency. Nearly 50% of food produced is lost through crop loss or waste during storage, distribution, marketing and household use. Some of these inefficiencies—especially crop and storage losses—can be addressed with small investments in simple farming and storage technologies.

The sea level rise recorded in Indian seas during 1970–2010 was more than 8 cm (Unnikrishnan and Shankar 2007), and the projection for 2050 and 2100 is more than 20–40 and 50–70 cm, respectively. Under the influence of climate change, beaches, sand dunes and cliffs currently eroding will continue to do so under increasing sea level. The human settlements (especially those of the fisher folks), transportation and tourism infrastructure at or near the coast are vulnerable to more frequent flooding, rising sea levels and possible increase in the magnitude and frequency of tropical storms and other natural calamities due to climate change. Main threats could be salinity ingress in water bodies and inundation of low-lying areas, with resultant loss of fertile agricultural lands in coastal areas. However, aquatic biodiversity like fishery resources may provide opportunity for adaptive livelihood measures for food security in the coastal area (Anon. 2010). Conventionally coastal defences have relied upon 'hard defence' structures such as sea walls. However, evidence suggest that resilient coastal ecosystems such as mangroves, coral reefs, sand dunes and salt marshes can play an effective role in coastal protection as buffer against extreme events. Further, they also ensure resources for the livelihoods of local farmers and fishermen, thereby allowing more flexibility to adaption process. Considering the fact that coastal ecosystems alone would not ultimately help in

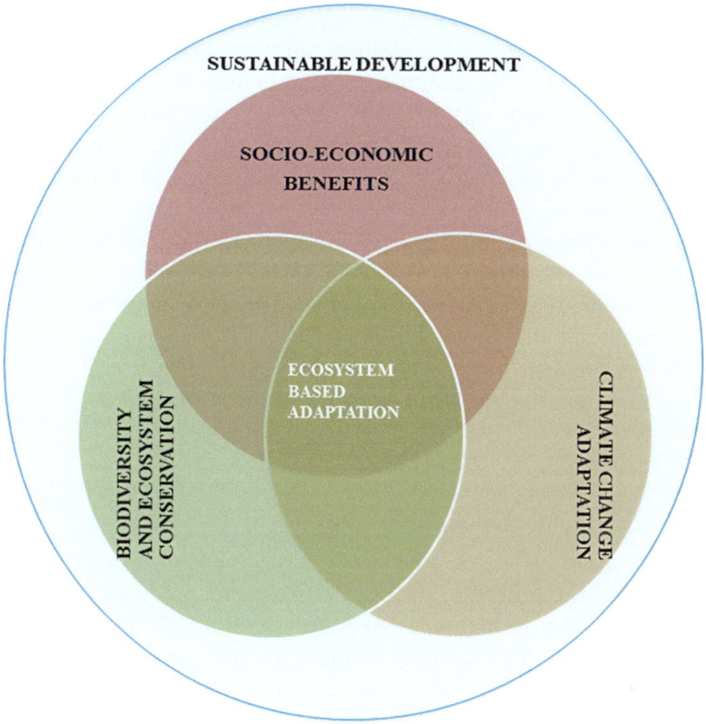

Fig. 5.4 Linkages of ecosystem-based adaptation and sustainable development

addressing the climate change impacts, integration of 'hard defence' measures with proper land-use planning and ecosystem management is being popularised.

As with coastal ecosystems, natural freshwater ecosystems provide vital water regulation services and can play a role in adaptation to water scarcity as well as flooding. Actions to reduce degradation of watersheds, through reduced deforestation, afforestation and soil conservation, can lower vulnerability to drought, and the maintenance and restoration of water-regulating services of wetlands are important for flood control. The need of the hour is to integrate conventional and modern watershed management practices in order to make adaptation a practical reality at the grass roots.

Ecosystem-based adaptation (EbA) is an important approach for achieving multiple benefits in the context of sustainable development. EbA has been defined by the CBD as 'the use of biodiversity and ecosystem services as part of an overall adaptation strategy to help people to adapt to the adverse effects of climate change'. This definition clearly identifies a strong link between biodiversity, ecosystem services, climate change adaptation and societal resilience. In order to achieve sustainable development, a synergy between biodiversity and ecosystem conservation, socio-economic benefits and climate change adaptation is necessary (Fig. 5.4). In addition to protection from climate change impacts, EbA also provides many other benefits to communities, for example, through the maintenance and enhancement of ecosystem services crucial for livelihoods and human well-being, such as clean water and food. Appropriately designed ecosystem management initiatives can also contribute to climate change mitigation by reducing emissions from ecosystem loss and degradation and enhancing carbon sequestration.

The contribution of biodiversity to societal adaptation varies according to the local situations. However, the available evidences make it clear that integrated management strategies incorporating ecosystem approach into broader cross-sectoral adaptation policies as a compliment to structural and technological measures would assist in more sustainable adaptation strategies. However, these efforts in developing countries would require more institutional and financial support, which at present are not fully available.

5.4.2 Mitigation

'Mitigation', in the context of climate change, is a human intervention to reduce the sources or enhance the sinks of greenhouse gases (GHGs). Because mitigation is intended to reduce the harmful effects of climate change, it is part of a broader policy framework that also includes adaptation to climate impacts. Mitigation can mean using new technologies and renewable energies, making older equipment more energy efficient, or changing management practices or consumer behaviour. Protecting natural carbon sinks like forests and oceans and creating new sinks through silviculture or green agriculture are also elements of mitigation.

The IPCC provided growing evidence of the importance of natural ecosystems in the carbon cycle (especially in carbon storage and sequestration) and therefore in mitigation policies. Climate mitigation polices focused on reducing CO_2 emissions can have both positive and negative impacts on biodiversity. A series of renewable energy projects such as biofuel or wind farms and monoculture plantations are being planned without consideration of their biodiversity impacts. Mitigation strategies have the potential to be directly beneficial to biodiversity by focusing on protecting carbon-rich land, such as forests, on managing or restoring sensitive ecosystems such as wetlands, or by promoting best practice management.

India is a party to the United Nations Framework Convention on Climate Change (UNFCCC). The Paris Agreement agreed that future global warming should be limited to below 1.5 °C relative to pre-industrial levels. India has already prepared an Intended Nationally Determined Contributions (INDCs) to plan for mitigation measures. Better land-use management, including conserved habitats, can reduce carbon emissions from land-use change and help remove carbon dioxide from the atmosphere, thus helping to mitigate global warming (e.g.,, reducing emissions from deforestation and degradation, improving soil carbon storage in soils). A portfolio of land-use management activities, including the protection of natural forest and wetland carbon stocks, the sustainable management of forests, the use of native assemblages of forest species in reforestation activities, sustainable wetland management, restoration of degraded wetlands and carbon-friendly agricultural practices, can contribute to the objectives of both the UNFCCC and CBD. Reducing emissions from deforestation and forest degradation (REDD) activities should take biodiversity conservation and sustainable use into account, as this helps maintain forest ecosystem resilience and the long-term stability of the carbon pool as well as providing co-benefits in terms of the delivery of other ecosystem services, including supporting sustainable livelihoods.

Agricultural soils are important carbon sinks with great potential to mitigate climate change, and soil biodiversity plays an important role in soil carbon cycles. There are a large number of agricultural management activities (e.g., conservation tillage, erosion control practices and irrigation) that will sequester carbon in soils and which may have positive or negative effects on biodiversity, depending on the practice and the context in which they are applied. Better understanding and management of soils have potential to bring important conservation and use benefits, mitigate climate change, avoid land degradation and improve water retention and productivity. Bio-based agricultural systems that require less external energy input could also contribute to mitigating climate change.

The use of erosion control practices, which include water conservation structures, vegetative strips used as filters for riparian zone management and agroforestry shelterbelts for wind erosion control, can reduce the displacement of soil organic carbon and provide opportunities to increase biodiversity. The irrigation can increase crop production, but will degrade water resources and aquatic ecosystems. It is important to include farmer-centred participatory approaches and consideration of local or indigenous knowledge and technologies, promote cycling and use of organic materials in low-input farming systems and use a diverse array of locally adapted crop varieties voiding degradation of wetlands. These are beneficial mitigation options. Revegetation activities that increase plant cover on eroded, severely degraded or otherwise disturbed lands have a high potential to increase carbon sequestration and enhance biodiversity. Sequestration rates will depend on various factors, including revegetation method, plant selection, soil characteristics and site preparation and climate.

Some marine ecosystems, such as mangroves, salt marshes and seaweed ecosystems, beyond having high biodiversity values and providing breeding grounds and nurseries for fisheries, can also play a key role in mitigating global climate change through their ability to store carbon. These blue carbon ecosystems are being degraded at a very high level in Kerala, and the current stretch of mangrove forests in Kerala is only 663.09 ha (Anon. 2012) as against 70,000 ha reported by Blasco (1975) in earlier period, and climate change-related changes would seriously hamper their sustainability. Climate change assessment report documents that sea level rise along the Indian coast would submerge the mangroves as well as increase the salinity of the wetland, and this would favour mangrove plants that tolerate higher salinity (Anon. 2010).

Marine ecosystems may offer mitigation opportunities, but the potential implications for ecosystem function and biodiversity are not well understood. Oceans are substantial reservoirs of carbon with approximately 50 times more carbon than is presently in the atmosphere. There have been suggestions to fertilise the ocean to promote greater biomass production and thereby sequester carbon and to mechanically store carbon deep in the ocean. However, the potential for either of these approaches to be effective for carbon storage is poorly understood, and their potentially large negative impacts on ocean and marine ecosystems and their associated biodiversity are unknown.

Renewable energy sources (crop waste, solar and wind power) may have positive or negative effects on biodiversity depending upon site selection and management practices. Replacement of fuel wood by crop waste, the use of more efficient wood stoves and solar energy and improved techniques to produce charcoal can also take pressure from forests, woodlots and hedgerows. Most studies have demonstrated low rates of bird collision with windmills, but the mortality may be significant for rare species. Hydropower has been promoted as a technology with significant potential to mitigate climate change by reducing the greenhouse gas intensity of energy production but has potential adverse effects on biodiversity. In some cases, emissions of carbon dioxide and methane caused by dams and reservoirs may be a limiting factor on the use of hydropower to mitigate climate change.

Large-scale hydropower development can also have other high environmental and social costs such as loss of biodiversity and land, disruption of migratory pathways and displacement of local communities. The ecosystem impacts of specific hydropower projects vary widely and may be minimised depending on factors including type and condition of pre-dam ecosystems, type and operation of the dam (e.g., water flow management) and the depth, area and length of the reservoir.

It is clear from the foregoing that climate mitigation policies have the potential to impact biodiversity in both positive and negative manner. Therefore mitigation efforts, especially those involving larger infrastructure and investment, should be implemented with proper impact assessment so as to develop 'win-win' mitigation policies that are beneficial for both climate change mitigation and biodiversity.

5.5 Research and Awareness

In order to enable informed decisions on the appropriate mitigation and adaptation strategies, there is a growing need to strengthen the wealth of climate data and information and transform the knowledge into action. Further, a fundamental restructuring of the way energy, land, water and biological resources are managed is needed to achieve a cost-effective transition to low-carbon economy and society. Research and innovation will play an important role in defining cost-effective decarbonisation pathways, developing alternative technological and socio-economic solutions for decision-makers and for the society as a whole, while informing them on related risks and costs, besides planning suitable adaptation and mitigation programmes.

In the case of biodiversity, the responses of individual species to climate variables, role of smaller biomes in regulating climate, adaptation required for specific cases and integrated mitigation programmes with due consideration of biodiversity are the areas that requires more intensive research. At the species level, there are numerous correlative modelling studies simulating the potential impact of climate change on their distribution. Experimental studies are extremely useful in determining the effect of climate change on aspects of ecosystem composition, structure and function. Improvement of regional scale climate models coupled with transient ecosystem models that deal with multiple pressures with appropriate spatial and temporal resolution and that include spatial interactions between ecosystem and landscapes is also needed (IPCC 2002). Further, at each local ecosystem, identification of biodiversity conservation and sustainable use activities and policies that would beneficially affect climate change adaptation and mitigation options is yet another priority area of research.

The Conference of the Parties to the United Nations Framework Convention on Climate Change held at Paris in 2015 has decided for Green Climate Fund (GCF) to expedite support for the least developed countries and parties of other developing countries for the formulation of national adaptation plans and for the subsequent implementation of policies, projects and programmes identified by them. The GCF is a unique global initiative to respond to climate change by investing into low-emission and climate-resilient development. Given the urgency and seriousness of the challenge, the fund is mandated to make an ambitious contribution to the united global response to climate change.

India's National Action Plan on Climate Change Assessment (NAPCC) envisages National Mission on Strategic Knowledge for Climate Change as one of the missions with the objective of vulnerability assessment, research and observation and data management implemented through the Department of Science and Technology. Considering that climate change is a relatively new challenge, the focus of the conservation action plan should be on generating awareness and building capacity, and this should be done across all levels of the government and external stakeholders involved in the different sectors. This strong drive towards building capacity will result in empowering people and organisations to be able to address, manage and respond to climate change concerns. Information, education and communication (IEC) technologies should be better utilised for the purpose. Climate change should also be included as a topic in academic curricula in education.

There is a need to involve stakeholders, particularly coastal community, in a more proactive way in the climate action plan implementation. This involvement will relate to:

1. Promoting much greater climate change awareness within community
2. Identifying problematic issues relevant to climate change
3. Support in monitoring of climate-induced problems
4. Ensuring greater accountability to the people on climate change issues

Stakeholder involvement will be an effective tool with stakeholders, who play an important part in bringing out the solutions.

Climate change is an interdisciplinary subject that cuts across physics, chemistry, biology, earth sciences, economics, technology development, etc. Therefore, multiple data sets are required even to simulate the current situations by different models. So, current data on observations on climate, natural ecosystems, soils, water from different sources, agricultural productivity and inputs and socio-economic parameters amongst others are continuously required. In this context, it is essential to have accessibility to databases at micro levels prepared with various agencies and to continuously monitor the impacts. Further, there is also a need for capacity building to co-ordinate world-class climate change research in India.

5.6 Policy

IPCC reports make it clear that climate change and associated impacts will continue for centuries, even if anthropogenic emissions of greenhouse gases are stopped. There is a clear opportunity to implement mutually beneficial activities (policies and projects) that take advantage of the synergies between the United Nations Framework Convention on Climate Change and its Paris Climate Agreement (2015), the Convention on Biological Diversity and broader national development objectives. However, these opportunities are rarely being realised due to a lack of national coordination amongst sectoral agencies to design policy measures that exploit potential synergies between national economic development objectives and environmentally focused projects and policies.

Many countries have established specific policies and strategies on climate change adaptation, mitigation and adaptation. However, there is a need for greater integration between the CBD and UNFCCC to maximise the synergies in environmental protection, not only for climate change but for poverty alleviation, food and water security and long-term sustainability, besides addressing the impacts of climate change. Developing a sound and pragmatic Climate Change Action Plan and its subsequent implementation requires strengthening/evolving supportive institutions, information, finance, technology and public support. The strategy has to be developed in consonance with national and regional developmental objectives and contexts, based on available sound scientific information and by following a participatory approach including consultation with local stakeholders and communities.

India's National Action Plan on Climate Change (NAPCC) and its subsequent eight missions such as National Solar Mission, National Mission for Enhanced Energy Efficiency, National Mission on Sustainable Habitat, National Water Mission, National Mission for Sustaining the Himalayan Ecosystem, National Mission for a 'Green India', National Mission for Sustainable Agriculture and National Mission on Strategic Knowledge for Climate Change represent 'multi-pronged, long-term and integrated strategies for achieving key goals in the context of climate change. NAPCC outlines a strategy by which India will adapt to and mitigate challenges of climate change, while maintaining a high growth rate, protecting poor and vulnerable sections of society and achieving national growth objectives. The government's proposals, while void of specific targets and details, put forward action plans that could be easily developed through regional and state-level strategies as well as to incorporate into current and future developmental plans.

Integration of biodiversity into developmental plans and policies would pave the way for ecosystem-based approach for climate change mitigation and adaptation, besides ensuring sustainability of resources.

5.7 Conclusions

The impacts of climate change and its linkage with biodiversity are better understood, and there are growing demands for reducing the ecological footprint in human species to protect, preserve and restore our rich biodiversity for posterity. In order to avoid escalating threats on biodiversity with changing climate and to plan and implement mitigation and adaptation programmes, there are growing demands to mainstream the strategies and action into policy framework.

References

Adhavan D, Kamboj RD, Marimuthu N, Bhalodi MM (2014) Seasonal variation and climate change influence coral bleaching in Pirotan Island, Gulf of Kachchh Marine National Park, Gujarat. Curr Sci 107:1780–1781

Aggarwal PK, Pathak H, Kumar N (2009) Global climate change and Indian agriculture: a review of adaptation strategies, trust for advancement of agricultural sciences, New Delhi

Aggarwal, PK, Kumar SN, Chauhan N, Saxena R, Jain S, Rani S (2011) Impact of climate change on crop productivity in Western Ghats, coastal and north-eastern regions of India. Curr Sci, Special Issue, 101(3) 332–341

Anonymous (2010) Climate change and India: a 4x4 assessment. A sectoral and regional analysis for 2030s. Ministry of Environment and Forests, Government of India, p 24

Anonymous (2012) Coastal zones of India. Space Application Centre, ISRO, Ahmedabad, p 597

Ateweberhan M, McClanahan TR (2010) Historical sea-surface temperature variability predicts climate change-induced coral mortality. Mar Pollut Bull 60:964–970

Arthur R (2000) Coral bleaching and mortality in three Indian reef regions during an El Niño southern oscillation event. Curr Sci 79:1723–1729

Barnosky AD, Matzke N, Tomiya S, Wogan GO, Swartz B, Quental TB, Marshall C, McGuire JL, Lindsey EL, Maguire KC, Mersey B, Ferrer EA (2011) Has the Earth's sixth mass extinction already arrived? Nature 471:51–57

Biju kumar A (2012) Shifting baselines and jellyfish blooms: importance of research in relation to jelly fish export potential. Fishing Chimes 32(7):30–31

Blasco F (1975) The mangroves in India. Institut Francais de Pondicherry, Pondicherry, p 175

Brotz L, Cheung WL, Kleisner K, Pakhomov E, Pauly D (2012) Increasing jellyfish populations: trends in Large Marine Ecosystems. Hydrobiologia 690:3–20

Catenazzi A, Lehr E, Vredenburg VT (2014) Thermal physiology, disease, and amphibian declines on the eastern slopes of the Andes. Conserv Biol 28:509–517

CBD (2009) Connecting biodiversity and climate change mitigation and adaptation. Report of the Second Ad Hoc Technical Expert Group on Biodiversity and Climate Change. Montreal, Technical series no. 41, p 126

CBD (2014) An updated synthesis of the impacts of ocean acidification on marine biodiversity, In: Hennige S, Roberts JM, Williamson P (eds) Montreal, Technical series no. 75, p 99

Chaturvedi RK, Gopalakrishnan R, Jayaraman M, Bala G, Joshi NV, Sukumar R, Ravindranath NH (2011) Impact of climate change on Indian forests: a dynamic vegetation modelling approach. Mitig Adapt Strateg Glob Chang 16(2):119–142

Chaturvedi RK, Kattumuri R, Ravindranath D (2014) Mainstreaming adaptation to climate change in Indian policy planning. Int J Appl Econ Econ 22(1):23–56

Cheung WWL, Lam VWY, Sarmiento JL et al (2009) Projecting global marine biodiversity impacts under climate change scenarios. Fish Fish 10:235–251

Cheung WWL, Watson R, Pauly D (2013) Significance of ocean warming in global fisheries in catch. Nature. doi:10.1038/nature12156

Chitale VS, Behera MD, Roy PS (2014) Future of endemic flora of biodiversity hotspots in India. PLoS One 9(12):e115264, 1–15

Darwall WRT, Holland RA, Smith KG, Allen D, Brooks EGE, Katarya V, Pollock CM, Shi Y, Clausnitzer V, Cumberlidge N, Cuttelod A, Dijkstra KDB, Diop MD, Garcia N, Seddon MB, Skelton PH, Snoeks J, Tweddle D, Christophe J (2011) Implications of bias in conservation research and investment for freshwater species. Conserv Lett 4:474–482

Dhanya CT, Kumar A (2015) Making a case for estimating environmental flow under climate change. Curr Sci 109:1019–1020

Dulvy NK, Rogers SI, Jennings S, Stelzenmüller V, Dye SR, Skjoldal HR (2008) Climate change and deepening of the North Sea fish assemblage: a biotic indicator of warming seas. J Appl Ecol 45:1029–1039

Ellison JC (2012) Climate change vulnerability assessment and adaptation planning for mangrove systems. World Wildlife Fund (WWF), Washington, DC

Field CB, Barros VR, Mastrandrea MD et al (2014) Summary for policymakers. In: IPCC. Climate Change 2014: impacts, adaptation, and vulnerability. Working group II contribution to the IPCC 5th assessment report. Cambridge University Press, Cambridge/New York

Gautam HR, Bhardwaj ML, Kumar R (2013) Climate change and its impact on plant diseases. Curr Sci 105(12) 25: 1685–1691

Gosain AK, Sandhya R, Anamika A (2011) Climate change impact assessment of water resources of India. Curr Sci 101(3):356–371

Hawkes LA, Broderick AC, Godfrey MH, Godley BJ (2009) Climate change and marine turtles. Endanger Species Res 7:137–154

Hawkins JP, Roberts CM, Clark V (2000) The threatened status of restricted-range coral reef fish species. Anim Conserv 3:81–88

INCCA (2010) Climate change and India: a 4X4 assessment – a sectoral and regional analysis for 2030s. Ministry of Environment and Forests, Govt of India, 160pp

IPCC (2002) Climate change and biodiversity. IPCC Technical paper V. Intergovernmental panel on climate change. Geneva, p 77

IPCC (2014a) Climate Change 2014. In: Pachauri RK, Myer LA (eds) Synthesis report. Contribution of working groups I, II and III to the Fifth Assessment Report of the Intergovernmental Panel on climate change. IPCC, Geneva, p 151

IPCC (2014b) Summary for policymakers. In: Field CB, Barros VR, Dokken DJ, Mach KJ, Mastrandrea MD, Bilir TE, Chatterjee M, Ebi KL, Estrada YO, Genova RC, Girma B, Kissel ES, Levy AN, MacCracken S, Mastrandrea PR, White LL (eds) Climate change 2014: impacts, adaptation, and vulnerability. Part A: Global and sectoral aspects. Contribution of Working Group II to the Fifth Assessment Report of the Intergovernmental Panel on Climate Change. Cambridge University Press, Cambridge/New York, USA, 32pp

Jagtap TG, Nagle VL (2007) Response and adaptability of mangrove habitats from the Indian subcontinent to changing climate. Ambio 36:328–334

Kumar NS, Aggarwal PK, Rani S et al (2011) Impact of climate change on crop productivity in Western Ghats, coastal and northeastern regions of India. Curr Sci 101:332–341

Laloë JO, Cozens J, Renom B, Taxonera A, Hays GC (2014) Effects of rising temperature on the viability of an important sea turtle rookery. Nat Clim Chang 4:513–518

Lee PLM, Dawson MN, Neill SP, Robins PE, Houghton JDR, Doyle TK, Hays GC (2013) Identification of genetically and oceanographically distinct blooms of jellyfish. J R Soc Interface 10(80):1–11

MoEF (2014) India's fifth national report to the convention on biological diversity. Ministry of Environment and Forests, New Delhi, 100p

Mohamed KS, Sathianandan TV, Kripa V, Zacharia PU (2013) Puffer fish menace in Kerala: a case of decline in predatory control in the south-eastern Arabian Sea. Curr Sci 104:426–429

O'Briena K, Leichenko R, Kelkar U, Venemad H, Aandahla G, Tompkinsa H, Javedc A, Bhadwalc S, Bargd S, Nygaarda L, Westa J (2004) Mapping vulnerability to multiple stressors: climate change and globalization in India. Glob Environ Chang 14(4):303–313

Padmakumar KB, Menon NR, Sanjeevan VN (2012) Is occurrence of harmful algal blooms in the exclusive economic zone of India on the rise? Int J Oceanograph, Article ID 263946, 7 pp. doi:10.1155/2012/263946

Parry ML, Canziani OF, Palutikof JP, Linden PJ Hanson CE (eds) (2007) Climate change 2007. Impacts, adaptation and vulnerability. Contribution of Working Group II to Change 2007: Impacts, Adaptation and Vulnerability. Contribution of Working Group II to the Fourth Assessment Report of the Intergovernmental Panel on Climate Change. Cambridge University Press, Cambridge, p 982

Perry AL, Low PJ, Ellis JR, Reynolds JD (2005) Climate change and distribution shifts in marine fishes. Science 308:1912–1915

Portner HO, Farrell AP (2008) Physiology and climate change. Science 322:690–692

Purcell JE, Uye SI, Lo WT (2007) Anthropogenic causes of jellyfish blooms and their direct consequences for humans: a review. Mar Ecol Prog Ser 350:153–174

Raha A, Das S, Banerjee K, Mitra A (2012) Climate change impacts on Indian Sunderbans: a time series analysis (1924–2008). Biodivers Conserv 21:1289–1307

Rajendran K, Sajani S, Jayasankar CB, Kitoh A (2013) How dependent is climate change projection of Indian summer monsoon rainfall and extreme events on model resolution? Curr Sci 104:1409–1418

Richardson AJ, Bakun A, Hays GC, Gibbons MJ (2009) The jellyfish joyride: causes, consequences and management responses to a more gelatinous future. Trends Ecol Evol 24(6):312–322

Rosenfield JA (2002) Pattern and process in the geographical ranges of freshwater fishes. Glob Ecol Biogeogr 11:323–332

Roxy MK, Modi A, Murtugudde R, Valsala V, Panickal S, Prasanna Kumar S, Ravichandran M, Vichi M, Lévy M (2016) A reduction in marine primary productivity driven by rapid warming over the tropical Indian Ocean. Geophys Res Lett 43:1–8. doi:10.1002/2015GL066979

SCBD (2010) Strategic action plan for biodiversity 2011–2020 and the Aichi targets. Secretariat of the Convention on Biological Diversity, Montreal

Syvitski JPM, Kettner AJ, Overeem I, Hutton EWH, Hannon MT, Brakenridge GR, Day J, Vörösmarty C, Saito Y, Giosan L, Nicholls RJ (2009) Sinking deltas due to human activities. Nat Geosci 2:681–686

Unnikrishnan AS, Shankar D (2007) Are sea level rise trends along the coasts of northern Indian Ocean consistent with global estimates? Glob Planet Chang 57:301–307

Venter O, Fuller RA, Segan DB, Carwardine J, Brooks T, Butchart SHM, Marco M, Iwamura T, Joseph L, O'Grady D, Possingham HP, Rondinini C, Smith RJ, Venter M, Watson JEM (2014) Targeting global protected area expansion for imperilled biodiversity. PLoS Biol 12(6):e1001891, 1–7

Victor DG, Zhou D, Ahmed EHM, Dadhich PK, Olivier JGJ, Rogner HH, Sheikho K, Yamaguchi M (2014) Introductory chapter. In: Edenhofer O, Pichs-Madruga R, Sokona Y, Farahani E, Kadner S, Seyboth K, Adler A, Baum I, Brunner S, Eickemeier P, Kriemann B, Savolainen J, Schlömer S, Stechow C, Zwickel T, Minx JC (eds) Climate change 2014: mitigation of climate change. Contribution of Working Group III to the Fifth Assessment Report of the Intergovernmental Panel on Climate Change, Cambridge University Press, Cambridge/New York, USA, pp 113–150

Vivekanandan E (2013) Climate change: challenging the sustainability of marine fisheries and ecosystems. J Aquat Biol Fish 1(1 & 2):54–67

Vivekanandan E, Rajagopalan M, Pillai NGK (2009) Recent trends in sea surface temperature and its impact on oil sardine. In: Aggarwal PK (ed) Global climate change and Indian agriculture. Indian Council of Agriculture Research, New Delhi, pp 89–92

Wells ML, Trainer VL, Smayda TJ (2015) Harmful algal blooms and climate change: learning from the past and present to forecast the future. Harmful Algae 49:68. doi:10.1016/j.hal.2015.07.009

Williams JW, Blois JL, Shuman BN (2011) Extrinsic and intrinsic forcing of abrupt ecological change: case studies from the late quaternary. J Ecol 99(3):664–677

WWF (2014) Living Planet Report 2014. Species and spaces, people and places. WWF International, Gland, p 36

Indigenous Knowledge and Intellectual Property Rights

6

P.E. Rajasekharan and K. Souravi

Abstract

Indigenous knowledge refers to what indigenous people know and do and what they have known and done for generations – practices that evolved through trial and error and proved flexible enough to cope with change. Intellectual property, very broadly, means the legal rights, which result from intellectual activity in the industrial, scientific, literary and artistic fields. Intellectual property law aims at safeguarding creators and other producers of intellectual goods and services by granting them certain time-limited rights to control the use made of those productions. Intellectual property has increasingly assumed a vital role with the rapid pace of technological, scientific and medical innovation that we are witnessing today. Moreover, changes in the global economic environment have influenced the development of business models where intellectual property is a central element establishing value and potential growth. The complicated contemporary issues especially the ethical, cultural, historical, political, religious/spiritual and moral dimensions rather legal involved in the relationship between indigenous/traditional knowledge and intellectual property rights with examples and case studies are detailed in this chapter.

Indigenous peoples have the right to maintain, control, protect and develop their cultural heritage, traditional knowledge and traditional cultural expressions, as well as the manifestations of their sciences, technologies and cultures, including human and genetic resources, seeds medicines, knowledge of the properties of fauna and flora, oral traditions, literatures, designs, sports and traditional games visual and performing arts. They also have the right to maintain, control, protect and develop their intellectual property over such cultural heritage, traditional knowledge, and traditional cultural expressions. (United Nations Declaration on the rights of Indigenous peoples, Article 31, 2007.)

P.E. Rajasekharan (✉) • K. Souravi
Division of Plant Genetic Resources, ICAR-Indian Institute of Horticultural Research, Hessaraghatta Lake post, Bangalore 560 089, India
e-mail: rajasekharan.pe@gmail.com; souravi.karpakal81@gmail.com

S. Abdulhameed et al. (eds.), *Bioresources and Bioprocess in Biotechnology*, DOI 10.1007/978-981-10-3573-9_6

Keywords
Traditional knowledge • TK • Intellectual property rights • IPR • CBD

6.1 Introduction

Indigenous knowledge or traditional knowledge (TK) has been in use from centuries and centuries by the indigenous people and local communities; it is a quintessential knowledge that is generated, improved upon and passed on from generation to generation, thus becoming a part of their cultural identities. Traditional knowledge is the backbone in vital areas such as healthcare and food security. About 80% of the population in developing countries such as India and China still largely depend on the traditional medicines; they provide the most affordable mode of treatment for people from lower economic backgrounds and have also been the source of undeniable knowledge for development of modern medicines. Also TK helps in the development of the newer plant varieties by local farmers and thereby creating an upscale market for these indigenous varieties and the associated knowledge in the areas of agriculture in many of agriculture-dependent developing countries in Asia and Africa.

Overall the traditional knowledge systems are an absolute requirement for the well-being and sustainable development of the communities involved and also help portray and showcase their intellectual and cultural vitality. Which in turn means that it is not just vital to only sustain the mere possessed knowledge, but there is a requirement to pay equal importance to the paraphernalia of social and physical environment, of which it is a part of, i.e. the indigenous people, their well-being and their interests, as they are the very custodians of such unquenchable knowledge. The various sectors such as biotechnology and human healthcare industries that are involved in the creation and production of essential requirements such as chemical, medicinal and agro-products have looked upon the TK in the recent past, thus leading to a sudden surge in TK and biodiversity bioprospecting (Reid et al. 1993). The prime intellectual response to the very mere necessities of life is supplied from the most practical component, i.e. the traditional knowledge and its prospecting; some of the classic examples of knowledge-derived products are Jeevani, the wonder drug from India, and the *Hoodia* from South Africa. Other less known examples include the use of Plao-noi used by the Thai traditional healers to treat ulcers and the Western Amazon indigenous healers who use the Ayahuasca vine to prepare various medications.

However, the significant value of TK or the obligations in regard to its use have only led to the loss of TK through the destruction of the physical, social and cultural values of the associated communities. When TK-based benefits are sought out for industrial or commercial advantage, there is a strong concern that the associated knowledge may be misappropriated, and the rightful custodians are not accordingly rewarded. In the current scenario, the need of the hour is to nurture with sustainable use of TK and also protect and respect the beliefs and values of the TK holders. This led to the question whether TK is also a part of the intellectual property's hornets nest. The IP system being legally and practically a protective cover to the interests of the involved communities, along with providing significant contributions towards

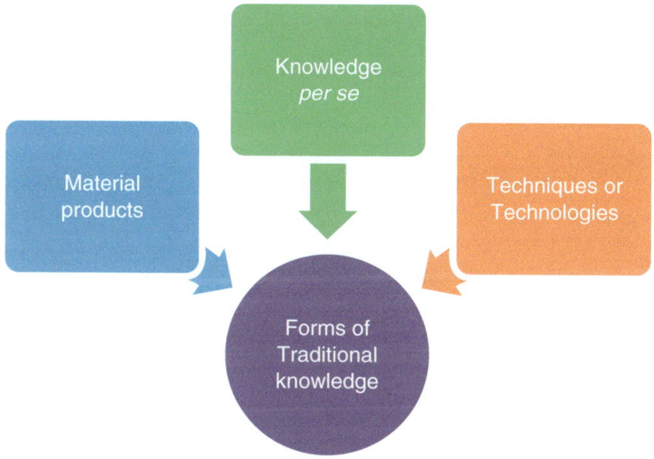

Fig. 6.1 Three different forms of traditional knowledge (TK)

sustainable growth and development, is an issue that needs to be debated and discussed upon. In this context, this chapter tries to address the role of IP in the effective protection and promotion of TK.

6.2 TK Protection and the Involved National Framework

The very need for the protection arises as TK manifests itself in various dimensions such as tangible (material products), intangible resources (knowledge per se) and also in the form of processes and procedures (techniques or technologies in various descriptive forms) (Muller 2013) (Fig. 6.1).

TK is often defined in roles of cultural, environmental, social and religious context. This holistic nature of TK is what that gives it a distinctive nature, and it is this rationale that has made its protection in terms of IP policies and framework very complex. Overall, the need to protect stems from the reasons such as equity among all the involved stakeholders, conservation of the biological resources, preservation of the traditional cultures, prevention of unauthorized use and finally promotion of its various uses (Correa 2001).

Equity Inequity is expressed when the true TK holders are normally not recognized and thus not well compensated. This scenario can be seen in case of traditional farmers, wherein they are the true custodians of native landraces; they would have developed the best varieties through continuous selection and in the process also conserved the traditional plant varieties over generations; however, the monopoly comes into play when the seed companies protect the rebranded improved varieties that are developed by the traditional farmers or protect the new varieties that were developed using these traditional varieties through plant breeder's rights.

Table 6.1 Some of the controversial cases that involved the wrongful patenting of biological resources

Sl.no	Biopirated cases	TK country	Patented country	Status
1.	Turmeric patent*1	India	United States	Patent revoked
2.	Coloured cotton*2	Latin America	United States	No relief
3.	Barley gene*3	Ethiopia	United States	No relief
4.	Neem products*4	India	United States	Few of the patents revoked
5.	Ayahuasca*5	South America	United States	Patent upheld
6.	Hoodia cactus*6	South Africa	United States and others	Benefits shared
7.	Basmati*7	India	United States	Patent revoked
8.	Ginger*8	India	United Kingdom	Patent revoked
9.	Kava*9	Pacific	United States	No relief
10.	Piper*9	India	United Kingdom	No relief
11.	Phyllanthus*9	India	United States	No relief
12.	Quinoa*9	Andes	United States	No relief

*1 http://www.ias.ac.in/currsci/oct252001/955.pdf

*2 https://www.idrc.ca/en/book/people-plants-and-patents-impact-intellectual-property-trade-plant-biodiversity-and-rural

*3 Bryant (1994)

*4 Singh Nijar (1994)

*5 Ardhede (2006)

*6 Intellectual Property as a Lever for economic growth (2003)

*7 http://ssrn.com/abstract=1143209

*8 http://www.slideshare.net/patent_unitedipr/protection-of-traditional-knowledg

*9 Biopiracy Newsletter (2011)

Conservation of Biological Resources Protection of TK helps conserve and promote sustainability of the biological resources as it is the TK of the indigenous communities, their tradition, culture and practices. Most of these communities are dependent on nature such as flora and fauna; hence, protecting their values contributes significantly towards conservation of biological resources.

Preservation of Traditional Cultures Many of the TKs are nothing but a manifestation of the lifestyle and self-identities of the TK holders, i.e. the indigenous communities; hence, protection of the associated TKs helps protect the identities of the TK holders and in their sustenance. Although IPR doesn't seem to play a role, it needs to be understood that the protection of TK through appropriate IPRs can give a better profile and more the reason for protection.

Prevention of Biopiracy The misappropriation of TK can be termed as "biopiracy" – exploitation of indigenous knowledge and biodiversity without benefit sharing or in a derogatory manner. There are many such case studies that can be quoted to best suit in this context (Table 6.1.)

Cases of Biopiracy in India (http://www.tkdl.res.in/tkdl/Langdefault/common/BioPiracy.asp?GL)

Turmeric (Curcuma longa Linn.):

In 1995, two expatriate Indians were granted US patent for use of turmeric in wound healing. The CSIR argued that it was a TK of India and was in used from centuries ago, and therefore its medicinal use was not a novel invention. Their claim was supported by documentary evidence of traditional knowledge, including ancient Sanskrit text and a paper published in 1953 in the Journal of the Indian Medical Association. The US PTO upheld the CSIR objections and revoked this patent in 1997.

Neem (Azadirachta indica A. Juss.):

In 1994, EPO granted a patent to the US Corporation W.R. Grace Company and US Department of Agriculture for a method for controlling fungi on plants by the aid of hydrophobic extracted neem oil. In 1995, a group of international NGOs and representatives of Indian farmers represented a legal opposition. Evidence was submitted that the fungicidal effect of extracts of neem seeds had been known and used for centuries in Indian agriculture to protect crops and, therefore, was not something novel. In 2000 EPO revoked the patent, which was again challenged in 2001 and further rejected by EPO in 2006.

Promotion of TK If the TK of the holders are properly used, promoted and commercialized, the benefits that arise would also be manyfolds and through IPR laws if properly used would help in effective sharing of benefits with the TK holders and can also help further conserve and promote the particular TK.

The conceptual understanding of IP-based protection which is in the eyes of some is the mere exclusion of authorized use by the third parties, whereas for others it is a tool to protect the existing TK and their holders. The former being the defensive protection and the latter positive protection, defensive protection is to fill in the loopholes in the IP system which may allow IPR claims that emanate from TK without the consent or even prior information to the people who actually hold the TK (Ansong 2013). The Biodiversity Act 2000 is a form of defensive protection regime against unauthorized use of biological resources and TK (Dutfield 2003). In order to take up IPR protection for any invention originating from a biological resource or TK from India, it is mandatory to have NBA permission. However under positive protection the TK holder, i.e. the indigenous communities or individuals, can themselves take up the protection of their TK or TK-associated biological resources using various protective mechanisms such as patents (Cottier, Panizzon 2004).

Limitations of the Current IPR Regimes There are two major limitations of the IPR systems that are making them inadequate to protect the TK and the TK holders (Moni 2006).

- Notion of property ownership: IPR rights are largely individualistic rights; they do not recognize or cover on community rights, wherein knowledge is largely held by communities' rather than single individuals. Also in a simpler context, the TK is largely passed on from one generation to the next; hardly any innovation or modification happens during the course of time.

- One of the main requirements of an effective IPR regime such as patents requires three parameters to be fulfilled, namely, novelty, inventive step and commercial viability. However, TK does not pass these requirements as TK is not something new or novel, and it is a knowledge that has been in the public domain for a very long period of time. Further most of these TKs do not contain any inventive step as they would have been passed down the generations without any modifications or changes.

6.3 International Efforts to Protect TK

The initial interests to protect TK related to biological resources can be traced back to the early 1980s by the FAO, the mandate of which has been included in the International Treaty on the Plant Genetic Resources for Food and Agriculture (ITPGRFA) 2001 (http://data.iucn.org/dbtw-wpd/htm/EPL057-expguide-international-treaty/Article9.html). In the year 1992, the CBD recognized the importance of conservation, sustainable development and finally access and benefit sharing (ABS) of the associated TK. It called for Prior Informed Consent (PIC), participation of indigenous and local communities and benefit sharing as priorities for any use of TK. The Nagoya Protocol, a resultant of the CBD discussions, states that the access and use of TK is possible only through PIC, and the local communities must be included in benefit sharing (Biber-Klemm 2006). The CBD along with the WTO debated on TRIPS for a better functioning IPR system. In 2007 a regime addressing the rights of indigenous people and local commutates in relation to their IP, values and innovations was bought into effect by the UN Declaration (Jorge et al. 1999; Thomas 2010).

6.4 TK Protection in India Under the IPR Regime

6.4.1 Trade-Related Aspects of Intellectual Property (TRIPS Agreement)

Article 1 states:

"Members may, but shall not be obliged to, implement in their law more extensive protection than is required by this Agreement, provided that such protection does not contravene the provisions of this Agreement."

India as a signatory of the TRIPS agreement follows the IPR prescriptions as provided by the TRIPS; some of the main provisions in relation to TK are discussed below:

- Article 27.3(b) provides for either the use of *sui generis* systems or a combination of patents and *sui generis* systems to protect plant varieties. Such a system can be termed as defensive protection. Plant breeders of indigenous communities and related TK holders of biological resources are benefitted from such a protection.

- According to Article 22.1 of the TRIPS Agreement, Geographical Indications of Origin (GIOs) may help protect products of certain plants bred in traditional communities, wherein specific quality or characteristic of such plants can be attributed to the particular geographical location of the traditional community.

- Article 39.2 provides for the protection from disclosure of information lawfully if such information is secret, has commercial value and reasonable steps have been taken by the person to keep it secret. Some forms of TK are not in the public domain of the community and may be held exclusively by TK holders like herbalists and healers. For example, in The Nana Drobo case, an exclusive knowledge of the medicinal value of certain herbs, which had commercial value, was protected under this clause. Hence, although TK is not specifically mentioned in the TRIPS agreement, the flexibility in the TRIPS may offer policy choices for some forms of TK to be protected as undisclosed information.

***Sui Generis* System** At times the existing norms, laws and regulations may not be enough to get what we aspire for, so is the case with TK, although we have a stable framework for IPR laws; yet these may not be sufficient to protect the TK and rights of its holders, this can be attributed to the very holistic nature of TK, the larger than life TK resources to be protected and importantly its highly attributable diverse nature. In order to overcome these, countries such as India and the Philippines have come up with a *sui generis* legislation and enacted the same, in respect to the nature of the resources that have to be protected. In simple words, *sui generis* simply means "its own" or "one that is of its own kind", a system that is specifically formulated for a particular country keeping in mind the requirements and needs of the country; however, upholding the primary objectives, i.e. to identify, protect and promote the rights of the indigenous people and the local communities who are the holders of the TK and also to recognize the potential for using or exploiting these resources.

Moreover, it is important for countries to have somewhat similar *sui generis* systems as these will enable the formation of a platform for the establishment of an international *sui generis* system as articulated recently by the G-15 group of developing countries (Barton et al. 2002). A number of countries such as Peru, the Philippines, Costa Rica, Thailand and Venezuela have already put forth effective *sui generis* systems. Further, an international treaty such as that discussed by the megadiverse countries, viz. India, Indonesia, Peru, Costa Rica, China, Ecuador, Kenya, Costa Rica, Mexico, Brazil, Venezuela and South Africa, will focus on policies attributable to the three pillars of CBD, namely, conservation, sustainable use of the biological/genetic resources and benefit sharing. These policies are framed in order to bring in an effective regulatory mechanism for the effective use of biodiversity and not to make any amendments or be contradictory in relation to the international laws such as IP polices which would in turn dilute the IP laws and create more loopholes and hinder the effective working of these laws.

Costa Rica (http://www.servat.unibe.ch/icl/cs00000_.html)

Article 82 states: "The state expressly recognizes and protects, under the common denomination of sui generis community intellectual rights, the knowledge, practices and innovations of indigenous peoples and local communities related to the use of components of biodiversity and associated knowledge. This right exists and is legally recognized by the mere existence of the cultural or knowledge related to genetic resources and biochemicals; it does not require prior declaration, explicit recognition nor official registration; therefore it can include practices which in future acquire such status..... no form of intellectual or industrial property rights protection.... shall affect such historic practices."

Thailand (http://www.thailawforum.com/articles/Thai-traditional-medicine-protection-part1-5.html)

"Thai Traditional Medicinal Intelligence Act": Formula of Thai traditional medicine accorded "national status", known as national formula if it is of commercial use for research and development and production of drugs then it is subject to permission from the government, violation of this act is punishable.

In case of private formula, third parties must seek permission from the private rights holder. The rights conferred is ownership rights throughout the life of the TK owner and up to 50 years after his/her death.

Article 8(j) states ".... respect, preserve and maintain knowledge, innovations and practices of indigenous and local communities embodying traditional lifestyles relevant for the conservation and sustainable use of biological diversity and promote their wider application with the approval and involvement of the holders of such knowledge, innovations and practices and encourage the equitable sharing of the benefits arising from the utilization of such knowledge, innovations and practices".

6.5 Convention of Biological Diversity (CBD Article 18(j))

This international agreement was signed in June 1992 with 191 countries (India inclusive) being parties to the convention, with three main goals:

- Conservation of biodiversity
- Sustainable use of the components
- Sharing of benefits arising in a fair and equitable way

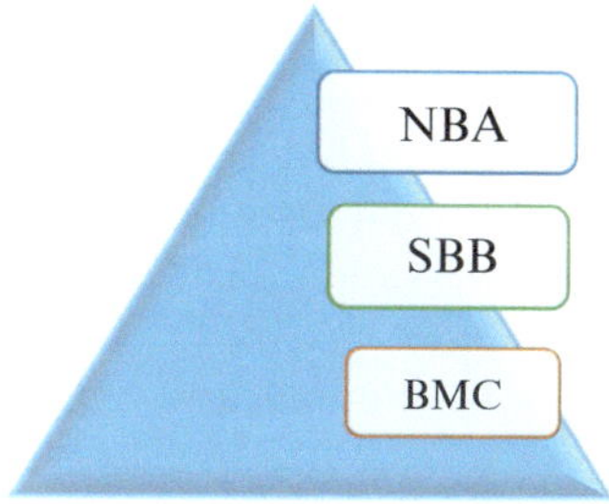

Fig. 6.2 Three-tier system under Biodiversity Act 2000

It further provided its member countries a set of protocols that were framed on strong underlying principles that aimed to answer two important questions: Firstly, how should access to genetic resources be provided? Secondly, how should benefits arising from such resources be shared? However, under this agreement, choice of protection and benefit sharing with TK holders is relative. There is no prescribed form of binding norms that are to be adhered to.

In order to comply with the CBD agreement, India came up with the Biodiversity Act 2000; under this regime, a three-tier system functions, namely, the National Biodiversity Authority (NBA), State Biodiversity Board (SBB) and the Biodiversity Management Committee (BMC); these systems not only help in documentation of the existing traditional knowledge but also include in their mandate conservation, preservation, sustainable use and finally fair and equitable benefit sharing among all the stakeholders (Fig. 6.2).

NBA: Functions and Activates

- Access to biological resources and associated TK by foreign nationals, institutions or companies.
- All matters relating to transfer of the results of the research to any foreign national.
- Imposition of terms and conditions to obtain equitable sharing of benefits.
- Establish sovereign rights over the bioresources of India.
- Approvals for seeking any form of IPRs in/outside India for an invention based on research or information pertaining to biological resource and associated TK obtained from India.
- Coordinates ABS activities of SBB and BMC.
- Select and notify areas as biodiversity heritage sites.
- Finally the NBA acts as national component authority to take any and all the decisions pertaining to ABS.
- Fees and royalties paid to the NBA for approvals are deposited in the national biodiversity fund used for conservation and development of resource areas accessed.

SBB: Functions and Activates

- Access to biological resources to Indians for commercial purposes.
- Governing body to restrict any activity that violates the common objectives of conservation, sustainable use and equitable benefit sharing.

BMC: Functions and Activities

- BMC mainly functions for conservation, sustainable use and documentation of biodiversity and chronicling of knowledge related to biodiversity.
- NBA and SBB consult the respective BMCs relating the use of biological resources and related knowledge within their jurisdiction.

Access and Benefit Sharing Mechanism A mechanism to address the roles of commercial entities, TK holders and other genetic resource developers as well as the specific roles of the countries and communities from which the genetic resources had their origin led to the urgent requirement of formulation and implementation of an efficient system (Lucia and Marin 2002). The quest for such an efficient system that would provide an equitable return to developing countries and its numerous indigenous communities and TK holders was coalesced into what was called as access and benefit sharing (ABS). Access and benefit sharing are completely two different entities but linked concepts: access to genetic resources and equitable sharing of the benefits arising from the utilization of genetic resources. The ABS started functioning within the negotiations of the CBD, with add-on in the form of Bonn Guidelines and much recently the Nagoya protocol, which tried to make the system of ABS much more efficient.

With the advent of a number of guidelines that propose effective ABS system in respect to medicinal plant biodiversity, numerous questions arises such as "Are these proposals feasible especially in respect to the allocation of property rights to such resources? Can the problems caused by transgressing acceptable or sustainable environmental limits be solved or even be controlled by traditional market mechanisms and property rights? Or will we, by commodifying the global commons, end up with the commercial enclosure or privatization of all the ecosystems and all the knowledge that make up the earth's living space, both physically and intellectually?"

Further India is taking cautious but steady steps towards heralding ABS regime into medicinal biodiversity. ABS is already in practice with the help of governing agencies such as the NBA; it is also undergoing a sea of change to meet up all the requirements of fair and equitable benefit sharing for which a number of propositions are in the pipeline. A debate on the interplay between ABS and biological resources and TK is the most relevant at this juncture; this interplay will determine the contours of ABS in years to come. How this interaction will shape the portfolio of development in India and the implications thereof is the important question that needs to be addressed.

Table 6.2 Examples of TK Protection under the existing IP laws in other countries

Sl.no	Country	Brief discription of TK protection and legislation
1.	Canada	Relies on Copyright Act, Trademark Act and Trade secret Act for protection of TK based creations.
2.	France	GI and origin (for patent claims) are protected, also fraud sanctions are imposed, if creations or products of some particular TK holding community is sold fraudulently.
3.	Portugal	Industrial property code protects trade, certification and collective marks, origin appellations and GIs
4.	Vietnam	TK based products or creations have been long protected under various covers such as patents, trademarks and appellations of origin.
5.	New Zealand	"Maori Made Mark" was established to authenticate and mark products or creations of indigenous communities, The Royal Commission on Genetic Modification was established to help in assistance for patent applications in consultation with the indigenous community.
		Have also developed IP guidelines for patent examiners in respect to patent applications in significance to the TK and TK holders and even the indigenous resources like micro-organisms or flora and fauna.

The draft guidelines on ABS were the first step taken by the NBA to put in an objective and a highly regulated system for the conduct of ABS (http://www.entetelegale.com/PDF/DRAFT_ABS_GUIDELINES_NBA_INDIA.pdf). They provide a non-discriminatory framework for granting approvals for access to biological resources and knowledge associated thereto and the fair and equitable sharing of the benefits arising from the utilization, in conformity with the Biological Diversity Act 2002. The ABS guidelines are normally divided into four parts:

- Preliminary
- User obligations
- Provider obligations
- NBA approvals and benefit sharing principles.

The guidelines lay out the conditions under which biological resources and associated knowledge thereto shall be granted and under which the sharing of benefits arising out from the utilization shall be qualified as fair and equitable. The ABS guidelines are effectively administered by the CBD and NBA, to ensure conservation and sustainable use (Table 6.2).

There are a few benefit sharing principles that have been mentioned in the ABS protocols that pave way to equitable and fair benefit sharing procedures which are as follows (https://cmsdata.iucn.org/downloads/an_explanatory_guide_to_the_nagoya_protocol.pdf):

- In case of access to biological resources for commercial utilization, the benefit sharing shall be in the form of ensuring sustainable livelihoods to providers, and the providers shall be shared with the knowledge of best practices to ensure conservation and sustainable use of the biological resources. This will hold true also in case the biological resource is cultivated or domesticated.

- If the biological resource is collected from public land, then the user in addition to ensuring sustainable livelihood of the provider will have to also deposit a certain amount to the NBA.
- It further provides benefit sharing principles in case the access is for research purposes. The research can further be divided as non-commercial and commercial. Commercial research is further divided into collaborative and non-collaborative. The non-collaborative research is further divided as cases where the results of research are shared and where they are not shared, and accordingly for each of these classes the benefit sharing formula is prescribed.
- It also asks that the application is made to seek NBA approval before seeking any IPRs.

The Kani Story (www.lead-journal.org/content/07001.pd)
The Kani case study is one of the initial models of TK protection using IPR and also the recognition of the TK holders even before a confirmative legislative framework on was formed. The subject of this case study is the role of benefit sharing arrangements concerning the "Jeevani" drug, which was developed by scientists at the Tropical Botanic Garden and Research Institute (TBGRI), based on the tribal medicinal knowledge of the Kani tribe in Kerala, South India.

Within the Kani tribe the customary rights to transfer and practice certain traditional medicinal knowledge are held by tribal healers, known as Plathis. The knowledge was divulged by three Kani tribal members to the Indian scientists who isolated 12 active compounds from Arogyapacha, developed the drug "Jeevani" and filed two patent applications on the drug (and another patent based on the same plant but for a different use). The technology was then licensed to the Arya Vaidya Pharmacy, Ltd., (Coimbatore), an Indian pharmaceutical manufacturer pursuing the commercialization of Ayurvedic herbal formulations. A trust fund was established to share the benefits arising from the commercialization of the TK-based drug "Jeevani" with the Kani tribes.

Other Success Stories *Herbavate ointment is based on the knowledge of seven innovators from six districts Sabarkantha, Panchmahal, Dang, Mehsana, Patan and* ***Bhavnagar*** *of Gujarat. Herbavate exhibits remarkable properties against eczema and a variety of inflammatory and infectious skin conditions. Technology involved in the production of the ointment was developed by Sristi lab, by pooling the associated TK of farmers and herbalists, licensed to matrix agrochemicals. The raw materials required were sourced from the common people. Benefits were shared in accordance to NBA procedure between the local communities and the company in the year 2007 (http://www.entetelegale.com/PDF/DRAFT_ABS_ GUIDELINES_ NBA_INDIA.pdf).*

Azadirachta indica or otherwise commonly known as neem have been exported to the tune of 2000 kgs to Japan by an Indian company Bio-India Biologicals. These neem leaves have been collected from Amarchinta village Biodiversity Management

Committee (BMC) of Mahbubnagar district, Andhra Pradesh. The neem leaves are collected and dried by undertaking a few special operations by the villagers of Amarchinta before handing over to the exporter. The NBA was paid a royalty of 5% of FOB to the tune of Rs. 55,035.00 by the exporter and transferred part of the royalty amount to Amarchinta BMC for planting neem saplings and creation of awareness about biodiversity conservation and employment of the local communities (http://www.eurogeographics.org/WorkGroups/WG1/eu_directive.pdf).

The Hoodia Model *(http://www.ipngos.org/NGO%20Briefings/Hoodia%20 case%20of%20 benefit%20sharing.pdf) One of the oldest indigenous civilizations is the San tribe of South Africa. This tribe was found to be the true owners of a very valuable TK, namely, the Hoodia plant or locally called as the Xhoba. They used this plant quench or quell hunger when they used to head out for hunting that prevailed for several days. In the year 1963, CSIR based on this TK went on to research this plant and isolated the proactive compound P57; they licensed this compound to a large pharmaceutical company, Phytopharma, which in turn sublicensed it to Pfizer. Commercialization was on huge note and a lot of money was made; however, the San tribes who were the actual TK holders were not recognized. In 1998 the CSIR agreed upon the acknowledgement of the owners of TK to be benefitted, followed by a memorandum of understanding in the year 2003 that said that the benefits that aroused from this plant will be shared with the respective indigenous community, San tribes, the true TK holders.*

6.6 Traditional Knowledge Digital Library (TKDL)

After facing biopiracy cases of neem and turmeric patents, the National Institute of Science Communication (NISCOM) and Department of Indian System of Medicine and Homoeopathy (ISM&H) in 1999 decided to establish the TKDL. The main purpose of this digital library was to facilitate the systematic arrangement, dissemination and retrieval of information in relation to TK. This database helps to provide legitimate records in respect to TK, which is currently being used by the various patent offices, to obtain prior art on their patent filings (http://www.tkdl.res.in/tkdl/langdefault/Common/Abouttkdl.asp?GL=Eng) (Table 6.3).

Implications of This Interplay The dynamics of TK and IPR have manifold impact due to their magnitude. Their implications are not just restricted to a single outcome but are multifaceted. The IP regime not only influences research portfolio but also the contours of development. The underlying motive of placing monetary value on TK and its associated biodiversity should encourage people to protect it. Therefore, the economic implications are not only predominant but also obvious. The other two implications are on social and ecological dimensions. TK benefit sharing is an embracing complex, interlinking notions of social and ecological processes that highlight the gains from ecosystem services that accrue to the stakeholders through multilevel governance processes. These three impacts are not mutually exclusive but often overlap (Fig. 6.3).

Table 6.3 Some of the recent patent application/patent grants that were abandoned/terminated/revoked based on the TKDL prior art evidence

Sl.no	Title/plants used/used for	Applicant	Status
1.	Compositions for the encapsulation of natural product extracts in oil medium in hard gelatin capsules and a method of encapsulation (Use of *kutaja, haritaki, bibhitaki, amla* and *coconut* for the management of **psoriasis**)	Vilambi NRK Reddy, United States	Application abandoned
2.	Composition for the treatment of diabetes mellitus and metabolic syndrome (Use of *haritaki, baheda, amla, turmeric/haldi, gurmar buti, karela, kutaja, methi, and konch* for the treatment of **diabetes mellitus**)	Morrison and Foerster, United States	Application abandoned
3.	Oral hygiene composition comprising myrtle (Use of *myrtle* for the treatment of **gingivitis** and **toothache**)	Mars Incorporated, United States	Application abandoned
4.	Herbal composition and method for treatment of airway inflammation using the same (Use of *peppermint, almond and lotus* for the treatment of **dry cough, bronchial asthma** and **rhinitis**)	Taichung Veterans General Hospital, Taiwan	Application abandoned
5.	Novel anticancer agent, methods for obtaining the same and pharmaceutical compositions thereof (Use of *Sphaeranthus indicus* as **immunopotentiators/anti-ageing/antioxidant**)	MMI Corporation, United States	Application abandoned
6.	Phytochemical combinations that regulate pathological immunity (Use of *cinnamon, ginger, turmeric and wine grape* for the treatment of **itching, eczema** and **pitta dominance** conditions)	Cowan; Fred M., United States	Application abandoned
7.	Pharmaceutical composition for the treatment of excess mucous production (Use of *ginger* and *Picrorhiza* for the treatment of **phlegmatic cough, bronchial asthma, cough/bronchitis**, diseases of **phlegm** and as **phlegmagogue**)	Nicholas John Larkins, United Kingdom	Application terminated 2 weeks before grant
8.	Cannabigerol for use in treatment of diseases benefiting from agonism of CB1 and CB2 cannabinoid receptors (Use of *bhang (Cannabis sativa)* for the treatment of **pain, bronchial asthma, rheumatism, opisthotonus, delirium, tetanus, psychosis, convulsion, nephritis, eczema, psoriasis, arthralgia, lumbago, diabetes** and **cancer**)	GW Pharma Ltd., United Kingdom	Application terminated 78 weeks before grant
9.	New use for cannabinoid-containing plant extracts (Use of *bhang (Cannabis sativa)* for the treatment of **amnesia, paralysis, Bell's palsy, epilepsy, diabetes mellitus, psoriasis, alopecia, rheumatism, cancer/carcinoma, nephritis, bronchial asthma** and **colitis**)	GW Pharma Ltd., United Kingdom	Application terminated 84 weeks before grant

(continued)

Table 6.3 (continued)

Sl.no	Title/plants used/used for	Applicant	Status
10.	A synergistic ayurvedic/functional food bioactive composition (cincata) and a process of preparation thereof (Use of *Eugenia*, *Salacia* and *Cinnamomum* for treatment of **diabetes** and **related disorders**)	M/s Avesthagen Limited, India	Deemed to be revoked

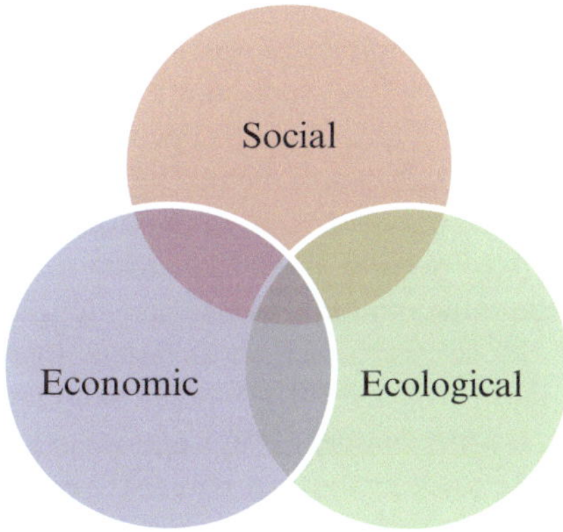

Fig. 6.3 Overlapping implications of TK and IPR

The interplay between TK and IPR will help fuel future research contours and portfolios. This interplay is a dynamic scenario with one to one, one to many and finally many to one interaction. There is no deterministic "cause and effect" relationship. Availability of monetary benefits is a critical determinant of protection and conservation of the existing biodiversity and also the economic upliftment of the stakeholders, thereby their social life. The magnitude of availability of TK, its sustainable utilization, fair and equitable sharing of the benefits arising from the utilization of the biological resource and other possible actors are influenced by the nature of the IPR regime (http://www.ecologyandsociety.org/vol17/iss1/art17/).

In order to visualize the likely scenario in the area of TK conservation and sustainable utilization, gaining more insights into the IPR workings and related aspects is inevitable. Core competencies of and synergies between these sections are of equal vital importance. The interplay and the outcome will have manifold impact on the respective stakeholders like the pharmaceutical sectors or the TK holders. The economic impact will be most obvious and explicit (https://www.thecommonsjournal.org/index.php/ijc/article/view/330).

Considering the peculiarities of our agrarian economy, social and ecological implications are of utmost importance. Often these implications tend to transgress and overlap their respective domains. The extent to which negative impact can be minimized will depend on the degree of imagination that will go into the making of a *sui generis* TK regime. Given the already existing and potential instruments for policy making in India, the aforesaid task is not insurmountable. Against such a backdrop, the dynamics of interplay between the IPR and TK needs thorough understanding.

IPR is an important tool for conservation, sustainable use and monetary monopoly if rightly used; importantly an egalitarian approach needs to be showcased when the decisions pertaining to TK are taken into account keeping in mind the local communities and their associated TK (Blankeney 1999). Appreciation and understanding the intricacies in its manifold dimensions is the first step towards the change.

The Way Forward The best way forward would be to reach out to the TK holders and work in the grass root level. IPR is a complex regime; an understanding of this level of complexness will not be easily understood and accepted by all the stakeholders involved such as the TK holders or the indigenous communities; hence, there is a requirement to decimate the right-hand information, in the best possible way to reach out to them. The purpose of networking is required. The help of government organizations, institutions and NGOs can be taken for this very purpose.

Also priority should be made to set up regional centres that would gather and accumulate the traditional knowledge records of the particular areas as to provide easier access to its records and maintenance of the respective TKs in their areas of control. These need to be then supplied to central networks such as the TKDL to be accessed by larger groups and in national and international platforms.

Accessible and transparent frameworks need to be developed in order to help the TK holders to understand the reality in the amount of biological resources that can be accessed and the kind of realistic benefits that they will be getting.

Further an in-depth understanding of the IPR laws is necessary; this would enable to develop newer ways to develop an improved TK protection keeping in the interests of the indigenous communities. Also appropriate industrial codes of conduct and ethical guidelines such as PIC need to be stringently implemented in any areas involving TK and its holders. Finally and most importantly, there is an urgent need for effective international alternative dispute resolution mechanisms to resolve disputes involving IPR and TK in international platforms.

References

Ansong A (2013) Developing countries and decision-making in the WTO: reconciling the principles of special and differential Treatment and sovereign Equality of states. A PhD thesis submitted to Middlesex University https://eprints.mdx.ac.uk/12448/1/AAnsong_thesis.pdf

Barton J, Alexander D, Correa C, Mashelkar RA, Samuels G, Thomas S (2002) Joint Declaration G-15 Experts meeting on Science & Technology, "Sustainable use of Biodiversity, traditional knowledge and protection systems". 3–5 April, 2002, Carcas

Biber-Klemm S (2006) The current Law on Plant Genetic Resources and Traditional knowledge: basic issues and perspectives. In: Biber-Klemm, Susette and Cottier, Thomas W (eds) Chapter 2: In rights tp plant genetic resources and traditional knowledge: basic issues and prospectives. CABI. UK. pp 56–110

Biopiracy Newsletter, Envis Center on Environmental Biotechnology, vol 19, Dec 2011, pp 1–16

Blankeney (1999) Perspectives on intellectual property – IP aspects of ethnobiology, vol 6. Sweet and Maxwell Publications, London

Bryant E (1994) Corporate patents or global piracy? Africans oppose privatization of genetic innovations. African Farmer, April, page 37
CBD Article 18(j), https://www.cbd.int/traditional
Correa CM (2001) Traditional knowledge and Intellectual property, issues and options surrounding the protection of traditional knowledge – a discussion paper. Quaker Peace and service, Geneva, November 2001
Cottier T, Panizzon M (2004) Legal perspectives on traditional knowledge: the case for intellectual property protection. J Int Econ Law 7(2):371–400
Dutfield G (2003) Protecting traditional knowledge and folklore, UNCTAD-ICTSD issue paper no. 1, http://www.ictsd.org/pubs/ictsd_series/iprs/CS_dutfield
Henrik Ardhede (2006) Traditional knowledge and the patent system – irreconcilable differences or a simple case of mistaken identity? Autumn, Master thesis, Faculty of Law University of Lund
http://data.iucn.org/dbtw-wpd/htm/EPL057-expguide-international-treaty/Article9.html. Last seen on 12-11-2015
http://ssrn.com/abstract=1143209. Last seen on 11-11-2015
http://www.ecologyandsociety.org/vol17/iss1/art17/. Last seen on 11-11-2015
http://www.entetelegale.com/PDF/DRAFT_ABS_GUIDELINES_NBA_INDIA.pdf. Last seen on 15-11-2015
http://www.entetelegale.com/PDF/DRAFT_ABS_GUIDELINES_NBA_INDIA.pdf. Last seen on 11-10-2015
http://www.eurogeographics.org/WorkGroups/WG1/eu_directive.pdf. Last seen on 11-11-2015
http://www.ias.ac.in/currsci/oct252001/955.pdf. Last seen on 17-11-2015
https://www.idrc.ca/en/book/people-plants-and-patents-impact-intellectual-property-trade-plant-biodiversity-and-rural. Last seen on 17-11-2015
http://www.ipngos.org/NGO%20Briefings/Hoodia%20case%20of%20benefit%20sharing.pdf. Last seen on 17-11-2015
http://www.servat.unibe.ch/icl/cs00000_.html. Last seen on 07-11-2015
http://www.slideshare.net/patent_unitedipr/protection-of-traditional-knowledge. Last seen on 11-11-2015
http://ssrn.com/abstract=1143209. Last seen on 17-11-2015
http://www.thailawforum.com/articles/Thai-traditional-medicine-protection-part1-5.html. Last seen on 15-11-2015
http://www.tkdl.res.in/tkdl/langdefault/Common/Abouttkdl.asp?GL=Eng. Last seen on 19-11-2015
http://www.tkdl.res.in/tkdl/Langdefault/common/BioPiracy.asp?GL=. Last seen on 11-11-2015
http://www.tkdl.res.in/tkdl/LangGerman/common/Outcome.asp. Last seen on 27-11-2015
http://www.venezuelaemb.or.kr/english/ConstitutionoftheBolivarianingles.pdf. Last seen on 07-11-2015
https://cmsdata.iucn.org/downloads/an_explanatory_guide_to_the_nagoya_protocol.pdf. Last seen on 17-11-2015
https://www.thecommonsjournal.org/index.php/ijc/article/view/330. Last seen on 11-11-2015
Intellectual Property as a Lever for economic growth, The African Experience, Part II, WIPO Magazine, November/December 2003, pages 6 and 7
Jorge C, Brenden T, Manuel R (1999) Acceso a Recursos Geneticos. Lecciones Y Experiencias. WRI, SPDA. Lima, Peru
Lucia P, Marin C (2002) Providing protection for plant genetic resources patents, sui generis systems & bio partnerships. Kluwer Law International, New York
Moni W (2006) What is *sui generis* system of intellectual property? Tecno-policy brief 13. African Technology Policy Network Studies, Nairobi
Muller MR (2013) Protecting shared traditional knowledge: issues, challenges and opinions, ICTSD program on Innovation, Technology and Intellectual Property, October 2013, issue paper no: 39. ICTSD, Geneva
Reid WV, Laird SA, Meyer CA, Gámez R, Sittenfeld A, Janzen DH, Gollin MA, Juma C (eds) (1993) Biodiversity Prospecting: Using Genetic Resources for Sustainable Development. World Resources Institute, Washington, DC

Singh Nijar G (1994) A conceptual framework and essential elements of a rights regime for the protection of indigenous rights and biodiversity. In: Biodiversity convention briefings. Third World Network, Penang, page 4

Thomas H (2010) Disclosure requirements in patent law and related measures: a comparative overview of existing national and regional legislation on IP and biodiversity. In: Werth A, Reyes S (eds) Triggering the synergies between intellectual property rights and biodiversity. GIZ Eschborn, Frankfurt am Main, pp 293–226

www.lead-journal.org/content/07001.pdf. Last seen on 17-11-2015

Educating Biodiversity

7

Lekshmi K. Edison, S. Pradeep Kumar, and N.S. Pradeep

Abstract

Biodiversity is life in all variety, vital to the continued existence of all ecosystems, life forms, and microorganisms on earth. There are more than 14 million species found from mountaintop to deep-sea vent, connecting and interacting with each other and creating communities and systems. Everyone in the world directly depends on natural ecosystems and its resources for a healthy and protected life. Currently the world is facing an immense biodiversity crisis. The increasing rate of biodiversity loss indicates that current developments are not sustainable. Industrialization and urbanization of the last century inversely affect our natural ecosystems, which in turn increases the number of threatened species. The stress on water, air, and natural ecosystems will rise globally in the coming years if human attitudes and actions do not change. If we wish to assure the continuing health of the planet, we must develop acknowledgeable and motivated community that realizes what biodiversity is and why it is so important. Education is one of the best tools to create citizens having skills and confidence for protecting biodiversity and feel empowered to do so. Education also challenges people to think seriously and productively about their world. The main focus of biodiversity education is to create knowledge, curiosity, and essential skills to solve a range of biodiversity problems with reference to the limited background. For developing biodiversity awareness among students, practical

L.K. Edison • N.S. Pradeep (✉)
Division of Microbiology, Jawaharlal Nehru Tropical Botanic Garden and Research Institute, Palode, Thiruvananthapuram 695562, Kerala, India
e-mail: lek_kedison@yahoo.co.in; drnspradeep@gmail.com

S. Pradeep Kumar
Kerala State Council for Science Technology and Environment, Government of Kerala, Sasthra Bhavan, Pattom, Thiruvananthapuram 695004, Kerala, India
e-mail: dr.s.pradeepkumar@gmail.com

S. Abdulhameed et al. (eds.), *Bioresources and Bioprocess in Biotechnology*,
DOI 10.1007/978-981-10-3573-9_7

oriented biodiversity education methods should be developed to attain sustainable biodiversity knowledge and motivate to conserve local biodiversity. This chapter discusses the relevance of biodiversity education for sustainable life.

Keywords
Biodiversity education • Sustainable development • Protection • Threats • Challenges

7.1 Introduction

Biodiversity education is considered as a model education for understanding sustainable development. All significant magnitudes of sustainability – ecological, economic, and social – can be embraced with biodiversity education. At present *Education for a Sustainable Development (ESD)* is an unknown concept for many. Teachers representing only one third are familiar with these concept. Biodiversity education is not like conventional teaching or studying species names. It is something more than that where naming of species is an initial point; go after knowledge of complex interrelations and connections with enough student activities. This way, all significant proportions of sustainability – ecological, economic, and social – can be embraced when teaching biodiversity. The UN Decade of Education for Sustainable Development engages people through innovative ways, by focusing on education and training to address the interlinking issues of ecosystems and livelihoods; the interactions of nature with culture, society, and economy; and the actions we can take without affecting our quality of life.

7.2 Why Biodiversity Education

During the coming 20–30 years, due to changes in environment induced by human intervention, the existence of more than a million species of plants and animals will be in danger. Extinction rate will be more than thousand times than the estimated normal rate (100 species per day). Both plants and animals will come in the list of lost, endangered, and threatened species. About 10% of temperate region plant species and 11% of the world's 9000 bird species are at some risk of extinction. In the tropics, the destruction of forests threatens 130,000 species which live nowhere else. This distressing rate of extinction is a global problem, leading to the unavoidable biodiversity education. Conservation biology is a part of science to understand the loss of biodiversity problems and discover solutions. These solutions can be implemented by environmental education (UNESCO).

7.3 Biodiversity Education in Institutions

The Indian National Policy on Education (1986) states that the *Protection of the Environment* must form an essential part of the curriculum at all phases of education. A national curricular framework visualizing the national system of education contains several elements having a common core including a straight attitude on the natural and social environment. This is included not only in the instructional materials but also in the classroom and out of school activities. The curricular area related to work experience can generate a socially useful productive work through direct participation of children in environment-associated field programs, like environmental sanitation, planting and nurturing of trees, etc.

In India, the University Grants Commission (UGC) is responsible for establishing or supporting environmental education (EE) courses at tertiary level. Financial assistance has been provided by UGC for introducing environmental education, and today more than 150 universities are offering courses related with environment at postgraduate and undergraduate level. At lower primary level, biodiversity ideas are introduced with outdoor activities. Children are encouraged to observe the immediate outdoor environmental features and observe birds, plants, and common animals along with nonliving objects like rocks, sand, pebbles, metallic objects, etc. The common local names of the living objects are also taught and encouraged in activities like classification based on similarities and differences. At upper primary level, the curriculum recommended introduction of preliminary concept of laws of nature, dynamics of nature, community functioning, relation between resources and daily activities of life, etc. Children are encouraged in making collections of plant parts, flowers, seeds, fruits; making notes of medicinal plants, their domestic uses, local names, time of flowering; etc. Very simple principles like genetic and species diversity may be demonstrated to children who are encouraged to prepare herbarium sheets out of such plant specimens. They may be guided to prepare lists of domesticated animals; indoor animals like lizards, spiders, cockroaches; etc.

At the high school level, different categories of biodiversity including genera, species, and unique ecosystems are introduced. The students are allowed to get involved in collecting and documenting information on biodiversity elements of their localities. At higher secondary level, more serious study of biodiversity is introduced. Students are assigned with the task of preparing biodiversity registers for specific localities under able supervision and guidance. These tasks are often linked to their optional studies.

At tertiary level, biodiversity should be offered as major or minor subjects involving both theory and practical field studies. Graduate level syllabi can be included with topics like distribution of biodiversity, ecological determinants of biodiversity, inventorying and monitoring biodiversity, human uses and manipulation of biodiversity, intellectual property rights, human impacts on biodiversity, and managing a nation's heritage of biodiversity.

In 1995, renowned environmentalist Professor Madhav Gadgil has suggested ten modules for biodiversity education for both formal and non-formal, and these modules are accepted worldwide. The modules are:

1. Identification of different groups of people in terms of their relationship with the living resources of the region from over which they meet the bulk of their resource requirements, such as fuel wood or grazing for their livestocks
2. Mapping the mosaic of ecological habitats of the study site
3. Recording the different species of plants and animals and their uses known to local people
4. Recording the abundance and distribution of these living organisms in the different habitats of the study site
5. Documenting the ecological history of the study site, especially for the last two decades for which people have excellent recall
6. Recording the ongoing patterns of utilization of the living resources of the study site, to meet subsistence as well as commercial demands and the extent to which different groups of local people as well as outsiders benefit or lose from these uses
7. Documenting ongoing attempts at regulation of uses of living resources, or their conservation, both on part of government agencies and in form of efforts by local communities
8. Recording the development aspirations of local people and how these relate to the diversity of living resources of the region
9. Documenting the agreements as well as differences in the approaches of the different sections of the local communities in their prescriptions for the management of living resources of the study sites
10. Documenting the various emerging options for managing the natural resources of the study sites with a particular focus on conservation of biodiversity

7.4 A Framework of Biodiversity Education Goals by UNEP

The prime goal *: to give support to society in becoming environmentally knowledgeable and, above all, skilled and dedicated citizens who are willing to work individually and collectively toward maintaining or enhancing biodiversity at local, regional, and global levels.*

According to the United Nations Environment Programme (UNEP), biodiversity curriculum has four goal levels, each providing sufficient knowledge about biodiversity to make ecologically reliable and sensible resolutions with respect to biodiversity issues.

Goal level 1: The ecological foundation level

A. Define biodiversity
B. Describe where biodiversity exists in the world
C. Describe the status of biodiversity in their own local and regional area
D. Describe the ecological importance of biodiversity
E. Describe the factors which can reduce biodiversity

Goal level 2: Conceptual awareness level Biodiversity issues and values

A. Current and future biodiversity issues that may exist locally, regionally, or globally
B. The nature of biodiversity issues
C. The nature of human responses to stressful issues
D. The range of values which interrupt on biodiversity issues
E. The values or benefits which can be derived from protecting our (local/global) biodiversity
F. The relationship between human cultural mores and activities (e.g., religious, economic, political, social, etc.) and biodiversity problems and issues
G. The important role of individual behaviors in creating or resolving biodiversity problems or issues
H. The scientific and technological means which exist for managing or protecting biodiversity resources
I. The sociopolitical processes and institutions involved in managing biodiversity problems and issues
J. The local, regional, and global actions which are being implemented and planned to manage biodiversity resources

Goal level 3: The investigation and evaluation level

A. The ability and willingness to monitor trends in their natural and social world and anticipate future implications for biodiversity
B. The knowledge and skills needed to identify and investigate biodiversity issues and to synthesize the gathered information
C. The ability to analyze the critical components of a biodiversity issue including stakeholders, status of science/technology, and values and beliefs of stakeholders
D. The ability to identify alternative solutions and to evaluate the solutions in terms of relevant information about science, stakeholder perceptions (values and beliefs), and available (political, social, economic, natural) resources of implementation of alternatives
E. The ability to clarify their own values and priorities related to a biodiversity issue
F. The ability to identify and tolerate alternative values and priorities held by others in biodiversity issues
G. The ability to participate effectively in group processes to evaluate and select effective solutions which would resolve biodiversity issues and problems

Goal level 4: Action skills level – Training and application

A. The ability to participate effectively in group processes to evaluate, select, and take effective actions which would implement selected solutions to biodiversity issues and problems

B. The ability to identify, evaluate, and select actions which would implement effective solutions to biodiversity problems and issues
C. The ability to effectively implement selected actions regarding biodiversity problem solutions (UNESCO)

7.5 Milestones in the History of Global Biodiversity Education

Several important conferences in relation with biodiversity crisis due to human population growth and lethal environmental changes were held for more than three decades. The outcome from these conferences supported a noticeable development in the theoretical, methodological, and social awareness for sustainable development. The Stockholm Conference on "The Human Environment" in 1972 and the 1977 Tbilisi principles on "Environmental Education" are considered as a starting point. These conferences provided some of the first fundamental proposals on actions to battle against pollution of the environment after focusing the political statements as a reaction to the environmental crisis of the 1970s. In 1983, "World Commission on Environment and Development," commonly referred to as the "Brundtland Commission" and was established by secretary-general of the UN, proposed a global agenda for dealing with the world's environmental problems. Afterwards, they recognized that environmental issues were entangled with social and economic issues and addressed the subject environment and development in large in UN conferences. In 1992, The Rio Earth Summit United Nations Conference on Environment and Development (UNCED) was held, followed by the World Summit on Sustainable Development (WSSD) at Johannesburg in 2002. According to the Rio document "Agenda 21," Chapter 36, education is critical for promoting sustainable development and improving the capacity of the people to address environment and development issues. The United Nations declared the World Decade of Education for Sustainable Development in 2002 for the years 2005 to 2014 to establish the role of education in facilitating the sustainable development. The United Nations Educational, Scientific and Cultural Organization (UNESCO) became the lead agency responsible to define education for sustainable development and promoting national education systems to direct on the way to sustainability (Ulbrich et al. 2010).

7.6 What Is Biodiversity?

> Earth provides enough to satisfy every man's needs, but not every man's greed (Mahatma Gandhi).

Earth is distinct from any other places in the universe because only earth is known to support life. It happens everywhere: on the slopes of high mountains and on the floors of the oceans, in scorching deserts, and at the frigid poles. Life which has been sustained on planet earth appears in countless shapes, forms, and sizes ranging from microscopic bacteria to more complex animals. This assortment of life is known as biodiversity. Biodiversity is both a measure of the variety of life and a sign of the overall health of our planet.

Biodiversity is a contracted form of the term biological diversity. The word biodiversity is relatively new, but the term biological diversity was popularized from 1985 by a number of authors (Chen and Chen 2004). Biodiversity or the variety of life on the earth is the most incredible thing on the planet, which explains how living things interact with each other. It consists of all the species of animals and plants and other life forms and the varieties that exist within each species, also includes range of ecosystems such as deserts, oceans, and tropical rain forests. The genetic variations that make each individual unique are the part of biodiversity.

Biodiversity is the foundation of life on earth and is extremely complex, dynamic, and varied. It is very crucial for the functioning of all ecosystems, life forms, and microorganisms because they depend on each other and cannot exist without its services and products. It is an interdependent, closely interlinked life system that binds all life on earth together in a delicate and intricate web where each strand plays a crucial role and depends upon the other for survival. A loss of even one strand robustly affects human well-being and well-being of every other living creature that leads to endangering life on earth.

All the fundamentals of biodiversity have led to a number of proper definitions. Eighty-five dissimilar definitions were reviewed by Delong (1996). But mainly important and widely accepted definition is included within the *Convention on Biological Diversity* (CBD). This treaty was signed by over 150 nations at the United Nations Conference on Environment and Development, held in Rio de Janeiro, Brazil, in 1992. CBD defines biodiversity as "the variability among living organisms from all sources including, inter alia [among other things], terrestrial, marine and other aquatic ecosystems and the ecological complexes of which they are part; this includes diversity within species, between species and of ecosystems."

Estimation of biodiversity is one of the supreme challenges to mankind and requires millions of days with highly skilled biologists. Current biodiversity is estimated at 10–100 million species, including 1.7 million formally catalogued. Of these around one million is animal species (dominated by insects – around 5–10 million); 2,48,000 higher plants, 69,000 fungi, 31,000 protozoa, 27,000 algae, 5000 bacteria, and 1000 viruses. Table 7.1 describes the relative number of described species in major taxa.

Importance of biodiversity measurement arises from the fundamental questions related with biodiversity such as where it occurs, how fast it is disappearing, or how it can be maintained. Although biodiversity can be measured in different ways, the most commonly used measure is that of species richness. Species richness (number of species in a given area) represents a single but important metric that is valuable as the common currency of the diversity of life, but to fully detain biodiversity, it must be integrated with other metrics. According to Purvis and Hector (2000) and

Table 7.1 Relative number of described species and estimated number in major taxa

No.	Taxon	No. of described species	Estimated no. of species
1.	Viruses	4000	4,00,000
2.	Bacteria	4000	1,00,000
3.	Fungi	72,000	15,00,000
4.	Protozoa	40,000	2,00,000
5.	Algae	40,000	4,00,000
6.	Plants	2,70,000	4,00,000
7.	Nematodes	25,000	4,00,000
8.	Crustaceans	40,000	1,50,000
9.	Arachnids	75,000	7,50,000
10.	Insects	9,50,000	80,00,000
11.	Mollusks	70,000	2,00,000
12.	Vertebrates	45,000	50,000
13.	Others	1,15,000	2,50,000

Gaston and Spicer (2004), there are a number of reasons mentioned as a primary measure for biodiversity:

1. *Species often keep their genes to themselves and thus can have independent evolutionary trajectories and unique histories; it thus makes biological sense to measure species richness rather than a higher taxonomic grouping.*
2. *It is often easier to count the number of species compared to other measures of biodiversity. Humans tend to be able to recognize species, and these are the units typically used in folk knowledge, practical management, and political discourse. Humans can visualize variation in biodiversity as variation in species richness.*
3. *There is a substantial body of information already available on species, for example, in museums and herbaria.*
4. *Species richness can act as a "surrogate" for other measures of biodiversity. In general as long as the number of species involved is moderate, greater numbers of species will tend to have more genetic diversity and will tend to have greater ecological diversity as more niches, habitats, or biomes will be represented.*

7.7 Levels of Biodiversity

Biodiversity refers to life consisting of three levels genetic, species, and ecosystem diversity. *Genetic diversity* is the discrepancy of genetic composition of individuals within a species. Genes are the fundamental unit of heredity passed from parents to offspring that contains special codes and instructions to determine individual's characteristics. Because they have distinctive genotype, each individual showed some dissimilarity from others in some of the characteristics, which has given us beautiful nature around us. Greater genetic diversity has been seen in larger populations in comparison with smaller populations. Similarly, species composed of many populations typically have more genetic diversity than do species limited to one or few populations. Although we have many techniques necessary to monitor the amount of genetic variability in a population or species, we still lack the resources to distinguish the genotype of single individuals. Genetic diversity is generally compared among populations by characterizing the amount of heterozygosity within every population. Heterozygosity is calculated by investigating specific traits to determine the presence of multiple alleles in that population for that trait. If all individuals in a population show the same genetic blueprint for a particular trait, then the population is known as homozygous for that trait, and heterozygous means more than one genetic pattern is evident in the population for that trait. Increased heterozygosity increases genetic diversity.

Species diversity can be defined as the number of species within a particular ecological area. Even though the number of species increases, species diversity also functions as a representation of the size of populations. More varied ecosystems have numerous species with the equal population sizes. Another type of ecosystem having the same number of species but having a few species with large populations and the rest with very small populations would be less diverse. It is understood that all species have significant role in the structure and function of ecosystems.

Ecosystem diversity refers to different types of habitats. Ecosystems have both biotic and abiotic components, composed of structural and functional relationships. The ecosystem structure is provided by living organisms, and the functional relationship is something like the pathways by which energy and nutrients move through the ecosystem. Variety of ecosystems and its abundance in a particular area will define ecosystem diversity. Even as the concept of an ecosystem is easy to imagine, it may be hard to identify where one ecosystem ends and another begins, because the areas of unique ecosystems overlap to form important ecological areas known as *ecotones*. *Ecotones* contain a diversity of species because they enclose representatives from each neighboring ecosystem as well as species. The diversity of ecosystems constitutes a biome. For example, coral reefs are the most diverse aquatic biome and may even support a greater diversity of species than do tropical rain forests.

Biodiversity that we see around us today is the outcome of over 3.5 billion years of evolutionary development, shaped by natural processes and increasingly by human influence. It sustains the web of life and we fully depend on it to meet our food, healthcare, and other needs. Conserving biodiversity is basic to our survival and well-being and using it sustainably forms part of our culture and lifestyle. Biodiversity along with ecosystem services contributes poverty eradication and national development.

Biodiversity is not distributed evenly across the globe. Certain countries, lying mostly in the tropics, are characterized by majority of species richness and more number of endemic species. They account for 60–70 % of the world's biodiversity.

Table 7.2 The 17 megadiverse nations of the world with their biodiversity

Countries	No. of mammals	No. of birds	No. of reptiles	No. of amphibians	No. of angiosperms
Australia	255	740	686	197	23,000
Brazil	428	622	467	516	55,000
China	394	1195	270	265	27,000
Columbia	359	1721	383	407	45,000
Ecuador	280	1447	345	358	15,000
India	350	1200	353	197	15,000
Indonesia	515	1519	600	270	20,000
Madagascar	131	250	269	144	10,000
Malaysia	293	1200	294	171	15,000
Mexico	449	1010	717	282	25,000
Peru	361	1750	297	251	20,000
Congo	409	1086	280	216	10,000
South Africa	243	800	370	117	20,300
Papua New Guinea	250	762	293	197	9000
Philippines	240	556	186	74	8120
United States	479	654	8750	230	20,000
Venezuela	323	547	322	287	21,073

These are called *megadiverse countries*. Seventeen megadiverse countries are identified by Conservation International in 1998. In 2002, another organization was formed based on interests and prim concerns for the preservation and sustainable use of biodiversity and traditional knowledge. But this organization does not include all megadiverse countries recognized by Conservation International (www.conservation.org). Table 7.2 details the 17 megadiverse nations and their biological diversity.

7.7.1 Biodiversity Hotspots

Biodiversity hotspot concept was defined by British ecologist Norman Myers (1988). Hotspot regions enclose great diversity of endemic species, specific to geographical area and at the same time drastically impacted and distorted by human activities. It must support 1500 endemic plant species or 0.5% of the total global area. For consideration as a biodiversity hotspot, a region must be with a loss of more than 70% of its original habitat. The existing primary vegetation is the basis for evaluating human impact in the region. There are 35 hotspots all over the world covering only 1.4% of global area and 17.3% of earth's land surface containing 44% of plant species and 35% of terrestrial animal species (Fig. 7.1).

Fig. 7.1 Map indicating the 35 global biodiversity hotspots (Courtesy: https://www.e-education.psu.edu/geog030/node/393)

7.8 Relevance of Biodiversity

Biodiversity is essential because it generates benefits, related with the ecological functions of biodiversity in sustaining viable populations along with ecosystems. This knowledge has been very slow to recognize by human species. In fact the quality and existence of our own lives depend on the long-lasting functions of earth ecosystems. Biodiversity is one of a critical constituent of our environment. Every species in the ecosystem has some role and some are very crucial; not all species execute equally important functions. It is very urgent that we act conservatively to maintain the highest level of diversity because our ecological science has not yet recognized the relevant importance of all species in our ecosystem.

Earth really contains about 50 million different species. If some are lost or extinct, the total number would be affected only minutely. If so, why conservation of biodiversity has extreme importance today? The enlightenment of web of life gives us the answer for this question. The web of life or food chain links each and every form of life on earth. All life on earth is mutually supporting and man is only a thread in the delicate web of relationships. If a bond in the food chain is broken, it causes a series of counterblasts and man himself moves closer to his destruction. Thus the importance of each and every creature in the web is evident. Damaging the food chain only produces negative results, leading to the devastation of species. This shows why biodiversity and its constituents are crucial to sustain the balance of nature.

The opportunity for the production of new varieties is greater when the diversity of wild species is superior. Nearly all crops and livestock at present we depend for food production are hybrids; they are originating from wild species. These hybrid species are vulnerable when compared with wild ancestors; it may get totally wiped out due to any anomalous climatic variation. The greater the variety of wildlife we conserve, the more will be able to learn from it.

Beyond ecological functions, we also receive certain important benefits from biodiversity. These additional benefits we discover from biodiversity consist of medicines and drugs, alternative foods, energy, building supplies, and many other things that support ourselves economically. Many of these resources have not yet been exploited to their complete potential. The loss of these precious biodiversity resources would change our excellence of life and human survival.

The biodiversity resides greatly on the tropical rain forests. It contains the most diverse terrestrial ecosystems and also has the status of most threatened. However, the protection of biodiversity in less species-rich ecological unit is also equally significant (Reid and Miller 1989). These ecosystems supply the resources requisite by the local people for continued existence. The products used by humans like food, medicines, and other products have come from almost every ecosystem from each corner of the earth. At present researchers are exploring many naturally occurring edible food plants (approximately 75,000) in existence that can be replaced by predominate food plants like wheat, corn, rye, and soybeans (Wilson 1988). Due to the diversity of plant species, the tropics hold potential supply of alternative food sources. Coconuts, peanuts, and sugar are some examples of the food items which originated in the tropics. Researchers point out that a lot of underexploited tropical species will become more habitually used in the future (Plotkin 1988).

A great deal of the energy used on this planet is originated from living or long-dead plants. Approximately 1.5 billion people, across all cultures, used wood to meet 90% of their fuel needs (Eckholm 1975). If appropriately managed, wood can be used as a source of renewable energy. But, in many parts of the world, wood supplies cannot be reloaded fast enough to meet the increasing energy demands. In developed nations nonrenewable fossil fuels, product of the natural changes in the earth's diversity around tens of millions of years ago, are the primary supply of energy. Renewable energy sources excluding wood will reduce our need on nonrenewable resources that may incline the burden on our forests. For example, all around the world about 60 million tons of rice husks are produced annually that can be used as an alternative energy source (Beagle 1978). Another significant application of biodiversity is in the form of wood products for construction where it can be replaced by variety of other building materials which will reduce the utilization of traditional wood and wood products. An alternative method suggested by Darrow and Saxenian (1986) is the use of cement made from rice husk ashes combined with lime.

7.9 Threats to Biodiversity

Due to overexploitation and lack of knowledge, there is a drastic change in our environment which is necessary for the persistent survival of biodiversity. The damage to environment has arrived at a dangerous point and is addressed as biodiversity loss. Natural environment is changing due to human activity leading to mass extinctions of species. This is partly due to other environmental factors like ecological substitutions, biological factors, and pathological causes.

One of the chief threats to biodiversity is *habitat destruction*. Massive areas of woody lands are emptied annually. It has been estimated that annual deforestation is around 11.3 million hectares (Lanly 1982). Urbanization and desertification are the main reasons of agricultural land loss. Aquatic habitats are exhausted as a result of chemical fertilizer revolution. Contemporary styles of transportation, efficient machines to clear the land, and some government policies have speed up habitat destruction and fragmentation.

Habitat fragmentation happens when ecosystems are broken down into smaller parts, habitually isolated from each other. Such type of fragmentation may cause isolation and changes in climatic factors. By the end of this century, disappearance of thousands of species occurs due to fragmentation of large natural ecosystems (Lovejoy 1980). Destruction of habitat is starting from urban areas for widening the urban circles. One example is the ruin of the Atlantic Forest of Brazil that was once found in the area occupied by the city of Sao Paulo. When the city has grown, further areas of the forest have been cracked, and the species coupled with this region became extinct earlier than they were illustrated by scientists (Murphy 1988). In the tropical and temperate areas of the world, results of such type of habitat destruction will have shown contradictory effects on plant, mammal, and bird distributions. Many plants and birds are good colonizers and therefore may not be as unfavorably affected as mammals. Habitat destruction, predominantly deforestation and drainage of wetlands and urbanization of coasts, has been connected to landownership patterns and government policies. In many countries, a disproportionately large percentage of the land has been owned by a small percentage of the population. For example, in Guatemala, 63% of the agricultural land has been claimed by only 2% of the landowners (Brough 1991). The government provides homestead marginal agricultural land to landless individual to improve their livesout of frequent influence. Damming or draining in wetland ecosystems for crop production is another means of habitat destruction. In developed and undeveloped countries, governments provide financial assistances and tax credits for certain industries like logging industry and cattle ranching industry that take on in deforestation.

Many scientists predict that within the next 50–100 years, global warming is expected to increase quite a lot of degrees. When compared to normal fluctuations, a change in 2 °C is considered as large by scientists, and it is expected it would leave the planet warmer than at any time in the past 100,000 years (Schneider and Londer 1984). Precipitation and temperature due to global warming affected species ranges and mortality. The fossil records demonstrate that the people of the past have compacted with climate changes by shifting their ranges to match their most favorable climate (Peters 1988). Changes in the universal precipitation pattern may result in shifts in agriculture lands. As a result of climate changes, many of agricultural cultivars may not be as productive or even viable because most of the agricultural products have a particular geographic area for their cultivation. This defines the need for the preservation of the genetic diversity of food species that may get adapted to a new changed atmosphere. A major climatic change may create new plant and animal associations by mixing species of natural communities (Van Devender and Spaulding 1979). Due to stressful interactions between species, these new associations do not normally occur together and could lead to the extinction of species.

Chemical spills, discharge of sewages, and inappropriate use of pesticides and herbicides can result in the destruction of ecosystems and species. For example, residual wastes from industries contain dioxin and other toxic substances. However, detectable effects on an organisms' health in natural ecosystems have not been found (Anthony and Kozlowski 1982, Dressler et al. 1986). Polychlorinated biphenyls, sulfur dioxide, and pesticides are the major toxic by-products of urban areas which may cause a threat to biodiversity (Ehrlich and Ehrlich 1981). Air and water pollution may afflict ecosystems and diminish the size of susceptible populations. In Europe and the USA, air pollution has been associated to forest diebacks. Air pollution along with acid rainfall may weaken tree species by getting infected with pathogens diminishing the productivity of that area (Reid and Miller 1989). Acid precipitation also troubles aquatic ecosystems by altering the pH of the water, resulting in the death of pH-sensitive species.

The introduction of foreign species substituting local and native species is mentioned as one of the largest reasons of biodiversity loss. Foreign species replace and frequently affect the extinction of inhabitant species. Introduction of invasive plant and animal species caused an estimated annual economic damage of about US$ 1.4 trillion. The effects of introducing tigers, lions, and elephants into New York City would positively enhance the diversity of the city, but it causes certain significant problems to the native species by competing with them for food and shelter, by breeding with them and destroying unique genetic types, or by altering predator-prey interactions. Organization for biodiversity does not mean biologists should introduce as many species as feasible into an area. The most disruptive species included mammalian herbivores, vertebrate and invertebrate predators, and species that can simply attack new ecosystems (Vitousek 1988). An example is the mongoose which was introduced to Caribbean islands to manage snake populations and the Old World rat, which was introduced coincidently. The result of this introduction was that snake numbers were only partially controlled. Mongooses preyed greatly on resident species of birds, mammals, and reptiles and also removed a number of species (Robinson and Bolen 1984).

Ecosystem degradation and killing of wildlife populations through direct human intervention are one of the major factors of biodiversity loss. These have been destroyed as a part of forest management, agricultural practices, and urbanization. Consequently, many of the ecosystems and the species associated with them may not get replaced. A large number of species are threatened by hunting, poaching, and illegal trade. In 1979, 1.3 million elephants were estimated in Africa, but today the numbers came down drastically to 6,50,000. Natural disturbances and destructions have always been a dynamic force of all natural system. Occasionally biodiversity can be conserved by allocating natural changes, rather than by preventing them.

According to current estimates, an average of 9,00,000 vertebrate species got extinct after every one million years, means of 90 species in every century. In the case of higher plants, it is estimated that one species has become extinct every 27 years. In recent years, human interventions such as destruction of lowland forests and introduction of predators, diseases, and weedy plants in oceanic islands have undergone mass extinctions, because their area is small and the population of species is also limited. About 75% island-dwelling birds and mammals become extinct in the past several years. Coral reefs are the home for larger variety of marine species than any other marine environment. These are the marine equivalents of rain forests; the movement of trawler to them is equivalent to that of bulldozer through rain forests. Rapid destruction of the tropics and other diverse ecosystems raise a genuine fear that the world's biodiversity is under risk of extinction in the next 20–30 years. In the last 400 years, 120 species of mammals and 225 species of birds have become extinct in the world.

7.10 IUCN Categories of Threat

All species in plants and animals form a closed web in a stable ecosystem and form a state of dynamic equilibrium. However some evolutionary and geological changes under natural conditions lead to habitat loss and sequentially lead to the extinction of species. Accordingly a good number of beautiful and interesting plant and animal species are vanishing forever. The International Union for the Conservation of Nature and Natural Resources (IUCN) is the world's chief authority on the conservation status of species. The IUCN Red List of Threatened Species is established in 1964, differently known as the *IUCN Red List* or *Red Data List*. It is the world's most complete record of the global conservation status of biological species. Species are categorized by the IUCN Red List into nine groups, based on the criteria such as rate of decline, population size, area of geographic distribution, and degree of population and distribution fragmentation.

The latest update of the IUCN Red List of Threatened Species shows that 17,291 species out of the 47,677 assessed are threatened with extinction. The results reveal that 21% of all known mammals, 30% of all known amphibians, 12% of all known birds, 28% of reptiles, 37% of fresh water fishes, 70% of plants, and 35% of invertebrates assessed so far are under threat. The accepted categories of existence or extinction of groups of plants and animal species are given below.

Extinct (EX) – An extinct taxon means those species which are not found after searches in known and or expected habitat at appropriate times (diurnal, seasonal, annual); its historic range have failed to record, and there is no reasonable doubt that the last individual has died. Surveys should be over a time frame appropriate to the taxon's life cycle and life form.

Extinct in the wild (EW) – A taxon is extinct in the wild when it is known only to survive in cultivation, in captivity, or as a naturalized population well outside; its historic range and exhaustive surveys in known and or expected habitat, at appropriate times (diurnal, seasonal, annual), have failed to record an individual. Surveys should be over a time frame appropriate to the taxon's life cycle and life form.

Critically endangered (CR) – A taxon is critically endangered when the best available evidence indicates that it meets any of the criteria for critically endangered and is therefore considered to be facing an extremely high risk of extinction in the wild.

Endangered (EN) – A taxon is endangered when the best available evidence indicates that it meets any of the criteria for endangered, and it is therefore considered to be facing a very high risk of extinction in the wild.

Vulnerable (VU) – A taxon is vulnerable when the best available evidence indicates that it meets any of the criteria for vulnerable, and it is therefore considered to be facing a very high risk of extinction in the wild.

Near threatened (NT) – A taxon is near threatened when it has been evaluated against the criteria but does not qualify for critically endangered, endangered, and vulnerable now, but is close to qualifying for or is likely to qualify for a threatened category in the near future.

Least concern (LC) – A taxon is least concern when it has been evaluated against the criteria and does not qualify for critically endangered, endangered, vulnerable, or near threatened. Widespread and abundant taxa are included in this category.

Data deficient (DD) – A taxon is data deficient when there is inadequate information to make a direct, or indirect, assessment of its risk of extinction based on its distribution and/or population status. A taxon in this category may be well studied, and its biology well known, but appropriate data on abundance and/or distribution are lacking. Data deficient is therefore not a category of threat. Listing of taxa in this category indicates that more information is required and acknowledges the possibility that future research will show that threatened classification is appropriate. It is important to make positive use of whatever data are available. In many cases great care should be exercised in choosing between DD and a threatened status. If the range of a taxon is suspected to be relatively circumscribed, and a considerable period of time has elapsed since the last record of the taxon, threatened status may well be justified.

Not evaluated (NE) – A taxon is not evaluated when it is has not yet been evaluated against the criteria (http://www.iucnredlist.org).

The equilibrium of nature is depended on the wildlife population. Wildlife becomes extinct or its number declines when equilibrium of nature is cracked. The vanishing of wild species can alter web of life, disturbing the balance between predator and prey thus in turn causes the multiplication of insect pests. Bullfrogs are the main predators of malaria-spreading mosquitoes, and its overharvesting causes the upsurge of disease.

Out of 350 species of mammals, 116 are generally endangered. There are about 2100 species and subspecies of birds which are endangered. Of the reptiles, the endangered ones are a species of tortoises, five species of lizards, and three species of crocodiles and turtles. IUCN has estimated that about 10% of the world's vascular plant species are in varying levels of threat.

7.11 Convention on Biological Diversity

The Convention on Biological Diversity (CBD) also known as Biodiversity Convention is a multilateral treaty, for protecting biological resources. This agreement was put forward at the time of the Earth Summit in Rio de Janeiro on 5 June 1992, where native groups and NGOs from all around the world assembled for the largest international environment meeting. The agreement came into effect on

29 December 1993. All ecosystems, species, and genetic resources come under this contract. The chief objectives of the treaty are:

1. Conservation of biological diversity or biodiversity
2. Sustainable use of its components
3. Fair and equitable sharing of benefits arising from genetic resources

In other words, its aim is to expand national policies for the conservation and sustainable use of biological diversity. It is frequently observed as the authenticate document concerning sustainable development. One hundred ninety-five states and the European Union are parties (governments of the countries) to join the convention. That means almost all of the world's governments have dedicated to achieving the goals of the convention. All UN member states with the exception of the USA have approved the treaty. The convention explains how parties should turn these goals into action. It also recommends parties to help each other by sharing resources and technologies.

Cartagena Protocol on Biosafety, a biosafety protocol, is an additional agreement to the CBD, adopted in January 2000. It deals with technology development and transfer, benefit sharing, and biosafety issues. This protocol is to protect biological diversity from the potential risks caused by living modified organisms arising through modern biotechnology.

Central body of the convention is the *Conference of the Parties* (COP), containing of all governments and organizations that have approved the treaty. This vital authority evaluates growth and developments of the convention, recognizes new priorities, and sets work strategies for members. The other functions of COP are making amendments to the convention, creating skilled advisory bodies, reviewing progress reports by member nations, and collaborating with other international organizations and agreements. The COP uses proficiency and support from supplementary bodies that are established by the convention.

CBD's office known as *Secretariat* is located in Montreal, Canada, and it is controlled by the United Nations Environment Programme (UNEP). Its key functions are to help the parties by organizing meetings, preparing documents, assisting the implementation of the program of work, coordinating with other international organizations, and collecting and publishing information.

Another organization comes under CBD is the *Subsidiary Body on Scientific, Technical and Technological Advice* (SBSTTA), a committee consisting of professionals from member governments proficient in relevant areas and performs important role in creating recommendations to the COP on scientific and technical issues.

National Biodiversity Strategies and Action Plans (NBSAPs) are the chief agency for executing the convention at the national level, which come under Article 6. The main objective of this body is the preparation of national biodiversity strategy at country level and make sure that this strategy is in a streamline to scheduling the activities of all sectors and ensuring that these activities should have a significant effect on biodiversity.

7.12 Issues Under the CBD

Some issues under the convention include:

1. Measures and incentives for the conservation and sustainable use of biological diversity
2. Regulated use to genetic resources and traditional knowledge, including prior informed consent of the party providing resources
3. Sharing, in a fair and equitable way, the results of research and development and the benefits arising from the commercial and other utilization of genetic resources with the contracting party providing such resources (governments and/or local communities that provided the traditional knowledge or biodiversity resources utilized)
4. Access to and transfer of technology, including biotechnology, to the governments and/or local communities that provided traditional knowledge and/or biodiversity resources
5. Technical and scientific cooperation
6. Coordination of a global directory of taxonomic expertise (Global Taxonomy Initiative)
7. Impact assessment
8. Education and public awareness
9. Provision of financial resources
10. National reporting on efforts to implement treaty commitments

7.13 Challenges of Biodiversity Education

Education has been recognized as a major tool to attain sustainability and biodiversity protection by changing human approaches in connection with nature (Ehrlich and Pringle 2008). Education has great opportunities to contribute by serving citizens to become well informed, decisive, and competent and as a result become able to work in favor of biodiversity (Dreyfus et al. 1999).

Since the realization of the CBD, a number of assessments have been carried out to understand the intensity of awareness on biodiversity in different countries, suggesting that education, extensions, and public awareness approaches are failing to attain the interest and inspiration for people to act in favor of biodiversity conservation, and also the significance of the values of sustaining biodiversity is not getting across. One of the current international surveys conducted for the Secretariat of CBD discloses the need of informing and empowering the future generations (Airbus Report 2010). The survey was conducted in 2010, sampled 10,000 children between the ages of 5 and 18 from ten countries. First 40% of the sample is interested in watching TV or playing computer games, and only 4% is careful about the environment, and additionally 9% are considered as looking after animals as most important (CBD press release 2010). These results recommend that biodiversity education has not been successfully penetrating different parts of society. Public and governmental authorities should take action considering biological resources as a relevant issue.

Low-level awareness regarding biodiversity conservation has increasingly achieved more relevance in national and international programs. International agreements like CBD and organizations like the International Union for the Conservation of Nature and Natural Resources (IUCN) or the World Wildlife Fund have been established and continually worked for protecting and sustaining the natural biodiversity. According to CBD the effective actions to prevent biodiversity loss should depend on communication, education, and awareness strategies to make sure that "everyone understands the value of biodiversity and what steps they can take to protect it, including thorough changes in personal consumption and behavior" (SCBD 2010).

To fulfill the CBD's objectives, the CBD recognizes public education and awareness as a crucial tool. Specifically, Article 13 recommends to the contracting parties media and conservation-related topics as part of educational programs which helps in promoting and encouraging biodiversity protection (CBD – Article 13 2006). As a part of the country's national biodiversity strategy, the CBD has established the *Communication, Education and Public Awareness* (CEPA) program. Its chief objective is communicating and raising awareness about biodiversity by incorporating it into the education systems of CBD's all participants. Environmental education (EE) and Education for Sustainable Development (ESD) were recognized as strategies to deal with environmental concerns through education, each emerged at different times and situations. Environmental education was first defined by Stapp (1969) as a new approach in generating scientific awareness and imparting knowledge on nature and life.

The fundamental reasons of biodiversity loss come from social, economic, political, cultural, and historical features of each civilization (WEHAB working group 2002). This is due to underprivileged governance to lack of knowledge and awareness regarding the significance of biodiversity. Both EE and ESD recognize the importance of biodiversity education with relations and interdependencies between environmental and socioeconomic issues.

According to Navarro-Perez and Tidball (2012), biodiversity educational field faces four main challenges. The first challenge is defining the approach for biodiversity education and understanding how the nature and approaches of both EE and ESD programs can potentially manipulate biodiversity education. Many educators have approved and considered EE as a multidisciplinary approach of education that focuses on nature, environment, and society as mutually supporting and indivisible bodies, although EE has been very environmentally focused, failing to illustrate the combined effects that lead to environmental change. ESD also highlights interconnections between society, economy, and the environment and has been considered as a good approach by including problems of ethics and equity along with new forms of thinking and learning (Hesselink et al. 2000). Even though the implementation of sustainability education can be challenging for both schools and teacher education, some type of biodiversity education mixed the activities and machineries from both EE and ESD.

The second challenge refers to the complexities to create both educators and learners in managing an idea that is considered as ill defined. In addition, its multi-dimensional nature connecting to social, economic, and environmental relations makes it a hard concept to pass on effortlessly and significantly to learners (Wals 2001). Integration of biodiversity as an educational subject will also depend on the theoretical framework of the educators.

The third challenge refers to the significance of getting different and broad listeners through important messages. Different surveys on public approaches around the world show that the importance of biodiversity conservation messages is not getting across. This means that the community requests to be more engaged. The awareness can be raised through non-formal educations, and communications related with biodiversity also motivate all levels of society.

Lastly, the fourth main challenge for biodiversity education is to reconnect people and nature. Because of urbanization most people live in urban areas having less connection with nature. Some authors have suggested that education must focus on increasing contact with nature in childhood and youth through a variety of activities (Louv 2005; van Weelie and Wals 2002; Lindemann-Mathies et al. 2009). These early contacts influence people to increase their curiosity in nature (Chawla and Cushing 2007).

These challenges can be overcome by increasing the public knowledge and awareness about biodiversity. Along with these challenges, biodiversity education should educate biodiversity in its different meanings and dimensions. It would enable the learners to develop critical thinking skills about biodiversity and its protection through multilevel understanding and analyzing. These abilities can authorize learners and help them understand their prospective actions according to their own interests and concerns. These four challenges detailed above indicated that biodiversity educational programs should highlight experimental and social learning so as to promote a new "concern" in relationship with nature.

References

Airbus (2010) Bio-index Report. Published by CBD

Anthony RG, Kozlowski R (1982) Heavy metals in tissues of small mammals inhabiting wastewater irrigated habitats. J Environ Qual 11:20–22

Beagle E (1978) Rice husks conversion to energy, FAO agricultural services Bull. No. 31. UNIPUB, New York

Brough HB (1991) A new lay of the land. World Watch 4(1):12–19

CBD (2006) Article 13. Public education and awareness. Retrieved from http://www.cbd.int/convention/articles.shtml?a=cbd-13

CBD Press Release (2010) Alarming global survey on children's perceptions on nature. Retrieved from http://www.cbd.int

Chawla L, Cushing FD (2007) Education for strategic environmental behavior. Environ Educ Res 13(4):437–452

Chen ZX, Chen SY (2004) Nematology: advances and perspectives, Vol 1. Donald Ward Dickson publishers, p 439

Convention on Biological Diversity. Retrieved from www.cbd.int

Convention on Biological Diversity: Global Taxonomy Initiative (1996) Retrieved from www.cbd.int/gti
Darrow K, Saxenian M (1986) Appropriate technology sourcebook. By Volunteers in Asia. Stanford, p 800
Dressler RL, Storm G, Tzilkowski WM, Sopper WE (1986) Heavy metals in cottontail rabbits on mined lands treated with sewage sludge. J Environ Qual 15:278–281
Dreyfus A, Wals AEJ, van Weelie D (1999) Biodiversity as a postmodern theme for environmental education. Can J Environ Educ 4:155–175
Eckholm E (1975) The other energy crisis: firewood, World watch Paper 1. World watch Institute, Washington, DC
Ehrlich PR, Ehrlich AH (1981) Extinction: the causes and consequences of the disappearance of species. Random House, New York, p 305
Ehrlich PR, Pringle RM (2008) Where does biodiversity go from here? A grim business as-usual forecast and a hopeful portfolio of partial solutions. PNAS 105(1):11579–11586
Gaston KJ, Spicer JI (2004) Biodiversity: an introduction, 2nd edn. Blackwell publishers, Malden
Hesselink F, van Kempen PP, Wals AEJ (2000) ES Debate, International debate on education for sustainable development. IUCN. Gland
Issues under CBD. Retrieved from https://www.cbd.int/gbo1/chap-02.shtml
IUCN Red list of threatened species. Retrieved from http://www.iucnredlist.org
Lanly JP (1982) Tropical forest resources, FAO Forestry Paper 30. Food and Agricultural Organisation, Rome, p 106
Lindemann-Matthies P, Constantinou C, Junge X (2009) The integration of biodiversity education in the initial education of primary school teachers: 4 comparative studies from Europe. Educ Environ Res 15(1):17–37
Louv R (2005) Last child in the woods: saving our children from nature-deficit disorder. Algonquin Books of Chapel Hill, Chapel Hill
Lovejoy TE (1980) A projection of species extinctions. Journal of Council on Environmental Quality and the U.S. Dept. of State. The global 2000 report to the President: entering the twenty first century. U.S. Govt. Printing Office, Washington, DC, pp 328–331
Megadiverse countries. Retrieved from http://www.conservation.org/documentaries/Pages/mega-diversityaspxMegadiversity: The 17 Biodiversity Superstars
Murphy DD (1988) Challenges to biological diversity in urban areas. In: Wilson EO (ed) Biodiversity. National Academy Press, Washington, DC, pp 71–78
Navarro-Perez M, Tidball KG (2012) Challenges of biodiversity education: a review of education strategies for biodiversity education. Int Electron J Environ Educ 2(1):13–30
Peters RL (1988) The effect of global climatic change on natural communities. In: Wilson EO (ed) Biodiversity. National Academy Press, Washington, DC, pp 450–464
Plotkin MJ (1988) The outlook for new agricultural and industrial products from the tropics. In: Wilson EO (ed) Biodiversity. National Academy Press, Washington, DC, pp 106–116
Purvis A, Hector A (2000) Getting the measure of biodiversity. Nature 405:212–219
Reid WV, Miller KR (1989) Keeping options alive: the scientific basis for conserving biodiversity. World Resources Institute, Washington, DC, p 128
Robinson WL, Bolen EG (1984) Wildlife ecology and management. Macmillan Publ. Co, New York, p 478
Schneider SH, Londer R (1984) The co-evolution of climate and life. Sierra Club Books, San Francisco, p 563
Secretariat of the CBD (2010) Global biodiversity outlook 3. Montreal, 94 pages. Retrieved from http://www.cbd.int/gbo3/ebook
Stapp WB (1969) The concept of environmental education. J Environ Educ 1(30):33–36
Ulbrich K, Settele J, Benedict FF (2010) Biodiversity in education for sustainable development – reflection on school-research cooperation. Pensoft Publishers, Sofia/Moscow

UNESCO – Environmental Education Module: Biological Diversity. Retrieved from http://portal.unesco.org/education/en/ev.phpURL_ID=41439&URL_DO=DO_TOPIC&URL_SCTION=201.html

Van Devender TR, Spaulding WG (1979) The development of vegetation and climate in the southwestern United States. Science 204:701–710

Van Weelie D, Wals EJ (2002) Making biodiversity meaningful through environmental education. Int J Sci Educ 24(11):1143–1156

Vitousek PM (1988) Diversity and biological invasions of oceanic islands. In: Wilson EO (ed) Biodiversity. National Academy Press, Washington, DC, pp 181–189

Wals AEJ (2001) Biodiversity as a bridge between nature conservation education and education for sustainability. Roots 23

WEHAB Working group (2002) A framework for action on biodiversity and ecosystem management. World summit on sustainable development

Wilson EO (1988) The current state of biological diversity. In: Biodiversity. National Academy Press, Washington, DC, pp 3–20

Trends in Biodiversity Documentation and Assessment

8

A.R. Sivu and N.S. Pradeep

Abstract

Biodiversity is the variability among living organisms from all sources, including terrestrial, marine, and other aquatic ecosystems and the ecological complexes of which they are part; this includes diversity within species, between species, and of ecosystems. This diversity of living creatures forms a support system, which has been used by each civilization for its growth and development. The rich biodiversity has been instrumental in providing humanity with food security, health care, and industrial goods that has led to high standard of living in the modern world. The diversity of life on earth is very rich that if we use it sustainably and judiciously, we can go on developing new products for generations. For this reason, we must realize the importance of biodiversity as a major resource. Many traditional societies have played significant roles in conserving biodiversity. They value biodiversity as a part of their livelihood through cultural and even sometimes religious sentiments. Assessing the status and trends of biodiversity is essential for sustainable development strategies at all levels, from village to nation to region. Ecological communities maintain the ecological and evolutionary processes that sustain life. These are necessary to help maintain the planet's chemical balance, moderate climate, renew soil, and conserve species diversity. They are also the source of all biological wealth-supplying food, raw materials, medicines, recreational resources, and a store of other goods and services worth many billions of dollars per year. The genetic stocks within crop varieties, livestock breeds, and their wild relatives provide essential traits for increasing and

A.R. Sivu (✉)
Department of Botany, NSS College, Nilamel, Kollam 691535, Kerala, India
e-mail: sivuar@gmail.com

N.S. Pradeep
Division of Microbiology, Jawaharlal Nehru Tropical Botanic Garden and Research Institute, Palode, Thiruvananthapuram 695562, Kerala, India
e-mail: drnspradeep@gmail.com

S. Abdulhameed et al. (eds.), *Bioresources and Bioprocess in Biotechnology*,
DOI 10.1007/978-981-10-3573-9_8

improving agricultural production and the development of biotechnologies. Biodiversity assessment is the first step in protecting the complete range of morphological and genetic diversity of species on Earth and in reaching the goals of conservation biology. Assessment begins with identifying organisms that make up biological communities and understanding evolutionary histories. Scientific advancements in molecular sequencing can help clarify and provide support for identifications. This chapter discusses the value of biodiversity, trends in documentation, and the tools and techniques commonly adopted for biodiversity.

Keywords
Biodiversity • Trends • Characterization • Assessment

8.1 Introduction

Biodiversity cannot be defined linearly as diversity of species. It is on the contrary, concerned much more about how the environment is diversely distributed through with biotic and abiotic (i.e., physical, chemical, and other nonliving environmental factors) components and how they sustain themselves through various processes of ecological interaction. Biodiversity is defined as "the variability among living organisms from all sources including terrestrial, marine and other aquatic ecosystems and the ecological complexes of which they are part and this includes diversity within species, between species and of ecosystems" or "biodiversity is that part of nature which includes, the different genes among the individuals of a species; the variety and richness of all the plant and animal species at different scales in space, locally, in a region, in the country and the world; and the types of ecosystems, both terrestrial and aquatic, within a defined area." Biodiversity is essentially a term used to cover all of nature's variety, including its life-forms and processes. Biodiversity can be classified into three fundamental levels in a biological hierarchy.

8.1.1 Genetic Diversity

Genetic diversity is the variety of genetic information contained among individuals within a species. Gene is the most basic structure of species. It influences the characteristics of all species. Within any given species, there can be several "varieties" and "races" which differs from one another in certain characteristics. There are multiple characteristics in each species. To name a few of these characteristics are size, shape, color, aggressiveness, boldness, resistance against disease, hot and cold weather, etc. Thus a species may be smaller or bigger in size, black or brown in color, or may be aggressive or calm in nature, etc. Such different characteristics of same species are termed as "genetic diversity."

8.1.2 Species Diversity

Species diversity refers to the number of different species found at a place or in an ecosystem. It also refers to the studies to know how these kinds of species vary from season to season at the same place or different places. Species diversity is the most common level of diversity.

8.1.3 Ecosystem Diversity

Ecosystem diversity relates to variety of habitat, biotic communities, and ecological processes in the biosphere. There exist different varieties of ecosystem, for example, a lake ecosystem is different from an ocean ecosystem. This diversity is assessed at the continental or global level. Depending upon the availability of biotic resources, an ecosystem develops its own community of living organisms. Ecosystem diversity is a complex of species diversity.

Biodiversity can be understood better by classifying its components into three interdependent hierarchical levels. They are composition, structure, and function (Franklin 1988, Noss 1990). Composition comprises the genetic and species diversity here as the structure denotes the physical aspects of a system; let it be a gene, species, community, or an ecosystem. Function, on the other hand, explores the various processes involved in a biodiversity environment lie gene flow, population fluctuations, evolutionary processes, etc.

8.2 Documentation of Biodiversity

Fleming and Aagaard in 1993 expressed that documentation is the fundamental procedure to conserve biodiversity. Documentation does not simply mean mapping and making catalogues, but a detailed repertoire of documents enumerating species distributions, habitat distributions, and the various natural and anthropogenic process involved. At a time when resources like money and skilled labor needed for conservation procedures are limited and taking into consideration actualities like the impossibility of putting up full inventories for conservation and the fact that any rapid step toward sampling and collection of data might lead to statistical errors and consequent loss of some biodiversity, it is imperative that there must be some standardized, quantitative, and repeatable documentation protocols that are universal by nature. For instance, a standard method of documentation must be user-friendly as well as precise, as far as the information and location are concerned. It would also be remarkable if such databases are accessible through local area network (LAN), wide area network (WAN), or other such service. The World Conservation Monitoring Centre (WCMC), United Nations Environmental Programme (UNEP), World Wildlife Fund (WWF), etc. are examples of such repositories accessible from anywhere around the world. In the context, it is advisable that planners and decision-makers must employ a decision framework for establishing priorities for conserving biodiversity. McNeely et al. (1990) suggests three elements useful on the formation of such a decision framework. They are (i) distinctiveness, (ii) threat, and (iii) utility.

8.2.1 Distinctiveness

Distinctiveness aims at prioritizing distinctive characteristics or unique elements of biodiversity. The goal is to sustain diverse forms of life and processes. For instance, such an approach would give optimum interest to endemic species, habitats, etc.

8.2.2 Threat

It concentrates on the chances of losing particular elements of biodiversity due to anthropogenic activities. Accordingly the priority must fall on those areas where the magnitude of human intrusion is at its peak. However, McNeely et al. (1990) pinpoints a considerable weakness. The imminence of threat is dependent on our knowledge of the system, and thus, the adequacy of this approach declines less we know about the system.

8.2.3 Utility

Utility is an approach hooked on anthropogenic perspective. The method prioritizes those elements of biodiversity that are of materialistic importance. An important drawback of this approach is that those elements that are of no direct economic value but nonetheless endemic and ecologically significant might get overlooked.

Finally, documentation of biodiversity is incomplete if there is the absence of trained personnel. The service of taxonomist and para-taxonomist are of utmost importance in order to identify and classify biodiversity (Janzen 1992). The Committee on Research Priorities in Tropical Biology asserts the point by recommending a fivefold increase in the number of systematists to get on with the procedures. Hence researchers in the domain need to be encouraged. Social scientists, including economists, should be involved to address anthropogenic influences on patterns of biodiversity.

8.3 Genetic Diversity

Gene is the most basic structure of species. It influences the characteristics of all species. Within any given species, there can be several verities and races which differ from one another in certain characteristics. There are multiple characteristics in each species. Few of them are size, shape, color, boldness, resistance against diseases, hot and cold weather, etc. So a species may be smaller or bigger in size, black or white in color, etc.; such different characteristics of same species are termed as genetic diversity. The importance of within species or genetic diversity is often overlooked. Yet, diversity within species is the ultimate source of biodiversity at higher levels. Genetic and life-history variation and population structure and dynamics shape the way species respond to their environment. The potential for subsequent evolutionary change is determined, in large part, by the genetic variation. Domestic crops and animals, for example, are derived from and modified using genetic diversity from within wild species.

The composition of genetic diversity within species can be quantified and monitored as (i) single-locus variation, (ii) quantitative variation, (iii) chromosomal polymorphisms, and (iv) inbreeding. Each one of these are briefly discussed below. Lande and Barrowclough (1987) provide more detailed descriptions on this.

8.3.1 Single-Locus Genetic Variation

Variation at single chromosome location is most easily measured at the protein level but may also be examined at other levels including DNA. Electrophoresis is a relatively inexpensive and widely used technique for surveying protein polymorphisms. The proportion of individuals carrying different alleles, or different forms of the gene, at a locus (heterozygote's) is the measure of genetic variation derived from such surveys. It can be computed directly from the observed frequencies of actual heterozygotes at each locus or can be estimated from allele frequencies. A problem with heterozygosity as a measure of genetic variation is that there is no objective standard to compare calculated values, as they vary naturally among species (Fleming and Aagaard 1993).

Various molecular techniques have been used for measuring genetic diversity and bar coding the same. The most common methods include isozymes, restriction fragment length polymorphism (RFLP), and numerous genetic marker assays based on polymerase chain reaction such as random amplified polymorphic DNA (RAPD), simple sequence repeats (SSR), and amplified fragment length polymorphism (AFLP) for diversity studies and specific markers, i.e., *ITS*, *Rbcl*, *Matk*, etc. for sequence analysis and bar coding.

8.3.1.1 Isozyme Analysis

Isozyme analysis (or starch gel electrophoresis SGE) is the oldest of the molecular marker techniques still widely in use today having been developed some 30 years ago. Immediately after its development, there was a fantastic explosion of studies using isozyme analysis to investigate the genetic variation within and among natural populations (Avise 1994). The high level of variability within species uncovered by isozyme analysis has a major effect on evolutionary theory. Isozyme analysis still remains as the standard tool for surveys of genetic variability within and among conspecific populations as evidenced by the large numbers of population structure surveys published regularly using this technique (Bacilieri et al. 1994; Hsiao and Rieseberg 1994) and the comprehensive periodic reviews of plant population structure solely of studies employing isozyme analysis (Hamrick and Godt 1989, 1996).

Isozyme analysis is relatively inexpensive when compared to DNA-based techniques. Costly apparatus such as PCR machines and expensive materials such as primers, *taq* polymerase, and radioactive isotopes are not required. This is much more important in population structure studies than other investigations as large sample sizes are required. Isozyme analysis also requires much shorter development times to produce suitable markers. The long history of its use across a wide range of taxa has left a legacy of methodologies and recipes which should enable suitable markers to be found in almost every species within a reasonable time scale. Moreover, isozyme analysis does not require the same level of lab expertise as DNA-based techniques and is consequently much more accessible to non-laboratory specialists.

8.3.1.2 Restriction Fragment Length Polymorphism

DNA nucleotide variation resulting from point mutations occasionally leads to loss or gain of restricted endonuclease site. DNA alterations involving larger regions as a result of insertions, deletions, inversions, and translocations can result in changes in the distribution of several restriction endonuclease site simultaneously. The size distribution of DNA fragments affected by such changes will be different from that of unaffected individual resulting in a unique RFLP pattern. RFLP between two or more individuals may be detected in the DNA by digestion with an appropriate restriction enzyme followed by electrophoresis and Southern blotting. The Southern blot is with a labeled nucleic acid fragment that is homologous to the affected region. Single or low copy sequence in the genome are ideal to be used as RFLP probes, and the source may vary from cDNA sequences to anonymous sequences such as noncoding parts of the gene. Nuclear RFLP probes have been shown to be stably inherited, segregated in a Mendelian fashion and exhibit codominance (Swati et al. 1999).

8.3.1.3 Hybridization Using Multi-locus Probes

This technique is a derivative of RFLP analysis but differs from the latter by the type of hybridization probe applied to reveal genetic polymorphisms. To obtain a multi-locus DNA fingerprint, probes are used to create complex banding patterns by recognizing multiple genomic loci simultaneously. Each of these loci are characterized by more or less regular arrays of tandemly repeated DNA motifs that occur in different numbers at different loci. Two categories of such multi-locus probes are mainly used. The first category comprises cloned DNA fragments which are complimentary to so called mini-satellites, tandem repeats of a basic motif of about 10–60 bp in length. Genetic fingerprinting using mini-satellites (also known as VNTR fingerprinting) was first developed by Jeffreys et al. (1985). Since then, mini-satellites have been cloned from many organisms including plants (e.g., rice, *Arabidopsis thaliana*, etc.), and synthetic mini-satellite probes have been designed. The second category of probe is exemplified by short, synthetic oligonucleotides which are complimentary to microsatellites, i.e., tandem repeats of about one to two base pairs (synonyms: simple sequence repeats or SSRs, short tandem repeats STRs). This technique was first developed by Weber and May (1989) and is also known as "oligonucleotide fingerprinting" or "hybridization-based microsatellite fingerprinting." The use of oligonucleotide probes is compatible with an in-gel hybridization approach that is faster and more efficient than conventional Southern blotting. In recent years, both mini-satellite and microsatellite probes have been applied to RFLP fingerprinting of numerous animals, plants, and fungal species (Swati et al. 1999).

8.3.1.4 RFLP Analysis of mtDNA

RFLP analysis involves cutting up the mtDNA molecule into a number of fragments using a variety of restriction enzymes, separating the resultant fragments by size using electrophoresis, then visualizing the banding profile of fragments using ethidium bromide. In such a way, a fragment profile for each mitochondrial genome is generated. The presence or absence of bands can then be scored allowing the creation of a matrix from which genetic distance calculations are easily made. The banding profile directly reflects the position and frequency of restriction sites within the *mt*DNA molecule. The number and position of restriction sites are determined both by mutation and by recombination events. Plant *mt*DNA differs significantly from animal *mt*DNA (Bachmann 1994). Plant *mt*DNA is highly variable in size existing as a collection of different sized circles with extremely complex recombination processes. While animal *mt*DNA has a rate of evolution commensurate with studies of conspecific populations, plant *mt*DNA evolves much more slowly with respect to nucleotide substitution making this technique largely unsuitable for analysis of conspecific populations (Bachmann 1994; Avise 1994). The reasons for the slow rate of nucleotide substitution are still obscure but may include error-free DNA replication mechanisms or efficient DNA repair systems. The complicated recombination of plant *mt*DNA, technical difficulties, and the unsuitable rate of evolution have limited the use of *mt*DNA in the analysis of plant populations although there are some examples (Strauss et al. 1993; Swati et al. 1999).

Chloroplast DNA (*cp*DNA) has been used much more extensively than mtDNA in studies of plants. However *cp*DNA evolves at a rate only three to four times faster than plant mtDNA rendering it similarly unsuitable for analysis of nonspecific plant populations, and its use is generally restricted to systematic studies at a higher taxonomic level than the population although there have been a few exceptions.

8.3.1.5 RFLP Analysis of scnDNA

This technique screens for restriction site polymorphisms among sequences that are only represented once throughout the haploid nuclear genome. The data produced have similarities with those produced by isozyme analysis as the RFLP "alleles" are detected codominantly; homozygotes and heterozygotes can therefore be identified for particular restriction site or loci. Suitable genetic variation for population studies can be found in almost any species making this technique potentially very useful (Avise 1994). However, this technique is very expensive in time and resources.

8.3.1.6 RAPD Analysis

RAPD analysis has been used in studies estimating the genetic variation within and among populations (Bachmann 1994; Lynch and Milligan 1994; Chong et al. 1994). RAPD analysis is a PCR-based technique that uses short arbitrary primers to generate short DNA fragments from whole genome DNA. The technique relies on the statistical chance that complementary primer sites occur elsewhere on the same strand of DNA as inverted repeats enclosing only a relatively short sequence (up to a few thousand base pairs). PCR amplifies the sequence enclosed between the primer sites producing a range of fragments from similar sites throughout the genome. Electrophoresis is used to separate fragments by size, and the bands are visualized using ethidium bromide. The particular type of genetic variation sampled involves the number of primer sites throughout the genome and the numbers of base pairs between them. The variation in the position and frequency of these primer sites is like restriction sites generated by mutation and recombination events.

In common with other DNA-based techniques, genetic variation can almost always be found with RAPD (Bachmann 1994). As there are potentially thousands of primer combinations, it may only be a matter of persistence (and expense) to uncover potentially informative genetic polymorphisms. RAPD analysis has uncovered genetic variation where other techniques, such as isozyme analysis, have failed (Brauner et al. 1992). Because ethidium bromide staining is used to visualize the RAPD fragment profiles, the need for costly autoradiography is eliminated. Furthermore, as the primers consist of random sequences, it is reasonable to assume that the whole genome is sampled more randomly than with other methods (Bachmann 1994). There are significant drawbacks with this technique that seriously limit its applicability for investigating population structure. The most important disadvantage concerns the dominant expression of RAPD alleles. As both scnDNA RFLPs and isozyme loci are codominantly expressed, both homozygotes and heterozygotes can be easily scored directly from the gel. However, RAPD alleles are dominantly expressed making the estimation of allele frequency much less accurate especially when recessive alleles are at low frequency (Swati et al. 1999).

Lynch and Milligan (1994) have endeavored to produce a methodology that allows estimation of allele frequency from RAPD loci. However, they concede that completely unbiased estimates of allele frequency are not possible with this type of dominantly expressed marker. For estimates of allele frequency in a population of accuracy comparable with estimates made with codominant markers, sample sizes must be increased by between two and ten times depending on the frequency of rare alleles (Lynch and Milligan 1994). Furthermore, loci at which there are very rare alleles must be excluded from analyses simply because allelic frequency is too low to be accurately estimated. Consequently many more loci must be included in initial screenings to allow for later disqualification. Divergence between populations has been successfully estimated using RAPD analysis (Chong et al. 1994). However, in the analysis of their data set, one had to rely on an existing data set derived from previous isozyme analysis to provide an unbiased estimate of *F1* to compensate for the inherent biases in their RAPD marker system. The advisability of screening populations first with isozyme then with RAPDs to achieve an accurate estimate of population divergence seems dubious. A further shortcoming of RAPD analysis concerns the anonymity of RAPD bands (Bachmann 1994 ; Swati et al. 1999).

RAPD analyses are perhaps of less importance as the technique is more widely used and refined, and still the method is being considered as unreliable and of allowing very limited comparison between studies (Bachmann 1994). At present, the theoretical tools available for population structure analysis using RAPDs are not sufficiently sophisticated to provide acceptable estimates of the standard genetic parameters in widespread use. The variability uncovered by RAPDs is such that the technique could be used much more profitably in clonal identification. There are many variations in RAPD technique developed for the sake of convenience in specific cases, which include:

(a) *DNA amplification fingerprinting* (*DAF*): Caetano-Anolles et al. (1991) employed single arbitrary primers as short as five bases to amplify DNA using polymerase chain reaction. In a spectrum of products obtained, simple patterns are useful as genetic markers for mapping, while more complex patterns are useful for DNA fingerprinting. Band patterns are reproducible and can be analyzed using polyacrylamide gel electrophoresis and silver staining. DAF requires careful optimization of parameters; however, it is extremely amenable to automation and fluorescent tagging of primers for early and easy determination of amplified products. DAF profiles can be tailored by employing various modifications such as predigesting of template. This technique has been useful in genetic typing and mapping.
(b) *Arbitrary primed polymerase chain reaction* (*AP-PCR*): This is a special case of RAPD, wherein discrete amplification patterns are generated by employing single primers of 10–50 bases in length for PCR of genomic DNA (Welsh et al. 1991). In the first two cycles, annealing is under non-stringent conditions. The final products are structurally similar to RAPD products. Compared to DAF, this variant of RAPD is not very popular as it involves autoradiography. Recently, however, it has been simplified by separating the fragments on agarose gels and using ethidium bromide staining for visualization.
(c) *Sequence characterized amplified regions for amplification of specific band* (*SCAR*): Michelmore et al. (1991) and Martin et al. (1991) introduced this technique wherein the RAPD marker termini are sequenced and longer primers are designed (22–24 nucleotide bases long) for specific amplification of a particular locus. These are similar to STS markers in construction and application. The presence or absence of the band indicates variation in sequence. These are with better reproducibility than RAPDs. SCARs are usually dominant markers; however, some of them can be converted into codominant markers by digesting them with tetra cutting restriction enzymes, and polymorphism can be deduced by either denaturing gel electrophoresis or SSCP (Rafalski et al. 1993). Compared to arbitrary primers, SCARs exhibit several advantages in mapping studies (codominant SCARs are informative for genetic mapping than domi-

nant RAPDs), map-based cloning as they can be used to screen pooled genomic libraries by PCR, physical mapping, locus specificity, etc.

(d) *Cleaved amplified polymorphic sequences* (*CAPs*): These polymorphic patterns are generated by restriction enzyme digestion of PCR products. Such digests are compared for their differential migration during electrophoresis (Koniecyzn and Ausubel 1993, Jarvis et al. 1994). PCR primer for this process can be synthesized based on the sequence information available in databank of genomic or cDNA sequences or cloned RAPD bands. These markers are codominant in nature.

(e) *Randomly amplified microsatellite polymorphisms* (*RAMPO*): In this PCR-based strategy, genomic DNA is first amplified using arbitrary (RAPD) primers. The amplified products are then electrophoretically separated, and the dried gel is hybridized with microsatellite oligonucleotide probes. There are several advantages for oligonucleotide fingerprinting (Epplen 1992). RAPD (Williams et al. 1990) and microsatellite-primed PCR (Weising et al. 1995; Gupta et al. 1994) are thus combined to get the speed of the assay, high sensitivity, high level of variability detected, and the non-requirement of prior DNA sequence information (Richardson et al. 1995).

(f) *Sequence-tagged microsatellite site markers* (*STMS*): This method includes DNA polymorphism using specific primers designed from the sequence data of a specific locus. Primers complementary to the flanking regions of the simple sequence repeat loci (Weber et al. 1989) yield highly polymorphic amplification products. Polymorphisms appear because of variation in the number of tandem repeats (VNTR loci) in a given repeat motif. Tri- and tetranucleotide microsatellites are more popular for STMS analysis because they present a clear banding pattern after PCR and gel electrophoresis (Hearne et al. 1992). However, dinucleotides are generally abundant in genomes and have been used as markers. Examples include $(CA)n$ $(AG)n$ and $(AT)n$ (Rafalski et al. 1993). The di- and tetranucleotide repeats are present mostly in the noncoding regions of the genome, while 57% of trinucleotide repeats are shown to reside in or around the genes. A very good relationship between the number of alleles detected and the total number of simple repeats within the targeted microsatellite DNA has been observed. Thus larger the repeat number in the microsatellite DNA, the greater is the number of alleles detected in a large population (Yang et al. 1994).

(g) *Direct amplification of mini-satellite DNA markers* (*DAMD-PCR*): Introduced by Heath et al., (1993), DAMD-PCR has been explored as a means of generating DNA probes useful for detecting polymorphism. DAMD-PCR clones can yield individual-specific DNA fingerprinting pattern and thus have the potential as markers for species differentiation and cultivar identification (Somers et al. 1996).

8.3.1.7 Inter Simple Sequence Repeat Markers (ISSR)

In this technique, reported by Zietkiewicz et al. (1994), primers based on microsatellites are utilized to amplify inter-SSR DNA sequences. Here, various microsatellites anchored at the 3′ end are used for amplifying genomic DNA which increases their specificity. These are mostly dominant markers, though occasionally few of them exhibit codominance. An unlimited number of primers can be synthesized for various combinations of di-, tri-, tetra-, and pentanucleotides [(4)3 = 64, (4)4 = 256] with an anchor made up of a few bases which can be exploited for a broad range of applications in plant species. Inter-simple sequence repeat or ISSR markers are more and more in demand, because they are known to be abundant, very reproducible, highly polymorphic, highly informative, and quick to use (Zietkiewicz et al. 1994). ISSR uses the presence throughout the genome of simple sequence repeats (SSR) which is ubiquitous, abundant, and highly polymorphic tandem repeat motifs composed of 1–7 nucleotides. A unique PCR primer composed of a few SSR repeats (with or without anchored end) was used to amplify DNA sequences between two inverted SSR composed of the same units and spaced out by less than around 4 Kbp.

8.3.1.8 Amplified Restriction Fragment Length Polymorphism (AFLP)

AFLP technique was originally used to fingerprint genomic DNA samples. These DNA fingerprints could be used as a source of genetic markers to generate linkage DNA maps and as markers that are linked to a trait or locus of interest. AFLP is a highly procedural assay. Briefly, it involves restriction of genomic DNA with a combination of two enzymes, one of which is a rare cutter such as *Eco R1* and the other being a frequent cutter such as *Mse 1*. Enzyme-specific adapters are then ligated to the restricted fragments. Adapter ligation step results in the genesis of a huge set of restricted fragments that are different with respect to the length and sequence. However, they are identical with respect to the flanking adapter sequences. The next step involves selective pre-amplification of restricted fragments leading to the generation of template DNA that is used in the AFLP reaction. In pre-amplification, primers specific to *Eco R1* and *Mse 1* adapters and carrying an extra nucleotide at the 3′ end are utilized. This ensures selective amplification of only subset of the adapter ligated restriction fragments, all of which carry the nucleotide complementary to the +1 selective nucleotides present in both the primers. This step is crucial as it simultaneously reduces the complexity of fingerprint by eliminating many restriction fragments and generates an unlimited source of template DNA.

AFLP has several other advantages that make it applicable in assessment of genetic diversity, genetic mapping, and tagging studies. The assay is highly reproducible owing to the stringent PCR reaction conditions. It has a high multiplex ratio that ensures a thorough genome scanning both for polymorphism as well as for markers linked to traits of interest. Since AFLP generates both shared and unique bands when tested across many accessions, they become ideal tools for inferring genetic relationships, particularly at inter- and intraspecies level.

8.3.1.9 Sequencing-Based Markers

DNA sequencing can also be used as a definitive means for identifying species. Variations due to transversion, insertion, or deletion can be accessed directly and information on a defined locus can be obtained. Genetic variation occurs extensively at the single nucleotide level. Direct sequencing can efficiently identify such single nucleotide polymorphisms that usually depend on how closely related are the organisms being compared. Other sequencing-based strategies include analysis of the variable internal transcribed spacer (ITS) sequences of ribosomal DNA (rDNA). The ITS region of 18 s–26 s rDNA has proved to be a useful sequence for phylogenetic studies in many angiosperm families. The level of ITS sequence variation suitable for phylogenetic analysis is found at various taxonomic levels within families, depending on the linkage. A number of researchers have also sequenced other regions of DNA such as *trn*K of chloroplast and spacer region of 5 s rDNA as diagnostic tools for authentication purpose.

8.3.2 Quantitative Variation

Quantitative variation is the variation in continuous traits that are controlled by several interacting genes. It can be measured and monitored with heritability studies. Heritability measures the portion of the total or phenotypic variability in a trait that is genetically based. Narrow-sense heritability, the ratio of additive genetic variance to the total phenotypic variance, is the form commonly used to monitor changes in quantitative genetic variation (Fleming and Aagaard 1993).

8.3.3 Chromosomal Polymorphisms

Chromosomal polymorphism is another method of examining genetic variation within species by providing the arrangement of chromosomes. Some species are known to contain more than one chromosomal sequence so karyotypic analyses may be used to look for such polymorphisms. However, chromosomal polymorphisms are not common in all organisms. Furthermore, different types of chromosomal variation require differing interpretations, and lack of variability may be unimportant. Karyotypic analysis is thus of limited use for quantifying genetic diversity.

8.3.4 Inbreeding

Inbreeding is the production of offspring from the mating/breeding of individuals or organisms that are genetically more closely related and have deleterious effects on the variability of populations and species. Inbreeding leads to homozygosity which can increase the chances of offspring being affected by recessive traits, and it ill leads to a decreased biological fitness of population which is called inbreeding depression. Quantifying the exact levels of inbreeding in a population is difficult or requires some knowledge of the pedigree of the individuals of interest, which is difficult to obtain for natural populations. Alternatively, it is possible to monitor inbreeding indirectly by examining indicators of genetic stress, such as variation in morphology, but such variation is confounded by phenotypic effects (Fleming and Aagaard 1993).

There are several methods available for quantifying and monitoring the composition of genetic variation; their use will be constrained. Monitoring quantitative genetic variation will provide the most important information to understand the long-term adaptability of populations, but it is difficult and expensive to undertake.

8.4 Species Diversity

Species diversity refers to the number of different species found at a place or in an ecosystem. It also refers to the studies to know as to how these kinds of species vary from season to season at the same place or different at different places. Species diversity is the most common level of diversity. The simplest and most obvious measure of species diversity is the number of species. This is referred to as species richness and involves the counting of species in an area that is well delimited in space and time. We have to work with samples rather than a total inventory of species in a particular area. This creates problems in determining the number of species because counts depend on sampling intensity. Species-area or species-abundance relationships can be used to get around this and derive measures of diversity. In the former case, species density is studied and in the latter case, species richness.

8.5 Ecosystem Diversity

Ecosystem diversity is the organization of assemblages of microorganisms, plants, and animals, together with the nonliving components of the environment that influence energy flow and nutrient cycling ,or it relates to a variety of habitat, biotic communities, and ecological processes in the biosphere. There exist different varieties of ecosystem. A lake ecosystem is different from an ocean ecosystem. This diversity is assessed at the continental or global level. Depending upon the availability of biotic resources, an ecosystem develops its own community of living organisms. Ecosystem diversity is a complex species diversity. Within this category also included community and landscape diversity. Landscape diversity refers to the pattern of land forms, vegetation types, and land uses and community diversity to the assemblages of organisms (Urban et al. 1987). Rapid human-induced alterations of the earth's environment have resulted in an urgent need to understand the structure and function of local ecosystems, as well as the global biosphere. Effects on ecosystem diversity have cascading influences on diversity at all lower levels in the biological hierarchy. Hence, global problems such as greenhouse warming and stratospheric ozone depletion will have fundamental effects on the conservation of species diversity. Landscape features such as size of patches, and their heterogeneity, perimeter-area ratios, position, and connectivity have important influences on species composition, abundance, and viability. Landscape structure can be inventoried and monitored primarily by remote sensing using aerial photography and satellite imagery. It is important that interpretation of such images be verified by on-the-ground observations, known as "ground truthing." Data from remote sensing can also be combined with information from other sources, organized and displayed with a geographical information system (GIS) to allow analysis of land use and habitat modification. Time series analysis of this data is a powerful monitoring technique to track abiotic and biotic changes and disturbances in the environment. GIS, in some cases, may also allow analysis of spatial relationships between different landscapes and thus contribute to a better understanding of causes behind changes in biodiversity (Fleming and Aagaard 1993).

Combining investigations of landscape patterns with community-ecosystem structure and function will be particularly profitable, landscape variables when combined with knowledge of species assemblages and processes can be used to provide inventories of community and ecosystem diversity. Scott et al. (1990), used such a process to develop a method of identifying centers of species richness and endemism to locate gaps in the distribution of protected areas. Because of cascading effects, monitoring landscape function and composition can also provide a means of understanding ecosystem structure. Landscape disturbance, such as fire frequency and seasonality may be key determinants of biodiversity in some ecosystems. Total diversity of native species at the landscape level is maximized when disturbance occurs at its historical frequency and in its historical pattern (Hobbs and Huenneke 1992). If altered, key species may be lost and new spaces may invade (Fleming and Aagaard 1993).

Finally, combined inventories of landscape and community-ecosystem diversity will aid monitoring programs considerably. Repeated, extensive inventories of species will often be impractical; however, periodic inventories of vegetation by remote sensing can be used to effectively monitor changes in biodiversity at other levels. Knowing the relationship between vegetation patterns and species assemblages, inferences can be drawn about species distributions. Thus, combining knowledge of landscape diversity with that of community-ecosystem assemblages may be our most cost-effective means of monitoring environmental degradation both on local and global scales (Fleming and Aagaard 1993).

8.6 Conclusions

Conserving biodiversity is a daunting task. The sheer number of living organisms precludes complete enumeration and geographic referencing before crucial conservation decisions have to be implemented. It is generally recognized that a species-by-species approach is too arduous. It must also be recognized that emphasis on single-species conservation can be dangerous, especially when directed toward the rarest organisms instead of species assemblages. Yet, species richness and endemism have attracted the attention of biologists likely due, in part, to a bias toward preserving tangible life-forms rather than poorly understood life processes. On both practical and theoretical grounds, we probably have no choice but to use an ecosystem-community perspective in conservation, which incorporates both the diversity of life-forms and processes.

Given current circumstances, the best way to minimize species loss is likely to be to maintain the integrity of ecosystems. Loss of natural habitat is almost universally acknowledged as the primary threat to biodiversity. There is a need to focus on habitat diversity with the intention to maintain the range of ecosystem processes (e.g., selection pressures), the number of species, and the evolutionary potential of the organisms. There are several reasons for focusing on habitat conservation. (1) We lack knowledge about ecosystems and the needs of species and how to meet these needs. (2) Ecosystem protection will save both species and ecosystems. (3) Conservation of ecosystems will have positive effects on the species within them. (4) The ecosystem approach focuses attention on the long-range problem.

The rapid and ever increasing loss of biodiversity has forced our hand. Documentation and measurement of biodiversity must proceed quickly and thus focus on the need to conserve particular habitats, recognizing their contribution to biodiversity at multiple levels of biological organization.

References

Avise JC (1994) Molecular markers, natural history and evolution. Chapman and Hall, New York

Bachmann K (1994) Molecular markers in plant ecology. New Phytol 126:403–418

Bacilieri R, Labbe T, Kremer A (1994) Intraspecific genetic structure in a mixed population of *Quercus petraea* (Matt) Liebl and Q robur L. Heredity 73:130–141

Brauner S, Daniel JC, Stuessy TF (1992) Ribosomal DNA and RAPD variation in the rare plant family Lactoridaceae. Am J Bot 79(12):1436–1439

Caetano Anolles G, Bassam BJ, Gresshoff PM (1991) Bio Technol 9:553–557

Chong LD, Traynor-Kaplan A, Bokoch GM, Schwartz MA (1994) The small GTP-binding protein Rho regulates a phosphatidylinositol 4-phosphate 5-kinase in mammalian cells. Cell 79:507–513

Epplen JT (1992) Advances in electrophoresis, vol. 5. In: Chrambach A, NJ Dunn and BJ Radola (eds), VCH, Cambridge, pp 59–112

Fleming IA, Aagaard K (1993) Documentation and measurement of biodiversity. NINA Utred 50:1–23

Franklin JF (1988) Structural and functional diversity in temperate forests. In: Wilson EO, Peter FM (eds) Biodiversity. National Academy Press, Washington, DC, pp 166–175

Gupta M, Chyi YS, Romero-Severson J, Owen JL (1994) Amplification of DNA markers from evolutionarily diverse genomes using single primers of simple-sequence repeats. Theor Appl Genet 89:998–1006

Hamrick JL, Godt MJW (1989) Allozyme diversity in plant species. In: Brown AHD, Clegg MT, Kahler AL, Weir BS (eds) Plant population genetics, breeding and germplasm resources. Sinauer Associates, Sunderland, pp 43–63

Hamrick JL, Godt MJW (1996) Effects of life history traits on genetic diversity in plant species. Phil Trans R Soc Lond B 351(1345):1291–1298

Hearne CM, Ghosh S, Todd JA (1992) Microsatellites for linkage analysis of genetic traits. Trends Genet 8:288–294

Heath DD, Iwama GK, Devlin RH (1993) PCR primed with VNTR core sequences yields species specific patterns and hyper variable probes. Nucleic Acids Res 21:5782–5785

Hobbs RJ, Huenneke LF (1992) Disturbance, diversity, and invasion: implications for conservation. Conserv Biol 6:324–337

Hsaio C, Chatterton NJ, Asay KH, Jensen KB (1994) Phylogenetic relationships of 10 grass species: an assessment of phylogenetic utility of the internal transcribed spacer region in nuclear ribosomal DNA in monocots. Genome 37:112–120

Janzen DH (1992) A south-north perspective on science in the management, use, and economic development of biodiversity. In: Sandlund OT, Hindar K, Brown AHD (eds) *Conservation of biodiversity for sustainable development*. Scandinavian University Press, Oslo, pp 27–52

Jarvis P, Lister C, Szabo V, Dean C (1994) Integration of CAPS markers into the RFLP map generated using recombinant inbred lines of *Arabidopsis thaliana*. Plant Mol Biol 24:685–687

Jeffreys AJ, Wilson V, Thein SL (1985) Hypervariable 'minisatellite' regions in human DNA. Nature 314:67–73

Koniecyzn A, Ausubel FM (1993) Plant J 4:403–410

Lande R, Barrowclough GF (1987) Effective population size, genetic variation, and their use in population management. In: Soulé ME (ed) Viable populations for conservation. Cambridge University Press, Cambridge, pp 87–123

Lynch M, Milligan B (1994) Analysis of population-genetic structure using RAPD markers. Mol Ecol 3:91–99

Martin GB, Williams JGK, Tanksley SD (1991) Rapid identification of markers linked to a *Pseudomonas* resistance gene in tomato by using random primers and near-isogenic lines. Proc Natl Acad Sci U S A 88:2336–2340

McNeely JA, Miller KR, Reid WV, Mittermeier RA, Werner TB (1990) Conserving the world's biological diversity. IUCN, World Resources Institute, World Bank, WWF-US and Conservation International, Washington DC

Michelmore RW, Paran I, Kesseli RV (1991) Identification of markers linked to disease-resistance genes by bulked segregant analysis: a rapid method to detect markers in specific genomic regions by using segregating populations. Proc Natl Acad Sci U S A 88:9828–9832

Noss RF (1990) Indicators for monitoring biodiversity: a hierarchical approach. Conserv Biol 4:355–364

Rafalski JA, Tingey SV (1993) Genetic diagnostics in plant breeding: RAPDs, microsatellites and machines. Trends Genet 9:275–280

Richardson T, Cato S, Ramser J, Kahl G, Weising K (1995) Hybridization of micro- satellites to RAPD: a new source of polymorphic markers. Nucleic Acids Res 23:3798–3799

Scott JM, Csuti B, Smith K, Estes JE, Caicco S (1990) Gap analysis of species richness and vegetation cover: an integrated conservation strategy for the preservation of biological diversity. In: Kohm K (ed) Balancingon the brink: a retrospective on the endangered species act. Island Press, Washington DC

Somers DJ, Zhou Z, Bebeli P, Gustafson JP (1996) Repetitive, genome-specific probes in wheat (*Triticum aestivum* L. em Thell) amplified with minisatellite core sequences. Theor Appl Genet 93:983–989

Strauss SH, Hong YP, Hipkins VD (1993) High levels of population differentiation for mitochondrial DNA haplotypes in *Pinus radiate, P. muricata* and *P. attenuate*. Theor Appl Genet 86:605–611

Swati PJ, Prabhakar K, Ranjekar V, Gupta S (1999) Molecular markers in plant genome analysis. Curr Sci 77:25–27

Urban DL, RV O'N, Shugart HH (1987) Landscape ecology. Bio Sci 37:119–127

Weber JL, May PE (1989) Abundant class of human DNA polymorphisms which can be typed using the polymerase chain reaction. Am J Hum Genet 44:388–396

Weising K, Nybom H, Wolff K, Meyer W (1995) In: Arbor A (ed) DNA fingerprinting in plants and fungi. CRC Press, Boca Raton, pp 1–3

Welsh J, Mc Clelland M (1991) Distribution of repetitive DNA sequences in eubacteria and application to fingerprinting of bacterial genomes. Nucleic Acids Res 19:6823–6831

Williams JGK, Kubelik AR, Livak KJ, Rafalski JA, Tingey SV (1990) DNA polymorphisms amplified by arbitrary primers are useful as genetic markers. Nucleic Acids Res 18:6531–6535

Yang GP, Saghai Maroof MA, Xu CG, Zhang Q, Biyashev RM (1994) Comparative analysis of microsatellite DNA polymorphism in landraces and cultivars of rice. Mol Genet 245:187–194

Zietkiewicz E, Rafalski A, Labuda D (1994) Genome fingerprinting by simple sequence repeat (SSR)-anchored polymerase chain reaction amplification. Genomics 20:176–183

Biopiracy

9

P.K. Satheeshkumar and Anoop Narayanan

Abstract

In the nineteenth century, globalization due to industrial revolution made significant impact in the intra- and international exchange of knowledge and goods. Even though there were many constructive features, globalization propelled few detrimental events like monopolizing the economy by a few countries and biopiracy. In simple words, the term biopiracy refers to the unethical and illegal exploitation of one's knowledge or resource without prior consent or benefit-sharing agreement by a third party through intellectual property right claims. The phenomenon is present in almost all commercial fields, especially in pharmaceutical, food and cosmetic industry. In a general modus operandi, the information gained will be skilfully manipulated to a patent, through which the information will become the intellectual property of the pirate. Superficially, there may not be any similarity with the prior knowledge, but either the resource or the product will be monopolized by the biopirate, thus preventing the actual inventor from utilizing it. It was noted that the benefits were always accounted to the developed countries even though 90 % of the information was from the developing/underdeveloped nations. Heavy exploitation accompanied with the inevitable destruction of biodiversity forces the organizations like UN to enforce strict means to protect the universe. At present, there are 170 countries that agreed to the fair sharing of information and profit and to the sustainable utilization of natural resources.

P.K. Satheeshkumar (✉)
Division of Microbiology, Jawaharlal Nehru Tropical Botanic Garden and Research Institute, Palode, Thiruvananthapuram 695562, Kerala, India
e-mail: pksat_74@yahoo.com

A. Narayanan
Department of Biological Sciences, Purdue University,
915 W. State Street, West Lafayette, IN 47907-2054, USA

S. Abdulhameed et al. (eds.), *Bioresources and Bioprocess in Biotechnology*,
DOI 10.1007/978-981-10-3573-9_9

Keywords
Bioprospecting • Traditional knowledge • Benefit sharing • Intellectual property • Biodiversity

9.1 Introduction

Last century witnessed a giant leap in the field of scientific knowledge mainly due to the invention of modern techniques and machines. While giving utmost significance to this, it is not possible to sideline the knowledge the old civilizations generated by experimenting with nature through thousands of years. It passed from one generation to the other mainly through the verbal route, so there was very little written documentation as evidence of this traditional knowledge in many of the old civilizations. So it was rightly said, when an old man dies, a library is burned indicating the loss of an unimaginable quantity of knowledge he possess. Along with the mechanization in the last century, the New World countries start exploring the indigenous knowledge on different aspects of life including mainly pharmaceutical, nutritional and genetic resources. This lead to the unauthorized exploitation and monopoly on traditional knowledge by a very few, and they never allowed others to get a share of it due to property rights. The so-called bioprospecting worldwide led to heavy loss of biodiversity and caused severe concern among different nations, and the Convention on Biological Diversity (1992) held at "Rio Earth Summit" framed few regulations to control it. The convention proposed a system where the biodiversity and traditional knowledge have to be exploited with the consent of local population and government and the profit is shared between the parties which may prevent the unsustainable exploitation of biodiversity and illegal access to the traditional knowledge. Biopiracy is a term that refers to the use of intellectual property systems to legitimize the exclusive ownership and control over biological resources and biological products and processes that have been used over centuries in non-industrialized cultures. Patent claims over biodiversity and indigenous knowledge that are based on the innovation, creativity and genius of the people of the Third World (Shiva 1997, Nazarea 2006). It has become very difficult to differentiate between bioprospecting and biopiracy due to the increased number of factors and ethical issues involved (Soejarto et al. 2005).

9.2 Pharmaceutical Biopiracy

Plants and animals were used as source of medicines for thousands of years in old civilizations. The nomadic culture gradually turned to settlements, and man started cultivating these plants, which they found to have medicinal properties. Different civilizations developed their own ways to treat diseases using products from different natural sources. In the past, except against severe endemics, which lessened the human populations from time to time, almost all other diseases could be treated with products of natural origin. As in the old civilizations, the duties were distributed among different groups of people; it would be possible that the ailments in the members were treated by a specific group of individuals and their families. Along with the very common methods of medication that almost everybody knew, these families developed products and methods to treat diseases, which were not very common. The experience gained through generations made them masters in the field. The monopoly along with a need to protect their knowledge would have made the foundations of intellectual rights as early in the sixth century.

For example, in India, there are more than 4,600 ethnic communities (source: ASI), and almost all of them have their own medical traditions, which are accessible very easily. It was estimated that nearly 7,500 plant species and several hundreds of animals, metals and minerals were used in the folklore medical practices of Indian communities. The traditions followed by both tribal and nontribal populations consist of people from all the strata of society including herbal healers, monks, midwives, bonesetters, physicians and housewives. The primary objective of this knowledge was not economic gain. The known tradition of Indian folklore dates back to nearly two millennia. It is even known that in the past, these traditions were even shared with neighbouring countries (Shankar 1997).

With the economic gain, the genuine search and research started taking shortcuts, and the investors started to explore traditional knowledge in the indigenous populations to get access to the products, which are patentable. Many of the old civilizations now prevailing in the Third World countries face a serious threat of biopiracy as multinational companies are patenting their traditional knowledge. The companies target the countries like India, Africa and other developing countries, which are well-off in traditional knowledge and bio-resources. They used the intellectual property system of their country to legitimize the restricted ownership and unrestricted usage over biological resources, products and processes that were the property of non-industrialized cultures for centuries. The cases of basmati rice, neem and turmeric are few of the examples of biopiracy in India.

Pharmaceutical biopiracy is the condition where the multinational pharmaceutical companies take advantage of local or traditional knowledge without acknowledging its origin (Nigh 2002; Hayden 2007). The patent rights they obtain for the product/knowledge render them the monopoly to make and commercialize it, but the indigenous people who are the original inventors receive little or nothing from the company. Even though the treaties and regulations and even the term biopiracy did not exist, one of the earlier reports of biopiracy and the issues tagged to it was from Norway. The compound cyclosporin A (trade name: Sandimmune Neoral) used for the treatment of patients having undergone organ transplantation was reported to be isolated from samples originally collected from Norway by a scientist based in Basel, Switzerland, on a vacation in 1962 (Christoffersen and Fish 1999). According to one of the reports released by the company (Sandoz became Novartis) Novartis, the gross revenue obtained by selling the drug in 1996 was 1.2 billion USD. Even though that time there was no agreement between nations for profit sharing, this example is an indicator of the huge economic loss to one of the partners if there are no systems for a genuine and legal bioprospecting.

Cancer and pathogenic infections, the two deadly diseases, account for the use of nearly 60% of drugs either directly derived or modelled from natural resources (Kurien and Das 2011). From the history of major drug molecules, it is clear that natural products are the most consistently successful source of drug leads. Natural products continue to provide greater structural diversity than standard combinatorial chemistry, and so they offer major opportunities for finding novel low-molecular-weight lead structures that are active against a wide range of assay targets (Gibbons 2008; Koehn and Carter 2005). The natural selection process exists among different species susceptible to pathogens always developing new variants of defence components which can never be produced by man-made techniques. The evolutionary process preserves those compounds, and in turn the resistance shown by the organism can be an inspiration to explore for the molecule to use against the same/related species of human pathogens (Walters 2011). As less than 10% of the world's biodiversity has been tested for biological activity, many more useful natural lead compounds are awaiting discovery. During 2001–2002, about 97 novel natural compounds with antitumour activity were identified and isolated from marine organisms. Among the three major groups of living beings (plant, animals and microbes), microbes are the most exploited group as a large number of commercially valuable therapeutic compounds are identified from them (Chaudhuri 2003; Gurang et al. 2013; Gupta et al. 2014).

The normal route to a new drug and its validation pursue the following steps. As a first step, metabolites or proteins are extracted from the samples. This is followed by the screening process where the extracts are screened for its effectiveness against various diseases or pathogens. The potential extracts are analysed for its cytotoxicity against mammalian cells, and the components are identified by various separation methods. This first step can take a few months' time to years together according to the availability and complexity of starting material. Once the compound is found non-toxic, the pharmacodynamics and pharmacokinetics studies are undertaken. The mechanism of action of the chemical entity will be explored in this stage. It may be possible to use the synthetic chemistry methods to modify the molecule to make them more active once the nature of interaction with the target and the mode of action are identified. This is a crucial point as the investors may take a decision on investing more on the drug to conduct the preclinical and clinical studies based on the profit possibilities in the future (Amir-Aslani and Mangematin 2009). In the next stage, the investor has to get the permission for preclinical and clinical studies in accordance with the existing legislations of the country. The safety and quality of the drug are evaluated continuously throughout the preclinical and clinical trials and even after its release as a drug in the market. In conclusion, it is estimated that an average of approximately USD800 million to USD2 billion has to be spent by the investor from the discovery phase to the evaluation and regulatory phase (Rose et al. 2012). The unexpected failure of many drugs in the past always put the investors in caution and is a reason for looking at the traditional knowledge of indigenous people to obtain the molecules with therapeutic value against various diseases. So, if a bioprospector get the information from the local people of the area who are accustomed to using the natural sources for therapeutics, their knowledge will act as a filter to pin down the plants or animals with medicinal potential.

The ethnobotanical route to identify a drug molecule has already proven worth and resulted in the discovery of many high-profile drug molecules (Koehn and Carter 2005). The examples include aspirin (from the bark of willow plant), digitoxin (from foxglove) and quinine (from the cinchona) used against pain, heart failure and malaria, respectively. It was observed that major bioethical issues usually occur in the initial stages of bioprospecting, i.e. when the researcher and the local people interact. The ethical negligence of the scholars ultimately results to the condition that the rights of the local stakeholders are at risk. The communication gap further widens when the issue is in between the local community and the multinationals who invest on the programme, due to many obstacles like language, culture and difficulties in accessing the real individuals in the decision-making group. The property rights in bioprospecting are ambiguous and often support the corporate interests (Edelman 1992; Cunha et al. 2010). The Convention on Biological Diversity (CBD) and Nagoya protocol have effectively inducted steps to legitimize the bioprospecting and thus reducing the risk of biopiracy in pharmaceutical industry.

9.3 Agri-food Biopiracy

Agri-food industry holds the second position in utilizing the knowledge on traditional food from the indigenous civilizations. But there is no way by which we can differentiate the present systems and traditions of culinary practices in terms of its originality and authenticity. It is known that in old civilizations, many of the diseases were treated with certain dietary formulations using very common fruits and vegetables, which were part of our routine diet. There are written documents in the epics of old Indus valley civilizations indicating the knowledge of our forefathers about what food is exactly needed by the body for proper growth and development. For example, in Ayurveda, there are clear indications of the food and the mode of cooking, to maintain the balance of human body (which was supposed to be made of three fundamental forces, fire, ether and water) and to evade the attack of diseases. So in this personalized cooking, according to the physical state of each individual, the mode of cooking, which includes frying, steaming, roasting and blanching, has to be adjusted.

Even though there are less reports compared to medicinal biopiracy, this is more important as it directly affects the food security of millions of people. More than 70% of the world population depend on the indigenous food materials like edible plant resources, mainly wheat, maize, rice and potato. Keeping these resources away from being monopolized by companies/individuals will be a threat to the world food security. Patenting of varieties, which are high yielding and adapted to various stress conditions by multinationals, will encourage their control by a few individuals. According to the socio-ecologists, the green revolution has already made the farmers heavily dependent on the agrochemicals and the seed companies, which can supply seeds resistant to these chemicals. The traditional means of cultivation using different varieties to suit the weather and environment, which was a highly successful practice in the past, had been abolished completely and in turn made the farmers more dependent on the seed and agrichemical companies. Moreover, when the genetically modified crops and high-yielding varieties raised through research (both and patent protected) were promoted globally during the green revolution, the highly adapted but less productive local varieties outcompeted (Yusuf 2010; Bhattacharya et al. 2013).

The piracy with regard to food industry is also applicable on issues where the indigenous methodology was later claimed by somebody as their own, even if there is documentary evidence to prove it. There are written documents from the sixteenth century about a prime food of Javanese people known as tempeh. Tempeh is prepared by the fermentation of soybeans and is one of the oldest known processes for food making. Consumed by both poor and rich, the food is a rich source of vitamin B12. Tempeh has gained popularity among the Western countries recently as a nutritional supplement, as an equivalent or as a substitute of animal protein. Many Japanese companies gained patents on tempeh, which clearly disregards the Javanese origin of this food process technology.

One of the finest examples of agri-food biopiracy is by Monsanto (Mahyco) when they introduced the Bt (genetically modified with the bacterial toxin from *Bacillus thuringiensis* against the lepidopteran insects) brinjal in India. The story becomes awkward when we look into the fact that India harbours more than 2,500 varieties of brinjal suitable to different regions and climate (Samuels 2012). One of the major setbacks observed in the field trials by Monsanto is that, if the GM plants are planted along non-GM plants, the non-GM plants are severely affected, forcing the farmers to purchase the Bt brinjal seeds, if somebody in the near field is cultivating GM brinjal. Somehow, farmer's protest against Bt brinjal made the government to put a ban on Bt brinjal for an infinite period of time. The committee constituted by the government of India has made strong recommendations to prevent the use of Bt brinjal in the Indian fields (Ministry of Environment and forests. 2010). Meantime, Monsanto in their test fields in India is using the native varieties of brinjal to generate high-yielding brinjal varieties through hybridization experiments. The new varieties containing the favourable characters of both the parents will be patented and used in genetic engineering programmes.

It is reported that six multinational Gene Giants are in control of the global agricultural research. The firms Syngenta, Monsanto, BASF, Dow, Bayer and DuPont share 60% of commercial seed business and 76.1% of agrochemical business. These companies also own more than 76% of private sector R&D in the above-said fields.

9.4 Genes and Genetic Resources

The investment for innovation can be different with regard to the target groups which utilize the results of the research. For example, the investments in the human medical field are very high, and the potential for regaining the investment along with good profit is more promising. Either the molecules (small molecules or RNA/protein molecules) or the methods (diagnostic methods to determine or differentiate) of high potential in the human medical field will always prompt the investors to spend money on innovative R&D (Jefferson et al. 2015). With regard to plant science research, the profit and potential for commercialization are comparatively low. It happens when the target species of research is cultivated/used by a small population. Compared to the previous generation, patents on specific traits incorporated into the plants (Bt resistance, herbicide resistance, etc.) and patents related to that have been changed to the patenting of nucleotide and amino acid sequences which will have application in a broad spectrum of target group plants. Another form of patents is on the plant variety. These high-yielding varieties have been developed by either hybridization experiments or through different biotechnological approach.

With regard to the plants, there are two forms of intellectual property rights, namely, utility patents and plant variety protection (PVP). While the patents grant the inventor the right to prevent all others from making, using and/or retailing the product invented for 15–20 years, PVP rights given to the plant breeders are intended to protect the distinctness, stability, uniformity and novelty of the variety. Like any other IPR, both patents and PVP give exclusive domination of the holder over an invention/creation for commercial exploitation over a defined period. Farmers are allowed to access the germplasm of the variety to improve it through experimentation. The table gives a glimpse of IPRs and PVP filed by major investors during 2004–2008 (Fig. 9.1).

The Convention on Biological Diversity considered plant genetic resources as a tradable commodity subjected to national sovereignty rights (Aoki 2009). It was proposed that the biopiracy on genes and genetic resources has taken a distinct shape by the induction of IRRI. This institution gives a lot of opportunities to both public and private sector players to get access to the germplasm of a large number of varieties which acted as a resource for their research. With the invention of molecular techniques, IRRI started using the marker-assisted selection (MAS) among the varieties to choose them easily for a particular trait using simple molecular technique (Wang 2000). They have developed nearly 35 markers by 2000 to tag different traits in rice. This marker-assisted selection has given a lot of opportunity to the private sector companies to protect the genes as a proprietary item and to strengthen their claims. In 1999, a patent on soybean with increased yield was filed by Monsanto in 81 countries. It claims that the entire cultivating varieties of soybean containing specific sequences or fragments of sequences of DNA which is supposed to enhance the yield identified through MAS are protected by intellectual property rights to Monsanto. It gives Monsanto the monopoly on the wild soybean (*Glycine soja*) of the southern China province (working report, GRAIN).

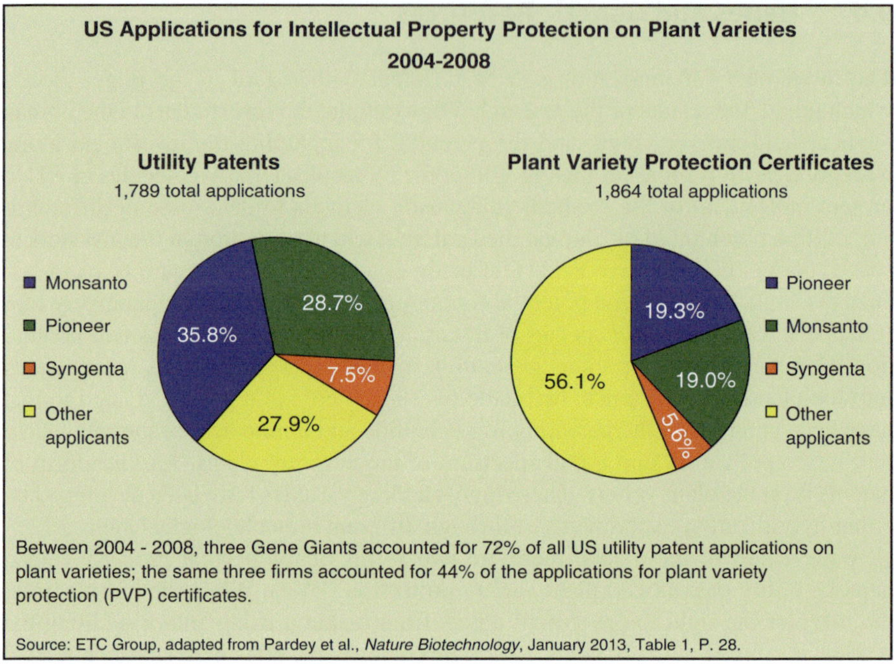

Fig. 9.1 Patents on plants varieties filed between 2004 and 2008

Monsanto made headlines in 2000 with an announcement that it will share a working draft of the rice genome with the public sector, International Rice Genome Sequencing Project (IRGSP). This was proclaimed as a generous movement by a multinational company for the benefit of the society by many. But there are a few who pointed out the fact that Monsanto would have completed a certain percentage of sequencing and may need to just fill the gaps. Once the specific genes are identified, it is very easy to file a patent for the same with a basic set of experiments. The table below indicates the patents on rice filed in the US/European and Japan patent offices by September 2000 (Table 9.1).

9.5 Indigenous Intellectual Property vs. Patent Rights

United Nations Educational, Scientific and Cultural Organization (UNESCO) and WHO have accepted long back the necessity of having an integrated indigenous knowledge platform (consists of all forms of knowledge particular to communities of different origin accumulated over hundreds of years) to improve the global health. The ethnomedicine or traditional medicine system was well established in all ancient civilizations, and it was possible for them to control many life-threatening diseases specific to their area. The geographic restrictions are vanished due to the increased mobilization between populations, and thus the traditional knowledge

Table 9.1 Biotech patents on rice genes, transgenic rice plants or methods to obtain them worldwide

Company/organization	Country	Number of patents
DuPont	USA	95
Mitsui	Japan	45
Monsanto	USA	36
Aventis	France-Germany	28
Syngenta	Switzerland	20
Japan Tobacco	Japan	19
Cornell Research Foundation	USA	18
Mitsubishi	Japan	14
CSIRO	Australia	9
University of California	USA	9
Sumitomo	Japan	8
Purdue Research Foundation	USA	8
Dow AgroSciences	USA	8
US Department of Agriculture	USA	6
Advanced Technologies	USA	5
DNA Plant Technologies	USA	5
Hokko	Japan	5
Total biotech patents on rice: 609		
Top 17 patent holders: 56%		

Source: Compiled by GRAIN from Derwent Biotechnology Abstracts

also gets expanded. In fact, the corporate sector recognized the commercial factor in the traditional knowledge than the government/public sector and started extorting the information from these communities by all means possible. They obtained IPR protection in many countries on many products developed purely on indigenous knowledge.

Globalization and increased awareness of their own intellectual rights have made the Third World countries to question the disparities as they were just the consumers of goods made of their own knowledge and they were deprived of using their indigenous knowledge to make the goods anymore. Organizations like WTO have made great contributions in this by developing an interdependent system of trade along with multilateral agreements between nations and normalization of standards of property rights. The internationalization of property claim system has evoked serious questions on many IPRs owned by different corporates and on bioprospecting and biopiracy. However since few years back, the voice of indigenous communities for justice and access to their legal rights was not getting any response from the system which in fact controlled by the corporates. The bioprospecting was formalized, and the companies have filed IPRs on whatever they want, even when the innovation was initially developed by the local people, which is biopiracy. The affected communities were exploited for years without giving them any financial or social benefits. Examples of biopiracy included the patented seeds, trees, plant species, cell lines, hybrid plant and crops, etc.

The early voices against bioprospecting and biopiracy from the Third World countries have evoked serious attempts by organizations like UN and WTO which led to the revocation of few patents issued to different corporates. The major concern is that the biopiracy activities are not restricted to corporates. National governments, academia and other organizations working in the field of bioprospecting also reported to help the corporates in obtaining the traditional knowledge of indigenous communities without their consent. One of the famous examples is the sale of the cactus "*Hoodia*" by the South Africa's Council for Scientific and Industrial Research (CSIR) to the pharma company Phytopharm. The indigenous community was neither given access to the plant for their health needs nor got any rewards for their intellectual property. For Phytopharm, it was a high-profit business as they have patented and later sold the technology to Pfizer for USD 21 million. This clearly indicates the need for a rethinking on the systems and regulations operating in the field to make sure that the indigenous population is getting a decent share for their knowledge.

9.6 Legalized Bioprospecting and Ownership Claims

The key factors involved in regulating the indigenous knowledge distribution in any country are the government officials, representatives of indigenous groups, lawmakers and persons involved in regulatory organizations. In some cases, government specifically constitutes an institution to oversee the bioprospecting activities taking place in the country. These institutions are responsible for sustainable resource exploitation and in generating benefit-sharing contracts. When it comes to the development, science plays a crucial role. To do science, the government needs the infrastructure, co-operation of public and private sectors to invest for basic research ventures, good opportunities for quality education and up-to-date technology to compete with the global market. The changes which can happen through scientific and technological interference in the industry, economy and international interaction will always keep the government vigilant about the resource utilized and the outcome.

According to the Bonn Guidelines (2002), there are five main issues which decide the government's interest in bioprospecting including:

1. The institutional infrastructure for dealing with international matters and for providing in-country research and valuation methods
2. The legality, scope and uses of contracts
3. The inclusion of the public in the source country and the process of informing any potential property sharers
4. The actual manner in which benefits are shared and distributed
5. The implementation of the contractual agreement and fidelity of the parties involved

It is the responsibility of the government to make the resource available for researchers, which will propel the basic research on many value-added products irrespective of the country barrier. The legal procedure for procuring the resources for R&D between nations can be transparent and simple, and the source countries can have laws to protect their rights on the intellectual rights and the profit sharing. So it is clear that the national development cannot be compensated in the name of biopiracy. At the same time, the dealings of the researchers and the investors across different countries have to follow the legal routes of the source country from where the natural resource is procured. Taking together the conservation of the resource and exploiting it in a sustainable way can facilitate bioprospecting and scientific research for valuable products.

9.7 Socio-environmental Perspective and Biopiracy

According to Vandana Shiva, the present-day patent rights are the modified version of the charter given to Columbus while he started his journey to explore and conquer in the fourteenth century. In the charter, the dual verb, discover and conquer, was repeated seven times to assert rights to all "islands and mainlands" before their discovery. The religious colonization is successfully replaced with economic colonization. Religious colonization needed territory or wealth in the form of gold or other precious metals which is no longer considered in the economic colonization as the knowledge can be converted into property/money equal to what gold/land did in the religious colonization. When the biodiversity of a country is patented, the whole system of socioeconomic existence is at stake. Before the Uruguay Round on GATT (General Agreement on Tariffs and Trade), there was a system existing in most of the countries protecting the intellectual property regimes of their own. The consortium of 123 countries involved in the Uruguay Round has agreed to change the IPR laws through TRIPS (Trade-Related Aspects of Intellectual Property Rights), which has redefined the patentable property, thus disturbing the balance of socioeconomic and ethical equilibrium existing in that countries. The scope of TRIPS agreement to cover life forms in fact augmented the piracy on indigenous knowledge.

The Convention on Biological Diversity (1992) has recognized the supreme rights of the nations over its natural resources. It is the responsibility of the government to utilize the resources for its population in a sustainable manner. The claims of CBD on natural resources have evoked questions on the authority of nations on it against the individual groups within and ultimately the individual himself. In a small country, many of the indigenous knowledge have been unified and nationalized among various groups (Hayden 2003). But in large countries with many individual groups, the nationalization and further profit-sharing-based knowledge disposal may not work out. This can lead to the fighting of indigenous groups among themselves, and some may prove their authenticity over the other. Granting special rights to the groups further create indigenous clashes (Harvey 2001). At the end, the question is still unclear; who is the actual owner of the natural resources, the government, the community or the individual?

Either bioprospecting or biopiracy, if not controlled, will leave a scar on the ecosystem as the natural resource may get exhausted in no time. On the view of profit and to reduce the expensive R&D, the corporates will always try to bypass the regulations and may result in excessive exploitation, which harms the environment very badly. It is thus necessary to put a system into act to prevent the unethical exploitation of both nature and the source group. The socio-environmental responsibilities involved in bioprospecting are mainly done by a diverse group of individuals, namely, non-governmental organizations (NGOs), activists, specific interest groups, local collaborations and, probably, academic researchers in different fields like biology, ecology, political science, sociology or anthropology. They all will have a common objective of protecting the environment and the interests of the local population along with the sustainable exploitation of the natural resource (Arenas et al. 2009; Peterson 2001). Social, moral and ethical norms are the driving force of these groups and will stand for the interests of the local communities. They can better communicate with the locals and convey the message to the researchers and the representatives of the company and make sure that they protect the rights of the local community. Following the research ethics and the regulations of profit sharing by the company can also ensure that the nature and the people are equally considered. Studies have shown that the conservation of biodiversity has a positive impact on the improved social life in terms of better health, quality of life and improved livelihoods for the local groups (Barrett et al. 2011).

9.8 Anti-biopiracy Movements

The anti-biopiracy movements have started after the Uruguay Rounds and the interests of the Third World countries were not taken care of. It was also considered as one of the major components of the anti-globalization movement. It witnessed the participation of indigenous communities, farmers, ecologists, socioeconomic workers, etc. The biodiversity-rich Third World countries resisted the domination of the developed north and made a review on the charter of TRIPS. There were number of outbursts around the globe after the implementation of TRIPS, and many patent rights granted to multinationals in the USA and Europe were challenged (detailed information is included in the case study section).

The first Convention on Biodiversity (1992) put forth a number of concerns over the exploration and use of natural resources, benefit sharing and intellectual property rights. After 10 years in 2002, a group of intellects called the ETC group (Action Group on Erosion, Technology and Concentration) concluded that the CBD could not establish reliable methods to track down the issue of biopiracy. ETC works closely in the socioeconomic and ecological issues around the globe and interacts with the people from different sections of society. There are ecologists, economists, anthropologists, biologists and so on, who keep a close watch on the corporate movements and trade technologies. They have noticed that there are groups around the globe who realize and resist biopiracy. So from 2002 onwards, the ETC group has started presenting the Captain Hook Award to "infamous and outstanding achievements in biopiracy" and Cog Awards (cogs were ships designed to repel pirate attacks) for the rebels.

9.9 International Treaties on Biopiracy

There are three treaties internationally which safeguard the interest of traditional communities from biopirates. The treaties are:

1. Convention on Biological Diversity (CBD) evolved of the Rio Earth Summit (1992)
2. The Cartagena Protocol (2000)
3. The Nagoya Protocol (2010)

Convention on Biological Diversity

Restoration of biodiversity was in cards for major countries since the middle of the nineteenth century. The disastrous effects generated by uncontrolled destruction of live forms and their environment led to significant climate changes. In the past, there were few conferences and meetings under the mentorship of the United Nations Environment Programme (UNEP) that started in Vienna and Switzerland in 1988. In 1992, UNEP started a campaign to protect the Mother Earth, and the convention seeks support from all over the globe. One hundred sixty-eight nations supported the programme and signed the treaty. The treaty focused on the loss of diversity and the measures to restore it. As the developing countries possess the major chunk of biodiversity, the treaty gave ample importance to the knowledge and benefit sharing among the nations arising out of the commercial and other utilizations of genetic material. The national sovereignty over its resources must be recognized, and the benefits should be shared between the inventor and the country of origin of the raw material. The CBD will act as a bridge in between nations to formulate legislations in common interest (CBD website). With regard to the traditional knowledge, the CBD and the member countries have undertaken "to respect, preserve and maintain" such knowledge and practices, to promote their wider application with the approval and involvement of the communities concerned and to encourage the equitable sharing of the benefits derived from their utilization (CBD 1992).

The Cartagena Protocol

In January 2000, the member nations of CBD have adopted a supplementary agreement known as The Cartagena Protocol on Biosafety, which was on the safe handling, transport and use of LMOs (living modified organisms) developed by artificial techniques and procedures by the biotechnologists and may have severe adverse effects on the existing biodiversity and also to the human health.

The Nagoya Protocol
The Nagoya Protocol on Access to Genetic Resources and the Fair and Equitable Sharing of Benefits Arising from Their Utilization to the Convention on Biological Diversity is an international agreement between the nations involved in CBD on the sharing of benefits in terms of cash, sharing of technology, resources or manpower. It enhances the legal transparency among the nations on access, utilization and benefit sharing.

9.10 Case Studies

The realization that the illegal bioprospecting and exploitation harm the ecological and socioeconomic balance has made certain non-profit organizations like RAFI (Rural Advancement Foundation International), the ETC group, the Third World Network (TWN) and International Centre for Trade and Sustainable Development (ICTSD) to keep a close watch on the claims of biopiracy and report it from time to time to the public through its various reports. The pirates have explored almost all possible space on Earth including all the climatic zones of land and marine ecosystems. In the aquatic environment, especially the costal explorations for new threads, the researchers never accepted the ownership of the source country from where the samples were collected. There were very serious cases of biopiracy wherein prestigious academic institutions, multinational companies and eminent scholars were a part of. Few examples of the piracy reports are given below.

Pirates of the Sea As everybody knows, the mother sea is a source of endless number of valuable products including food and medicine. The fauna and flora of marine ecosystem are reported to be rich in highly variable secondary metabolite compounds with potential therapeutic applications. According to Arrieta et al. (2010), there has been a rapid growth in the human appropriation of marine genetic resources (MGRs), with over 18,000 natural products and 4,900 patents associated with genes of marine organisms, and the latter component is escalating at a rate of 12% per year. The marine collections are present in the sovereign territory of the country and often unique of its kind. As per the reports of the organization, Tambuyog, a Philippine organization working in the field of marine biodiversity, there are many patent seekers who collect the specimens from local people along with important knowledge on its medicinal value. *Pseudopterogorgia elisabethae* is a sea whip, which produces pseudopterosin, an anti-inflammatory agent. The compound was developed by a professor of California University and included as an ingredient in the skin cream, and the same compound was licensed to two more companies as pharmaceutical ingredient. The skin cream "Resilience" alone generated revenue of USD 7,50,000/– in the year 1998. Another group of scientists with the aid of American Cancer Society collected *Diazona chinensis* and identified an anticancer compound from the organism. Eleutherobin, a medicine given to patients with solid tumours, is developed by the University of California's Scripps Institute of Oceanography and is licensed to Bristol-Myers Squibb. Another compound manoalide isolated from a marine sponge by again the Scripps Institute scientists generated 30 patents and licensed to Allergan Pharmaceuticals. In all these cases, the inventors diplomatically avoided giving the exact place from where the specimen is collected, thus preventing the source country from getting any benefits (RAFI communiqué).

Grain Biopiracy Rice is the staple food for most of the Asian countries. The cultivation spans around 150 million hectares and produces nearly 90% of the world's total rice production. It energizes 80% of the people in Asia on a daily basis. The agriculture history of Asia is nothing without rice, and it goes back to hundreds of thousands of years. Over the years, the farmers of different parts of Asia have developed varieties best suited to their region, cultivated and utilized. An estimate of nearly 140,000 varieties of rice according to scientists has existed in Asia. The farmers choose on a specific variety depending on the climate, soil conditions, availability of water and presence of any biotic or abiotic harm doers in an area. Certain rice varieties have slender grains, and others are with short or round ones. The nature of the rice varies from aromatic, medicinal, sticky, slow cooking, etc. and was developed by the farmers through years of observation and experimentation. The knowledge on the variety like the preparation of land, seeding time, water requirements, intervals of land maintenance and harvesting was transferred from generation to generation.

The green revolution has made significant changes in the system. Only few varieties were preferred over the other on certain features. As a result, the last 30 or 40 years have witnessed a sharp decline in the indigenous knowledge and the varieties cultivated. It was estimated that only half of the cultivated area with irrigation facilities in Asia showed a marginal increase in production and the agriculture becomes too dependent on agrochemicals. The soil fertility and the ecological balance were disturbed to the extent that the productivity declined in most of the areas. The so-called high-yielding varieties needed heavy agrochemical support which in turn destroyed many useful pests. These issues are directly linked to the loss of biodiversity.

Induction of the International Rice Research Institute (IRRI) in 1960 funded through the Consultative Group on International Agricultural Research (CGIAR), which operates from the World Bank, can be considered as one of the biggest biopiracy movements in the twentieth century. IRRI took over rice cultivation in Asia as it collected the farmer's wealth (the rice varieties) and created a gene bank (nearly 80,000 Asian rice varieties are conserved in IRRI gene bank). The gene bank represents almost all possible genome/genes available in the rice population, which acts as a goldmine for the researchers to look for new combinations to create new high-yielding varieties. While the intention to constitute a system like IRRI was to help the developing countries to cope up with the population explosion by increasing the yield, more than the poor countries, the industrialized countries benefited tremendously from IRRI. It was reported that three-quarters of the rice cultivation in the USA is dependent on the germplasm provided by IRRI, which earn the country nearly USD 1 billion since 1970. In fact, every year, the four countries Australia, the USA, Canada and New Zealand benefit an economic gain of nearly USD 655 million using the rice varieties obtained from IRRI. The huge monetary gain by the developed countries using the genetic resources of the Third World countries was challenged in the CBD 1992, and the regulations were put forth to access the germplasm at IRRI, and the varieties are no more patentable.

Livestock Biopiracy It is not only the plants of medicinal and commercial importance but also the animals that are prone to biopiracy. One of those incidents has been reported from Australia with regard to a breed of cattle. A joint venture between the Commonwealth Scientific and Industrial Research Organization (CSIRO) Australia and a consortium named Boran and Tuli Producers in 1987 collected Tuli embryos from Zimbabwe and Boran from Zambia. They have taken the embryos to Cocos Island and implanted into surrogate mothers. The live calves were introduced to Australia in 1990 (CSIRO media release). Commutech, a civil society organization (CSO) working with the biodiversity development and the intellectual rights of Africa, condemned this move. The Tuli breed of cattle was developed by the government of Zimbabwe in 1945 using the cows and bull (Tswana bread) from the lowland region of Southern Zimbabwe. Even though the breed was initially intended to improve the livestock of African farmers, the commercial benefits are realized very soon. The breed was sold to the European farmers for many years.

The Tuli breed brought to Australia made significant changes to the beef industry. The breed was characterized by high fertility, disease resistance, and adaptation to environmental stress with excellent meat properties. The consortium believed that the indigenous livestock can be improved with the introduction of Tuli breed by the breeding with British breed cows. The economical estimate of this induction is reported as there was a 30% increase in the production of beef, and this change is converted to USD2.4 million dollars per year. Further to this the selling of purebred Boran and Tuli embryos to farmers of the USA is a heavily profitable business to the consortium. It was reported that the Australians collect the purebred from Zimbabwe and sell it to the USA. If a fair share business has to be offered to Africans, a minimum of USD 40 million dollars (5 %) has to be spared to both Zimbabwe and Zambia if the economic gain from the beef industry is USD 800 million dollars.

Plant Biopiracy *Pentadiplandra brazzeana* is a West African plant characterized by the presence of a super-sweet protein "brazzein". Scientists from the University of Wisconsin have identified and isolated this protein from the berries of this plant, which was consumed for centuries by the people of Gabon, Africa. The university earned three commercial patents (Table 9.2) with the genomic sequence of brazzein

Table 9.2 University of Wisconsin's exclusive monopoly patents on West Africa's sweet genes

S no	US patent number	Approval date
1	5,741,537	April 21, 1998
2	5,527,555	June 18, 1996
3	5,346,998	September 13, 1994

Source: Rafi communiqué report (2000)

with which they have developed artificial systems to produce the protein in high-tech laboratories. This avoided the need for collecting the plants from the natural habitat and extracting the berries for the super-sweet protein (Rafi communiqué 1995). The brazzein is 200 times sweeter than sugar, which makes it a promising low-calorie candidate instead of sugar. Wisconsin has sold the technology to many companies and programmes on the way to express large quantities of brazzein using genetically modified maize plants. The low-calorie diet sweetener holds a market value of nearly USD 1.4 million in 2000.

Even though there was an agreement between the indigenous people and the exploiters, if it is not fulfilled to the maximum, then it also can be termed as biopiracy. One such incident was reported in the Philippines. The project involving the researchers from the University of Illinois, Chicago (UIC), was aimed to collect and evaluate the plants from the Palawan region of the Philippines for any potential therapeutic compound. In the memorandum of agreement (MOA), the partners MacArthur Foundation and the National Cancer Institute, Philippine National Herbarium, University of the Philippines and the Palawan Council for Sustainable Development (PCSD) have agreed to share the benefits.

Later, the Palawan NGO Network, Inc., an organization working against biopiracy, has discovered that one of the conditions, i.e. get the consent from the people of Palawan to collect the plant sample, was never satisfied. UIC later established a trust fund, which made an agreement to share the benefits of the potential drug. The table (Table 9.3) below indicates the disparity existing in the profit sharing even after the agreement.

East Asian countries, especially Thailand, Philippines and Indonesia, reported many incidents of biopiracy. *Cananga odorata* (ilang-ilang in the Philippines) is an ornamental plant, and the varieties present in the Philippines are highly useful in the perfume industry. French companies which used these flowers for the making of perfume imported these flowers from the Philippines for more than two decades. Later on they put up their own plantations in South Africa for *Cananga odorata* and stopped its import from the Philippines. They further secured a patent for the perfume formula which is based on the native Filipino species. There are other plants like *Lagerstroemia speciosa*, *Momordica spp.* and *Centella asiatica*, which were conventionally used as medicinal source by the native people of Indonesia and Thailand and were patented for its medicinal properties by many Japanese companies.

India is also a victim of biopiracy by the Western countries. The major instances include the cases of turmeric, neem and the basmati rice. In all these three, European-/US-based companies earned patents on the commercial products from

Table 9.3 Benefit sharing in a mutual agreement on paper vs. real

	Philippine trust fund	Royalties to inventors	Funds for institutions
On paper	47%	14.5%	4.5%
In reality	2.3%	0.73%	0.23%

these plants. The legislations and the diplomatic movements from the government of India have forced the patent authorities to overturn the patents later.

9.11 Future Perspectives

With the changing scenario of socioeconomic relations among the countries and the easy access of information from one end to the other by the expansion of technology, it has become very difficult to conceal the under-table trade how it was a few years back. The pirates are now taking new means to achieve their goals, through few indigenous so-called non-profit organizations and consortia, which help the MNC to collect the knowledge and resource. It is a fact that once the information and source organism are revealed, it may be possible that the benefits may or may not be shared. A wider collaboration between the Third World countries that are rich in heritage is needed to protect their knowledge, and this association should be able to control the biopiracy and illegal bioprospecting. There should be a databank on the available indigenous knowledge which can act as a resource to prevent the MNC from taking out any specific information and making rights on it. More than bilateral agreements, the multilateral agreements including the parties of different dimensions (there may be more than one country in an area that would have developed same information and can get benefited only if all of them were included in the agreement) will facilitate the dissemination of knowledge to the prospector, and the benefit sharing will be more meaningful. There must be a consortium involving all heritage-rich countries, and the revenue generated should be divided among the members for the conservation of their knowledge and resources. It should be noted that the medical formulations developed (based on the indigenous knowledge gained in the past) by the Western companies still control 80% of the developing world's medical field, even though their forefathers used somewhat similar medications from the same source for centuries to treat the same disease successfully. So if the indigenous knowledge is not preserved and practised, the developing countries have to spend more on their medical needs than what they acquire through benefit sharing. Even though the illegal practice of biopiracy becomes legitimate through bilateral free trade and benefit-sharing agreements, the time is not so far when the intellectuals from the developing countries start resisting the free trade agreements between developed and developing countries. Rather, an alliance of developing countries to generate commercial products of international value using traditional knowledge and the resources will be more appealing than the agreements between developed and Third World countries.

References

Amir-Aslani A, Mangematin V (2009) The future of drug discovery and development: shifting emphasis towards personalized medicine. Technol Forecast Soc Change. Doi: 10.1016/j.techfore. 2009.09.005

Arenas D, Lozano JP, Albareda L (2009) The role of NGOs in CSR: mutual perceptions among stakeholders. J Business Ethics 88:175–197
Arrieta JM, Arnaud-Haond S, Duarte CM (2010) What lies underneath: conserving the oceans' genetic resources. Proc Natl Acad Sci USA 107(43):18318–18324
Aoki K (2009) Symposium: when worlds collide: intellectual property at the interface between systems of knowledge creation: panel II: knowledge creations systems on the national stage: "Free Seeds, Not Free Beer": participatory plant breeding, open source seeds, and acknowledging user innovation in agriculture. 77 Fordham L. Rev. 2275
Barrett CB, Travis AJ, Dasgupta DP (2011) On Biodiversity conservation and poverty traps. Proc Natl Acad Sci USA 108:13907–13912
Bhattacharya S, Chattopadhyay DJ, Mukhopadhyay A (2013) Changing dimensions of food security in a globalized world: a review of the perspectives for environment, economy and health. Int Res J Environ Sci 2(3):67–73
Bonn Guidelines (2002) Bonn guidelines on access to genetic resources and fair and equitable sharing of the benefits arising out of their utilization. United National Environmental Programme (UNEP) Convention on Biological Diversity. http://www.cbd.int/abs/bonn.shtml
Chaudhuri SK (2003) Microbial biopiracy in India; How to fight back? J Intellect Prop Rights 8:389–399
Christoffersen LP, Fish SC (1999) Standing up to Biopiracy foresting sustainable development through bioprospecting. Resour Afr 7:23–24
Cunha MP, Guimarães-Costa N, Rego A, Clegg SR (2010) Leading and following (Un) ethically in Limen. J Business Ethics 97:189–206
Edelman LB (1992) Legal ambiguity and symbolic structures: organizational mediation of civil right law. Am J Sociol 97:1531–1576
Gibbons S (2008) Phytochemicals for bacterial resistance--strengths, weaknesses and opportunities. Planta Medica 74:594–602
Gupta C, Prakash D, Gupta S (2014) Natural useful therapeutic products from microbes. J Microbiol Exp 1(1):00006. DOI: 10.15406/ jmen.2014.01.00006
Gurang N, Ray S, Bose S, Rai V (2013) A broader view: microbial enzymes and their relevance in industries, medicine, and beyond. BioMed Res Int. dx.doi.org/10.1155/2013/329121
Harvey N (2001) Globalisation and resistance in post-cold war Mexico: difference, citizenship, and biodiversity conflicts in Chiapas. Third World Q 22:1045–1061
Hayden C (2003) When nature goes public: the making and unmaking of bioprospecting in Mexico. Princeton University Press, Princeton
Hayden C (2007) Taking as giving: bioscience, exchange, and the politics of benefit-sharing. Soc Stud Sci 37:729–758
Jefferson OA, Köllhofer D, Ehrich TH, Jefferson RA (2015) Gene patent practice across plant and human genomes. Nat Biotechnol 33(10):33–38
Koehn F, Carter GT (2005) The evolving role of natural products in drug discovery. Nature 4:206–220
Kurien K, Das A (2011) Nagoya Protocol and its implications on pharmaceutical industry. Beroe Inc. Publication, pp 3–6
Ministry of Environment and forests (2010) National consultations on BT brinjals: a primer on concerns, issues and prospects. Centre for environment education, India.
Nazarea VD (2006) Local knowledge and memory in biodiversity conservation. Annu Rev Anthropol 35:317–335
Nigh R (2002) Maya medicine in the biological gaze: bioprospecting research as herbal fetishism. Curr Anthropol 43:451–477
Peterson K (2001) Benefit sharing for All?: bioprospecting NGOs, intellectual property rights, New Governmentalities. PoLAR 24:78–91
RAFI Communiqué, "Biopiracy Update: A Global Pandemic," Sept-October, 1995.
Rose J, Quave CL, Islam G (2012) The four-sided triangle of ethics in bioprospecting: pharmaceutical business, international politics, socio-environmental responsibility and the importance of local stakeholders. Ethnobiol Conserv 1:1–25

Samuels J (2012) Genetically engineered Bt brinjal and the implications for plant biodiversity – revisited. Greenpeace, India

Shankar D (1997) Traditional medicine and biopiracy. Ancient Sci Life 17(1):67–71

Shiva V (1997) Biopiracy: the Plunder of nature and knowledge. South End Press, Cambridge, MA

Soejarto DD, Fong HH, Tan GT, Zhang HJ, Ma CY, Franzblau SG, Gyllenhaal C, Riley MC, Kadushin MR, Pezzuto JM, Xuan LT, Hiep NT, Hung NV, Vu BM, Loc PK, Dac LX, Binh LT, Chien NQ, Hai NV, Bich TQ, Cuong NM, Southavong B, Sydara K, Bouamanivong S, Ly HM, Thuy TV, Rose WC, Dietzman GR (2005) Ethnobotany/ethnopharmacology and mass bioprospecting: issues on intellectual property and benefit-sharing. J Ethnopharmacol 100:15–22

Walters D (2011) Plant defence: Warding off attack by pathogens, herbivores and parasitic plants. Wiley-Blackwell, London

Wang R (2000) Strategic rice research: Impossible problems, possible solutions. Presentation at CGIAR International Centers Week, Washington, DC, 26 October 2000

Yusuf M (2010) Ethical issues in the use of the terminator seed technology. Afr J Agric Res 9(52):8901–8904

Online Resources

https://www.cbd.int/convention/guide/default.shtml?id=intaction

http://www.etcgroup.org/files/publication/327/01/com_biopiracy.pdf

https://www.cbd.int/abs

http://www.greenpeace.org/india/PageFiles/446445/GE-Bt-brinjal-revisited.pdf.

http://bch.cbd.int/protocol/background/

https://www.wto.org/english/docs_e/legal_e/gatt47_e.pdf

https://www.wto.org/english/docs_e/legal_e/27-trips.pdf

https://www.grain.org/article/archive/categories/14-reports

https://www.grain.org/article/entries/91-ftas-trading-away-traditional-knowledge

https://www.grain.org/article/entries/30-intellectual-property-rights-ultimate-control-of-agricultural-r-d-in-asia#55

https://www.csiro.gov.au.

https://www.questia.com/magazine/1G1-111800832/the-new-scramble-for-africa-biopiracy-africa-is

Part II

Strategies for Sustainable Utilization

Biodiversity Mining Through Transcriptome Analysis

10

K.K. Sabu and F. Nadiya

Abstract

Secondary metabolites or chemical constituents of plants are functional products synthesized through more complex multienzymatic secondary metabolite pathways. The release of few secondary metabolites is related to the exertion of morphological differentiation and maturation during plant development. Secondary metabolite compounds designate huge chemical diversity, *i.e.*, each plant possesses its own complex diverse set of metabolites. This diversity imparts analytical challenges, characterization of a number of metabolites in parallel, and quantification of particular compound of interest. These compounds itself are related to significant plant traits (color and fragrance of flowers, taste and color of food, and resistance toward pests and pathogens) and also for the synthesis of fine chemicals such as medicines (anticancer, anti-inflammatory, antioxidants, etc.), flavors, fragrances, dyes, insecticides, pheromones, and antimicrobial agents. The emergence of cost-effective high throughput sequencing or next-generation sequencing (NGS) technology quickly expanded transcriptome information of several plant species, which could be analyzed for quick identification of previously unknown genes and enzymes and elucidation of biosynthetic pathways. The present chapter details diverse secondary metabolites, gene expression patterns, molecular basis of chemical diversity of the compounds, and application of NGS based transcriptome profiling for biosynthetic pathway elucidation.

Keywords

Transcriptome • Metabolomics • Phytochemicals • Gene sequencing • Biomolecules • Biodiversity mining

K.K. Sabu (✉) • F. Nadiya
Biotechnology and Bioinformatics Division, Jawaharlal Nehru Tropical Botanic Garden and Research Institute, Palode, Thiruvananthapuram 695562, Kerala, India
e-mail: sabu@jntbgri.res.in

S. Abdulhameed et al. (eds.), *Bioresources and Bioprocess in Biotechnology*,
DOI 10.1007/978-981-10-3573-9_10

10.1 Introduction

From several years, plant metabolomics has great importance due to its huge chemical diversity when compared to animals and microbes. The extensive chemical biodiversity of plants was explored to develop well-targeted therapeutic strategy for better protection, less side effects, and decreased relapse. These chemical constituents are small organic molecules with specific biological activity restricted to particular plant species, genus, or family and are termed as secondary metabolites which comprises the major physiologically active chemical constituents distributed to distinct taxonomic group, and many of which are applied to several prime valued chemicals, pharmaceuticals, flavors, and fragrances (Fabricant and Farnsworth 2001; Oksman-Caldentey and Saito 2005). Different from the function of primary metabolites in fundamental processes, like plant growth and development, secondary metabolites chiefly participates in mediating plant environment interactions, including the pollinator attraction and defense response to plant pathogens (Xiao et al. 2013). These specialized metabolites are distinguished by diverse chemical structures and can be grouped into five major categories depending on the nature of biosynthesis: polyketides, isoprenoids, alkaloids, phenylpropanoids, and flavonoids. Biosynthesis of polyketides is elucidated through the acetate-mevalonate pathway; the isoprenoids such as terpenoids and steroids are produced from a five-carbon precursor isopentenyl diphosphate (IPP), through mevalonate pathway or the novel MEP (non-mevalonate or Rohmer) pathway; the various amino acids present produce alkaloids, and a combination of phenylpropanoids and polyketides synthesizes the flavonoids (Robert Verpoorte 2000). Broadly, most categories possess thousands of known and novel compounds with more emerging discoveries and interpretations. Both primary and secondary metabolite production is initiated through complex, enzymatic, multistep biosynthetic pathways (Nicolaou and Chen 2011). Plant phenolics such as flavonoids and phenols are synthesized through various routes and thus comprise a heterogeneous group. The two fundamental pathways associated are the shikimic acid and the malonic acid pathways. The shikimic acid pathway participates in the biosynthesis of most plant phenolics, and the soluble carbohydrates serve as the basic factor in producing phenolic component. The conversion of simple carbohydrate precursors derived from glycolysis and pentose phosphate pathway to aromatic amino acids is mediated through shikimic acid pathway (Herrmann and Weaver 1999). However, the shikimic acid and acetate-malonate pathways are the prominent metabolic routes of polyphenolic synthesis in plants, and two precursors, acetate and phenylalanine, are required for flavonoid synthesis (Van Soest 1982; Jung and Fahey 1983). The potential application of diverse secondary metabolites in plant breeding technology was extensively studied (Wink 1988). Figure 10.1 outlines how metabolites from the process of photosynthesis, glycolysis, and Krebs cycle lead to synthesis of secondary metabolites (Ghasemzadeh and Jaafar 2011).

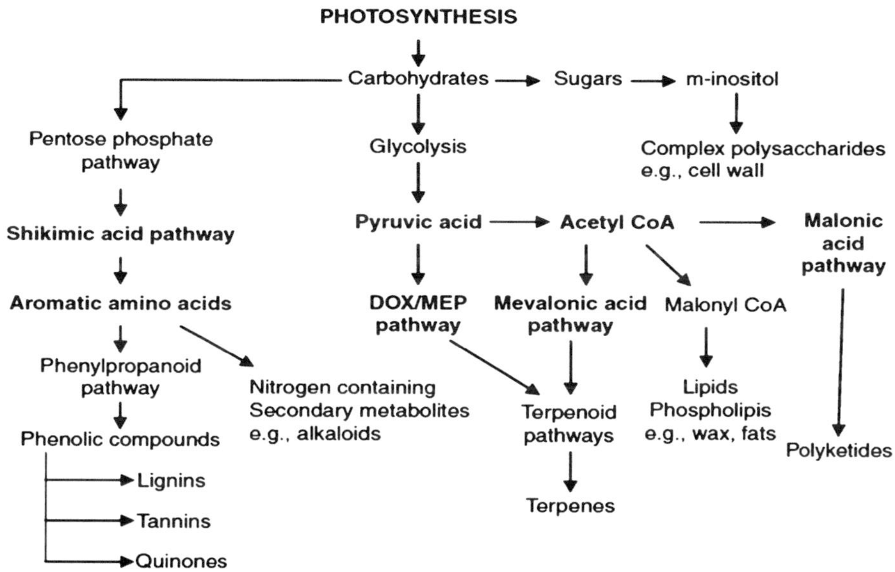

Fig. 10.1 Elucidation of biosynthetic pathways synthesizing secondary metabolites (Adopted from Ghasemzadeh and Jaafar 2011)

10.2 Functions and Applications of Secondary Metabolites

Essential Oils Essential oils also known as volatile oils are complex mixtures, constituted by terpenoid hydrocarbons, oxygenated terpenes, and sesquiterpenes responsible for their characteristic aroma in plants, and also serve as internal messengers Harrewijn et al. (2001). Due to their powerful fragrance, they are involved in attracting pollinating insects and act as defense substances or plant volatiles against diseases and predators (Evans and Mitch 1982). Location of essential oil varies in different parts with different plants, for example, they can be observed in leaves like in eucalyptus trees (*Eucalyptus citriodora*) and citronella (*Cymbopogon nardus*); in roots such as that of the vetiver grass (*Vetiveria zizanioides*); in stems such as peteribi wood (*Cordia trichotoma*) and incense, chinchilla (*Tagetes minuta*), and lemongrass (*Cymbopogon citratus*); in flowers like lavenders (*Lavandula officinalis*); and in fruits of citrus species (lemon, orange), and also they are available in seeds such as anise (*Pimpinella anisum*), coriander (*Coriandrum sativum*), and pepper (*Piper nigrum*), among others (Baser and Buchbauer 2015). The major representatives of essential oil involve α-pinene, β-pinene, limonene, α-ocimene, geraniol, anethole, germacrene D, α-terpenoil, γ-cadinene, δ-cadinene, myrcene, etc. in varying concentrations in different plants (Shexia Ma 2012) (Fig. 10.2).

Several research findings showed the multipurpose application of essential oil in perfumery, food industry, sweets and beverage preparation, and aroma therapeutic products of plant origin (Bernáth and Fuleky 2009). The role of essential oil in

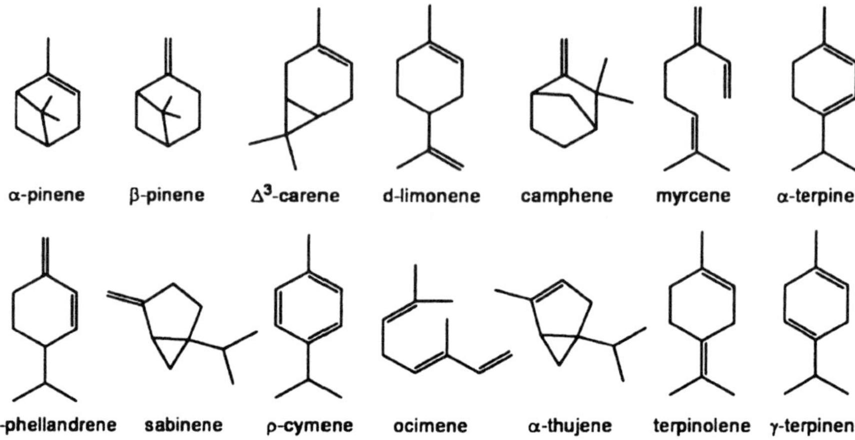

Fig. 10.2 Chemical structure of major essential oil components in plant species (Adopted from Shexia Ma 2012)

pharmacological effects such as antibacterial, antifungal, and antiviral properties was reported (Böhme et al. 2014). Several reports demonstrated that plants, particularly diverse bioactive components of essential oils, protect crops from contamination by various mold species (Kitic et al. 2013).

Terpenoids The terpenes, or isoprenoids, constitute one of the most diverse classes of metabolites, and are isoprene derivatives synthesized from acetate through mevolanate pathway. Steroids, gibberlic acid, and carotenoids represent few members of this class. Terpenoid biosynthesis proceeds through the condensation of isoprene units (C5) and is categorized by the number of five carbon units found in the core structure (Mahmoud and Croteau 2002). Biosynthesis of terpenes involves two pathways: the mevalonate pathway in the cytosol and the MEP pathway (2-C-methyl-D-erythritol-4-phosphate) in plastids. The major building blocks of terpene are an isoprene unit that is derived from isopentenyl diphosphate (IPP) and dimethylallyl diphosphate (DMAPP) (McGarvey and Croteau 1995). Terpenes include monoterpenes (constituents of the essential oils, floral scents, and defensive resins particularly in aromatic plants), sesquiterpenes (less volatile than monoterpenes), diterpenes which contribute large group of terpenoids with a wide range of biological activities (e.g., phytol, reduced form of geraniol), and sesterpenes (least common group of terpenoids) (Wang et al. 2005). Plants utilize terpenoid compounds for various purposes such as growth and development and predominantly more specific chemical interactions and plant response to abiotic and biotic environment stress. Usually, plant-derived terpenoids have been used in the food, pharmaceutical, and chemical industries, and shortly they have been explored in producing certain biofuel products. Besides, the ecological priority of terpenoid metabolites has achieved great attention in developing novel strategies for viable pest control and plant stress protection (Tholl 2015). Drugs derived from terpenoids such as artemisinin and its

Table 10.1 Classification and application of terpenes according to number of isoprene units

Class	Isoprene units/ C-atoms	Pharmacological importance	References
Monoterpenes	2/10	Antitumor properties	Crowell (1999)
Sesquiterpenes	3/15	Anti-inflammatory and antimalarial properties	Brossi et al. (1988) and Safayhi et al. (1994)
Diterpenes	4/20	Anticancer activity	Gogas and Fountzilas (2003)
Sesterpenes	5/25	Antimicrobial activity	Giannini et al. (2001)
Triterpenes	6/30	Analgesic and anti-inflammatory properties	Mohamed et al. (2009)
Tetraterpenes	8/40	Antioxidant and cardiovascular properties	Wagner and Elmadfa (2003)

derivatives provided significant effects on human disease therapy and prevention, and enormous benefits to patients and for the pharmaceutical industries (Wang et al. 2005). Reports also showed the anticancer property of terpenoid compounds which promises to open new more opportunities for cancer therapy (Huang et al. 2012) (Table 10.1).

Alkaloids They can be defined as pharmacologically active, nitrogen-containing basic compounds of plant origin which blocks ion channels, obstructs enzymes, or hinders with neurotransmission and exert hallucinations, coordination loss, convulsions, vomiting, and death. Previous studies demonstrated that alkaloids do have major role in ecochemical functions either as in plant defense mechanism against invading pathogens and herbivores or, another case like pyrrolizidine alkaloids, as pro-toxins for insects, which precede additional modification of alkaloids and then integrate them to its own defense secretions (Hartmann 1991). The first alkaloid isolated is morphine from *Papaver somniferum*, quinine from *Cinchona* species, and coniine (poison hemlock) from *Conium maculatum*, and reserpine purified from *Rauwolfia serpentina* (the prominent reported true alkaloids). Usually alkaloid biosynthesis occurs from either of the common amino acids, specifically, aspartic acid, lysine, tyrosine, and tryptophan (Pearce et al. 1991) (Table 10.2).

Flavonoids Flavonoids represent low molecular weight (Heim et al. 2002; Fernández et al. 2006) bioactive polyphenols (Hollman and Katan 1999) which play pivotal role in photosynthesizing cells synthesized by phenylpropanoid pathway. They are huge class of polyphenolic compounds bearing a γ-benzopyrone structure and are ubiquitous in plants. Numerous studies demonstrated the key role of secondary metabolites particularly possessing phenolic nature such as flavonoids in wide variety of pharmacological activities (Pandey 2007; Mahomoodally et al. 2008). It has been shown that flavonoids and hydroxylated phenolic substances are produced by plants in response to pathogenic infection particularly to microbial infections (Dixon et al. 1983). Though flavonoids hold numerous biochemical

Table 10.2 Source and pharmacological properties of major alkaloids

Alkaloid name	Source	Pharmacological activity	References
Atropine	*Atropa belladonna*, *Datura stramonium*, *Mandragora officinarum*	Competitive antagonist of muscarinic acetylcholine receptors, anticholinergic, anti-myopia effects	McBrien et al. (2013) and Gu et al. (2011)
Berberine	*Berberis* species, *Hydrastis*, *Canadensis*, *Xanthorhiza simplicissima*, *Phellodendron amurense*, *Coptis chinensis*, *Tinospora cordifolia*, *Argemone mexicana*, and *Eschscholzia californica*	Anti-inflammatory, antibacterial/viral, recent experiments showed antidiabetic and beneficial effects on cardiovascular system and various types of cancer such as intestinal	Agyapong et al. (2013), Kim et al. (2010), Zha et al. (2010) and Zhang et al. (2010)
Coniine	*Conium maculatum*, *Sarracenia flava*	Neurotoxin, poisonous	Hajek et al. (2013) and Panter et al. (2013)
Morphine	*Papaver somniferum* and poppy derivatives	Act on CNS (central nervous system), on myenteric plexus, and acute pulmonary edema and reduce the shortness of breath	Clarke et al. (1988), Rozov-Ung et al. (2014) and Takita et al. (2000)
Nicotine	Solanaceae plants family	Stimulant, antiherbivore, insecticide, anti-inflammatory	Achan et al. (2011) and Melton (2006)
Quinine	*Cinchona succirubra*, *C. calisya*, *C. ledgeriana*, plants of Rubiaceae family	Antimalarial, antipyretic, analgesic, anti-inflammatory, antiarrhythmic, bacteriostatic	Adnyana et al. (2013), Clifford et al. (2012) andEl-Tawil et al. (2010)
Solanine	*Solanum tuberosum*, *S. lycopersiam*, *S. nigrum*, plants of Solanaceae family	Antifungal, antipesticide, sedative, anticonvulsant, anticarcinogenic, anti-inflammatory	Kenny et al. (2013), Lu et al. (2010) and Mohsenikia et al. (2013)

Source: Kabera et al. (2014)

properties, the frequently studied feature of most groups of flavonoids is its antioxidant property. The flavonoid-based antioxidant property depends on the arrangement of functional groups in the nuclear structure. The total number of hydroxyl groups, substitution, and configuration critically influences the mechanisms involved in the antioxidant activity like the ability to metal ion chelation and free-radical scavenging (Heim et al. 2002; Pandey et al. 2012). Flavonoids such as catechin, apigenin, quercetin, naringenin, rutin, and venoruton were illustrated for their potential hepatoprotective properties (Tapas et al. 2008). Various clinical findings showed the significance and safety of flavonoids in treating hepatobiliary dysfunction and digestive disorders including fullness sensation, loss of appetite, abdominal pain, and nausea. *Equisetum arvense* flavonoids, hirustrin, and avicularin obtained from different sources were observed to accord protection toward

Table 10.3 Flavonoid classes and its dietary source

Class	Flavonoid	Dietary source	References
Flavanol	(+)-Catechin (−)-Epicatechin Epigallocatechin	Tea	Lopez et al. (2001)
Flavone	Chrysin, apigenin, rutin, luteolin, and luteolin glucosides	Fruit skins, red wine, buckwheat, red pepper, and tomato skin	Hara et al. (1995), Kreft et al. (1999) and Stewart et al. (2000)
Flavonol	Kaempferol, quercetin, myricetin, and tamarixetin	Onion, red wine, olive oil, berries, and grapefruit	Stewart et al. (2000)
Flavanone	Naringin, naringenin, taxifolin, and hesperidin	Citrus fruits, grapefruits, lemons, and oranges	Miyake et al. (2000)
Isoflavone	Genistin, daidzin	Soya bean	Reinli and Block (1996)
Anthocyanidin	Apigenidin, cyanidin	Cherry, raspberry, and strawberry	Hertog et al. (1992) and Stewart et al. (2000)

Source: Kumar and Pandey (2013)

chemically prompted hepatotoxicity in HepG2 cells (Spencer et al. 2009; Kim et al. 2011). Flavonoid extracts from various plant species have been observed to exert antibacterial activity (Mishra et al. 2013). Other flavonoids such as flavones, apigenin, galangin, isoflavones and flavanones, flavonol glycosides, and chalcones were also reported to possess potent antibacterial activity (Cushnie and Lamb 2005). Flavonoids such as luteolin, apigenin, hesperidin, and quercetin were observed to exert anti-inflammatory and analgesic effects. Flavonoids act particularly on enzyme systems significantly associated with inflammatory processes, specifically tyrosine and serine-threonine protein kinases (Nishizuka 1988; Hunter 1995) (Table 10.3).

Phenolics Phenolic compounds are plant secondary metabolites that constitute the largest group of substances synthesized through shikimate-phenylpropanoid pathway, deriving phenylpropanoids, or the polyketide and acetate-malonate pathway, which leads to produce simple phenols. They comprise a big reservoir of natural chemical diversity that holds a wide variety of compounds and enzymes and regulate the mechanisms of gene regulation and transport of metabolites and enzymes. Phenols exhibit antioxidant, anti-inflammatory, anticarcinogenic, and other biological properties and may provide protection to oxidative stress and some infectious diseases (Park et al. 2001). Two main phenolic compounds found in most plants are hydroxybenzoic and hydroxycinnamic acids.

Phenolic acids are present in plants as esters or glycosides attached with other natural compounds like flavonoids, alcohols, hydroxyfatty acids, sterols, and glucosides (Dai and Mumper 2010). The total phenolic compounds in tea, coffee, berries, and fruits were estimated up to 103 mg/100 g fresh weight (Manach et al. 2004).

Tannins Tannins are phenolic compounds made up of diverse group of oligomers and polymers synthesized during shikimic acid pathway (phenylpropanoid pathway) and can be grouped into two major categories: hydrolysable tannins and condensed tannins (non-hydrolysable) (Athanasiadou et al. 2001; Chaichisemsari et al. 2011; Hassanpour et al. 2011; Maheri-Sis et al. 2011). Condensed tannins are the usual tannins found in forage legumes, trees, and stems (Barry and McNabb 1999) and widely distributed in legume pasture species including *Lotus corniculatus* and in some types of acacia and numerous other plant species (Degen et al. 1995). Tannins differ from other secondary metabolites in forming complex with proteins, starch, minerals, other basic and large molecular compounds, pigments, and metallic ions. While evaluating polyphenols, some tannin, ellagitannins, and their oxidized analogs, pentagalloyl glucose and EGCG, exhibited effective suppression of tumor initiation in second stage of two-stage chemical carcinogenesis (Ito et al. 1999). Cancer-preventing activity of tannins through inhibition was substantially validated on EGCG, disclosing its positive and potential effects in this area (Yoshizawa et al. 1987). Tannins also possess beneficial effects on protein metabolism in ruminants, lowering rumen degradation of dietary protein and enhancing amino acids absorption in the small intestine.

Glycosides Glycosides can be phenol, alcohol, or sulfur compounds largely distributed in the plant kingdom. The sugar which can be observed in cyclic structure consists of a sugar part known as glycon and non sugar group called aglycon. The sugar part is covalently attached to the aglycon through hydroxyl group. The non-sugar part aglycon may be a terpene, flavonoid, or any other natural product. Many plants store chemicals in the form of inactive glycosides, which can be activated by enzyme hydrolysis (Polt 1995). Glycosides found major therapeutic applications such as anticancer (Zhou et al. 2013), expectorant (Fernández et al. 2006), sedative, and digestive activities (Galvano et al. 2004).

Saponins Saponins are high molecular weight glycosides and contain sugar unit(s) attached to a triterpene or a steroid aglycone. Saponins mostly exhibit detergent properties and lower the surface tension of aqueous solutions and so provide stable foams when in contact with water. Saponins were reported to cause hemolysis, often with bitter taste, and may exert toxic effects to cold-blooded animals (Guçlu-Ustundag and Mazza 2007). Several plant-derived drugs and traditional medicines, particularly distributed in Asia, consist of saponins, and hence there is a prominent focus in characterization and elucidation of their pharmacological and biological properties (Hostettmann and Marston 2005). Saponins possess major application in physicochemical properties (emulsification, sweetness, and bitterness), biological properties (antimicrobial, antioxidant, insecticide, and ichthycide), food, cosmetics, pharmaceutical industries, and soil bioremediation (Kabera et al. 2014).

10.3 Bioresources Mining Through Gene Manipulation

Available genomic resources and developing tools in synthetic biology promote the metabolic engineering of prime valued secondary metabolites in plants. Exploring the molecular biology of complex multistep biosynthetic pathways may reveal more opportunities for developing plants bearing better disease resistance, altering the levels of human health-promoting compounds (nutraceuticals) in food crops, and introducing genetically modified plants with improved secondary metabolites for the production of pharmaceuticals (Dixon and Arntzen 1997). The molecular characterization of numerous classes of plant natural product biosynthetic enzymes, like terpene cyclases and polyketide synthase, leads in developing transgenic plants with the aid of specific plant gene cloning through PCR strategies or through EST database information (Dixon 1999). More recent development in transgenic technology toward potential pharmacological studies includes the application of transgenic organ cultures for achieving increased production and biotransformation of vital secondary metabolites. Induction of Ri plasmid of *Agrobacterium rhizogenes* in hairy roots has been demonstrated as an effective way in eliciting various secondary metabolites which are synthesized normally in differentiated plant roots (Lam et al. 1984). *De novo* biosynthesis and biotransformation of particular secondary products that are normally produced in leaves of plants can be achieved either through the shooty teratomas stimulated by the tumor forming Ti plasmid or through *Agrobacterium tumefaciens* shooty mutant (Escobar and Dandekar 2003). Another more precise means to alter secondary metabolite producing pathways is accorded through transferring and expressing significant modified genes to specific plant cells using vector systems such as *Agrobacterium* (Chung et al. 2006). Transcriptional control of metabolism within the cells in response to developmental and environmental signals is a key factor of plant metabolic regulations (Gaudinier et al. 2015). Genetic engineering has become a potential strategy for producing improved secondary metabolite production and developing pharmaceuticals such as interferon, growth hormones, and growth factors (Goddijn and Pen 1995). This approach can also be applied in obtaining rare compounds restricted to wild plants, which are difficult to culture, or in low concentration through introducing key genes to biosynthetic pathways of interest to cultivar plant species. Another application is the utilization of transcriptional regulation of metabolic pathways for increased production of secondary metabolites through overexpression of specific genes or hindering undesirable step of metabolic pathways (Giulietti and Ertola 1997). Genetic engineering of plant genomes has led to direct alteration of plant metabolism and the ability to manipulate the amount and nature of plant secondary metabolites of commercial value. Hence, plants are nowadays thought to be a potential factory for producing diverse useful products (Kishore and Somerville 1993; Ap Rees 1995). The major aim of generating transgenic plants utilizing these novel and powerful approaches modifies plant secondary metabolic pathways to increase low molecular weight compounds and polymers, enhanced resistance towards biotic and abiotic stresses, modified food quality (comprising altered levels in carbohydrate, protein and lipid concentration), and formation of polypeptides for pharmaceutical, medical, and industrial use (Ap Rees 1995; Herbers and Sonnewald 1999; Miflin 2000; Ye et al. 2000; Kumar 2001; Veronese et al. 2001; Rohini and Rao 2001; Lessard et al. 2002; Sharma et al. 2002). Recently, several plant derived products such as taxol (anticancer), vinblastine/vincristine (anticancer), artemisinin (antimalarial), reserpine (antihypertension), and quinine (antimalaria) have emerged as potent drugs for treating several human disorders. Genetic manipulation can be attributed through decreasing the production of either an unwanted compound or group of compounds in the secondary metabolic pathway (Verpoorte and Memelink 2002). Another investigation on modification of terpenoid biosynthesis in *Arabidopsis* has shown that a gene encoding 1-deoxy-D-xylulose-5-phosphate synthase (DXPS) enzyme has been continuously overexpressed in the initial enzymatic step of terpenoid metabolism. Enhanced gene expression leads to increased enzymatic activity and resulted in increased production of terpenoids in *Arabidopsis* (Estévez et al. 2001). Antioxidants such as flavonoids and anthocyanins are beneficial to human health and hence used as key targets for genetic manipulation. Based on this attempt on increasing flavonoid levels in tomato by overexpressing, the *Petunia* chalcone isomerase (CHI) gene showed a 78-fold increase of flavonoid levels in tomato fruits (Muir et al. 2001). Genetic improvement on alkaloid levels in *Catharanthus roseus* by overexpressing the genes encoding tryptophan decarboxylase (TDC) produced increased levels of tryptamine alone without changing the total alkaloid content (Verpoorte and Memelink 2002), while overexpression of STR (strictosidine synthase) resulted in increased total alkaloid contents (Canel et al. 1998). Whereas, the overexpression of a transcription regulatory factor ORCA3, regulating major steps in alkaloid metabolic pathway in *C. roseus*, cannot induce enhanced alkaloid production (van der Fits and Memelink 2000). A study was reported on anthocyanin biosynthesis in maize which is regulated through combinatorial action of two transcription factors R and C1. The whole flavonoid biosynthetic pathway has been induced through overexpressing of the transcription factors R and C1 in *in vitro* undifferentiated maize cell cultures (Grotewold et al. 1998). In rice activation of anthocyanin, biosynthesis was achieved through overexpressing the maize transcription factors C1 and R in combination with the chalcone synthase gene resulting in increased fungal resistance (Gandikota et al. 2001). In *Arabidopsis* the overexpression of a single MYB-type transcription factor (PAP1) resulted in plants bearing intense purple pigmentation during entire plant development (Borevitz et al. 2000). Based on these reports, it is evident that strict and specific genetic control of natural product accumulation during natural plant development can be altered by overexpression of either of the few transcription factors, even in heterologous plant species for better production and resistance (Table 10.4).

In some cases transcription factors inhibit the plants from accumulating natural compounds. In *Arabidopsis* inhibition of gene MYB4 resulted with increased sinapate ester levels in leaves and enhances the tolerance to UV-B irradiation (Jin et al. 2000). Likewise, the similar expression manipulation in strawberry led to reduced flower pigmentation and lower anthocyanin and flavonol levels, proving that in strawberry fruit MYB-type transcription factor acts as a repressor in particular steps of flavonoid metabolic pathway (Aharoni et al. 2001). The application of

Table 10.4 Plant-derived secondary metabolites and their pharmaceutical properties

Secondary metabolites	Compounds	Plant source	Therapeutic property
Alkaloids	Atropine, hyoscyamine, scopolamine	*Solanaceous* spp.	Anticholinergic
	Vinblastine, vincristine	*Catharanthus roseus* L.	Antineoplastic
	Nicotine	*Nicotiana* spp.	Smoking cessation
	Codeine, morphine	*Papaver somniferum* L.	Analgesic, antitussive
	Quinine	*Cinchona* spp.	Antimalarial
	Quinidine	*Cinchona* spp.	Cardiac depressant
	Artemisinin	*Artemisia annua* L.	Antimalarial
Terpenes and steroids	Diosgenin, hecogenin, stigmasterol	*Dioscorea* spp.	Oral contraceptives, hormonal drugs
	Taxol and other taxoids	*Taxus brevifolia* Nutt.	Antineoplastic
	Digoxin, digitoxin Sennosides A and B	*Digitalis* spp., *Cassia angustifolia* Vahl.	Cardiotonic, laxative
Others	Ipecac, podophyllotoxin	*Cephaelis ipecacuanha* (Brot.) A. Rich, *Podophyllum peltatum* L.	Emetic, antineoplastic

Adopted from Teli and Timko (2004)

β-carotene biosynthesis into rice through overexpressing genes encoding phytoene synthase, phytoene desaturase, and lycopene β-cyclase is a major achievement of transcription-based metabolic engineering (Ye et al. 2000).

A study demonstrates the manipulation of entire secondary metabolite biosynthesis pathway in a heterologous plant species through overexpression of cyanogenic glucoside biosynthesis genes from *Sorghum bicolor* to *Arabidopsis* (Tattersall et al. 2001). Upon tissue damage cyanogenic glucoside dhurrin present in sorghum got hydrolyzed by a β-glucosidase, and the generating cyanide can act as a significant pest deterrent and insecticide. Dhurrin is synthesized from tyrosine catalyzed by two multifunctional cytochrome P450 enzymes (CYPs) in the presence of UDPG glucosyltransferase. Overexpression of the first enzyme cytochrome P450 in the *Arabidopsis* pathway resulted in the formation of p-hydroxybenzylglucosinolates, which are normally absent in this plant species (Bak et al. 1999; Petersen et al. 2001). Hence the overexpression of the *Sorghum* specific glucosyltransferase in the presence of two CYP genes induces dhurrin synthesis in *Arabidopsis* and resulted in dhurrin producing transgenic *Arabidopsis* which release increased levels of cyanide during tissue damage. As a result transgenic *Arabidopsis* leaf tissue was rejected by larvae of the flea beetle *Phyllotreta nemorum*, and it was observed that larvae feeding on transgenic *Arabidopsis* leaves got expired. This study effectively proved that enhanced levels of a foreign metabolite were induced in a plant species without affecting its growth and development and gained positive effects on resistance against invading pests. Another example suggests the effective expression of synthetic three gene clusters in *Escherichia coli* to produce plant flavanones such as pinocembrin and naringenin using amino acids phenylalanine and tyrosine (Hwang et al. 2003). Reports also suggest the successful manipulation of yeast with two plant specific genes to generate stilbene resveratrol acts as antimicrobial agent from fed 4-coumaric acid (Becker et al. 2003). *E. coli* strains were developed by the overexpression of artificial amorpha-4, 11-diene synthase gene and the mevalonate isoprenoid pathway from *Saccharomyces cerevisiae* to increase the production of artemisinin precursor amorphadiene (a sesquiterpene olefin). The strains developed in such a way can serve as good manifest hosts for the synthesis of all terpenoid compounds when a terpene synthase gene is available because isopentenyl and dimethylallyl pyrophosphates function as the universal precursors to all isoprenoid biosynthesis (Martin et al. 2003) (Table 10.5).

Several approaches can be taken into account such as knocking down an enzymatic step in the pathway through lowering the extent of corresponding mRNA via antisense, co-suppression or RNA interference (RNAi) technologies, or overexpression of an antibody toward the enzyme. It has been reported that essential oil in mint plants was improved by downregulating synthesis of the undesirable content methanofuran through overexpressing an antisense derivative of the methanofuran synthase gene (Mahmoud and Croteau 2001). In *Papaver somniferum* a terminal enzyme, codeinone reductase, involved in morphine biosynthesis was knocked out by RNAi targeting all members of the gene family and resulted in notable morphine and codeine level reduction. The genetically modified poppy latex was observed with a drastic change in the alkaloid pattern, with the accumulation of rare alkaloids (Allen et al. 2004). RNAi technology was applied to California poppy, (*E. californica*) by hindering berberine bridge enzyme (BBE) activity which results in the accumulation of (S)-reticuline, a major intermediate compound produced during isoquinoline alkaloid biosynthesis (Fujii et al. 2007). In transgenic *Panax ginseng*, the gene which encodes for dammarenediol synthase (DDS) was silenced to lower the DDS expression responsible for ginsenoside production to 84.5% in roots using pK7GWIWG2 vector through *Agrobacterium tumefaciens* based genetic transformation. From the study it has been inferred that the DDS expression has major role in synthesizing ginsenosides in *P. ginseng* and showed that no evident morphological changes were observed in DDS-RNAi transgenic plants with respect to its wild variety (Han et al. 2006). Till now RNAi has been utilized as a rapid reverse genetic tool for generating valuable crops and medicinal plants bearing new chemical phenotypes and also to understand the genes coding for synthesizing several pharmacologically relevant secondary metabolites. The advent of more RNAi techniques for genome-wide screening may aid rapid gene identification involved in novel compound production and may facilitate the development of specific genes to explore them for commercially valuable plant derived products such as drugs, flavoring agents, etc (Table 10.6).

In Western countries plant derived drugs gained huge market price. Some illustrations include the significance of Madagascar periwinkle (*Catharanthus roseus*) which derived two drugs vinblastine and vincristine (Mukherjee et al. 2001), anticancer drugs paclitaxel (Taxol), analgesic drug morphine, podophyllotoxin and camptothecin (Mukherjee et al. 2001), or semisynthetic drugs of steroid nature

Table 10.5 List of some factors that can be utilized for enhanced secondary metabolite production

Metabolite	Plant species	Factor manipulated	References
Morphine, codeine	*Papaver somniferum*	*Vertcillium dahliae*	Heinstein (1985)
Indole alkaloids	*Catharanthus roseus*	Fungal elicitor	Eilert et al. (1987); Rijhwani and Shanks (1998)
Indole alkaloids	*Catharanthus roseus*	Diethyl amino ethyl dichloro phenyl ether	Lee et al. (1998)
N-acetyl-tryptamine	*Catharanthus roseus*	*Pythium aphanidermatum*	Eilert et al. (1986)
Diterpenoid tanshinones	*Salvia miltiorrhiza*	Yeast elicitor	Yan et al. (2005)
Saponins	*Panax ginseng*	Low-energy ultrasound	Wu and Lin (2002)
Sanguinarine	*Papaver bracteatum*	Dendryphion	Cline and Coscia (1988)
Taxol	*Taxus chinensis*	Fungal elicitor	Wang et al. (2001)
Tropane alkaloids	*Brugmansia suaveolens*, *Spodoptera frugiperda*	Methyl jasmonate	Alves et al. (2007)
Sesquiterpenoids	*Datura stramonium*	Metal ions	Threlfall and Whitehead (1988)
Capsidiol, debneyol, scopoletin, nicotine	*Nicotiana tabacum*, *Phytopthora cryptogea*	Yeast extract, cryptogein cellulase, methyl jasmonate	Lecourieux et al. (2002), Milat et al. (1991) and Wibberley et al. (1994)
Raucaffrincine	*Rauwolfia canescens*	Yeast elicitor, methyl jasmonate	Gundlach et al. (1992) and Parchmann et al. (1997)
Isoflavonoids	*Lotus corniculatus*	Glutathione	Robbins et al. (1995)
Digoxin, purpurea glycoside A	*Digitalis lanata*	Temperature	Kreis and Reinhard (1992)
Ubiquinone	*Nicotiana tabacum*	Temperature	Ikeda et al. (1977)
Crude alkaloids	*Catharanthus roseus*	Temperature	Courtois and Guern (1980)
Anthocyanin	*Daucus carota*	Light	Gläßgen et al. (1992)
Sesquiterpenes	*Matricaria chamomilla*	Light	Mulder-Krieger et al. (1988)
Monoterpenes	*Citrus limo*	Light	Leisso et al. (2015)
Flavonoids, tannins	*Cyrtanthus contractus*, *C. guthrieae*	Sodium chloride	Böhme et al. (2014)
Digitoxin	*Digitalis purpurea*	Phosphate	Hagimori et al. (1982)

Source: Ncube and Van Staden (2015)

Table 10.6 RNAi-mediated regulations reported in some plants

Target gene	Host	Active product	References
Cytochrome P450 (+) menthofuran synthase	*Mentha x piperita*	Menthofuran	Mahmoud and Croteau (2001)
Berberine bridge enzyme	*Eschscholzia californica*	Benzophenanthridine alkaloids	Park et al. (2002)
Putrescine N-methyltransferase	*Nicotiana tabacum*	Pyridine and tropane alkaloids	Chintapakorn and Hamill (2003)
Codeinone reductase	*Papaver somniferum*	Codeine and morphine	Allen et al. (2004)
Limonene-3-hydroxylase gene	*Mentha x piperita*	Limonene	Mahmoud et al. (2004)
Berberine bridge enzyme, N-methylcoclaurine 3′- hydroxylase (CYP80B1)	*Papaver somniferum*	Morphine, codeine, sanguinarine	Frick et al. (2004)
De-etiolated1 (DET1)	*Solanum lycopersicum*	Carotenoid and flavonoid in fruits	Davuluri et al. (2005)
Benzoic acid/salicylic acid carboxyl methyltransferase PhBSMT1	*Petunia x hybrida*	Methylbenzoate	Underwood et al. (2005)
Glutatione S-transferase 1 and 2, ubiquitin conjugating enzyme, putative cullin, anthocyanidin synthase, putative flowering-time gene, GPT, and the glutathione conjugate transporter	*Torenia fournieri*	Anthocyanin	Nagira et al. (2006)
Phenylacetaldehyde synthase (PAAS)	*Petunia hybrida*	Whole suppression of phenylacetaldehyde and 2-phenylethanol	Kaminaga et al. (2006)
Cinnamoyl-CoA Reductase	*Solanum lycopersicum*	Phenolics	Van der Rest et al. (2006)
Codeinone reductase	*Papaver somniferum*	Morphinan alkaloid	Larkin et al. (2007)
BBE	*Eschscholzia californica*	(S)-reticuline	Fujii et al. (2007)
Norcoclaurine 6-O-methyltransferase	*Coptis japonica*	Benzylisoquinoline alkaloid	Inui et al. (2007)
Coniferyl alcohol acyltransferase	*Petunia x hybrida*	Coniferyl aldehyde and homovanillic acid	Dexter et al. (2007)
Cinnamyl alcohol dehydrogenase	*Linum usitatissimum*	Lignin reduction	Wróbel-Kwiatkowska et al. (2007)
Salutaridinol 7-O-acetyltransferase	*Papaver somniferum*	Morphinan alkaloids	Allen et al. (2008)
Salutaridinol 7-O-acetyltransferase	*Papaver somniferum*	Morphine	Kempe et al. (2009)
Tryptophan decarboxylase	*Catharanthus roseus*	Tryptamine	Runguphan et al. (2009)

Source: Borgio (2009)

hormones obtained from diosgenin (Robert Verpoorte 2000). Genetic engineering of the lignin biosynthesis pathway reached potential interest due to the selection of model plants in exploring metabolic pathways and also due to its biomass content, which is indirectly related to its forage digestibility and quality requirement in pulping industry (Eudes et al. 2014). Findings strongly suggest that removal of lignin from developing plants without affecting its development is crucial. Attempts failed in downregulating particular genes coding for enzymes in the lignin biosynthetic pathway in plant species such as poplar, maize, pine, and switch grass using natural mutants or gene silencing (RNAi) techniques. However some reports also showed restricted genetic modifications are successful in moderate reduction of lignin content, biomass content modification, and modestly increasing saccharification efficiency, forage digestibility, and pulping yield (Li et al. 2008). Genetically modified plants also have been utilized for the developing antibodies against dental caries, rheumatoid arthritis, malaria, viral related cancers, cholera, diarrhea, HIV, rhinovirus, influenza viruses, hepatitis B virus, herpes simplex virus, etc. (Thomas et al. 2002). Recently it has become clear that chemical diversity of plants has great potential than any other human made chemical library, and thus the plant kingdom holds extensive resource of pharmacologically valuable compounds which are yet to discover. The major challenge observed in plant metabolic profiling is its complexity and diversity of the chemical compounds (Oksman-Caldentey et al. 2004). The diverse chemical properties exhibited by each valuable compound restrict the analytical tools while proceeding with several secondary metabolites in a parallel manner (Trethewey 2004).

10.4 Transcriptome Sequencing for Exploring Secondary Metabolite Pathways

Transcriptomic data allows substantial opening for the identification of novel genes and assessing collection of ESTs (expressed sequence tags) present in the sample, which can be utilized for molecular marker development, particularly for non model organisms bearing no reference genome (Wang et al. 2009). The huge sequence data produced by transcriptome sequencing also gained a prominent success in illustrating metabolic pathways, global gene expression, and differential expression analysis. Due to the absence of immense genomic data for most of the economically significant plants, traditionally using microarray based transcriptome analysis in these plant systems becomes inconvenient because of the requirement of large EST or cDNA clone assemblies. The first generation Sanger technology or chain termination method based sequencing of model crops was successful; its throughput and huge cost exerted few limitations to sequence more plant species, particularly studies on complex genomes; and this motivated increased demand for novel and modified sequencing technologies. Besides, numerous non model plants are crucial assets for energy resource bearing unique characteristics to them, and thus use of a model plant to study them becomes a need (Carpentier et al. 2008). Consequently, genomics in non model species raised some challenges until the excellent progress made by second generation sequencing platforms with high throughput and comparatively low sequencing cost generally termed as NGS technologies. Hence with the advent of next generation sequencing (NGS) technology, RNA sequencing (RNA-seq) has been extensively applied to plants limiting complete genomic information accord the transcriptional evidence for various downstream applications (Duan et al. 2012). RNA-seq or transcriptome sequencing with the aid of several NGS platforms is usually less costly and facilitates *de novo* assembly due to its smaller size (10–100 times) when compared to genomes (Gayral et al. 2013). RNA-seq can be utilized for studies on population genetics (Neil et al. 2010; Renaut et al. 2012), phylogenetic research (Chiari et al. 2012; Timme et al. 2012), and molecular adaptation (Elmer et al. 2010; Künstner et al. 2010; Gayral et al. 2013). Expression level differences among tissues, genotypes, and different population help in understanding functional and evolutionary relationships (Wolf et al. 2010; Gayral et al. 2013; Nadiya et al. 2017).

The first and foremost factor before starting RNA-seq experiment relies in selection of a particular NGS platform because the data generated from different RNA sequencing platforms varies, and this variation can influence the experimental interpretations. Procedures for sample preparation change in each platform, and hence selection of proper platform with respect to the downstream application is a prerequisite to obtain experimental triumph. Different NGS platforms are available commercially, and most of them are under active development (Metzger et al. 2011). Many are based on sequencing by synthesis technology, using DNA polymerase or ligase as the major component. Roche 454, Illumina, Helicos, and PacBio (Pacific Biosciences) initiate their sequencing reaction with DNA polymerase, while SOLiD (Life Technologies) and Complete Genomics utilize DNA ligase. The sequencing platforms can additionally be classified as either single molecule based (sequencing a single molecule, such as Helicos and PacBio) or ensemble based (sequencing of multiple identical copies of a DNA molecule, such as Illumina and SOLiD). Recently, the NGS platform Illumina dominated the sequencing industry and becomes the leading company in the market which utilizes the sequencing by synthesis approach adapting fluorescently labeled reversible terminator nucleotides toward clonally amplified cDNA templates immobilized into an acrylamide coating on the surface of a glass flow cell (Bentley et al. 2008). The launching of Illumina Genome Analyzer and recent HiSeq 2000 has set the standard for massively parallel sequencing in a high throughput way, and in 2011 Illumina launched the MiSeq suitable for smaller laboratories and clinical diagnostics suggesting lower-throughput fast turnaround instrument. Another two new NGS platforms which revolutionized the genomic sequencing world include the Ion Torrent Personal Genome Machine (PGM) and the Pacific Biosciences (PacBio). The Ion Torrent PGM utilizes a different semiconductor technology which detects the protons released for each nucleotide incorporated during synthesis (Rothberg et al. 2011), and PacBio was developed with a process based on single-molecule real-time (SMRT) sequencing (Eid et al. 2009). Transcriptome sequencing can be preceded with single-end reading or paired-end reading. In single end method, the sequencer reads a fragment from only one end to the other, and in paired end reading, it starts at one read, finishes this direction at the specified read length, and then starts the second round of reading from the opposite end of the fragment. Paired end reads are expensive and time consuming to execute than single end reads. After generating the reads, transcriptome assembly is obligatory to convert individual reads into complete mRNA sequences or individual transcripts. The longer the individual reads, the simpler it is to assemble transcripts unambiguously, but recent leading NGS platforms usually produce short reads and should be assembled into contigs. Analysis of massive amount of RNA-seq data generated during experiments seems to be challenging and time consuming. For instance, Hiseq 2000 (Illumina) generates up to 200 million 100-nt reads (approximately 50 GB) of data in one sequencer run per lane. It is mandatory to analyze the data and not only detect transcriptome similarities but also for contig assembly, differential expression, and quantification for making findings to biological meaning (Chu and Corey 2012). For a plant species lacking good quality, reference genome requires *de novo* assembly, which is absolutely essential for downstream RNA-seq analyses to obtain an accurate overview of the transcriptome (Duan et al. 2012). For a model plant species for which genome information is available, the transcriptome assembly can be performed upon the reference genome, and the ultimate success of reference based assemblers depends on the quality of the reference genome being used. For a plant species without a well characterized reference genome, *de novo* transcriptome assembly should be performed for further analysis. Several *de novo* transcriptome assemblers have been developed such as SOAPdenovo-Trans (Xie et al. 2014), ABySs (Robertson et al. 2010), and Velvet (Zerbino and Birney 2008) followed by Oases (Schulz et al. 2012), Trinity (Haas et al. 2013), and MIRA (Mimicking Intelligent Read Assembly (Chevreux et al. 2004). RNA-seq has been utilized for hundreds of non model plants (Johnson et al. 2012; Schliesky et al. 2012). But still more extensive coverage for selected plant species is needed for better characterization of biosynthetic pathways of particularly important specialized metabolites. Hence, the PhytoMetaSyn Project (www.phytometasyn.ca) aimed at 75 non model plants which could produce terpenoids, alkaloids, and polyketides (Facchini et al. 2012), and among those six subgroups are mainly targeted such as sesquiterpenes, diterpenes, triterpenes, monoterpenoid indole alkaloids, benzylisoquinoline alkaloids, and polyketides for the identification of novel biosynthetic genes responsible for the diversity of specialized compounds present in the 75 species (Table 10.7).

Using NGS based transcriptome sequencing, several genetic and genomic studies related to the molecular mechanisms associated with the diverse chemical composition of commercially and pharmacologically relevant plant species were performed. A study demonstrated large scale unigene identification of *Polygala tenuifolia* was performed by Illumina sequencing and showed that several transcripts were involved in the biosynthesis of triterpene, saponins, and phenylpropanoid. The study also laid foundation to modify strategies to synthesize active compounds through marker assisted breeding or genetic engineering for *P. tenuifolia* and also inferred that the genes involved in the enrichment of secondary metabolite biosynthetic pathways could increase the prominent applications of *P. tenuifolia* in pharmaceutical industries (Tian et al. 2015). Transcriptome based study on pear

Table 10.7 List of 75 plant species included in PhytoMetaSyn Project for transcriptome analysis using NGS

Species	Common name	Family	Authority
Abies balsamea	Balsam fir	Pinaceae	(L.) Mill.
Acacia victoriae	Gundabluie	Fabaceae	Benth.
Amsonia hubrichtii	Hubricht's bluestar	Apocynaceae	Woodson
Argemone mexicana	Mexican prickly poppy	Papaveraceae	Linn.
Arnica chamissonis	*Chamisso arnica*	Asteraceae	Less.
Arnica montana	Leopard bane	Asteraceae	L.
Artemisia absinthium	Absinthe wormwood	Asteraceae	L.
Artemisia chamaemelifolia	Chamomile-leaved artemisia	Asteraceae	Vill.
Artemisia maritima	Sea wormwood	Asteraceae	L.
Berberis thunbergii	Japanese barberry	Berberidaceae	DC.
Camptotheca acuminata	Happy tree, tree of life	Nyssaceae	Decne.
Cannabis sativa	Cannabis	Cannabaceae	L.
Catharanthus ovalis		Apocyanaceae	Markgr.
Catharanthus roseus	Rosy periwinkle	Apocyanaceae	(L.) G.Don
Catharanthus longifolius	Periwinkle	Apocyanaceae	Pichon
Cedrela sinensis	Chinese mahogany	Meliaceae	(A.Juss.) M.Roem
Centella asiatica	Indian pennywort, gotu kola	Mackinlayoideae (Apiaceae)	(L.) Urban
Chelidonium majus	Greater celandine	Papaveraceae	L.
Chrysanthemum coronarium	Garland chrysanthemum	Asteraceae	L.
Cinchona ledgeriana	Jesuit's bark	Rubiaceae	(Howard) Bern
Cissampelos mucronata	Abuta	Menispermaceae	A. Rich.
Citrus paradisi	Grapefruit	Rutaceae	Macfad.
Cocculus trilobus	Korean moonseed	Menispermaceae	(Thunb.) DC.
Coleus forskohlii	Coleus	Lamiaceae	Andrews
Corydalis cheilanthifolia	Fern-leaf corydalis	Papaveraceae	Hemsley
Dianthus superbus	Large pink	Caryophyllaceae	L.
Diospyros kaki	Japanese persimmon	Ebenaceae	Thunb.
Eschscholzia californica	California poppy	Papaveraceae	Cham.
Euphorbia peplus	Cancer weed	Euphorbiaceae	L.
Glaucium flavum	Yellow hornpoppy	Papaveraceae	Crantz
Grindelia integrifolia	Entire-leaved gumwood	Asteraceae	DC.
Helichrysum petiolare	Licorice plant	Asteraceae	Hilliard & B.L.Burtt
Humulus lupulus	Common hop	Cannabaceae	L.
Hydrangea macrophylla	Big leaf hydrangea	Hydrangeaceae	(Thunb.) Ser.
Hydrastis canadensis	Goldenseal	Ranunculaceae	L.
Hypericum perforatum	St. John's wort	Guttiferae	L.
Isodon rubescens		Lamiaceae	Hemsley
Jatropha gossypiifolia	Cotton leaf	Euphorbiaceae	L.

(continued)

Table 10.7 (continued)

Species	Common name	Family	Authority
Jeffersonia diphylla	Twin leaf	Berberidaceae	(L.) Pers
Lactuca sativa	Lettuce	Asteraceae	L.
Ligularia fischeri	Ligularia	Asteraceae	Turcz.
Lippia dulcis	Aztec sweet herb	Verbenaceae	Trevir.
Lonicera japonica	Japanese honeysuckle	Caprifoliaceae	Thunb.
Mahonia aquifolium	Oregon grape	Berberidaceae	(Pursh) Nutt
Marrubium vulgare	Common horehound	Lamiaceae	L.
Matricaria recutita	German chamomile	Asteraceae	L.
Menispermum canadense	Canadian moonseed	Menispermaceae	L.
Nandina domestica	Sacred bamboo	Berberidaceae	Thunb.
Nigella sativa	Black cumin	Ranunculaceae	L.
Papaver bracteatum	Persian poppy	Papaveraceae	Lindl.
Pelargonium x hortorum	Scented geranium	Geraniaceae	L.H. Bailey
Petasites hybridus	Butterbur	Asteraceae	(L.) G.Gaertn., B.Mey. & Scherb.
Piper methysticum	Kava	Piperaceae	G.Forst.
Platanus occidentalis	American sycamore	Platanaceae	L.
Primula obconica	Poison or German primrose	Primulaceae	Hance
Prunella vulgaris	Woundwort	Lamiaceae	L.
Pseudolarix amabilis	Golden larch	Pinaceae	(J.Nelson) Rehder
Rheum rhabarbarum	Garden rhubarb	Polygonaceae	L.
Rauwolfia serpentina	Indian snakeroot	Apocyanaceae	(L.) Benth. ex Kurz
Rosmarinus officinalis	Rosemary	Lamiaceae	L.
Sanguinaria canadensis	Bloodroot	Papaveraceae	L.
Saponaria vaccaria (Vaccaria hispanica)	Cowcockle	Caryophyllaceae	L.
Silene vulgaris	Bladder campion	Caryophyllaceae	(Moench) Garcke
Stylophorum diphyllum	Celandine poppy	Papaveraceae	(Michx.) Nutt.
Swietenia mahagoni	West Indian mahogany	Meliaceae	(L.) Jacq.
Tabernaemontana elegans	Milkwood, toad tree	Apocyanaceae	Stapf
Thalictrum flavum	Meadow-rue	Ranunculaceae	L.
Thapsia garganica	Drias plant	Apiaceae	L.
Tinospora cordifolia	Heartleaf moonseed	Menispermaceae	(Thunb.) Miers
Tripterygium wilfordii	Thunder god vine	Celastraceae	Hook f.
Valeriana officinalis	Valerian	Valerianaceae	L.
Vancouveria hexandra	Inside-out flower	Berberidaceae	(Hook) C.Morren & Decne.
Vinca minor	Lesser periwinkle, myrtle	Apocyanaceae	L.
Xanthium strumarium	Rough cocklebur	Asteraceae	L.
Xanthorhiza simplicissima	Yellowroot	Ranunculaceae	Marsh.

Adopted from Xiao et al. (2013)

plants revealed the identification of major genes expressed under salt stress to salt was significantly involved in fundamental biological processes, secondary metabolite biosynthetic pathways, and signal transduction mechanisms (Xu et al. 2015). The genus *Panax* (Xiao et al. 2013) possesses about nine species generally called ginsengs and reported to have antidiabetic, anticancer, anti-inflammatory and immunomodulatory, and anti-allergic compounds. Consequently, using different NGS platforms, various candidate genes responsible for secondary metabolite biosynthesis were identified from *Panax* species (Sun et al. 2010; Luo et al. 2011; Li et al. 2013; Jayakodi et al. 2014). Another report suggests that *Podophyllum* species possess podophyllotoxin and aryl tetralyn lignan which are extensively used in partial synthesis of anticancer drugs, and the biosynthetic pathway and the putative genes responsible for the release of these two compounds were identified through NGS based metabolomic analysis through various bioinformatics tools (Marques et al. 2014). Transcriptome sequencing using Illumina MiSeq platform of young and mature leaf tissue of *Cassia angustifolia* identified genes involved in various secondary metabolite pathways including terpenoids, sennosides, and polyketides metabolism (Reddy et al. 2015). A study illustrated the preparation of a high quality EST database from *Glycyrrhiza uralensis* using 454 GS FLX platform, and based on the developed ESTs, novel putative candidate genes related to glycyrrhizin secondary metabolite biosynthetic pathway and novel genes for cytochrome P450s and glycosyltransferases were identified. The study also inferred that with the aid of organ specific expression pattern, three unigenes were identified to be responsible for cytochrome P450s and six unigene codes for glycosyltransferases and were the most probable putative genes involved in glycyrrhizin biosynthetic pathway (Li et al. 2010). The medicinal plant, *Withania somnifera*, due to the presence of the bioactive molecules withanolides, is widely used in Ayurvedic and other native medical treatments. To understand the basic molecular mechanism of withanolide biosynthetic pathway, transcriptome sequencing of *Withania* leaf and root was conducted which particularly produce withaferin A and withanolide A, respectively. Transcript annotation, gene ontology, and KEGG analyses elucidated a comprehensive view of enzymes associated with withanolide production. The study also identified members of cytochrome P450, glycosyltransferase, and methyltransferase gene families which are differentially expressed in leaf and root suggesting the presence of tissue specific withanolides. Thus the transcriptome sequence data developed for *Withania* may found new ways to elucidate complete biosynthetic pathway of tissue-specific secondary metabolites in non model plant and to introduce modifications for increased biosynthesis of withanolides via various emerging biotechnological approaches (Gupta et al. 2013). Transcriptome data of *Calotropis gigantea* (Sodom apple) species, a significant medicinal shrub and well known fiber resource, identified abundant gene transcript resources for evaluating the molecular characteristics of fiber and secondary metabolite biosynthetic pathways. The putative fiber responsible genes were elucidated and were experimentally validated through real-time PCR techniques (Muriira et al. 2015). *Curcuma longa* L., most widely used herbal medicine and important spice, is a rich source of biologically active compound curcumin. Exploring the basic molecular mechanism underlying curcumin biosynthesis may aid in modification of curcumin content and keeping the growth stability in different ways of cultivation. A study was proposed to investigate the candidate genes responsible for curcuminoid biosynthesis by utilizing *de novo* transcriptome assembly of rhizome transcriptome of *C. longa* and *C. aromatica*. In the study, differential expression analysis of two new polyketide synthase genes (clpks1 and clpks2) showed increased expression in *C. longa* compared to *C. aromatica* and inferred the major roles of these transcripts in curcuminoid biosynthesis. The study also provided useful data for altering the curcumin biosynthetic pathway in curcuma and its related species for developing new turmeric traits (Sheeja et al. 2015). Using RNA-seq technology, *Hedychium coronarium* transcriptome provided an important resource for functional genomic studies. The study showed the existence of a number of candidate scent-related genes such as flower-specific HcDXS2A, HcGPPS, and HcTPSs which perform major role in regulating the biosynthetic pathway synthesizing floral volatile terpenes. The results of the study also suggested that through uncovering the molecular mechanism of floral scent formation and pathway regulation in *H. coronarium*, various breeding techniques and genetic manipulations can be developed to generate scent associated traits of *Hedychium* with higher commercial values (Yue et al. 2015).

Another study based on the anticancer, antioxidant, and antimalarial properties of curcumin utilized the transcriptome sequencing of rhizome of three *C. longa* varieties and *de novo* transcriptome assembly. The study elucidated terpenoid biosynthesis pathway, other secondary metabolite pathways, and genes associated with biosynthetic pathways of several anticancer compounds (taxol, curcumin, and vinblastine), antimalarial compounds (artemisinin), acridone alkaloids, and other prominent metabolites such as sesquiterpenes like capsidiol, gossypol, phaseic acid, bergamotene, germacrene, and farnesene. The assembled data of this significant phytochemically valid herb provide information to develop fast-growing cultivars with increased terpenoid profiles such as anticancer, antimalarial, and antioxidant properties (Annadurai et al. 2013). A study was reported to determine the varying status and quality of oil accumulation in different avocado cultivar fruits, through transcriptome sequencing and analysis. Transcriptomic data gained information on molecular genetics and functional genomics and helps in the identification of the pathways and genes associated with the release of a diverse essential nutrients and beneficial phytochemicals. Hence the study represented a detailed illustration of transcriptomic variations observed during ripening of the Mexican avocado fruit which allows an effectual view of genes associated with fatty acid biosynthesis and the fruit ripening process (Ibarra-Laclette et al. 2015). *Boesenbergia rotunda*, a food ingredient and medicinal plant, is a rich source of panduratin A which is a flavonoid having a wide range of medicinal properties such as anticancer, anti-dengue, anti-HIV, anti-inflammatory, and antioxidant properties. Transcriptome sequencing along with digital gene expression profiling of native and phenylalanine treated *B. rotunda* identified differentially expressed genes involved in the panduratin A biosynthetic pathway. From this study it has been shown that several genes were upregulated and some were downregulated which includes upregulation of two phenylalanine ammonia-lyase (PAL), three 4-coumaroyl-coenzyme A ligase (4CL), and one chalcone synthase (CHS) which plays significant role in the phenylpropanoid pathway results in the synthesis of panduratin A (Md-Mustafa et al. 2014).

Uncaria rhynchophylla plant produces two terpene indole alkaloids having great medical importance, namely, rhynchophylline (RIN) and isorhynchophylline (IRN), which hinder and destabilize the formation of pathological indicator of Alzheimer's disease (amyloid β-protein). NGS based transcriptome sequencing *de novo* assembly and differential expression analysis identified candidate genes coding for enzymes involved in the biosynthesis of reliable secondary metabolites. Cytochrome P450, methyltransferase, and isomerase were identified as putative genes involved in late biosynthesis of RIN and IRN. Hence the transcriptome data obtained from this study serves as an important source for understanding the way of altering specific bioactive compounds from the extracts of *Uncaria* (Guo et al. 2014). *Centella asiatica* (L.) is a medicinal herb used widely in traditional therapeutic systems particularly in Ayurveda. The plant exhibits memory enhancer, antiaging, antipyretic, diuretic, and wound healing properties (Mangas et al. 2009) and is also found to have a good role in healing varicose veins, ulcer, lupus, eczemas, and mental retardation (Brinkhaus et al. 2000; James and Dubery 2009). Phytochemical analysis suggests that leaves of *C. asiatica* possess diverse secondary metabolites including triterpenoids, volatile monoterpenes and sesquiterpenes, flavonoids, and alkaloids (Suntornsuk and Anurukvorakun 2005; Zainol et al. 2003; Zhu et al. 1997). Based on these observations, an attempt was made to develop a transcriptome assembly resource of *C. asiatica* using NGS based transcriptome sequencing and *de novo* assembly of pooled samples of *C. asiatica* leaves. Different secondary compounds including isoprenoids, genes responsible for various primary and secondary metabolites, and several cellular and molecular functions were characterized in the study. Hence the information provided by the study such as phytochemicals, responsible genes, and its metabolic pathways can be utilized in various biotechnological manipulations for the enhanced production of metabolites (Sangwan et al. 2013).

High throughput sequencing technologies become an exceptionally significant source for the generation of reference genome sequence in non model plants. Hence transcriptome data is being used to characterize candidate genes and networks associated with diverse secondary metabolite production in plants. Tea plants (*Camellia sinensis*) possess healthy nutritional properties due to the presence of chemical constituents such as polyphenols (chiefly catechins), theanine, and caffeine. In a study on demonstrating secondary metabolite in tea when compared to oil tea, it was observed that major secondary metabolites including theanine were lower in oil tea when compared with tea. The genes coding for key enzymes responsible for regulation of these pathways were comparatively highly expressed in tea (Tai et al. 2015). The differential expression of tea and oil tea revealed the status of secondary metabolite pattern lies between tea and oil tea which helps in describing the molecular information leading to the biosynthesis of specific metabolites in tea, and this illustration can be utilized for various gene manipulation techniques for the development of novel high-quality breeds. Roots of *Euphorbia fischeriana* contain 12-deoxyphorbol-13-acetate (prostratin, a phorbol ester of tigliane diterpene).

Prostratin was observed as a protein kinase C activator which can be used in human immunodeficiency virus (HIV) treatments. Transcriptome sequencing and *de novo* assembly of root transcriptome of the *E. fischeriana* identified 26 unigenes encoding enzymes responsible for different biosynthetic pathways including casbene biosynthesis pathway, which produces prostratin precursor. The study also revealed the increased expression of ent-kaurene oxidase and tRNA dimethylallyltransferase enzymes initiates the production of kaurenol and cis-zeatin-O-glucoside required for the casbene biosynthesis. The transcriptomic resources developed from this study may facilitate more functional studies to enhance prostratin production and other phorbol esters of interest for improving HIV research and therapeutics (Barrero et al. 2011).

A study demonstrated the exploration and characterization of temporal and spatial transcriptome, miRNA, and mRNA expression analysis in developing bamboo culms revealed the molecular mechanisms underlying the process of sequentially elongated internodes from the base to the top. Various significant pathways including environmental adaptation, signal transduction, translation, transport and other metabolism, and gene annotations like cell growth, hormone mediated signaling, protein modification, primary shoot apical meristem specification, xylem and phloem pattern formation, response to stimuli, metabolic process, and biological regulation were observed to be responsible for rapid generation culms in bamboo. Furthermore the combined analysis of transcriptome, miRNAs, and posttranslational and proteomic studies represents the overall characterization and other molecular insights, particularly the complex phenomenon during the rapid generation of culms in moso bamboo (He et al. 2013). Transcriptome sequencing and *de novo* assembly of *Suaeda fruticosa*, a non-conventional crop, provided information on differentially expressed and unique genes which were categorized using gene ontology terms and their corresponding pathways. The complex genetic mechanism of the plant was revealed through analysis of predicted genes by providing comprehensive information on mechanism of salt tolerance, novel genes discovery, association, and comparison of differential expression profiles without salt and normal salt concentration (Diray-Arce et al. 2015). *Nicotiana benthamiana*, a model plant, has been extensively used in gene expression studies, and exploring the genomic resources might enhance further studies on developmental, metabolic, and defense pathways associated with *N. benthamiana* and in understanding the molecular mechanisms underlying in such a frequently used model plant.

A study on transcriptome data generated from nine different tissues of *N. benthamiana* revealed putative genes, RNAi-associated pathways, genome coding capacity, high level transient transgene expression, and susceptibility to virus infections (Nakasugi et al. 2013). This study may provide information on developing viral resistant plants through RNA silencing pathway exploration. A study on comparison of ginger (*Zingiber officinale* Rosc.) and mango ginger (*Curcuma amada* Roxb.), toward bacterial wilt infection, elucidated the overexpression of genes associated with MEP (mevalonate) pathway for the terpene/isoprene biosynthesis in *C. amada* when compared with *Z. officinale.* The severalfold upregulation of MEP pathway-regulated genes in mango ginger was observed due to its resistance to *R.*

solanacearum (bacterial wilt) through the secretion of phenolic compounds and terpenoids, and other highly expressed transcripts were annotated as genes associated with pathogen recognition, biotic and abiotic stress resistance genes, transcription factors, and signaling. Transcripts that were observed as differentially expressed in *C. amada* were associated with plant defense response to pathogens, i.e., genes responsible for bacterial defense response, oxidative stress, introduction of a physical barrier to inhibit pathogen progression, and systemic resistance. Finally, genes observed to be participating in mango ginger resistance against bacterium codes for proteins associated with lignin synthesis including cytochrome P450, succinyl-CoA ligase, and S-adenosylmethionine synthase and suggested that the lignin accumulation during the interaction between *C. amada* and bacterium serves as a physical barrier for the pathogen invasion. The study also reported the overexpression of some other resistance genes involved in phenylpropanoid biosynthetic pathway in *C. amada* (Prasath et al. 2014). Using 454 pyrosequencing, the leaf and root transcriptome of *Avena barbata* (wild oat) identified root-specific genes involved in secondary metabolic pathways such as isoflavone 7-O-methyltransferase, cytochrome P450, protein catabolism including aspartic proteinase nepenthesin-2, vignain cystein endopeptidase, and serine carboxypeptidase. Since the wild variety of this crop was subjected to allelic change for desirable agronomic traits, *Avena barbata* has been the subject to investigations to explore more genetic mechanisms for adaptation in adverse environmental conditions (Swarbreck et al. 2011).

Several biosynthetic pathways were now successfully elucidated using NGS based transcriptome sequencing which illustrated several different plant specialized metabolic pathways in numerous plants such as the study on synthesis of mild sedative valerenic acid by the valerian plant, *Valeriana officinalis*, production of natural sweetener hernandulcin from *Lippia dulcis*, and the synthesis of anticancer drug thapsigargin from *Thapsia garganica* (Pickel et al. 2012; Pyle et al. 2012). The major genes responsible for the production of pharmacologically relevant monoterpenoid indole alkaloid through MIA pathway were elucidated in *Catharanthus roseus*, *Tabernaemontana elegans*, and *Amsonia hubrichtii* using transcriptome sequencing. The transcripts from *C. roseus*, *A. hubrichtii*, and *T. elegans* were annotated as geraniol-10- hydroxylase, 10-hydroxygeraniol oxidoreductase, loganic acid O-methyltransferase, secologanin synthase, tryptophan decarboxylase, and strictosidine synthase, while genes responsible for 16-methoxy-2,3-dihydro-3-hydroxytabersonine N-methyltransferase, desacetoxyvindoline 4-hydroxylase, and deacetylvindoline acetyltransferase required for the biosynthesis of vinodiline were exclusively observed in *Catharanthus* supporting the evidences for the unique origin of vindoline in the *Catharanthus* genus (Xiao et al. 2013). *Hypericum perforatum* contains polyketides hypericin and hyperforin used for depression treatments. Polyketides are another group of structurally diverse and biologically active metabolites.

The biosynthetic pathway of prenylated acylphloroglucinol hypericin, usually concentrated in leaf and flower glands, was elucidated partially (Karppinen and Hohtola 2008; Karppinen et al. 2008). The study revealed that the acylphloroglucinol component of hyperforin is catalyzed by polyketide synthase which condenses isobutyryl-CoA using three molecules of malonyl-CoA to generate phlorisobutyrophenone (PIBP) which is then prenylated in the presence of dimethylallyl diphosphate (DMAPP) which acts as the donor and uses geranyl diphosphate (GPP) to produce hyperforin. Due to the high content of flavonoids in blackberries, the transcriptome data may provide significant novel resource for biosynthetic pathway based molecular research. A study on blackberry transcriptome data identified putative genes encoding enzymes catalyzing flavonoid biosynthetic pathways, genes related to other metabolic processes, and transcripts responsible for RNA virus resistance genes and fungal and bacterial pathogens (Garcia-Seco et al. 2015). The study may provide more insights on gene discovery to develop genomic tools for breeding of improved varieties with enhanced health benefits. Banana (*Musa accuminata*), an economically important fruit crop, usually undergoes ethylene-induced ripening which causes heavy postharvest losses due to fruit over ripening to farmers and consumers. The fruit ripening mechanism includes diverse physiological and biochemical changes, and several genes representing various metabolic pathways result in ripe and edible fruit. For uncovering the molecular mechanisms underlying in banana ripening, a study was demonstrated through sequencing the transcriptome of ripe and unripe stages of banana fruit pulp and detected genes responsible for the ripening mechanisms such as softening and synthesis of aroma volatiles.

Transcripts were annotated as genes encoding acyltransferases from the transcriptome data which are involved in the production of aromatic volatiles and flavor components. Additionally, the present study also reveals the significance and role of expansions, PL, and XTH in fruit softening, genes responsible for cell wall degradation, and more differentially expressed novel genes which could play a major role in banana ripening may serve in the future as a good candidate for emerging gene manipulation methods in banana fruit ripening that could reduce postharvest loss (Asif et al. 2014). To demonstrate the expression profile during chickpea flower development, RNA-seq analysis was performed, and the transcriptome data generated identified differentially expressed transcripts during shoot apical meristem, floral development, and various other metabolic processes. The study also provided information on molecular mechanisms, regulatory networks, and particular metabolic pathway responsible genes during the developmental stages in the legume, chick pea (Singh and Jain 2014). A comprehensive transcriptome data set was developed for sugarcane induced by *S. scitamineum* as a good resource for exploring the molecular mechanisms associated with sugarcane responses to *S. scitamineum*. The data analysis identified differentially expressed genes associated with plant-pathogen interaction, hormone signal transduction, phenylalanine metabolism, flavonoid biosynthesis, phenylpropanoid biosynthesis, and other pathogenic response related metabolic pathways (Que et al. 2014). Saponins, the major amphipathic glycosides, exert several health benefits from their biological and medicinal properties (anticancer and antioxidant properties). Roots of *Asparagus racemosus* contain steroids responsible for the synthesis of saponins which led to the comparison studies utilizing high throughput transcriptome sequencing and *de novo* assembly of root and leaf tissues. The study identified novel transcripts involved in saponin biosynthetic pathway and unique expression of genes encoding enzymes associated with the MVA pathway for triterpene biosynthesis in root when compared with *A. racemosus* leaf transcriptome, and the data provided may serve as a source for functional characterization at biochemical, cellular, and molecular level to manipulate saponin biosynthetic pathways (Upadhyay et al. 2014). Transcriptome analyses also help in studies related to gene discovery, transcript quantification, molecular marker development, small RNA profiling, and negative gene regulation; thereby sequencing strategies dominate EST-based microarray experiments which were comparatively tedious. Transcriptome sequencing and *de novo* assembly of wild and five cultivar varieties of *Elettaria cardamomum* Maton (small cardamom) revealed differentially and uniquely expressed putative genes involved in various secondary metabolite synthesis pathways including terpenoid and flavonoid biosynthesis.

Challenges in Transcriptome Sequencing Although transcriptome sequencing has shown significant advances on molecular basis of the discovery of several putative genes, data analysis exerts several restrictions and challenges such as generation and *de novo* assembly of large number of short reads, gene annotation of big data, and low transcript abundance especially in non model organisms such as medicinal and phytochemically important plants. The first and foremost challenge in plant transcriptome sequencing is the requirement of high quality RNA which is very difficult to isolate in required quantities good for library construction before sequencing. Another challenging factor is drawbacks of reference based transcriptome assembly which mainly depends on the quality of preferred reference genome. Most of the cases the alignment provides hundreds to thousands of misassemblies and genomic deletions (Salzberg and Yorke 2005) which lead to incomplete transcriptome assembly.

In some cases it is reliable to depend on the reference of a closely related species such as the use of strawberry reference genome for raspberry transcriptome assembly (Ward and Weber 2011), but still divergent genomic regions could be missed in such cases. Finally, reference based assembly may also induce trans spliced gene misassemblies where pre-mRNAs are spliced to a single mature mRNA which is crucial for obtaining information on metabolic pathways. Reference based assembly can generate comprehensive transcriptome profiling in case a high quality reference genome is available. The biological sequences appear very complicated, and hence error rates might be possible up to 1±4% per nucleotide, which result in mismatches (Claros et al. 2012). For instance, it has been reported that the Illumina sequencing generates sequence specific miscalls, GC-biased errors (Nakamura et al. 2011), and more substitution-type miscalls than indel-type miscalls (Hoffmann et al. 2009), while 454 pyrosequencing releases more indel-type miscalls than substitution-type due to its homopolymer length inaccuracy concerns (Gilles et al. 2011). The assemblers designed for Sanger reads were observed to be not suitable for NGS data, and in response new assemblers with more sophisticated approaches were developed (Imelfort and Edwards 2009). The recently introduced assemblers require servers or clusters with >500 GB of RAM and several terabytes of disk space. Though low-cost servers, supercomputing centers, and the emerging cloud computing solved most the requirements, upcoming sequencing projects including loblolly pine (Neale et al. 2004) and maritime pine (Díaz-Sala and Cervera 2011), with 22±30 Gbp genomes, challenge the data analysis through more required computational demands. Another challenge is the specificity of the currently available *de novo* assemblers toward particular sequencing platforms, for example, Trinity *de novo* assembler (Haas et al. 2013) strongly supports Illumina paired-end reads for assembly, and MIRA assembler (Chevreux et al. 2004) is designed with error corrections of short reads generated from 454 pyrosequencing and ion torrent reads, although the assemblers work with almost all platforms but may exert some errors such as misassemblies or incomplete assemblies which hinders the downstream applications. Performing *de novo* assembly in eukaryotic transcriptomes is much more challenging, not only due to its huge data but also due to the presence of alternatively spliced variants (Kumar and Khurana 2014). Generally, *de novo* transcriptome assembly needs greater sequencing depth for full-length transcript assembly than with reference-guided assembly strategy. Moreover, *de novo* transcriptome assemblers are prone to sequencing errors, low transcript abundance, and chimeric molecules generally observed in data set of non-model plants.

10.5 Conclusions

Plants are important source of diverse specialized metabolites, most of which play significant roles in pharmaceuticals, flavors, fragrances, and other industrial values. Commercial value of some secondary metabolites gained great importance in its production and practical possibilities to enhance the production through biotechnological approaches. The major focus of the present review is the application of next-generation-based transcriptome sequencing toward the identification of putative genes involved in the synthesis of diverse secondary compounds present in varying concentrations in different plant species. But most of these secondary metabolites are extensively located in non model plants without a well characterized or well assembled reference genome sequence. However, generation of huge amount of sequence data through NGS technologies becomes rapid and relatively low cost particularly when dealing with 454 pyrosequencing, ion personal genome machine, and latest Illumina sequencers with better modified base calling efficiency. Hence the pattern of specialized secondary metabolite biosynthetic pathways and putative candidate genes involved in the synthesis can be studied through data mining framework utilizing NGS technology combined with various computational algorithms. Several studies reported the exploration of these bioresources through sequencing technologies to alter biosynthetic pathways in plant systems for enhanced production of existing phytochemicals and to design a biologically active product library which can be screened for new drug applications. Hence the advent of transcriptome sequencing adds the way to pathway discovery along with gene expression studies, functional annotation information, and plant genetic engineering to explore gene functions in various plant species which provides the most promising approach on various aspects of plant biodiversity for developing economically valuable products. The wider availability of metabolite-profiling techniques may increase our knowledge in metabolic networks by detecting unexpected correlations and relations among various metabolites. Thus studies should significantly improve and must be able to execute quantitative analysis of distinct categories of metabolite fluxes and pools exerting no or few disturbances in metabolism. It has been expected that with all this information, it will be applicable by metabolic engineers to produce predictive models of plant secondary metabolism.

References

Achan J, Talisuna AO, Erhart A, Yeka A, Tibenderana JK, Baliraine FN, Rosenthal PJ, D'Alessandro U (2011) Quinine, an old anti-malarial drug in a modern world: role in the treatment of malaria. Malar J 10:1

Adnyana IK, Sukandar EY, Setiawan F, Christanti Y (2013) Efficacy and safety O-desmethyl Quinine Compare to Quinine for Nocturnal Leg Cramp. J Med Sci 13:819

Agyapong VIO, Singh K, Savage M, Thekiso TB, Finn M, Farren CK, McLoughlin DM (2013) Use of codeine-containing medicines by Irish psychiatric inpatients before and after regulatory limitations on their supply. Ir J Psychol Med 30:7–12

Aharoni A, De Vos CH, Wein M, Sun Z, Greco R, Kroon A, Mol JNM, O'Connell AP (2001) The strawberry FaMYB1 transcription factor suppresses anthocyanin and flavonol accumulation in transgenic tobacco. Plant J 28:319–332

Allen RS, Millgate AG, Chitty JA, Thisleton J, Miller JAC, Fist AJ, Gerlach WL, Larkin PJ (2004) RNAi-mediated replacement of morphine with the nonnarcotic alkaloid reticuline in opium poppy. Nat Biotechnol 22:1559–1566

Allen RS, Miller JAC, Chitty JA, Fist AJ, Gerlach WL, Larkin PJ (2008) Metabolic engineering of morphinan alkaloids by over-expression and RNAi suppression of salutaridinol 7-O-acetyltransferase in opium poppy. Plant Biotechnol J 6:22–30

Alves MN, Sartoratto A, Trigo JR (2007) Scopolamine in *Brugmansia suaveolens* (Solanaceae): defense, allocation, costs, and induced response. J Chem Ecol 33:297–309

Annadurai RS, Neethiraj R, Jayakumar V, Damodaran AC, Rao SN, Katta MA, Gopinathan S, Sarma SP, Senthilkumar V, Niranjan V (2013) *De Novo* transcriptome assembly (NGS) of *Curcuma longa* L. rhizome reveals novel transcripts related to anticancer and antimalarial terpenoids. PLoS One 8:e56217

Ap Rees T (1995) Prospects of manipulating plant metabolism. Trends Biotechnol 13:375–378

Asif MH, Lakhwani D, Pathak S, Gupta P, Bag SK, Nath P, Trivedi PK (2014) Transcriptome analysis of ripe and unripe fruit tissue of banana identifies major metabolic networks involved in fruit ripening process. BMC Plant Biol 14:1

Athanasiadou S, Kyriazakis I, Jackson F, Coop RL (2001) Direct anthelmintic effects of condensed tannins towards different gastrointestinal nematodes of sheep: *in vitro* and *in vivo* studies. Vet Parasitol 99:205–219

Bak S, Olsen CE, Petersen BL, Møller BL, Halkier BA (1999) Metabolic engineering of p-hydroxybenzylglucosinolate in *Arabidopsis* by expression of the cyanogenic CYP79A1 from Sorghum bicolor. Plant J 20:663–671

Barrero RA, Chapman B, Yang Y, Moolhuijzen P, Keeble-Gagnère G, Zhang N, Tang Q, Bellgard MI, Qiu D (2011) *De novo* assembly of *Euphorbia fischeriana* root transcriptome identifies prostratin pathway related genes. BMC Genomics 12:1

Barry TN, McNabb WC (1999) The implications of condensed tannins on the nutritive value of temperate forages fed to ruminants. Br J Nutr 81:263–272

Baser KHC, Buchbauer G (2015) Handbook of essential oils: science, technology, and applications. CRC Press, Boca Raton

Becker JVW, Armstrong GO, van der Merwe MJ, Lambrechts MG, Vivier MA, Pretorius IS (2003) Metabolic engineering of Saccharomyces cerevisiae for the synthesis of the wine-related antioxidant resveratrol. FEMS Yeast Res 4:79–85

Bentley DR, Balasubramanian S, Swerdlow HP, Smith GP, Milton J, Brown CG, Hall KP, Evers DJ, Barnes CL, Bignell HR (2008) Accurate whole human genome sequencing using reversible terminator chemistry. Nature 456:53–59

Bernáth J, Fuleky G (2009) Aromatic plants. Cultivated Plants, Primarily as Food Sources 2:329–352

Böhme K, Barros-Velázquez J, Calo-Mata P, Aubourg SP (2014) Antibacterial, antiviral and antifungal activity of essential oils: mechanisms and applications. In: Antimicrobial compounds. Springer, Heidelberg, pp 51–81

Borevitz JO, Xia Y, Blount J, Dixon RA, Lamb C (2000) Activation tagging identifies a conserved MYB regulator of phenylpropanoid biosynthesis. Plant Cell 12:2383–2393

Borgio JF (2009) RNA interference (RNAi) technology: a promising tool for medicinal plant research. J Med Plant Res 3:1176–1183

Brinkhaus B, Lindner M, Schuppan D, Hahn EG (2000) Chemical, pharmacological and clinical profile of the East Asian medical plant *Centella asiatica*. Phytomedicine 7:427–448

Brossi A, Venugopalan B, Dominguez Gerpe L, Yeh HJC, Flippen-Anderson JL, Buchs P, Luo XD, Milhous W, Peters W (1988) Arteether, a new antimalarial drug: synthesis and antimalarial properties. J Med Chem 31:645–650

Canel C, Lopes-Cardoso MI, Whitmer S, van der Fits L, Pasquali G, van der Heijden R, Hoge JHC, Verpoorte R (1998) Effects of over-expression of strictosidine synthase and tryptophan decarboxylase on alkaloid production by cell cultures of *Catharanthus roseus*. Planta 205:414–419

Carpentier SC, Panis B, Vertommen A, Swennen R, Sergeant K, Renaut J, Laukens K, Witters E, Samyn B, Devreese B (2008) Proteome analysis of non-model plants: A challenging but powerful approach. Mass Spectrom Rev 27:354–377

Chaichisemsari M, Maherisis N, Sadaghian M, Eshratkhah B, Hassanpour S (2011) Effects of administration of industrial tannins on nutrient excretion parameters during naturally acquired mixed nematode infections in Moghani sheep. J Am Sci 7:245–248

Chevreux B, Pfisterer T, Drescher B, Driesel AJ, Müller WEG, Wetter T, Suhai S (2004) Using the miraEST assembler for reliable and automated mRNA transcript assembly and SNP detection in sequenced ESTs. Genome Res 14:1147–1159

Chiari Y, Cahais V, Galtier N, Delsuc F (2012) Phylogenomic analyses support the position of turtles as the sister group of birds and crocodiles (Archosauria). BMC Biol 10:1–15. doi:10.1186/1741-7007-10-65

Chintapakorn Y, Hamill JD (2003) Antisense-mediated down-regulation of putrescine N-methyltransferase activity in transgenic *Nicotiana tabacum* L. can lead to elevated levels of anatabine at the expense of nicotine. Plant Mol Biol 53:87–105

Chu Y, Corey DR (2012) RNA sequencing: platform selection, experimental design, and data interpretation. Nucleic Acid Ther 22:271–274

Chung S-M, Vaidya M, Tzfira T (2006) *Agrobacterium* is not alone: gene transfer to plants by viruses and other bacteria. Trends Plant Sci 11:1–4

Clarke PB, Fu DS, Jakubovic A, Fibiger HC (1988) Evidence that mesolimbic dopaminergic activation underlies the locomotor stimulant action of nicotine in rats. J Pharmacol Exp Ther 246:701–708

Claros MG, Bautista R, Guerrero-Fernández D, Benzerki H, Seoane P, Fernández-Pozo N (2012) Why assembling plant genome sequences is so challenging. Biology (Basel) 1:439–459.

Clifford M, Leah M, Charles N (2012) Antiepileptic properties of Quinine: A systematic review. Annu Rev Neurosci 19:14

Cline SD, Coscia CJ (1988) Stimulation of sanguinarine production by combined fungal elicitation and hormonal deprivation in cell suspension cultures of *Papaver bracteatum*. Plant Physiol 86:161–165

Courtois D, Guern J (1980) Temperature response of *Catharanthus roseus* cells cultivated in liquid medium. Plant Sci Lett 17:473–482

Crowell PL (1999) Prevention and therapy of cancer by dietary monoterpenes. J Nutr 129:775S–778S

Cushnie TPT, Lamb AJ (2005) Antimicrobial activity of flavonoids. Int J Antimicrob Agents 26:343–356

Dai J, Mumper RJ (2010) Plant phenolics: extraction, analysis and their antioxidant and anticancer properties. Molecules 15:7313–7352

Davuluri GR, Van Tuinen A, Fraser PD, Manfredonia A, Newman R, Burgess D, Brummell DA, King SR, Palys J, Uhlig J (2005) Fruit-specific RNAi-mediated suppression of DET1 enhances carotenoid and flavonoid content in tomatoes. Nat Biotechnol 23:890–895

Degen AA, Becker K, Makkar HPS, Borowy N (1995) *Acacia saligna* as a fodder tree for desert livestock and the interaction of its tannins with fibre fractions. J Sci Food Agric 68:65–71

Dexter R, Qualley A, Kish CM, Ma CJ, Koeduka T, Nagegowda DA, Dudareva N, Pichersky E, Clark D (2007) Characterization of a petunia acetyltransferase involved in the biosynthesis of the floral volatile isoeugenol. Plant J 49:265–275

Díaz-Sala C, Cervera T (2011) Promoting a functional and comparative understanding of the conifer genome-implementing applied aspects for more productive and adapted forests (ProCoGen). In: BMC proceedings. BioMed Central Ltd, p P158

Diray-Arce J, Clement M, Gul B, Khan MA, Nielsen BL (2015) Transcriptome assembly, profiling and differential gene expression analysis of the halophyte *Suaeda fruticosa* provides insights into salt tolerance. BMC Genomics 16:1

Dixon RA (1999) Plant natural products: the molecular genetic basis of biosynthetic diversity. Curr Opin Biotechnol 10:192–197

Dixon RA, Arntzen CJ (1997) Transgenic plant technology is entering the era of metabolic engineering. Trends Biotechnol 15:441–444

Dixon RA, Dey PM, Lamb CJ (1983) Phytoalexins: enzymology and molecular biology. Adv Enzymol Relat Areas Mol Biol 55:1–136

Duan J, Xia C, Zhao G, Jia J, Kong X (2012) Optimizing *de novo* common wheat transcriptome assembly using short-read RNA-Seq data. BMC Genomics 13:1

Eid J, Fehr A, Gray J, Luong K, Lyle J, Otto G, Peluso P, Rank D, Baybayan P, Bettman B (2009) Real-time DNA sequencing from single polymerase molecules. Science 323(80):133–138

Eilert U, Constabel F, Kurz WGW (1986) Elicitor-stimulation of monoterpene indole alkaloid formation in suspension cultures of *Catharanthus roseus*. J Plant Physiol 126:11–22

Eilert U, De Luca V, Constabel F, Kurz WGW (1987) Elicitor-mediated induction of tryptophan decarboxylase and strictosidine synthase activities in cell suspension cultures of *Catharanthus roseus*. Arch Biochem Biophys 254:491–497

Elmer KR, Fan S, Gunter HM, Jones JC, Boekhoff S, Kuraku S, Meyer A (2010) Rapid evolution and selection inferred from the transcriptomes of sympatric crater lake cichlid fishes. Mol Ecol 19:197–211

El-Tawil S, Al Musa T, Valli H, Lunn M, Brassington R, El-Tawil T, Weber M (2010) Quinine for muscle cramps. Cochrane Database Syst Rev 4:CD005044

Escobar MA, Dandekar AM (2003) *Agrobacterium tumefaciens* as an agent of disease. Trends Plant Sci 8:380–386

Estévez JM, Cantero A, Reindl A, Reichler S, León P (2001) 1-Deoxy-D-xylulose-5-phosphate synthase, a limiting enzyme for plastidic isoprenoid biosynthesis in plants. J Biol Chem 276:22901–22909

Eudes A, Liang Y, Mitra P, Loqué D (2014) Lignin bioengineering. Curr Opin Biotechnol 26:189–198

Evans DA, Mitch CH (1982) Studies directed towards the total synthesis of morphine alkaloids. Tetrahedron Lett 23:285–288

Fabricant DS, Farnsworth NR (2001) The value of plants used in traditional medicine for drug discovery. Environ Health Perspect 109:69–75

Facchini PJ, Bohlmann J, Covello PS, De Luca V, Mahadevan R, Page JE, Ro D-K, Sensen CW, Storms R, Martin VJJ (2012) Synthetic biosystems for the production of high-value plant metabolites. Trends Biotechnol 30:127–131

Fernández SP, Wasowski C, Loscalzo LM, Granger RE, Johnston GAR, Paladini AC, Marder M (2006) Central nervous system depressant action of flavonoid glycosides. Eur J Pharmacol 539:168–176

Frick S, Chitty JA, Kramell R, Schmidt J, Allen RS, Larkin PJ, Kutchan TM (2004) Transformation of opium poppy (*Papaver somniferum* L.) with antisense berberine bridge enzyme gene (anti-bbe) via somatic embryogenesis results in an altered ratio of alkaloids in latex but not in roots. Transgenic Res 13:607–613

Fujii N, Inui T, Iwasa K, Morishige T, Sato F (2007) Knockdown of berberine bridge enzyme by RNAi accumulates (S)-reticuline and activates a silent pathway in cultured California poppy cells. Transgenic Res 16:363–375

Galvano F, La Fauci L, Lazzarino G, Fogliano V, Ritieni A, Ciappellano S, Battistini NC, Tavazzi B, Galvano G (2004) Cyanidins: metabolism and biological properties. J Nutr Biochem 15:2–11

Gandikota M, de Kochko A, Chen L, Ithal N, Fauquet C, Reddy AR (2001) Development of transgenic rice plants expressing maize anthocyanin genes and increased blast resistance. Mol Breed 7:73–83

Garcia-Seco D, Zhang Y, Gutierrez-Mañero FJ, Martin C, Ramos-Solano B (2015) RNA-Seq analysis and transcriptome assembly for blackberry (*Rubus* sp. Var. Lochness) fruit. BMC Genomics 16:5

Gaudinier A, Tang M, Kliebenstein DJ (2015) Transcriptional networks governing plant metabolism. Current Plant Biology 3:56–64

Gayral P, Melo-Ferreira J, Glemin S, Bierne N, Carneiro M, Nabholz B, Lourenco JM, Alves PC, Ballenghien M, Faivre N, Belkhir K, Cahais V, Loire E, Bernard A, Galtier N (2013) Reference-Free Population Genomics from Next-Generation Transcriptome Data and the Vertebrate-Invertebrate Gap. PLoS Genet 9:e1003457

Ghasemzadeh A, Jaafar HZE (2011) Effect of CO2 enrichment on synthesis of some primary and secondary metabolites in ginger (*Zingiber officinale* Roscoe). Int J Mol Sci 12:1101–1114

Giannini C, Debitus C, Lucas R, Ubeda A, Payá M, Hooper JNA, D'Auria MV (2001) New sesquiterpene derivatives from the sponge *Dysidea* species with a selective inhibitor profile against human phospholipase A2 and other leukocyte functions. J Nat Prod 64:612–615

Gilles A, Meglécz E, Pech N, Ferreira S, Malausa T, Martin J-F (2011) Accuracy and quality assessment of 454 GS-FLX Titanium pyrosequencing. BMC Genomics 12:245

Giulietti AM, Ertola RJ (1997) Biotechnological strategies for production of plants and secondary metabolites of pharmaceutical interest. In: II WOCMAP congress medicinal and aromatic plants, Part 3: Agricultural production. Post Harvest Techniques, Biotechnology 502, pp 269–280

Gläßgen WE, Wray V, Strack D, Metzger JW, Seitz HU (1992) Anthocyanins from cell suspension cultures of *Daucus carota*. Phytochemistry 31:1593–1601

Goddijn OJM, Pen J (1995) Plants as bioreactors. Trends Biotechnol 13:379–387

Gogas H, Fountzilas G (2003) The role of taxanes as a component of neoadjuvant chemotherapy for breast cancer. Ann Oncol 14:667–674

Grotewold E, Chamberlin M, Snook M, Siame B, Butler L, Swenson J, Maddock S, Clair GS, Bowen B (1998) Engineering secondary metabolism in maize cells by ectopic expression of transcription factors. Plant Cell 10:721–740

Gu L, Li N, Gong J, Li Q, Zhu W, Li J (2011) Berberine ameliorates intestinal epithelial tight-junction damage and down-regulates myosin light chain kinase pathways in a mouse model of endotoxinemia. J Infect Dis 203:1602–1612

Guçlu-Ustundag O, Mazza G (2007) Saponins: properties, applications and processing. Crit Rev Food Sci Nutr 47:231–258

Gundlach H, Müller MJ, Kutchan TM, Zenk MH (1992) Jasmonic acid is a signal transducer in elicitor-induced plant cell cultures. Proc Natl Acad Sci 89:2389–2393

Guo Q, Ma X, Wei S, Qiu D, Wilson IW, Wu P, Tang Q, Liu L, Dong S, Zu W (2014) *De novo* transcriptome sequencing and digital gene expression analysis predict biosynthetic pathway of rhynchophylline and isorhynchophylline from *Uncaria rhynchophylla*, a non-model plant with potent anti-alzheimer's properties. BMC Genomics 15:676

Gupta P, Goel R, Pathak S, Srivastava A, Singh SP, Sangwan RS, Asif MH, Trivedi PK (2013) *De novo* assembly, functional annotation and comparative analysis of *Withania somnifera* leaf and root transcriptomes to identify putative genes involved in the withanolides biosynthesis. PLoS One 8:e62714

Haas BJ, Papanicolaou A, Yassour M, Grabherr M, Blood PD, Bowden J, Couger MB, Eccles D, Li B, Lieber M (2013) *De novo* transcript sequence reconstruction from RNA-seq using the Trinity platform for reference generation and analysis. Nat Protoc 8:1494–1512

Hagimori M, Matsumoto T, Obi Y (1982) Studies on the production of Digitalis cardenolides by plant tissue culture III. Effects of nutrients on digitoxin formation by shoot-forming cultures of *Digitalis purpurea* L. grown in liquid media. Plant Cell Physiol 23:1205–1211

Hajek P, McRobbie H, Myers K (2013) Efficacy of cytisine in helping smokers quit: systematic review and meta-analysis. Thorax 68:1037–1042

Han JY, Kwon YS, Yang DC, Jung YR, Choi YE (2006) Expression and RNA interference-induced silencing of the dammarenediol synthase gene in *Panax ginseng*. Plant Cell Physiol 47:1653–1662

Hara Y, Luo SJ, Wickremasinghe RL, Yamanishi T (1995) Special issue on tea. Food Rev Intl 11:371–542

Harrewijn P, van Oosten AM, Piron PGM (2001) Natural Terpenoids as Messengers: a multidisciplinary study of their production, biological functions, and practical applications. Kluwer Academic Press, Dordrecht, p 440

Hartmann T (1991) Alkaloids In herbivores; their interaction with secondary plant metabolites. In: Rosenthaland GA, Berenbaum MR (ed) The chemical participants, vol I, edn 2. Academic Press, San Diego

Hassanpour S, Sadaghian M, MaheriSis N, Eshratkhah B, ChaichiSemsari M (2011) Effect of condensed tannin on controlling faecal protein excretion in nematode-infected sheep: *in vivo* study. J Am Sci 7:896–900

He C, Cui K, Zhang J, Duan A, Zeng Y (2013) Next-generation sequencing-based mRNA and microRNA expression profiling analysis revealed pathways involved in the rapid growth of developing culms in Moso bamboo. BMC Plant Biol 13:119

Heim KE, Tagliaferro AR, Bobilya DJ (2002) Flavonoid antioxidants: chemistry, metabolism and structure-activity relationships. J Nutr Biochem 13:572–584

Heinstein PF (1985) Future approaches to the formation of secondary natural products in plant cell suspension cultures. J Nat Prod 48:1–9

Herbers K, Sonnewald U (1999) Production of new/modified proteins in transgenic plants. Curr Opin Biotechnol 10:163–168

Herrmann KM, Weaver LM (1999) The shikimate pathway. Annu Rev Plant Biol 50:473–503

Hertog MGL, Hollman PCH, Katan MB (1992) Content of potentially anticarcinogenic flavonoids of 28 vegetables and 9 fruits commonly consumed in the Netherlands. J Agric Food Chem 40:2379–2383

Hoffmann S, Otto C, Kurtz S, Sharma CM, Khaitovich P, Vogel J, Stadler PF, Hackermüller J (2009) Fast mapping of short sequences with mismatches, insertions and deletions using index structures. PLoS Comput Biol 5:e1000502

Hollman PH, Katan MB (1999) Dietary flavonoids: intake, health effects and bioavailability. Food Chem Toxicol 37:937–942

Hostettmann K, Marston A (2005) Saponins. Cambridge University Press, Cambridge

Huang M, Lu J-J, Huang M-Q, Bao J-L, Chen X-P, Wang Y-T (2012) Terpenoids: natural products for cancer therapy. Expert Opin Investig Drugs 21:1801–1818

Hunter T (1995) Protein kinases and phosphatases: the yin and yang of protein phosphorylation and signaling. Cell 80:225–236

Hwang EI, Kaneko M, Ohnishi Y, Horinouchi S (2003) Production of plant-specific flavanones by *Escherichia coli* containing an artificial gene cluster. Appl Environ Microbiol 69:2699–2706

Ibarra-Laclette E, Méndez-Bravo A, Pérez-Torres CA, Albert VA, Mockaitis K, Kilaru A, López-Gómez R, Cervantes-Luevano JI, Herrera-Estrella L (2015) Deep sequencing of the Mexican avocado transcriptome, an ancient angiosperm with a high content of fatty acids. BMC Genomics 16:599

Ikeda T, Matsumoto T, Noguchi M (1977) Effects of inorganic nitrogen sources and physical factors on the formation of ubiquinone by tobacco plant cells in suspension culture. Agric Biol Chem 41:1197–1201

Imelfort M, Edwards D (2009) *De novo* sequencing of plant genomes using second-generation technologies. Brief Bioinform 10:609–618

Inui T, Tamura K, Fujii N, Morishige T, Sato F (2007) Over expression of *Coptis japonica* norcoclaurine 6-O-methyltransferase overcomes the rate-limiting step in benzylisoquinoline alkaloid biosynthesis in cultured *Eschscholzia californica*. Plant Cell Physiol 48:252–262

Ito H, Miyake M, Nishitani E, Mori K, Hatano T, Okuda T, Konoshima T, Takasaki M, Kozuka M, Mukainaka T (1999) Anti-tumor promoting activity of polyphenols from *Cowania mexicana* and *Coleogyne ramosissima*. Cancer Lett 143:5–13

James JT, Dubery IA (2009) Pentacyclic triterpenoids from the medicinal herb, *Centella asiatica* (L.) Urban. Molecules 14:3922–3941

Jayakodi M, Lee S-C, Park H-S, Jang W, Lee YS, Choi B-S, Nah GJ, Kim D-S, Natesan S, Sun C (2014) Transcriptome profiling and comparative analysis of *Panax ginseng* adventitious roots. J Ginseng Res 38:278–288

Jin H, Cominelli E, Bailey P, Parr A, Mehrtens F, Jones J, Tonelli C, Weisshaar B, Martin C (2000) Transcriptional repression by AtMYB4 controls production of UV-protecting sunscreens in *Arabidopsis*. EMBO J 19:6150–6161

Johnson MTJ, Carpenter EJ, Tian Z, Bruskiewich R, Burris JN, Carrigan CT, Chase MW, Clarke ND, Covshoff S, Depamphilis CW, Edger PP, Goh F, Graham S, Greiner S, Hibberd JM, Jordon-Thaden I, Kutchan TM, Leebens-Mack J, Melkonian M, Miles N, Myburg H, Patterson J, Pires JC, Ralph P, Rolf M, Sage RF, Soltis D, Soltis P, Stevenson D, Stewart CNJ, Surek B, Thomsen CJM, Villarreal JC, Wu X, Zhang Y, Deyholos MK, Wong GK-S (2012) Evaluating methods for isolating total RNA and predicting the success of sequencing phylogenetically diverse plant transcriptomes. PLoS One 7:e50226

Jung HG, Fahey GC (1983) Nutritional implications of phenolic monomers and lignin: a review. J Anim Sci 57:206–219

Kabera JN, Semana E, Mussa AR, He X (2014) Plant secondary metabolites: biosynthesis, classification, function and pharmacological properties. J Pharm Pharmacol 2:377–392

Kaminaga Y, Schnepp J, Peel G, Kish CM, Ben-Nissan G, Weiss D, Orlova I, Lavie O, Rhodes D, Wood K (2006) Plant phenylacetaldehyde synthase is a bifunctional homotetrameric enzyme that catalyzes phenylalanine decarboxylation and oxidation. J Biol Chem 281:23357–23366

Karppinen K, Hohtola A (2008) Molecular cloning and tissue-specific expression of two cDNAs encoding polyketide synthases from *Hypericum perforatum*. J Plant Physiol 165:1079–1086

Karppinen K, Hokkanen J, Mattila S, Neubauer P, Hohtola A (2008) Octaketide-producing type III polyketide synthase from *Hypericum perforatum* is expressed in dark glands accumulating hypericins. Feder Eur Biochem Soc J 275:4329–4342

Kempe K, Higashi Y, Frick S, Sabarna K, Kutchan TM (2009) RNAi suppression of the morphine biosynthetic gene salAT and evidence of association of pathway enzymes. Phytochemistry 70:579–589

Kenny OM, McCarthy CM, Brunton NP, Hossain MB, Rai DK, Collins SG, Jones PW, Maguire AR, O'Brien NM (2013) Anti-inflammatory properties of potato glycoalkaloids in stimulated Jurkat and Raw 264.7 mouse macrophages. Life Sci 92:775–782

Kim JB, Yu J-H, Ko E, Lee K-W, Song AK, Park SY, Shin I, Han W, Noh DY (2010) The alkaloid Berberine inhibits the growth of Anoikis-resistant MCF-7 and MDA-MB-231 breast cancer cell lines by inducing cell cycle arrest. Phytomedicine 17:436–440

Kim SM, Kang K, Jho EH, Jung Y, Nho CW, Um B, Pan C (2011) Hepatoprotective effect of flavonoid glycosides from *Lespedeza cuneata* against oxidative stress induced by tert-butyl hyperoxide. Phytother Res 25:1011–1017

Kishore GM, Somerville CR (1993) Genetic engineering of commercially useful biosynthetic pathways in transgenic plants. Curr Opin Biotechnol 4:152–158

Kitic D, Pavlovic D, Brankovic S (2013) The role of essential oils and the biological detoxification in the prevention of aflatoxin borne diseases. Curr Top Med Chem 13:2767–2790

Kreft S, Knapp M, Kreft I (1999) Extraction of rutin from buckwheat (*Fagopyrum esculentum* Moench) seeds and determination by capillary electrophoresis. J Agric Food Chem 47:4649–4652

Kreis W, Reinhard E (1992) 12β-Hydroxylation of digitoxin by suspension-cultured *Digitalis lanata* cells: Production of digoxin in 20-litre and 300-litre air-lift bioreactors. J Biotechnol 26:257–273

Kumar PA (2001) Plant biotechnology: Future perspectives (Review Paper). Def Sci J 51:353

Kumar R, Khurana A (2014) Functional genomics of tomato: Opportunities and challenges in post-genome NGS era. J Biosci 39:917–929

Kumar S, Pandey AK (2013) Chemistry and biological activities of flavonoids: an overview. Sci World J 2013:1–16

Künstner A, Wolf JBW, Backström N, Whitney O, Balakrishnan CN, Day L, Edwards SV, Janes DE, Schlinger BA, Wilson RK (2010) Comparative genomics based on massive parallel transcriptome sequencing reveals patterns of substitution and selection across 10 bird species. Mol Ecol 19:266–276

Lam S, Lam B, Harrison L, Strobel G (1984) Genetic information on the Ri plasmid of *Agrobacterium rhizogenes* determines host specificity. Plant Sci Lett 34:345–352

Larkin PJ, Miller JAC, Allen RS, Chitty JA, Gerlach WL, Frick S, Kutchan TM, Fist AJ (2007) Increasing morphinan alkaloid production by over-expressing codeinone reductase in transgenic *Papaver somniferum*. Plant Biotechnol J 5:26–37

Lecourieux D, Mazars C, Pauly N, Ranjeva R, Pugin A (2002) Analysis and effects of cytosolic free calcium increases in response to elicitors in *Nicotiana plumbaginifolia* cells. Plant Cell 14:2627–2641

Lee K-T, Yamakawa T, Kodama T, Shimomura K (1998) Effects of chemicals on alkaloid production by transformed roots of belladonna. Phytochemistry 49:2343–2347

Leisso RS, Buchanan DA, Lee J, Mattheis JP, Sater C, Hanrahan I, Watkins CB, Gapper N, Johnston JW, Schaffer RJ (2015) Chilling-related cell damage of apple (*Malus*× *domestica* Borkh.) fruit cortical tissue impacts antioxidant, lipid and phenolic metabolism. Physiol Plant 153:204–220

Lessard PA, Kulaveerasingam H, York GM, Strong A, Sinskey AJ (2002) Manipulating gene expression for the metabolic engineering of plants. Metab Eng 4:67–79

Li C, Zhu Y, Guo X, Sun C, Luo H, Song J, Li Y, Wang L, Qian J, Chen S (2013) Transcriptome analysis reveals ginsenosides biosynthetic genes, microRNAs and simple sequence repeats in *Panax ginseng* CA Meyer. BMC Genomics 14:1

Li X, Weng J, Chapple C (2008) Improvement of biomass through lignin modification. Plant J 54:569–581

Li Y, Luo H-M, Sun C, Song J-Y, Sun Y-Z, Wu Q, Wang N, Yao H, Steinmetz A, Chen S-L (2010) EST analysis reveals putative genes involved in glycyrrhizin biosynthesis. BMC Genomics 11:268

Lopez M, Martınez F, Del Valle C, Orte C, Miro M (2001) Analysis of phenolic constituents of biological interest in red wines by high-performance liquid chromatography. J Chromatogr A 922:359–363

Lu M-K, Shih Y-W, Chang Chien T-T, Fang L-H, Huang H-C, Chen P-S (2010) ALPHA.-Solanine Inhibits Human Melanoma Cell Migration and Invasion by Reducing Matrix Metalloproteinase-2/9 Activities. Biol Pharm Bull 33:1685–1691
Luo H, Sun C, Sun Y, Wu Q, Li Y, Song J, Niu Y, Cheng X, Xu H, Li C (2011) Analysis of the transcriptome of *Panax notoginseng* root uncovers putative triterpene saponin-biosynthetic genes and genetic markers. BMC Genomics 12:1
Maheri-Sis N, Chaichi Semsari M, Eshratkhah B, Sadaghian M, Gorbani A, Hassanpour S (2011) Evaluation of the effects of Quebracho condensed tannin on faecal egg counts during naturally acquired mixed nematode infections in Moghani sheep. Ann Appl Biol Res 2:170–174
Mahmoud SS, Croteau RB (2001) Metabolic engineering of essential oil yield and composition in mint by altering expression of deoxyxylulose phosphate reductoisomerase and menthofuran synthase. Proc Natl Acad Sci 98:8915–8920
Mahmoud SS, Croteau RB (2002) Strategies for transgenic manipulation of monoterpene biosynthesis in plants. Trends Plant Sci 7:366–373
Mahmoud SS, Williams M, Croteau R (2004) Cosuppression of limonene-3-hydroxylase in peppermint promotes accumulation of limonene in the essential oil. Phytochemistry 65:547–554
Mahomoodally MF, Gurib-Fakim A, Subratty AH (2008) Antimicrobial activities and phytochemical profiles of endemic medicinal plants of Mauritius. Pharm Biol 43:237–242
Manach C, Scalbert A, Morand C, Rémésy C, Jiménez L (2004) Polyphenols: food sources and bioavailability. Am J Clin Nutr 79:727–747
Mangas S, Moyano E, Hernandez-Vazquez L, Bonfill M (2009) *Centella asiatica* (L) Urban: an updated approach. Plant Second terpenoids Research Signpost, Trivandrum, pp 55–74.
Marques JV, Dalisay DS, Yang H, Lee C, Davin LB, Lewis NG (2014) A multi-omics strategy resolves the elusive nature of alkaloids in *Podophyllum* species. Mol BioSyst 10:2838–2849
Martin VJJ, Pitera DJ, Withers ST, Newman JD, Keasling JD (2003) Engineering a mevalonate pathway in *Escherichia coli* for production of terpenoids. Nat Biotechnol 21:796–802
McBrien NA, Stell WK, Carr B (2013) How does atropine exert its anti-myopia effects? Ophthalmic Physiol Opt 33:373–378
McGarvey DJ, Croteau R (1995) Terpenoid metabolism. Plant Cell 7:1015
Md-Mustafa ND, Khalid N, Gao H, Peng Z, Alimin MF, Bujang N, Ming WS, Mohd-Yusuf Y, Harikrishna JA, Othman RY (2014) Transcriptome profiling shows gene regulation patterns in a flavonoid pathway in response to exogenous phenylalanine in *Boesenbergia rotunda* cell culture. BMC Genomics 15:1
Melton L (2006) Body Blazes. Sci Am 294:24
Metzger KJ, Klaper R, Thomas MA (2011) Implications of informatics approaches in ecological research. Eco Inform 6:4–12
Miflin B (2000) Crop improvement in the 21st century. J Exp Bot 51:1–8
Milat M-L, Ricci P, Bonnet P, Blein J-P (1991) Capsidiol and ethylene production by tobacco cells in response to cryptogein, an elicitor from *Phytophthora cryptogea*. Phytochemistry 30:2171–2173
Mishra A, Kumar S, Pandey AK (2013) Scientific validation of the medicinal efficacy of *Tinospora cordifolia*. Sci World J:1–8
Miyake Y, Shimoi K, Kumazawa S, Yamamoto K, Kinae N, Osawa T (2000) Identification and antioxidant activity of flavonoid metabolites in plasma and urine of eriocitrin-treated rats. J Agric Food Chem 48:3217–3224
Mohamed MA, Mammoud MR, Hayen H (2009) Evaluation of antinociceptive and anti-inflammatory activities of a new triterpene saponin from *Bauhinia variegata* leaves. Zeitschrift Fur Naturforschung 64:798–808
Mohsenikia M, Alizadeh AM, Khodayari S, Khodayari H, Karimi A, Zamani M, Azizian S, Mohagheghi MA (2013) The protective and therapeutic effects of alpha-solanine on mice breast cancer. Eur J Pharmacol 718:1–9

Muir SR, Collins GJ, Robinson S, Hughes S, Bovy A, De Vos CHR, van Tunen AJ, Verhoeyen ME (2001) Over expression of petunia chalcone isomerase in tomato results in fruit containing increased levels of flavonols. Nat Biotechnol 19:470–474

Mukherjee AK, Basu S, Sarkar N, Ghosh AC (2001) Advances in cancer therapy with plant based natural products. Curr Med Chem 8:1467–1486

Mulder-Krieger TH, Verpoorte R, Svendsen AB, Scheffer JJC (1988) Production of essential oils and flavours in plant cell and tissue cultures A review. Plant Cell Tissue Organ Cult 13:85–154

Muriira NG, Xu W, Muchugi A, Xu J, Liu A (2015) *De novo* sequencing and assembly analysis of transcriptome in the Sodom apple (*Calotropis gigantea*). BMC Genomics 16:1

Nadiya F, Anjali N, Jinu T, Gangaprasad A, Sabu KK (2017) Transcriptome profiling of Elettaria cardamomum (L.) Maton (small cardamom). Genomics Data 11:102–103. doi:10.1016/j.gdata.2016.12.013

Nagira Y, Shimamura K, Hirai S, Shimanuki M, Kodama H, Ozeki Y (2006) Identification and characterization of genes induced for anthocyanin synthesis and chlorophyll degradation in regenerated torenia shoots using suppression subtractive hybridization, cDNA microarrays, and RNAi techniques. J Plant Res 119:217–230

Nakamura K, Oshima T, Morimoto T, Ikeda S, Yoshikawa H, Shiwa Y, Ishikawa S, Linak MC, Hirai A, Takahashi H (2011) Sequence-specific error profile of Illumina sequencers. Nucleic Acids Res 39:e90–e90

Nakasugi K, Crowhurst RN, Bally J, Wood CC, Hellens RP, Waterhouse PM (2013) *De novo* transcriptome sequence assembly and analysis of RNA silencing genes of *Nicotiana benthamiana*. PLoS One 8:e59534

Ncube B, Van Staden J (2015) Tilting plant metabolism for improved metabolite biosynthesis and enhanced human benefit. Molecules 20:12698–12731

Neale DB, Wheeler NC, Centralia W (2004) The Loblolly Pine Genome Project. 2011 08 25]. http//dendrome. ucdavis. edu/NealeLab/lpgp

Nicolaou KC, Chen JS (2011) Classics in Total Synthesis III: Further Targets, Strategies, Methods 746 pages. Weinheim: Wiley-VCH.

Nishizuka Y (1988) The molecular heterogeneity of protein kinase C and its implications for cellular regulation. Nature 334:661–665

Oksman-Caldentey K-M, Inze D, Orešič M (2004) Connecting genes to metabolites by a systems biology approach. Proc Natl Acad Sci USA 101:9949–9950

Oksman-Caldentey K-M, Saito K (2005) Integrating genomics and metabolomics for engineering plant metabolic pathways. Curr Opin Biotechnol 16:174–179

O'Neil ST, Dzurisin JDK, Carmichael RD, Lobo NF, Emrich SJ, Hellmann JJ (2010) Population-level transcriptome sequencing of nonmodel organisms *Erynnis propertius* and *Papilio zelicaon*. BMC Genomics 11:1

Pandey AK (2007) Anti-staphylococcal activity of a pan-tropical aggressive and obnoxious weed *Parthenium histerophorus*: an *in vitro* study. Natl Acad Sci Lett 30:383–386

Pandey AK, Mishra AK, Mishra A (2012) Antifungal and antioxidative potential of oil and extracts derived from leaves of Indian spice plant *Cinnamomum tamala*. Cell Mol Biol 58:142–147

Panter KE, Welch KD, Gardner DR, Green BT (2013) Poisonous plants: Effects on embryo and fetal development. Birth Defects Res Part C: Embryo Today: Rev 99:223–234

Parchmann S, Gundlach H, Mueller MJ (1997) Induction of 12-oxo-phytodienoic acid in wounded plants and elicited plant cell cultures. Plant Physiol 115:1057–1064

Park E-S, Moon W-S, Song M-J, Kim M-N, Chung K-H, Yoon J-S (2001) Antimicrobial activity of phenol and benzoic acid derivatives. Int Biodeterior Biodegrad 47:209–214

Park S-U, Yu M, Facchini PJ (2002) Antisense RNA-mediated suppression of benzophenanthridine alkaloid biosynthesis in transgenic cell cultures of California poppy. Plant Physiol 128:696–706

Pearce G, Strydom D, Johnson S, Ryan CA (1991) A polypeptide from tomato leaves induces wound-inducible proteinase inhibitor proteins. Science 253(80):895–897

Petersen BL, Andréasson E, Bak S, Agerbirk N, Halkier BA (2001) Characterization of transgenic *Arabidopsis thaliana* with metabolically engineered high levels of p-hydroxybenzylglucosinolate. Planta 212:612–618

Pickel B, Drew DP, Manczak T, Weitzel C, Simonsen HT, Ro D-K (2012) Identification and characterization of a kunzeaol synthase from *Thapsia garganica*: implications for the biosynthesis of the pharmaceutical thapsigargin. Biochem J 448:261–271

Polt RL (1995) Method for making amino acid glycosides and glycopeptides. Google Patents Accessed from https://www.google.com/patents/US5470949

Prasath D, Karthika R, Habeeba NT, Suraby EJ, Rosana OB, Shaji A, Eapen SJ, Deshpande U, Anandaraj M (2014) Comparison of the transcriptomes of ginger (*Zingiber officinale* Rosc.) and mango ginger (Curcuma amada Roxb.) in response to the bacterial wilt infection. PLoS One 9:e99731

Pyle BW, Tran HT, Pickel B, Haslam TM, Gao Z, MacNevin G, Vederas JC, Kim S-U, Ro D-K (2012) Enzymatic synthesis of valerena-4,7(11)-diene by a unique sesquiterpene synthase from the valerian plant (*Valeriana officinalis*). Federation of European Biochemical Societies 279:3136–3146

Que Y, Su Y, Guo J, Wu Q, Xu L (2014) A global view of transcriptome dynamics during *Sporisorium scitamineum* challenge in sugarcane by RNA-seq. PLoS One 9:e106476

Reddy NRR, Mehta RH, Soni PH, Makasana J, Gajbhiye NA, Ponnuchamy M, Kumar J (2015) Next Generation Sequencing and Transcriptome Analysis Predicts Biosynthetic Pathway of Sennosides from Senna (*Cassia angustifolia* Vahl.), a Non-Model Plant with Potent Laxative Properties. PLoS One 10:e0129422

Reinli K, Block G (1996) Phytoestrogen content of foods—a compendium of literature values. Nutr Cancer 26:123–148

Renaut S, Grassa CJ, Moyers BT, Kane NC, Rieseberg LH (2012) The population genomics of sunflowers and genomic determinants of protein evolution revealed by RNAseq. Biology (Basel) 1:575–596.

Rijhwani SK, Shanks JV (1998) Effect of elicitor dosage and exposure time on biosynthesis of indole alkaloids by *Catharanthus roseus* hairy root cultures. Biotechnol Prog 14:442–449

Robbins MP, Thomas B, Morris P (1995) Phenylpropanoid defence responses in transgenic *Lotus corniculatus* II. Modelling plant defence responses in transgenic root cultures using thiol and carbohydrate elicitors. J Exp Bot 46:513–524

Robertson G, Schein J, Chiu R, Corbett R, Field M, Jackman SD, Mungall K, Lee S, Okada HM, Qian JQ (2010) *De novo* assembly and analysis of RNA-seq data. Nat Methods 7:909–912

Rohini VK, Rao KS (2001) Transformation of peanut (*Arachis hypogaea* L.) with tobacco chitinase gene: variable response of transformants to leaf spot disease. Plant Sci 160:889–898

Rothberg JM, Hinz W, Rearick TM, Schultz J, Mileski W, Davey M, Leamon JH, Johnson K, Milgrew MJ, Edwards M, Hoon J, Simons JF, Marran D, Myers JW, Davidson JF, Branting A, Nobile JR, Puc BP, Light D, Clark TA, Huber M, Branciforte JT, Stoner IB, Cawley SE, Lyons M, Fu Y, Homer N, Sedova M, Miao X, Reed B, Sabina J, Feierstein E, Schorn M, Alanjary M, Dimalanta E, Dressman D, Kasinskas R, Sokolsky T, Fidanza JA, Namsaraev E, McKernan KJ, Williams A, Roth GT, Bustillo J (2011) An integrated semiconductor device enabling non-optical genome sequencing. Nature 475:348–352

Rozov-Ung I, Mreyoud A, Moore J, Wilding GE, Khawam E, Lackner JM, Semler JR, Sitrin MD (2014) Detection of drug effects on gastric emptying and contractility using a wireless motility capsule. BMC Gastroenterol 14:1

Runguphan W, Maresh JJ, O'Connor SE (2009) Silencing of tryptamine biosynthesis for production of nonnatural alkaloids in plant culture. Proc Natl Acad Sci 106:13673–13678

Safayhi H, Sabieraj J, Sailer ER, Ammon HP (1994) Chamazulene: an antioxidant-type inhibitor of leukotriene B4 formation. Planta Med 60:410–413

Salzberg SL, Yorke JA (2005) Beware of mis-assembled genomes. Bioinformatics 21:4320–4321

Sangwan RS, Tripathi S, Singh J, Narnoliya LK, Sangwan NS (2013) *De novo* sequencing and assembly of *Centella asiatica* leaf transcriptome for mapping of structural, functional and regulatory genes with special reference to secondary metabolism. Gene 525:58–76
Schliesky S, Gowik U, Weber APM, Bräutigam A (2012) RNA-seq assembly–Are we there yet? Front Plant Sci 3:1–12
Schulz MH, Zerbino DR, Vingron M, Birney E (2012) Oases: robust *de novo* RNA-seq assembly across the dynamic range of expression levels. Bioinformatics 28:1086–1092
Sharma HC, Crouch JH, Sharma KK, Seetharama N, Hash CT (2002) Applications of biotechnology for crop improvement: prospects and constraints. Plant Sci 163:381–395
Sheeja TE, Deepa K, Santhi R, Sasikumar B (2015) Comparative Transcriptome Analysis of Two Species of *Curcuma* Contrasting in a High-Value Compound Curcumin: Insights into Genetic Basis and Regulation of Biosynthesis. Plant Mol Biol Report 33:1825–1836
Shexia Ma (2012). Production of secondary organic Aerosol from multiphase monoterpenes, atmospheric Aerosols – regional characteristics – chemistry and physics, (ed) Dr. Hayder Abdul-Razzak
Singh VK, Jain M (2014) Transcriptome profiling for discovery of genes involved in shoot apical meristem and flower development. Genomics data 2:135–138
Spencer JPE, Vauzour D, Rendeiro C (2009) Flavonoids and cognition: the molecular mechanisms underlying their behavioural effects. Arch Biochem Biophys 492:1–9
Stewart AJ, Bozonnet S, Mullen W, Jenkins GI, Lean MEJ, Crozier A (2000) Occurrence of flavonols in tomatoes and tomato-based products. J Agric Food Chem 48:2663–2669
Sun C, Li Y, Wu Q, Luo H, Sun Y, Song J, Lui EMK, Chen S (2010) *De novo* sequencing and analysis of the American ginseng root transcriptome using a GS FLX Titanium platform to discover putative genes involved in ginsenoside biosynthesis. BMC Genomics 11:262
Suntornsuk L, Anurukvorakun O (2005) Precision improvement for the analysis of flavonoids in selected Thai plants by capillary zone electrophoresis. Electrophoresis 26:648–660
Swarbreck SM, Lindquist EA, Ackerly DD, Andersen GL (2011) Analysis of leaf and root transcriptomes of soil-grown Avena barbata plants. Plant Cell Physiol 52:317–332
Tai Y, Wei C, Yang H, Zhang L, Chen Q, Deng W, Wei S, Zhang J, Fang C, Ho C (2015) Transcriptomic and phytochemical analysis of the biosynthesis of characteristic constituents in tea (*Camellia sinensis*) compared with oil tea (*Camellia oleifera*). BMC Plant Biol 15:1
Takita K, Herlenius E, Yamamoto Y, Lindahl SGE (2000) Effects of neuroactive substances on the morphine-induced respiratory depression; an *in vitro* study. Brain Res 884:201–205
Tapas AR, Sakarkar DM, Kakde RB (2008) Flavonoids as nutraceuticals: a review. Trop J Pharm Res 7:1089–1099
Tattersall DB, Bak S, Jones PR, Olsen CE, Nielsen JK, Hansen ML, Høj PB, Møller BL (2001) Resistance to an herbivore through engineered cyanogenic glucoside synthesis. Science 293(80):1826–1828
Teli NP, Timko MP (2004) Recent developments in the use of transgenic plants for the production of human therapeutics and biopharmaceuticals. Plant Cell Tissue Organ Cult 79:125–145
Tholl D (2015) Biosynthesis and biological functions of terpenoids in plants. Adv Biochem Eng Biotechnol 148:63–106
Thomas B, Van Deynze A, Bradford K (2002) Production of therapeutic proteins in plants. UCANR Publications, USA.
Threlfall DR, Whitehead IM (1988) The use of biotic and abiotic elicitors to induce the formation of secondary plant products in cell suspension cultures of solanaceous plants. Biochem Soc Trans 16:71–75
Tian H, Xu X, Zhang F, Wang Y, Guo S, Qin X, Du G (2015) Analysis of *Polygala tenuifolia* transcriptome and description of secondary metabolite biosynthetic pathways by illumina sequencing. Int J Genomics 2015:1–11.
Timme RE, Bachvaroff TR, Delwiche CF (2012) Broad Phylogenomic Sampling and the Sister Lineage of Land Plants. PLoS One 7:e29696

Trethewey RN (2004) Metabolite profiling as an aid to metabolic engineering in plants. Curr Opin Plant Biol 7:196–201

Underwood BA, Tieman DM, Shibuya K, Dexter RJ, Loucas HM, Simkin AJ, Sims CA, Schmelz EA, Klee HJ, Clark DG (2005) Ethylene-regulated floral volatile synthesis in petunia corollas. Plant Physiol 138:255–266

Upadhyay S, Phukan UJ, Mishra S, Shukla RK (2014) *De novo* leaf and root transcriptome analysis identified novel genes involved in steroidal sapogenin biosynthesis in *Asparagus racemosus*. BMC Genomics 15:1

van der Fits L, Memelink J (2000) ORCA3, a jasmonate-responsive transcriptional regulator of plant primary and secondary metabolism. Science 289(80):295–297

Van der Rest B, Danoun S, Boudet A-M, Rochange SF (2006) Down-regulation of cinnamoyl-CoA reductase in tomato (*Solanum lycopersicum* L.) induces dramatic changes in soluble phenolic pools. J Exp Bot 57:1399–1411

Van Soest PJ (1982) Nutritional ecology of the ruminant. Ruminant metabolism, nutritional strategies, the cellulolytic fermentation and the chemistry of forages and plant fibers. O & B Books, Inc.

Veronese P, Li X, Niu X, Weller SC, Bressan RA, Hasegawa PM (2001) Bioengineering mint crop improvement. Plant Cell Tissue Organ Cult 64:133–144

Verpoorte R (2000) Metabolic engineering of plant secondary metabolism. Springer Science & Business Media, Dordrecht

Verpoorte R, Memelink J (2002) Engineering secondary metabolite production in plants. Curr Opin Biotechnol 13:181–187

Wagner K-H, Elmadfa I (2003) Biological relevance of terpenoids. Overview focusing on mono, di- and tetraterpenes. Ann Nutr Metab 47:95–106

Wang C, Wu J, Mei X (2001) Enhancement of taxol production and excretion in *Taxus chinensis* cell culture by fungal elicitation and medium renewal. Appl Microbiol Biotechnol 55:404–410

Wang G, Tang W, Bidigare RR (2005) Terpenoids as therapeutic drugs and pharmaceutical agents. In: Natural products. Springer, pp 197–227

Wang Z, Gerstein M, Snyder M (2009) RNA-Seq: a revolutionary tool for transcriptomics. Nat Rev Genet 10:57–63

Ward JA, Weber CA (2011) Comparative RNA-seq for the investigation of resistance to *Phytophthora* root rot in the red raspberry'Latham'. In: X International Rubus and Ribes Symposium 946:67–72

Wibberley MS, Lenton JR, Neill SJ (1994) Sesquiterpenoid phytoalexins produced by hairy roots of *Nicotiana tabacum*. Phytochemistry 37:349–351

Wink M (1988) Plant breeding: importance of plant secondary metabolites for protection against pathogens and herbivores. Theor Appl Genet 75:225–233

Wolf JBW, Lindell J, Backström N (2010) Speciation genetics: current status and evolving approaches. Philos Trans R Soc B Biol Sci 365:1717–1733.

Wróbel-Kwiatkowska M, Starzycki M, Zebrowski J, Oszmiański J, Szopa J (2007) Lignin deficiency in transgenic flax resulted in plants with improved mechanical properties. J Biotechnol 128:919–934

Wu J, Lin L (2002) Elicitor-like effects of low-energy ultrasound on plant (*Panax ginseng*) cells: induction of plant defense responses and secondary metabolite production. Appl Microbiol Biotechnol 59:51–57

Xiao M, Zhang Y, Chen X, Lee E-J, Barber CJS, Chakrabarty R, Desgagné-Penix I, Haslam TM, Kim Y-B, Liu E (2013) Transcriptome analysis based on next-generation sequencing of non-model plants producing specialized metabolites of biotechnological interest. J Biotechnol 166:122–134

Xie Y, Wu G, Tang J, Luo R, Patterson J, Liu S, Huang W, He G, Gu S, Li S (2014) SOAPdenovo-Trans: *de novo* transcriptome assembly with short RNA-Seq reads. Bioinformatics 30:1660–1666

Xu Y, Li X, Lin J, Wang Z, Yang Q, Chang Y (2015) Transcriptome sequencing and analysis of major genes involved in calcium signaling pathways in pear plants (*Pyrus calleryana Decne.*). BMC Genomics 16:1

Yan Q, Hu Z, Tan RX, Wu J (2005) Efficient production and recovery of diterpenoid tanshinones in Salvia miltiorrhiza hairy root cultures with in situ adsorption, elicitation and semi-continuous operation. J Biotechnol 119:416–424

Ye X, Al-Babili S, Kloti A, Zhang J (2000) Engineering the provitamin A beta-carotene biosynthetic pathway into carotenoid free rice endosperm. Science 287(80):5451

Yoshizawa S, Horiuchi T, Fujiki H, Yoshida T, Okuda T, Sugimura T (1987) Antitumor promoting activity of (−)-epigallocatechin gallate, the main constituent of "Tannin" in green tea. Phytother Res 1:44–47

Yue Y, Yu R, Fan Y (2015) Transcriptome profiling provides new insights into the formation of floral scent in *Hedychium coronarium*. BMC Genomics 16:470

Zainol MK, Abd-Hamid A, Yusof S, Muse R (2003) Antioxidative activity and total phenolic compounds of leaf, root and petiole of four accessions of *Centella asiatica* (L.) Urban. Food Chem 81:575–581

Zerbino DR, Birney E (2008) Velvet: algorithms for *de novo* short read assembly using de Bruijn graphs. Genome Res 18:821–829

Zha W, Liang G, Xiao J, Studer EJ, Hylemon PB, Pandak WM Jr, Wang G, Li X, Zhou H (2010) Berberine inhibits HIV protease inhibitor-induced inflammatory response by modulating ER stress signaling pathways in murine macrophages. PLoS One 5:e9069

Zhang Q, Cai L, Zhong G, Luo W (2010) [Simultaneous determination of jatrorrhizine, palmatine, berberine, and obacunone in *Phellodendri amurensis* Cortex by RP-HPLC]. Zhongguo Zhong Yao Za Zhi= Zhongguo Zhongyao Zazhi= China J Chin Mater Med 35:2061–2064.

Zhou Q, Liang D, Deng A, Zhang J, Wu C, Nie Z, Wang Y (2013) Anti-tussive, expectorant and bronchodilating effects of ethanol extract of *Sorghum bicolor* (L.) Moench roots. J Ethnopharmacol 149:297–302

Zhu QY, Zhang A, Tsang D, Huang Y, Chen Z-Y (1997) Stability of green tea catechins. J Agric Food Chem 45:4624–4628

High-Throughput and *In Silico* Screening in Drug Discovery

11

Nandu Thrithamarassery Gangadharan,
Ananda Baskaran Venkatachalam, and Shiburaj Sugathan

Abstract

The process of drug discovery involves multiple branches of science. Discovery of novel molecule with biological modulation activity is a time-consuming and expensive process. High-throughput and *in silico* tools can reduce time and cost in drug discovery. The aim of high-throughput screening is to identify bioactive molecule from large compound collection and further development of active compounds to leads. There are two types of assay in high-throughput drug discovery: biochemical- and cell-based assays. Choice of assay depends on nature of target and assay feasibilities. Assay method should detect active compound from chemical library. Assay optimization and validation steps reduce false-positive and false-negative results. The assay results must be statistically validated to ensure reliability of results. The good assay design and implementation will give optimal results. *In silico* tools in drug discovery facilitate hit identification, hit to lead development, and optimization of druggability (improvement absorption, distribution, metabolism, excretion, and toxicity properties). High-throughput and *in silico* screening can be streamlined for hit identification and lead development. Streamlining of these methods reduces cost and time of drug discovery process. The wise use of these high-throughput methods can lead to discovery of drug with good potency and low toxicity profile.

N. Thrithamarassery Gangadharan • S. Sugathan (✉)
Division of Microbiology, Jawaharlal Nehru Tropical Botanic Garden and Research Institute, Palode, Thiruvananthapuram, Kerala 695562, India
e-mail: tg.nandu@gmail.com; drshiburaj@gmail.com

A.B. Venkatachalam
Atlantic Centre for Transplantation Research, Dalhousie University, Halifax B3H 4R2, Nova Scotia, Canada
e-mail: anandgeni@gmail.com

S. Abdulhameed et al. (eds.), *Bioresources and Bioprocess in Biotechnology*,
DOI 10.1007/978-981-10-3573-9_11

Keywords
Drug discovery • *In silico* • High throughput

11.1 Drug Discovery and High-Throughput Screening (HTS)

Drug discovery is a complicated and interdisciplinary process to identify novel drugs. First step of drug discovery is identification of competent drug target. Proteins are the main target class in drug discovery. Mostly proteins such as enzymes, receptors, and ion channels are targeted in drug discovery. Second step is target validation. Target validation is attained through biochemical assays and animal model experiments. After target validation, compounds modulating target will have to be identified. Next phase is assay development to screen modulators. Modulators that bring about dose-dependent target modulation are called lead compounds. Common pharmacophore can be developed from lead compounds showing common chemical properties. Structural activity relationship can be accessed and molecular descriptors can be optimized to improve selectivity and drug likeness of lead compounds; this process is known as lead optimization. Optimized lead compounds become potential candidates for drug development. These compounds are evaluated using animal models, and the successful compounds are selected for clinical trials (Carnero 2006).

The drug discovery process usually takes 10–12 years and costs over 1.5 billion USD (http://cen.acs.org/articles/92/web/2014/11/Tufts-Study-Finds-Big-Rise.html.). Among 1 million compounds screened, only 250 lead compounds enter into preclinical testing, 10 of them advance to clinical trial, and only 1 will be approved as drug by Food and Drug Administration (FDA).

High-throughput screening (HTS) involves testing of large number of chemical substances (natural products or synthetic compounds) against biological target. HTS identifies chemical compounds called hits that modulate drug target. Hit compounds are starting point of drug discovery. High-throughput screening is broadly defined as the testing of 10,000 to 100,000 compounds per day (Carroll et al. 2004; Wölcke and Ullmann 2001). HTS analyzes catalytic activity of enzymes, receptor ligand binding, opening and closing of ion channels, and phenotypic changes in cell (McDonald et al. 1999; Seville et al. 1996; Verma et al. 2004). HTS uses 96-well microtiter plate designed for serological studies (Sever 1962). Usually, HTS assays are carried out in 96-, 384-, or 1536-well microtiter plates.

In the early 1990s, advancement in combinatorial chemistry, genomics, and development of technologies like homogenous assays, high-density microplate, high-performance microliter dispensers, and imaging and laboratory automation was the driving force of HTS. In this period, colorimetry was used for enzymatic assay. ATP consumption readouts and bioluminescence technology were also developed. In the late 1990s, combinatorial chemistry focused on the enrichment of privileged motifs to create libraries of pure drug-like compounds, and scientists focused on assay development and quality controls. In the twenty-first century, availability

Fig. 11.1 Sequential steps involved in the high-throughput screening

of the human genome sequence provided opportunity to identify potential human drug targets (Carnero 2006).

The HTS process is a subset of the drug discovery process and can be described as the phase from target to lead (Fig. 11.1). The target of choice must be biologically relevant, and assay method must be robust. Compound showing desired biological activity should be detected by screening methods from the chemical library. False negatives and false positives can be minimized in the assay by setting low variability and high signal to background. There must be sufficient amount of cost-effective reagent available to run the entire screening campaign. Otherwise, it will impasse HTS operation (Macarron and Hertzberg 2011).

HTS assays can be divided into two categories, namely, biochemical assays and cell-based assays (An and Tolliday 2010). Biochemical assays include enzyme inhibition and receptor-ligand binding assays. The biochemical assay evaluates binding affinity or specific binding of compounds against biological target in an artificial environment. Purification is not possible for some biological targets. This limits application of biochemical assay in HTS. Besides, biochemical assays can't represent precise cell-specific responses against small molecules. Cell-based assays monitor drug effects on biochemical pathways or more specific targets (Sharma et al. 2010). Cell-based methods provide more biologically relevant microenvironment and direct information about cell permeability of compounds as well as acute cytotoxicity associated with the compounds.

11.2 Detection Methods in HTS

The most common assay readouts used in biochemical assay for HTS are optical, including absorbance, fluorescence, luminescence, and scintillation (Ramm 1999). Fluorescent or luminescent mechanism is highly amenable, which can be modified to different fluorescent assays (Inglese et al. 2007). Fluorescence happens on a timescale ranging from 10^{-9} s to 10^{-4} s, thereby allowing for light's many optical properties to be exploited by a number of detection methods (Lakowicz 2006).

Fluorescence Polarization or Fluorescence Anisotropy It is a solution-based homogeneous technique that allows rapid and quantitative analysis of diverse molecular interactions and enzymatic activities. FP uses polarized light to excite molecules in solution. Free fluorescent molecule shows fast tumbling, and it will be reoriented to prior light emission condition and consequently will show a low polarization value. But receptor-bound fluorophore that tumbles slowly will show a high polarization value (Trinquet and Mathis 2006). The fluorescent polarization approach has been adapted to almost every protein class like GPCRs, nuclear receptors, and enzymes. It can be used for the analysis of molecular interactions studies including protein-ligand, protein-protein, and protein-DNA binding events. It is also used for monitoring enzymatic reaction progress (Alpha et al. 1987).

Fluorescence Resonance Energy Transfer (FRET) It is an energy transfer between a fluorophore donor and a suitable fluorophore energy acceptor (Stryer 1978). The absorption spectrum of the acceptor should overlap with the emission spectrum of the donor. Two fluorophores are in close molecular proximity of each other to allow significant FRET. Most common FRET readouts are based on fluorescence variation between donor and acceptor. In cell-based HTS, genetically incorporated fluorescent indicators are used to understand signal transduction dependent on protein phosphorylation (Sato et al. 2002).

Bioluminescence Detection method is prevalent in HTS assays because of convenient detection and high sensitivity. In bioluminescence assay, ATP derived from enzyme-coupled reaction is detected by a reporter enzyme (luciferase from firefly *Photinus pyralis*) that acts on luciferin substrate to create a luminescent output. This method is free from compound interference and 1000 times sharper than equivalent fluorescence-based assays (Fan and Wood 2007). Luminescent-based assays have been used to access the activity of cytochromes P450, proteases, and monoamine oxidases (Sato et al. 2002). Luciferase can be mutated to develop multiple luminescent sensors in an assay. The development of dual luciferase reporter assays with different kinetics or emission maxima and substrate specificities has been used for identification of activities specific to the signaling pathway of interest (Davis et al. 2007). Luciferase reporters can be clumped with other detection formats, for example, combining green fluorescent protein (GFP) reporter (Bandyopadhyay et al. 2006), with β-galactosidase or alamarBlue (O'Boyle et al. 2005) to estimate cytotoxicity.

Bioluminescence Resonance Energy Transfer (BRET) It is a hybrid system involving the donor and acceptor (Pfleger and Eidne 2006). In BRET donor is a luminescent molecule excited by the enzyme (Renilla luciferase), and acceptor can be a fluorescent protein like a green fluorescent protein or yellow fluorescent protein. The use of an enzyme as an excitation source eliminates interferences from autofluorescent compounds and inner filter effects. BRET has been using as tool for the study of GPCRs by investigating receptor oligomerization or activation (Trinquet and Mathis 2006).

Disassociation-Enhanced Lanthanide Immunoassay (DELFIA) This method uses lanthanide ion chelates (europium or terbium), as fluorescent probe (Hemillä et al. 1984). The chelate, having longer fluorescent life, forms non-covalent association with a lanthanide ion and protects it from potential quenching by its environment. It acts as an antenna by transferring energy to lanthanides from the excitation source. UV-excited fluorescent probe emits their fluorescence 500 and 700, depending on lanthanide used, which has fluorescence more than 100 μs. Time-resolved fluorescent detection by pulsed excitation source rejects background produced by chemical compound, biological media, and instrumentation components.

Fluorescent lanthanide chelates are not stable in biological media. To overcome this limitation, two step-based heterogeneous assays were developed. In the initial step, nonfluorescent chelates associated with biomolecule conjugates were used to carry lanthanides during the biological reaction. In the second step, lanthanide ions are removed from nonfluorescent chelates by adding huge amount of different fluorescent chelating agents (Terpetschnig et al. 1995). DELFIA is free from fluorescent background, so very low detection limits can be achieved (Trinquet and Mathis 2006).

Homogenous Time-Resolved Fluorescence (HTRF) It combines FRET with time-resolved fluorescence detection. HTRF uses long lifetime FRET donor, cryptic lanthanide, which is formed by incorporating rare earth ion (e.g., europium) into ligand. This is called cryptand, it protects lanthanide ion from potential quenching by the environment and transfers energy to lanthanide ion from excitation source (Alpha et al. 1987). UV-excited europium has longer fluorescence (100–1000 μS) having wavelength range between 550 and 710 nm. Usually cross-linked allophycocyanin (XL665) is used as acceptor. Long lifetime of europium donor allows time-resolved fluorescence measurement. Clear distinction can be made between long-lived FRET signal of europium-associated acceptor and short-lived signal of freely diffusing acceptor, which allows separation of FRET signal from background.

Cryptic signal is independent of optical properties of test media at excitation wavelength. The signal ratio between acceptor signal and europium cryptate depends only on biological events under probe. High efficiency obtained in HTFR allows monitoring variety of biological interaction like enzyme activities and protein-protein interaction. It is independent of compound interference. It can be easily automated and miniaturized to wells format for the HTS to screen large chemical compound library (Kon et al. 2004; Mathis 1993).

Fluorescent Correlation Spectroscopy (FCS) It measures intensity fluctuations. The detection volume of this technique is around one femtoliter. Confocal microscope is needed to observe the reaction (Pope et al. 1999). In this assay, the fluorescent molecules are passed through signal and only through detection volume measured; therefore, signal is related to the diffusion coefficient of the sample. Autocorrelation techniques analyze fluorescent intensity fluctuations, which give information regarding the concentration of fluorescent molecule, their diffusion through detection volume, and the average brightness per molecule (Moore 1999). Receptor-bound fluorescent molecules show around tenfold increase in variation in diffusion time. FCS is free from compound interference. It is suitable for miniaturization in HTS (Trinquet and Mathis 2006).

Alpha (Amplified Luminescent Proximity Assay) Screen It uses a donor bead and acceptor bead (Seethala and Prabhavathi 2001). Both beads are bearing a biological partner under examination. When both beads under investigation come close together, the laser excitation of photosensitizer in the donor bead results in the generation of the singlet oxygen, which reacts with thioxene derivatives in the acceptor bead and generates chemiluminescence. This activates fluorophore, and activated fluorophore subsequently emits fluorescence. It can be used to investigate large molecule interaction and large number of targets. Low detection limit can be achieved in Alpha Screen. It can be easily miniaturized because of its homogenous format. Antioxidants and metal ions can affect measurement.

Beta-Lactamase Reporter System It can be used for study of mammalian expression system. It makes use of TEM-1 beta-lactamase which lacks 23 amino acids at the N-terminal end. This enzyme uses CCF2/CCF4 as substrate composed of dyes, 7-hydroxycoumarin-3-carboxamide and fluorescein, bridged by cephalosporin (Zlokarnik et al. 1998). De-esterified CCF2/4 trapped in the cell can easily be detected by FRET, which generates green-colored fluorescence. Cleavage CCF2/4 by beta-lactamase causes loss of FRET, which results in blue-colored fluorescence. CCF2/4 substrate enables ratiometric data analysis (the net blue fluorescence signal intensity and the net green fluorescence signal intensity ratio; em. 460/em. 530) that helps minimize variation in cell number or substrate concentration because the emission maxima of the cleaved and intact beta-lactamase substrates are distinct. Cytotoxicity can be assessed using uncleaved substrate (Qureshi 2007)

11.3 Assay Formats in HTS

There are two assay formats for HTS: biochemical and cell-based formats. Biochemical assays are designed to find compounds that interact with an isolated target *in vitro* environment. Cell-based assays assess phenotypic effect of compounds on the cell. HTS-compatible technology has developed to measure G protein-coupled receptor (GPCR) (Schroeder 1996) and ion channel function (González et al. 1999); confocal imaging platforms for rapid cellular and subcellular imaging and the continued development of reporter gene technology lead to development of user-friendly cell-based assays. For most drug discovery programs, both cell-based and biochemical assays are required for hit discovery, characterization, and subsequent lead optimization. If technical conceivability, expense, and throughput are equitable, cell-based assays are often preferred for HTS because compounds tested will be interacting with a more realistic mix of protein target conformations in their physiological milieu, i.e., with the right companions at the right concentration. Further, cell-based assays tend to avoid some common artifacts in biochemical assays such as aggregators (Shoichet 2006). Cell-based assays may wrongly identify hits that do not act on targets and may miss hits with low cell membrane permeability. If a cell-based assay is chosen for primary screening, a biochemical assay will often be used as a secondary screen to characterize hits and guide lead optimization. An effective HTS strategy considers both the primary and subsequent secondary assay designs carefully.

11.4 *In Vitro* Enzyme-Based HTS Assays

Developing HTS assays for enzymes necessitates recognition of accurate enzyme and substrate forms, purification methods, precise measurement of kinetic parameters, characterization of cofactors, choice of detection technology, and mode of action. HTS enzymes have been developed using detection of substrate consumption, product formation, and enzyme-ligand binding. Development of enzyme-based HTS begins with demonstration of catalytic activity on substrate *in vitro* environment. Available literature will provide information about initial test conditions. Preliminary experiments provide the Michaelis-Menten constant that is critical for assay optimization (Copeland 2003).

11.4.1 Components of Enzyme-Based Assays

Enzyme *In vitro* biochemical assays use enzyme isolated from cell. Absence of native *in vivo* conditions can significantly affect enzyme activity and stability. Enzymes may be expressed as truncated variants, or it may be expressed in alternatively tagged species. These artificial conditions may give compounds irrelevant in physiological conditions or miss compounds showing activity in physiological conditions. The choice of protein construct for HTS assay depends on stability and activity of enzyme construct. Besides, compounds with known modes of inhibition can be used for evaluation of constructs to use in a HTS based on the mode of inhibition. When multiple constructs of enzyme are used for same substrate, the activity can be compared by determining in kcat/Km. Very large difference in kcat/Km or subtle difference in Km value between constructs is indicative of difference in structure and/or stability. Comparison between different preparations of enzyme constructs is also possible using specific activity (Acker and Auld 2014).

Impure enzyme may lead to aberrant result. Mass spectrometry can be used to analyze enzyme purity. Enzymatic purity analysis is done by analyzing IC50 curves of known inhibitor or by determination of Michaelis-Menten parameters and comparing with previous results (Scott et al. 2004). Sometimes, vehicle-carrying compound may impair enzyme function. Vehicle sensitivity is estimated by titrating the vehicle in different concentrations against enzyme and analyzing its activity. Besides, poor interaction of enzyme with tubing and surfaces of dispensing unit results in a decrease of effective enzyme concentration. This can be prevented by adding BSA or small amount of detergent in the reaction buffer.

Substrate Concentration of substrates comparative to their Km will have influence on inhibitor type identified. In the case of competitive inhibitor, substrate concentration above Km value decreases the capability of inhibitor to bind enzyme active site. In uncompetitive inhibitor screening, high substrate concentration relative to Km value improves the binding of uncompetitive inhibitor. Allosteric (noncompetitive) inhibitors bind independent of the substrate molecule. Thus, it is unaffected by substrate concentration (Copeland 2003; Yang et al. 2009). While choosing an ideal substrate concentration, there must be correct balance between signal window and desired mode of inhibition (Iversen et al. 2006). When optimizing signal, it is better to keep substrate turned over low for identification weak inhibitors (Inglese et al. 2007). Insolubility of the substrate in assay buffer limits highest concentration possible. The substrate stability assays must be carried out to pinpoint stability effects, and modification should be made to resolve issues identified.

Cofactors Many enzymes require cofactors for structural integrity or to assist in the enzyme reaction. The enzyme may be purified with cofactor, or additional cofactor may need to be added to maximize enzymatic activity. The affinity of the cofactor will also influence whether a compound that competes with cofactor binding can be identified. The effects external of cofactor on biochemical enzyme assays can often be treated like substrate addition; the amount required depends on the level of activity needed and the necessity of the cofactor for the enzyme form one chooses to inhibit. It is usually best to use a saturating concentration of additional cofactor in the assay, when not specifically screening for cofactor-competitive compounds. The best possible signal to background ratio is identified by titration of cofactor. Finally, stability of a cofactor needs to be considered for the time and environment that the cofactor will be exposed to during an HTS run. For example, some cofactors are light sensitive (iron guanylylpyridinol), while others can change redox state in common buffers without reducing agents (iron salts). The timing of these modifications must be considered and tested to assure compatibility with the HTS process (Acker and Auld 2014).

Buffers *In vitro* biochemical assays are performed at near physiological pH in an attempt to mimic the intracellular environment of the native enzyme. For cytosolic proteins, pH=7.4 can be maintained by a number of buffers including Tris, HEPES, MOPS, and sodium or potassium phosphate buffers. However, simply because an enzyme is found in the cytosol does not guarantee that the activity will be optimal at pH=7.4. A range of pH values encompassing pH=7.4 should be tested in enzymatic activity assays, taking into account that differences in the local environment *in vitro* versus *in vivo* or changes in the protein construct from the native form could alter the optimum pH for reaction (Acker and Auld 2014).

The buffer choice also has significant influence on biochemical reaction because each buffer can have particular and compelling effects on a given enzyme target. Components of reaction mixture can interact poorly with certain buffers resulting in suboptimal assay conditions and affecting the reproducibility of an assay. For instance, phosphate buffer forms insoluble phosphate salt with Mg^{2+} and Ca^{2+} cofactors, this makes essential component unavailable for the reaction. Such an insoluble complex may result in poor liquid dispension in the automation equipment affecting the high-throughput screening. Free amine group in the tris buffer react with enzymes and/or substrates, altering the equilibrium of the system. Tris is also able to chelate metal ions, which could have deleterious effects on the activity of enzymes requiring metals for catalysis or structure (Desmarais et al. 2006).

11.5 Receptor-Ligand Binding-Based HTS Assays

There are two types of assay format used to investigate receptor-ligand binding: radioactive method and fluorescent-based methods. Radioactive methods like a filtration and scintillation proximity assay (SPA) are widely used for receptor binding assay. Both these methods use radiolabeled ligand and membrane. Reagents used for binding assays should have high specific activity, indicated by radioactivity/molecule of ligand, and its unit is Curies per millimole. Nonspecific binding of reagents used in binding assays can be reduced by coating filters with bovine serum albumin (BSA). Radiochemical purity of ligand should be above 90%. The purity of radiolabeled ligand decreases with time. Highly selective ligand will provide more reliable result. ^{125}I-labeled ligands will be stable for 1–2 months and tritated ligands have 3–6 months of stability. Beta energy released by ^{3}H produces proton on interacting with scintillates which can be measured by the detector. ^{125}I releases both gamma energy and beta energy. For flash plate assay ^{125}I will be ideal candidate. ^{3}H- or ^{125}I-labeled ligands can be used for SPA (Auld et al. 2012).

11.5.1 Assay Formats

SPA Ligand Binding Assay SPA ligand binding assay is a homogenous assay, which can be run in 96- or 384-well format. In this format, cell membranes are attached to SPA beads. When radio ligand binds to receptor, the proximate radio ligand transfers beta energy to scintillant in the bead and produces a signal that can be measured using microplate scintillation counter. The appropriate settling time needs to be determined experimentally (Auld et al. 2012).

Filtration Ligand Binding Assay In filtration assay, a separation of free radio ligand and radio ligand bound to the receptor is required for measurement. Filtration binding assay is carried out first in assay plate. After adding and incubating assay components, unbound ligand is removed by applying vacuum, while the bound ligand remains attached to the filter. The liquid scintillation cocktail or liquid scintillator is added to dried filter. Result is measured using microplate scintillation counter. It is more efficient than SPA (Auld et al. 2012).

Flash Plate Ligand Binding Assays In flash plate binding assay, a target receptor or receptor membrane is bound to wall of flash plate. Necessary compounds for flash plate assay were added. Energy from radio ligand interacts with scintillator bound in the wall of flash plate, producing signal. Unbound ligand far from plate scintillant can't produce signal, which distinguishes bound and unbound ligand (http://www.perkinelmer.com/Technologies/Assays-and-Reagents/Receptor-Ligand-Binding/default.xhtml.).

Fluorescent Polarization Assay for Receptor Binding Assay Receptor-ligand binding fluorescent polarization assays use a fluorescein-labeled ligand (tracer) and a large unlabeled receptor. Bound ligand will show high polarization value.

11.5.2 Assay Optimization

Ligand Concentration Bound ligand concentration can be determined using the equation given below:

$$\mathrm{BL} = \mathrm{Bmax} \times \mathrm{L} / (\mathrm{Kd} + \mathrm{L})$$

where BL = bound ligand concentration, Bmax = maximum binding capacity, L = total ligand concentration, and Kd = equilibrium affinity constant also known as dissociation constant.

IC_{50} for competitive binders can be calculated by the formula

$$\mathrm{IC}_{50} = (1 + \mathrm{L} / \mathrm{Kd}) \times \mathrm{Ki}$$

Functional assays should be performed for finding uncompetitive inhibitor. Keeping ligand concentration equal to dissociation constant (Kd) is ideal way to attain a good signal. This method is ideal for finding competitive binders (Auld et al. 2012).

Receptor Concentration The optimal concentration can be determined by calculating the proportion of bound ligand at the disassociation constant (Kd). Linearity of ligand binding with respect to membrane concentration must be evaluated. In radio filtration assay, it is better to keep membrane concentration at 10% of bound ligand at Kd concentration (Burt 1986). Determination of nonspecific binding (NSB) will give optimal assay result.

Preincubation and Equilibrium For slow receptor binders, preincubation is necessary. Maintaining equilibrium in HTS binding assays ensure a proper calculation of displacement by putative inhibitors.

11.6 Cell-Based HTS Assays

Cell-based assays for HTS can be divided into three types: reporter gene assays, second messenger assays, and cell proliferation assays (Sundberg 2000). Cell-based assays differentiate agonists and antagonists, pinpoint allosteric modulators, and provide direct information about intracellular permeability and stability compounds and cytotoxicity of the compound (Kunapuli et al. 2006). Besides, cell-based assays have been performed in a more biological relevant environment. They have been successfully used for early drug discovery in identifying high-quality leads. The main constituents of a cell-based HTS assays are cells, cell-culturing device, and detection system.

11.6.1 Cell Types Used for Cell-Based HTS Assays

Immortalized cell lines are inexpensive, easy to grow, dependable, and reproducible. Various types and sources of cells have been used in cell-based assays. Primary cells are capable of giving characteristic responses; even so, they are difficult to grow and transfect (Ebert and Svendsen 2010). Human cancer cells are widely used for anticancer drug screening assays. Recent advancements like cancer stem cells, which can differentiate, self-renew, and regenerate a phenocopy of the original tumor (Sabisz and Skladanowski 2009), are promising models in anticancer drug screening.

Embryonic stem cells (ESCs) can serve as better models for both drug efficacy and toxicity screening than primary or immortalized cells lines. ESCs are isolated from embryo and have unlimited capacity to self-renewal and can be differentiated into any cell type *in vivo*. IPSCs are pluripotent cells artificially derived from somatic cells by inducing a small set of powerful pluripotency genes. As IPSCs can be derived from patients with specific diseases, they have been considered as a new tool in drug discovery. Cell-based assays are mostly carried out in microtiter wells. It can be easily miniaturized to carry out assays at high-throughput scale and handled with a robotic system for automation (Sundberg 2000).

Common cell culture methods include single cells, monolayer cells on a two-dimensional (2D) surface, and multilayer cells or aggregate clusters in a 3D scaffold (Yang et al. 2008). 2D cell-based assays are generally used in drug screening because they are cheap and easy for operation. But the 3D cell culture generally shows similar *in vivo* morphology with intimate cell-cell and cell-extracellular matrix interactions, which are absent in 2D cell culture. Cell-based assays are widely used to assay the effects of compounds on cellular activities, cell number, intracellular calcium levels, nuclear size, mitochondrial membrane potential, and membrane permeability (Hughes et al. 2008).

11.6.2 Detection Methods in Cell-Based Assay

Online detection methods like electrochemical and optical techniques are used in HTS. Generally, optical sensation is effortless for miniaturization than electrochemical sensing (Kumar et al. 2004).

Electrochemical Methods Living cells act as an electrochemical system. Electrochemical changes in living cells have been used to define cell viability in a homogenous solution (Nonner and Eisenberg 2000). Metabolism in cells presided by modification in metabolic products or substrates. These changes can be detected by electrochemical methods based on cellular activities including potentiometry and amperometry (Zang et al. 2012). Conventional potentiometry cell-based sensors encompass gas-sensing electrode (GSE) or an ion-selective electrode (ISE) coated with a layer of cells (May et al. 2004). However, it requires very stable reference electrode, which limits its application.

Amperometric electrochemical method has been used for the determination of pH, DO, or glucose. The acidification rate in the proximity of cells can be calculated using a microphysiometer (Rabinowitz et al. 1997). Cellular biochemical reactions resulting from the accumulation of lactic acid and carbon dioxide can be relatively monitored using the pH value. Besides, heterogeneous pO2 distributions around tissues could be detected using a miniaturized system (Braun et al. 2001). Even so, many uncontrollable environmental factors affect metabolic activities which limit the applications of this approach in high-throughput cell-based assays.

Electrochemical Method Based on Barrier Behavior Electronic impedance significantly increases in the presence of cell. This property can be used for monitoring of biological status of the cell. Using electrochemical impedance spectroscopic techniques, real-time assessment of cytotoxicity and acute toxicity can be attained. Bioelectrical signal from electrogenic tissues can be used to test drugs against critical diseases such as cardiac arrhythmia, hypertension, Parkinson's disease, diabetes, depression, and neuropathic pain (Hogg et al. 2006). Single-wall carbon nanotubes (SWCNTs) are used for detection of membrane potential change of axon (Pui et al. 2010). This technique can be used to noninvasively track cellular activities for electrogenic cells with high throughput, high sensitivity, easy use, and the capacity of long-term cell culture.

Optical Methods Colorimetric methods are based on color change of the growth medium after cell metabolites react with chemical agents. Tetrazolium salts can differentiate living and nonliving cells based on the reduction of a tetrazolium salt by actively growing cells to a colored formazan product that can be quantified with a spectrophotometer. There is spectrum of HTS detection methods available using tetrazolium salts. Colorimetric methods are invasive, time-consuming, and laborious. Besides, they usually can only provide end point data, which can't provide more details about the effects of drugs on cells comparing to dynamic data. Automation of colorimetric end point assays is costly (Zang et al. 2012).

The detailed description of fluorescent and luminescent methods has been given in previous section.

Green fluorescent protein (GFP) has enabled online, noninvasive detection and quantification of cell proliferation and specific cellular functions. GFP-based cellular assays are inclinable to real-time, automated, and noninvasive analysis cellular events (Yang et al. 2008). Specific cellular function can be observed using reporter gene and can be expressed under control of promoter gene or regulatory DNA sequence. Activation of signal transduction pathway can be detected in this method

(Mahajan et al. 1999; Xu et al. 1998). Laser-scanning imaging systems with fluorescence microscopy can be used to investigate the context of living cells, quantify, intracellular proteins, and monitor the trafficking of proteins fused with fluorescent reporters and some subcellular structures (Abraham et al. 2004). Howbeit, their high costs and relatively low capacity limit their uses to the late phase compound characterization (Haney et al. 2006). Moreover, these imaging systems are limited to read planar images of cells cultured on 2D surfaces and are not suitable for 3D cell cultures (Zang et al. 2012).

11.7 Drug Target in HTS Assays

The major considerations for cherry picking therapeutic target for HTS are target validity, chemical tractability, and screen ability. The target must be disease relevant. There is reciprocal association between target novelty and validity. Some targets will have a high degree of validation, but low novelty and others will be highly novel but poorly linked to disease. Target validity can be evaluated by genetic method and/or compound-based experiments. Gene knockouts or RNA interference methods can be time engaging and sometimes lead to false inferences, but it is less expensive when comparing to compound-based experiments. Compound-based target validation approaches can be used for less-validated targets and to screen for tool compounds, followed by *in vivo* experiments. Most organizations are using combination of these methods (Macarrón and Hertzberg 2011).

The chemically amenable target increases the probability to find out drug-like compound producing therapeutically relevant effect. Certain target classes, like G protein-coupled receptors (GPCRs), ion channels, nuclear hormone receptors, and kinases, are more chemical amenable than others. About half of experimental and marketed drugs target five main protein families: G protein-coupled receptors (GPCRs), kinases, proteases, nuclear receptors (NRs), and ion channels (Hopkins and Groom 2002). The drawback of this approach is that it may entirely eliminate target classes that would otherwise be extremely attractive from a biological point of view.

A final factor to consider when choosing targets is screenability – the technical probability of developing a robust and high-quality screening assay. GPCRs, kinases, proteases, nuclear hormone receptors, and protein-protein interactions are relatively easy targets to establish screening assay. Ion channels are more difficult, although new technologies are being developed which make these more approachable from an HTS point of view (González et al. 1999). Approximately two thirds of therapeutic targets are comprised of enzymes and receptors (Zheng et al. 2006).

Phenotypic assays can be used for HTS if the target is inamenable, which measure cellular properties like a secretion of protein factors, chemotaxis, apoptosis, and cell shape change in response to test compound. Multiple targets have been screened in phenotypic assays. But these assays are difficult to configure and expensive (Macarrón and Hertzberg 2011).

All these factors should be evaluated before commencement of a HTS to make a choice to go forward. Chemically amenable, technically easy, inexpensive, and biologically relevant targets are ideal for drug screening, but these kinds of targets are rare (Macarrón and Hertzberg 2011).

11.7.1 Common Targets in HTS Assays

Protein Kinases Protein kinases are enzymes that phosphorylate the hydroxyl group present on serine, threonine, or tyrosine residues (Glickman et al. 2004) during posttranslational modification in cells and signaling pathways. The biochemical approaches to measure protein kinase activity can be divided into two categories: generic assays independent of subfamily and antibody-based formats that detect an epitope within the phosphorylated product. Measurement of ATP depletion via luciferase is an example of a generic assay format (Koresawa and Okabe 2004; Singh et al. 2004). The drawback of this assay is large quantity of enzyme required for sufficient signal to background ratio (Auld et al. 2013). ADP produced during the enzyme action can be detected using ADP specific antibody (Lowery and Kleman-Leyer 2006). Phosphorylated peptide is detected by radiometric filter binding assay. Immobilized metal-ion affinity-based fluorescence polarization (IMAP) is a homogeneous antibody-free method. The IMAP utilizes immobilized transition metals on nanoparticles as a "binding reagent" to make complexes with phosphate groups on phosphopeptides generated in a kinase reaction. Such a binding results increase in FP value (Sportsman et al. 2004). Antibody-based technologies and ALPHA, FP, and HTR-FRET are developed to detect phosphorylated peptide. These assays are ideal for kinase cascades and hierarchical phosphorylation.

Proteases Proteases are well-established drug targets (Leung et al. 2000). Proteases are measured using FRET-based or profluorescent substrates. The probes coupled to a protease cleaving site become fluorescent after protease cleavage (Karvinen et al. 2004). Inefficiency of fluorescent substrate to cover the entire binding site is the limitation of this assay. The AHTRF assay for endoproteases, "carboxypeptidase B," has been developed for HTS; in this assay peptide cleavage unmasks an epitope which is then recognized by an antibody (Ferrer et al. 2005).

Nuclear Receptor Nuclear receptors (NR) are a large family of ligand-activated transcription factors that act as transcriptional switches responding to lipophilic hormones, vitamins, dietary lipids, or other intracellular signals (Nagy and Schwabe 2004). NRs are regulated by hormones and metabolites (endobiotic ligands), or through xenobiotics (Olefsky 1999). Binding assays for NRs are divided into radiometric and fluorometric assays. GFP fused with NRs has been used to measure the glucocorticoid receptor (GR) translocation from nucleus to cytoplasm (Fung et al. 2006). NRs' agonist and antagonistic compounds can be detected using coregulator recruitment assays based on TR-FRET ligand (Gowda et al. 2006). Highly sensitive cell-based reporter gene assays fuse NR reaction components with reporter genes such as beta-lactamase, luciferase, and secreted alkaline phosphatase.

G Protein-Coupled Receptor GPCRs are the most targeted protein molecule in drug discovery. There are well-established assay methods for GPCR-targeted drug discovery (Eglen et al. 2007; Jacoby et al. 2006). HTS assays can be configured to monitor GPCR-orchestrated cellular events, such as protein phosphorylation, ion channel activity (Ferrer et al. 2003), modulation of secondary messenger (Chambers et al. 2003; Williams 2004), gene transcription (Dinger and Beck-Sickinger 2004), and cell proliferation. Intracellular Ca2+ stores activated by GPCRs can be analyzed using calcium-sensitive dyes such as fluo-3 and fura-4 and rapid inject imaging platform (Chambers et al. 2003). The GPCR secondary messengers have been analyzed by the direct measurement of inositol phosphate species (Eglen and Singh 2003; Trinquet et al. 2006) or cAMP (Williams 2004). GPCR internalization assays, independent of G protein subtype, can be applied to assess number of GPCRs using fluorescence microscopy (Inglese 2006; Taylor 2006) or microtiter plate reader (Hamdan et al. 2005).

Ion Channels Membrane-spanning ion channel proteins control the movement of inorganic ions such as Na^+, K^+, Ca^{2+}, and Cl into or out of cells. Ion channels are well-recognized therapeutic target for treating different diseases. Cell-based HTS assays have been developed for different types of ion channels using either fluorescence methods to monitor changes in membrane potential or microtiter plate readers to measure intracellular calcium levels (Inglese et al. 2007).

Ligand-receptor binding assays are established methodology for ion channel HTS. But inability to detect functional effect of compound on ion channels is the limiting factor of the binding assay. The calcium-sensing fluorescent probes such as fura-2, fluo-3, and fluo-4 can be used in HTS for ligand- and voltage-gated channels (Zheng et al. 2004). The fluorescent signal intensity of intracellular fluorophores increases proportionally with rise in intracellular free $Ca2^+$ concentration (Gee et al. 2000). Positional voltage sensors and FRET-based assays use voltage-sensing probes, which measure changes in membrane potential (González and Maher 2002). Atomic absorption spectroscopy, which has been used to measure ion transport, can be used in HTS ion flux assay. Automated patch-clamp instruments, with increase in throughput, have been used in HTS assays (Finkel et al. 2006).

11.8 Assay Validation and Evaluation of HTS

Hit and non-hit compound can be distinguished based on statistical evaluation. In enzyme-based HTS assays, blank, positive control, and negative control are used for primary evaluation. Blanks or NSB controls are prepared traditionally by adding an excess of unlabeled ligand used in ligand-receptor binding assays. Errors in sample processing and liquid handling can be distinguished by analysis of control plate.

Liquid handling errors are pinpointed by analysis of patterns. The analysis of performance starting with mean (M) and standard deviations (SD) for signal and background and combinations of these is as follows.

Signal to Background Signal to background ratio (S/B) furnishes information of the separation of positive and negative controls. S/B is useful to validate reagents in early assay development. But it is a poor indicator of assay quality (Zhang et al. 1999).

$$S/B = M\,\text{signal} / M\,\text{background}$$

Coefficient of Variation of Signal and Background It is the measure of variability, which is a function of the assay stability and precision of liquid handling and detection instruments (Macarrón and Hertzberg 2011).

$$CV = 100 \times SD / M(\%)$$

Z′ Factor The value of Z′ factor is a relative indication of the separation of the signal and background (Zhang et al. 1999). The Z factor's dimensionless scale ranging from 0 to 1 allows comparison of different assays and screens using the control wells (Z′) and sample wells (Z) of the plate (Inglese et al. 2007). Z′ must be assessed during assay development and validation and also throughout HTS campaigns on a per plate basis to assess the quality of dispensement and reject data from plates with errors (Coma et al. 2009b).

$$Z' = 1 - (3\sigma\,\text{max} + 3\sigma\,\text{min}) / \mu\,\text{max} - \mu\,\text{min}$$

Signal window and Z′ factor are used for assay validation. Signal window measures fold change in maximum and minimum output signals, but it is not as reliable as Z factor for predicting assay performance. Z′ factor measures the precision of this response within a plate and across plates. The minimum significance ratio (MSR) is used to track assay sensitivity variation between assay plates, which is often an indicator of reagent stability. The MSR can be calculated from titrations of control compounds on some or all assay plates (Inglese 2007).

After assay optimization, a pilot study must be done before starting HTS campaign. In this pilot step, representative sample must be tested from screening collection. It should be treated as the way HTS plates will be treated. The objectives of this pilot study are prediction of hit rate, assessment of screening compound interference, reproducibility of result, estimation of hit rate, and optimal sample concentration. The size of pilot screening compound collection can be 1–5% of total compound collection (Macarrón and Hertzberg 2011).

Apart from pilot screening, at least 10–20 assay plates should be run to test the HTS system in real action. Duplicates and triplicates of these samples should be run. Independently running duplicates evaluate the reproducibility of results. Triplicates access the rates of false positives and negatives (Coma et al. 2009a).

11.9 *In Silico* Screening in Drug Discovery

CADD is used to screen large compound library to identify biologically active compounds (hits), improving drug likeness of lead compound by increasing biological activity or optimizing pharmacokinetic properties and designing new compounds (Sliwoski et al. 2014).

CADD can be classified into two categories: structure or target based and ligand or pharmacophore based. Structure-based CADD (SBCADD) requires 3D structure of target molecule. Ligand-based CADD is done by superimposing active molecule to identify common chemical features or identifying interaction points between ligand and target molecule.

11.10 Pharmacophore-Based Drug Discovery

According to IUPAC, pharmacophore is defined as "the ensemble of steric and electronic features that is necessary to ensure the optimal supramolecular interactions with a specific biological target structure and to trigger (or to block) its biological response". Pharmacophoric descriptors include H-bond acceptors, H-bond donors, and aromatic, hydrophobic, positive, and negative ionizable groups. In case the macromolecular structure of target protein is not available, drug designing can be done using ligand-based drug discovery method. Chemical compounds with similar molecular fingerprints can be screened using molecular fingerprints of known ligands. A pharmacophore model can be generated either in a ligand-based method or in a structure-based manner (Yang 2010).

11.10.1 Ligand-Based Pharmacophore Modeling

Ligand-based pharmacophore modeling involves two steps: creation of conformational flexibility of ligands and alignment of multiple ligands in the training set. Mainly, various automated pharmacophore generators are available such as PHASE (Dixon et al. 2006) (Schrodinger Inc.), HypoGen (Li et al. 2000) (Accelrys Inc.), DISCO (Martin 2000), GALAHAD (Tripos Inc.), MOE (Chemical Computing Group), and HipHop (Barnum et al. 1996). There are two key challenges in ligand-based pharmacophore modeling. The first challenge is generating ligand flexibility. This problem can be solved either by precomputing conformations or by carrying out conformation analysis during the pharmacophore modeling process (Poptodorov et al. 2006).

Molecular alignment of multiple ligands is another difficult issue in ligand-based pharmacophore generation. There are two methods of alignment: property-based and point-based approaches (Wolber et al. 2008). The property-based approach uses molecular descriptors, generally represented by sets of Gaussian functions for alignment generation. In point-based algorithms, alignment is generated by super imposing atom pairs, chemical features or fragments by using the least-squares fitting.

Proper selection of training set compounds is necessary to ensure the accuracy of pharmacophore models (Poptodorov et al. 2006). Different training sets give different pharmacophore models of ligands that interact with same protein generated from same program.

11.10.2 Structure-Based Pharmacophore Modeling

Structure-based pharmacophore modeling depends on protein-ligand complex. A structure-based 3D pharmacophore reflects amino acids position and type of interaction in active site of protein. Major interaction sites between ligands and macromolecule can be determined by using the macromolecule-ligand-complex-based approach. The software used for macromolecule-ligand complex pharmacophore generation are GBPM (Ortuso et al. 2006), Ligand Scout (Wolber and Langer 2005), and Pocket v.2 (Chen and Lai 2006). This method needs macromolecule-ligand complex structure. Pharmacophore model with more than seven chemical features is not fit for practical applications, such as screening of 3D database (Pandit et al. 2006; Toba et al. 2006).

Pharmacophore building has been done using IC50 or Ki50 value of 15 diverse chemical structures. Alternatively, it can be done on the basis of three or four known active compound chemical structures from different chemical scaffolds (Toba et al. 2006; Funk et al. 2004).

11.11 Structure- or Target-Based CADD

This method is based on knowledge of 3D structure of target macromolecule (biological target). It depends on the hypothesis that molecules interacting with target protein can exert specific biologic effect. Therefore, novel compound with biological modulation activity can be screened by analysis of binding site. The 3D structure of macromolecules has been resolved by X-ray crystallography, NMR, and electron microscopy. The 3D coordinates of macromolecules are available in Protein Data Bank (PDB). The output of virtual screening depends on quality of PDB structure.

11.11.1 Homology Modeling of Protein

Complex protein molecules can't be solved by NMR analysis and X-ray crystallography techniques. In the absence of experimental protein structures, computational tools can be used for predicting 3D structure of proteins. Homology model of target protein is predicted based on template amino acid sequence. It is based on principle that distantly related primary structures of protein fold into similar tertiary structures. It involves the following steps: (1) template recognition and alignment of the target and template proteins, (2) alignment correction, (3) backbone generation, (4) loop modeling, (5) side-chain modeling, and (6) model optimization and validation. The software tools used for homology modeling are Swiss model, PSIPRED (Buchan et al. 2010), and MODELER (Marti-Renom et al. 2000). Using PSI-BLAST, template structure with similarity to target sequence can be found out (Altschul et al. 1990). Multiple alignment tools can be used for searching of template structure, which give more accurate result. Factors like template selection and resolution, alignment length and sequence identity between target and template influence output of homology model.

Gaps or insertions in sequence alignment are filled by connecting anchor residues. Missing residues are mostly seen in loop region. This can be filled by two modeling methods, either by knowledge-based approach or energy-based approach. Knowledge-based approach depends on an amino acid sequence present in target sequence. Energy-based methods create a large number of loop models, and quality has been assessed by energy-based function (Hillisch et al. 2004). Side-chain conformation prediction has been done based on Monte Carlo search (Rohl et al. 2004) and dead-end elimination (Desmet et al. 1992). These methods using side-chain conformation library are grouped by statistical methods (Krivov et al. 2009).

Models are refined by minimizing models using techniques such as Monte Carlo Metropolis minimization (Misura and Baker 2005), molecular dynamics (Raval et al. 2012), or genetic algorithms (Xiang 2006). Model evaluation is done by determining energy conformation of amino acids and by calculating root mean square difference between target and template.

11.11.2 Molecular Docking

Docking procedure predicts ligand conformation and orientation within a binding site of target macromolecule. Molecular docking studies give precise structural model and correct activity prediction. Docking is usually a multistep procedure in which one or more degrees of complication are added in each step (Brooijmans and Kuntz 2003). By applying docking algorithm, small molecule orients in the active site. Docking is a two-step process: first step is sampling conformations of the ligand in the active site of the protein, and the second is ranking these conformations via a scoring function. Ideal sampling algorithms should be reproducing the experimental binding mode, and the scoring function should also rank it highest among all generated conformations.

In molecular docking, three methods represent protein-ligand binding: atomic, grid, and surface (Halperin et al. 2002; Kitchen et al. 2004). Atomic method is used as a function of potential energy field. It is often used only during ranking. Surface methods work based on the topography of the molecule. These methods are guided by alignment of binding site and ligand, by minimizing the angle between the surfaces. It is usually used in protein-protein docking. The grid representation stores physicochemical features of receptor binding surface as energy potentials on grid points.

Docking methods have been classified as flexible docking and rigid-body docking (Halperin et al. 2002; Dias and de Azevedo 2008). Rigid-body docking considers physicochemical or geometrical complementarities as stationary state. Flexible docking considers a multiple conformations of a ligand and/or target. Treatment of ligand flexibility can be divided into three categories: systematic methods, random methods, and simulation methods. In systematic method algorithms try to investigate all possible conformation of the molecules. In this method ligands search the active site in stepwise manner. Docking programs, such as DOCK (Ewing et al. 2001), FLEXX (Rarey et al. 1996), GLIDE (Friesner et al. 2004), and FLOG (Miller et al. 1994), use this method. Stochastic or random methods operate by materializing arbitrary changes to either a single ligand or a population of ligands. The ligand evaluation is based on predefined probability function. The algorithms using random search methods are AutoDock (Morris et al. 1998) and GOLD (Jones et al. 1997). Molecular simulation methods calculate the system trajectory by the applying Newtonian mechanics.

11.11.3 Scoring Function in Molecular Docking

In molecular docking experiments, hundreds of thousands of macromolecule-ligand structures are generated. Incorrect conformation should be sorted out from correct macromolecule-ligand conformations. Reliable scoring function is necessary to distinguish valid binding conformation. Essentially, four types of scoring functions are currently applied: (1) force-field-based scoring functions, (2) empirical scoring functions, (3) knowledge-based scoring functions, and (4) consensus scoring functions (Sliwoski et al. 2014).

For force-field or molecular mechanics-based scoring functions, classic molecular mechanics are used. It usually calculates the internal ligand energy and the ligand-receptor interaction energy. These calculations are the sum of van der Waals and electrostatic interactions. AutoDock and DOCK use AMBER force field in ranking.

Empirical scoring functions fit parameters to experimental data, such as conformations and/or binding energy. The coefficients for various parameters are attained from regression analysis using experimentally calculated binding energies and X-ray crystallography information. Empirical scoring function terms are often easy to assess and are based on approximations. Empirical functions are used in several docking programs such as FLEXX (Rarey et al. 1996), SURFLEX (Jain 2003), and LUDI (Böhm 1992).

Information from experimentally determined structure is used in knowledge-based scoring function. In knowledge-based scoring function, protein-ligand complexes are modeled using interatomic distance. Several knowledge-based scoring functions are used in programs like DRUGSCORE (Velec et al. 2005), SMOG (DeWitte and Shakhnovich 1997), and BLEEP (Mitchell et al. 1999). Consensus scoring approach combines different scoring functions to balance the error in single method and improve scoring function.

11.12 QSAR and Drug Likeness Prediction

"Quantitative structure-activity relationships" (QSAR) relates chemical property of compound with its biological activity. Mathematical model can be constructed based on activity of set of ligands. Structure and activity can be quantified based on set of chemical properties called "molecular descriptors." QSAR modeling uses molecular descriptors as independent variables and activity as a dependent variable. QSAR model can predict biological activity of novel compounds from their molecular descriptors. Active compound can be screened from chemical database using QSAR model.

In silico drug likeness properties can be predicted using Lipinski rule of five. Lipinski rule of five is used to predict absorption, distribution, metabolism, and excretion (ADME) properties of lead compound. Lipinski's rule states that, in general, an orally active drug has no more than one violation of the following criteria:

- Not more than 5 hydrogen bond donors (nitrogen or oxygen atoms with one or more hydrogen atoms)
- Not more than 10 hydrogen bond acceptors (nitrogen or oxygen atoms)
- A molecular mass less than 500 daltons
- An octanol-water partition coefficient log P not greater than 5 (Lipinski 2000)

Prediction of drug likeness saves cost of pharmaceutical company for drug discovery. Lead compound can be optimized for improved pharmacokinetic properties.

11.13 Conclusions

The process of drug discovery can be accelerated by combining high-throughput and *in silico* screening. Using in silico screening, the number of compound screened for *in vitro* activity can be narrowed down. Compounds without biological activity and nondrug likeness can be eliminated *in silico* screening stage. This method reduces cost and time for HTS screening. Drug likeness of lead compound can be improved using *in silico* tools.

References

Abraham VC, Taylor DL, Haskins JR (2004) High content screening applied to large-scale cell biology. Trends Biotechnol 22:15–22

Acker MG, Auld DS (2014) Considerations for the design and reporting of enzyme assays in high-throughput screening applications. Perspect Sci 1:56–73

Alpha B, Lehn JM, Mathis G (1987) Angew Chem Int Ed 26:266

Altschul SF, Gish W, Miller W, Myers EW, Lipman DJ (1990) Basic local alignment search tool. J Mol Biol 5:403–410

An WF, Tolliday N (2010) Cell-based assays for high-throughput screening. Mol Biotechnol 45:180–186

Auld DS, Farmen MW, Kahl SD, Aidas K, Kevin LM, Chahrzad M, Jeffrey RW (2012) Receptor binding assays for HTS and drug discovery. In: Sittampalam GS, Coussens NP, Nelson H, Arkin M, Auld D, Austin C, Bejcek B, Glicksman M, Inglese J, Iversen PW, Li Z, McGee J, McManus O, Minor L, Napper A, Peltier JM, Riss T, Trask Jr OJ, Weidner J (eds) Assay guidance manual [Internet]. EliLilly & Company and the National Center for Advancing Translational Science, National Center for Biotechnology Information, Bethesda

Auld DS, Veith H, Cali JJ (2013) Bioluminescent assays for cytochrome P450 enzymes. Methods Mol Biol 987:1–9

Bandyopadhyay S, Ni J, Ruggiero A, Walshe K, Rogers MS, Chattopadhyay N, Glicksman MA, Rogers JT (2006) A high-throughput drug screen targeted to the 5'untranslated region of Alzheimer amyloid precursor protein mRNA. J Biomol Screen 11:469–480

Barnum D, Greene J, Smellie A, Sprague P (1996) Identification of common functional configurations among molecules. J Chem Inf Comput Sci 36:563–571

Böhm HJ (1992) LUDI: rule-based automatic design of new substituents for enzyme inhibitor leads. J Comput Aided Mol Des 6:593–606

Braun RD, Lanzen JL, Snyder SA, Dewhirst MW (2001) Comparison of tumor and normal tissue oxygen tension measurements using OxyLite or microelectrodes in rodents. Am J Physiol Heart Circ Physiol 280:H2533–H2544

Brooijmans N, Kuntz ID (2003) Molecular recognition and docking algorithms. Annu Rev Biophys Biomol Struct 32:335–373

Buchan DW, Ward SM, Lobley AE, Nugent TC, Bryson K, Jones DT (2010) Protein annotation and modelling servers at University College London. Nucleic Acids Res 38:W563–568

Burt DA (1986) Receptor binding methodology and analysis. In: O'Brien RA (ed) Receptor binding in drug research. Marcel Dekker, New York, pp 3–29

Carnero A (2006) High throughput screening in drug discovery. Clin Transl Oncol 8:482–490

Carroll SS, Inglese J, Mao SS, Olsen DB (2004) Drug screening: assay development issues. In: Prendergast GC (ed) Molecular cancer therapeutics: strategies for drug discovery and development. Wiley, Hoboken, pp 119–140

Chambers C, Smith F, Williams C, Marcos S, Liu ZH, Hayter P, Ciaramella G, Keighley W, Gribbon P, Sewing A (2003) Measuring intracellular calcium fluxes in high throughput mode. Comb Chem High Throughput Screen 6:355–362

Chen J, Lai L (2006) Pocket v.2: further developments on receptor-based pharmacophore modeling. J Chem Inf Model 46:2684–2691

Coma I, Clark L, Diez E, Harper G, Herranz J, Hofmann G, Lennon M, Richmond N, Valmaseda M, Macarron R (2009a) Process validation and screen reproducibility in high-throughput screening. J Biomol Screen 14:66–76

Coma I, Herranz J, Martin J (2009b) Statistics and decision making in high-throughput screening. Methods Mol Biol 565:69–106

Copeland RA (2003) Mechanistic considerations in high-throughput screening. Anal Biochem 320:1–12

Davis RE, Zhang YQ, Southall N, Staudt LM, Austin CP, Inglese J, Auld DS (2007) A cell-based assay for Ikappa Balpha stabilization using a two-color dual luciferase-based sensor. Assay Drug Dev Technol 5:85–103

Desmarais W, Bienvenue DL, Bzymek KP, Petsko GA, Ringe D, Holz RC (2006) The high-resolution structures of the neutral and the low pH crystals of aminopeptidase from Aeromonas proteolytica. J Biol Inorg Chem 11:398–408

Desmet J, De Maeyer M, Hazes B, Lasters I (1992) The dead-end elimination theorem and its use in protein side-chain positioning. Nature 356:539–542

DeWitte RS, Shakhnovich E (1997) SMoG: De novo design method based on simple, fast and accurate free energy estimates. J Am Chem Soc 119:4608–4617

Dias R, de Azevedo WF (2008) Molecular docking algorithms. Curr Drug Targets 9:1040–1047

Dinger MC, Beck-Sickinger AG (2004) Reporter gene assay systems for the investigation of G-protein-coupled receptors. In: Dingermann T, Steinhilber D, Folkers G (eds) Molecular biology in medicinal chemistry. Wiley-VCH Verlag GmbH & Co. KGaA, Weinheim, pp 73–94

Dixon SL, Smondyrev AM, Knoll EH, Rao SN, Shaw DE, Friesner RA (2006) PHASE: a new engine for pharmacophore perception, 3D QSAR model development, and 3D database screening: 1. Methodology and preliminary results. J Comput Aided Mol Des 20:647–671

Ebert AD, Svendsen CN (2010) Human stem cells and drug screening: opportunities and challenges. Nat Rev Drug Discov 9:367–372

Eglen RM, Singh R (2003) Beta galactosidase enzyme fragment complementation as a novel technology for high throughput screening. Comb Chem High Throughput Screen 6:381–387

Eglen RM, Bosse R, Reisine T (2007) Emerging concepts of guanine nucleotide-binding protein-coupled receptor (GPCR) function and implications for high throughput screening. Assay Drug Dev Technol 5:425–451

Ewing TJ, Makino S, Skillman AG, Kuntz ID (2001) DOCK 4.0: search strategies for automated molecular docking of flexible molecule databases. J Comput Aided Mol Des 15:411–428

Fan F, Wood KV (2007) Bioluminescent assays for high-throughput screening. Assay Drug Dev Technol 5:127–136

Ferrer M, Kolodin GD, Zuck P, Peltier R, Berry K, Mandala SM, Rosen H, Ota H, Ozaki S, Inglese J, Strulovici B (2003) A fully automated [35S]GTP gamma S scintillation proximity assay for the high-throughput screening of Gi-linked G protein-coupled receptors. Assay Drug Dev Technol 1:261–273

Ferrer M, Maiolo J, Kratz P, Jackowski JL, Murphy DJ, Delagrave S, Inglese J (2005) Directed evolution of PDZ variants to generate high-affinity detection reagents. Protein Eng Des Sel 18:165–173

Finkel A, Maiolo J, Kratz P, Jackowski JL, Murphy DJ, Delagrave S, Inglese J (2006) Population patch clamp improves data consistency and success rates in the measurement of ionic currents. J Biomol Screen 11:488–496

Friesner RA, Murphy RB, Friesner RA, Beard HS, Frye LL, Pollard WT, Banks JL (2004) Glide: a new approach for rapid, accurate docking and scoring. 2. Enrichment factors in database screening. J Med Chem 47:1750–1759

Fung P, Peng K, Kobel P, Dotimas H, Kauffman L, Olson K, Eglen RM (2006) A homogeneous cell-based assay to measure nuclear translocationusing beta-galactosidase enzyme fragment complementation. Assay Drug Dev Technol 4:263–72

Funk OF, Kettmann V, Drimal J, Langer T (2004) Chemical function based pharmacophore generation of endothelin-A selective receptor antagonists. J Med Chem 47:2750–2760

Gee KR, Brown KA, Chen WN, Bishop-Stewart J, Gray D, Johnson I (2000) Chemical and physiological characterization of fluo-4 Ca(2+)-indicator dyes. Cell Calcium 27:97–106

Glickman F, McGee J, Napper A (2004) Assay development for protein kinase enzymes. In: Sittampalam GS, Coussens NP, Nelson H, Arkin M, Auld D, Austin C, Bejcek B, Glicksman M, Inglese J, Iversen PW, Li Z, McGee J, McManus O, Minor L, Napper A, Peltier JM, Riss T, Trask Jr OJ, Weidner J (eds) *Assay guidance manual* [Internet]. EliLilly & Company and the National Center for Advancing Translational Science, National Center for Biotechnology Information, Bethesda

González JE, Maher MP (2002) Cellular fluorescent indicators and voltage/ion probe reader (VIPR TM): tools for ion channel and receptor drug discovery. Recept Channels 8:283–295

González JE, González J, Oades K, Leychkis Y (1999) Cell-based assays and instrumentation for screening ion-channel targets. Drug Discov Today 4:431–439

Gowda K, Marks BD, Zielinski TK, Ozers MS (2006) Development of a coactivator displacement assay for the orphan receptor estrogen-related receptor-gamma using time-resolved fluorescence resonance energy transfer. Anal Biochem 357:105–115

Halperin I, Ma B, Wolfson H, Nussinov R (2002) Principles of docking: an overview of search algorithms and a guide to scoring functions. Protein Eng Des Sel 47:409–443

Hamdan FF, Audet M, Garneau P, Pelletier J, Bouvier M (2005) High-throughput screening of G protein-coupled receptor antagonists using a bioluminescence resonance energy transfer 1-based beta-arrestin2 recruitment assay. J Biomol Screen 10:463–475
Haney SA, LaPan P, Pan J, Zhang J (2006) High-content screening moves to the front of the line. Drug Discov Today 11:889–894
Hemmilä I, Dakubu S, Mukkala VM et al (1984) Europium as a label in time-resolved immunofluorometric assays. Anal Biochem 137:335–343
Hillisch A, Pineda LF, Hilgenfeld R (2004) Utility of homology models in the drug discovery process. Drug Discov Today 9:659–669
Hogg DS, Boden P, Lawton G, Kozlowski RZ (2006) Ion channel drug targets – unlocking the potential. Drug Discov World 7:83–92
Hopkins AL, Groom CR (2002) The druggable genome. Nat Rev Drug Discov 1:727–730
Hughes JD, Blagg J, Price DA, Bailey S, Decrescenzo GA, Devraj RV, Ellsworth E, Fobian YM, Gibbs ME, Gilles RW, Greene N, Huang E, Krieger-Burke T, Loesel J, Wager T, Whiteley L, Zhang Y (2008) Physiochemical drug properties associated with in vivo toxicological outcomes. Bioorg Med Chem Lett 18:4872–4875
Inglese J (2006) Measuring biological responses with automated microscopy. Elsevier Academic, San Diego
Inglese J, Johnson RL, Simeonov A, Xia M, Zheng W, Austin CP, Auld DS (2007) High-throughput screening assays for the identification of chemical probes. Nat Chem Biol 3:466–479
Iversen PW, Eastwood BJ, Sittampalam GS, Cox KL (2006) A comparison of assay performance measures in screening assays: signal window, Z′ factor, and assay variability ratio. J Biomol Screen 11:247–252
Jacoby E, Bouhelal R, Gerspacher M, Seuwen K (2006) The 7 TM G-protein-coupled receptor target family. Chem Med Chem 1:761–782
Jain AN (2003) Surflex: fully automatic flexible molecular docking using a molecular similarity-based search engine. J Med Chem 46:499–511
Jones G, Willett P, Glen RC, Leach AR, Taylor R (1997) Development and validation of a genetic algorithm for flexible docking. J Mol Biol 267:727–748
Karvinen J, Elomaa A, Mäkinen ML, Hakala H, Mukkala VM, Peuralahti J, Hurskainen P, Hovinen J, Hemmilä I (2004) Caspase multiplexing: simultaneous homogeneous time-resolved quenching assay (TruPoint) for caspases 1, 3, and 6. Anal Biochem 325:317–325
Kitchen DB, Decornez H, Furr JR, Bajorath J (2004) Docking and scoring in virtual screening for drug discovery: methods and applications. Nat Rev Drug Discov 3:935–949
Kon T, Tanigawa T, Hayamizu K, Shen M, Tsuji T, Naito Y, Yoshikawa T (2004) Singlet oxygen quenching activity of human serum. Redox Rep 9:325–330
Koresawa M, Okabe T (2004) High-throughput screening with quantitation of ATP consumption: a universal non-radioisotope, homogeneous assay for protein kinase. Assay Drug Dev Technol 2:153–160
Krivov GG, Shapovalov MV, Dunbrack RL (2009) Improved prediction of protein side-chain conformations with SCWRL4. Proteins: Struct, Funct, Bioinf 77:778–795
Kumar S, Wittmann C, Heinzle E (2004) Minibioreactors. Biotechnol Lett 26:1–10
Kunapuli P, Lee S, Zheng W, Alberts M, Kornienko O, Mull R, Kreamer A, Hwang JI, Simon MI, Strulovici B (2006) Identification of small molecule antagonists of the human mas-related gene-X1 receptor. Anal Biochem 351:50–61
Lakowicz JR (2006) Principles of fluorescence spectroscopy. Springer, Heidelberg
Leung D, Abbenante G, Fairlie DP (2000) Protease inhibitors: current status and future prospects. J Med Chem 43:305–341
Li H, Sutter J, Hoffman R (2000) HypoGen: an automated system for generating 3D predictive pharmacophore models. In: Guner OF (ed) Pharmacophore perception, development, and use in drug design. International University Line, San Diego, pp 171–189
Lipinski CA (2000) Drug-like properties and the causes of poor solubility and poor permeability. J Pharmacol Toxicol Methods 44:235–249

Lowery RG, Kleman-Leyer K (2006) Transcreener: screening enzymes involved in covalent regulation. Expert Opin Ther Targets 10:179–190

Macarrón R, Hertzberg RP (2011) Design and implementation of high throughput screening assays. Mol Biotechnol 47:270–285

Mahajan NP, Harrison-Shostak DC, Michaux J, Herman B (1999) Novel mutant green fluorescent protein protease substrates reveal the activation of specific caspases during apoptosis. Chem Biol 6:401–409

Martin YC (2000) DISCO: what we did right and what we missed. In: Guner OF (ed) Pharmacophore perception, development, and use in drug design. International University Line, San Diego, pp 49–68

Martí-Renom M a, Stuart a C, Fiser a et al (2000) Comparative protein structure modeling of genes and genomes. Annu Rev Biophys Biomol Struct 29:291–325

Mathis G (1993) Rare earth cryptates and homogeneous fluoroimmunoassays with human sera. Clin Chem 39:1953–1959

May KML, Wang Y, Bachas LG, Anderson KW (2004) Development of a whole-cell-based biosensor for detecting histamine as a model toxin. Anal Chem 76:4156–4161

McDonald OB, Chen WJ, Ellis B, Hoffman C, Overton L, Rink M, Smith A, Marshall CJ, Wood ER (1999) A scintillation proximity assay for the Raf/MEK/ERK kinase cascade: high-throughput screening and identification of selective enzyme inhibitors. Anal Biochem 268:318–329

Miller MD, Kearsley SK, Underwood DJ, Sheridan RP (1994) FLOG: a system to select "quasi-flexible" ligands complementary to a receptor of known three-dimensional structure. J Comput Aided Mol Des 8:153–174

Misura KMS, Baker D (2005) Progress and challenges in high-resolution refinement of protein structure models. Proteins 59:15–29

Mitchell J, Laskowski R, Alex A, Thornton J (1999) BLEEP – potential of mean force describing protein-ligand interactions: I. Generating potential. J Comput Chem 20:1165–1176

Moore KJ (1999) Single molecule detection technologies in miniaturized high throughput screening: fluorescence correlation spectroscopy. J Biomol Screen 4:335–353

Morris GM, Goodsell DS, Halliday RS, Huey R, Hart WE, Belew RK, Olson AJ (1998) Automated docking using a Lamarckian genetic algorithm and an empirical binding free energy function. J Comput Chem 19:1639–1662

Nagy L, Schwabe JWR (2004) Mechanism of the nuclear receptor molecular switch. Trends Biochem Sci 29:317–324

Nonner W, Eisenberg B (2000) Electrodiffusion in ionic channels of biological membranes. J Mol Liq 87:149–162

O'Boyle DR, Nower PT, Lemm JA, Valera L, Sun JH, Rigat K, Colonno R, Gao M (2005) Development of a cell-based high-throughput specificityscreen using a hepatitis C virus-bovine viral diarrhea virus dual replicon assay. Antimicrob Agents Chemother 49:1346–1353

Olefsky JM (1999) Insulin-stimulated glucose transport mini review series. J Biol Chem 274:1863

Ortuso F, Langer T, Alcaro S (2006) GBPM: GRID-based pharmacophore model: concept and application studies to protein-protein recognition. Bioinformatics 22:1449–1455

Pandit D, So S-S, Sun H (2006) Enhancing specificity and sensitivity of pharmacophore-based virtual screening by incorporating chemical and shape features–a case study of HIV protease inhibitors. J Chem Inf Comput Sci 46:1236–1244

Pfleger KDG, Eidne KA (2006) Illuminating insights into protein-protein interactions using bioluminescence resonance energy transfer (BRET). Nat Methods 3:165–174

Pope A, Haupts U, Moore K (1999) Homogeneous fluorescence readouts for miniaturized high-throughput screening: theory and practice. Drug Discov Today 4:350–362

Poptodorov K, Luu T, Hoffmann RD (2006) Pharmacophore model generation software tools. In: Langer T, Hoffmann WD (eds) Pharmacophores and pharmacophore searches. Wiley-VCH Verlag GmbH & Co. KGaA, pp 15–47

Pui TS, Sudibya HG, Luan X, Zhang Q, Ye F, Huang Y, Chen P (2010) Non-invasive detection of cellular bioelectricity based on carbon nanotube devices for high-throughput drug screening. Adv Mater 22:3199–3203

Qureshi SA (2007) Lactamase: an ideal reporter system for monitoring gene expression in live eukaryotic cells. BioTechniques 42:91–95

Rabinowitz JD, Rigler P, Carswell-Crumpton C, Beeson C, McConnell HM (1997) Screening for novel drug effects with a microphysiometer: a potent effect of clofilium unrelated to potassium channel blockade. Life Sci 61:PL87–PL94

Ramm P (1999) Imaging systems in assay screening. Drug Discov Today 4:401–410

Rarey M, Kramer B, Lengauer T, Klebe G (1996) A fast flexible docking method using an incremental construction algorithm. J Mol Biol 261:470–489

Raval A, Piana S, Eastwood MP, Dror RO, Shaw DE (2012) Refinement of protein structure homology models via long, all-atom molecular dynamics simulations. Proteins: Struct, Funct, Bioinf 82:2071–2079

Rohl CA, Strauss CEM, Misura KM, Baker D (2004) Protein structure prediction using Rosetta. Methods Enzymol 383:66–93

Sabisz M, Skladanowski A (2009) Cancer stem cells and escape from drug-induced premature senescence in human lung tumor cells: implications for drug resistance and in vitro drug screening models. Cell Cycle 8:3208–3217

Sato M, Ozawa T, Inukai K, Asano T,Umezawa Y (2002) Fluorescent indicators for imaging protein phosphorylation in single living cells. Nature Biotechnol 20:287–94

Schroeder KS (1996) FLIPR: a new instrument for accurate, high throughput optical screening. J Biomol Screen 1:75–80

Scott JE, Williams KP (2004) Validating identity, mass purity and enzymatic purity of enzyme preparations

Seethala R, Prabhavathi F (2001) Handbook of drug screening. CRC Press, Hoboken, p 106

Sever JL (1962) Application of a microtechnique to viral serological investigations. J Immunol 88:320–329

Seville M, West AB, Cull MG, McHenry CS (1996) Fluorometric assay for DNA polymerases and reverse transcriptase. BioTechniques 21:664–672

Sharma SV, Da H, Settleman J (2010) Cell line-based platforms to evaluate the therapeutic efficacy of candidate anticancer agents. Nat Rev Cancer 10:241–253

Shoichet BK (2006) Screening in a spirit haunted world. Drug Discov Today 11:607–615

Singh P, Harden BJ, Lillywhite BJ, Broad PM (2004) Identification of kinase inhibitors by an ATP depletion method. Assay Drug Dev Technol 2:161–169

Sliwoski G, Kothiwale S, Meiler J, Lowe EW (2014) Computational methods in drug discovery. Pharmacol Rev 66:334–395

Sportsman JR, Gaudet EA, Boge A (2004) Immobilized metal ion affinity-based fluorescence polarization (IMAP): advances in kinase screening. Assay Drug Dev Technol 22:205–214

Sundberg S (2000) High-throughput and ultra-high-throughput screening: solution- and cell-based approaches. Curr Opin Biotechnol 11:47–53

Taylor DL (2006) In: Taylor D, Haskins JR, Giuliano KA (eds) High content screening. Humana, Totowa

Terpetschnig E, Szmacinski H, Malak H, Lakowicz JR (1995) Metal-ligand complexes as a new class of long-lived fluorophores for protein hydrodynamics. Biophys J 68:342–350

Toba S, Srinivasan J, Maynard AJ, Sutter J (2006) Using pharmacophore models to gain insight into structural binding and virtual screening: an application study with CDK2 and human DHFR. J Chem Inf Model 46:728–735

Trinquet E, Mathis G (2006) Fluorescence technologies for the investigation of chemical libraries. Mol BioSyst 2:380–387

Trinquet E, Fink M, Bazin H, Fink M, Bazin H, Grillet F, Maurin F, Bourrier E, Ansanay H, Leroy C, Michaud A, Durroux T, Maurel D, Malhaire F, Goudet C, Pin JP, Naval M, Hernout O, Chrétien F, Fink M, Bazin H, Grillet F, Maurin F, Bourrier E, Ansanay H, Leroy C, Michaud

A, Durroux T, Maurel D, Malhaire F, Goudet C, Pin JP, Naval M, Hernout O, Chrétien F, Chapleur Y, Mathis G (2006) d-myo-Inositol 1-phosphate as a surrogate of d-myo-inositol 1,4,5-tris phosphate to monitor G protein-coupled receptor activation. Anal Biochem 358:126–135
Velec HFG, Gohlke H, Klebe G (2005) Drug Score CSD-knowledge-based scoring function derived from small molecule crystal data with superior recognition rate of near-native ligand poses and better affinity prediction. J Med Chem 48:6296–6303
Verma R, Peters NR, D'Onofrio M, Tochtrop GP, Sakamoto KM, Varadan R, Zhang M, Coffino P, Fushman D, Deshaies RJ, King RW (2004) Ubistatins inhibit proteasome-dependent degradation by binding the ubiquitin chain. Science 306:117–120
Williams C (2004) cAMP detection methods in HTS: selecting the best from the rest. Nat Rev Drug Discov 3:125–135
Wolber G, Langer T (2005) Ligand Scout: 3-D pharmacophores derived from protein-bound ligands and their use as virtual screening filters. J Chem Inf Model 45:160–169
Wolber G, Seidel T, Bendix F, Langer T (2008) Molecule-pharmacophore super positioning and pattern matching in computational drug design. Drug Discov Today 13:23–29
Wölcke J, Ullmann D (2001) Miniaturized HTS technologies – uHTS. Drug Discov Today 6:637–646
Xiang Z (2006) Advances in homology protein structure modeling. Curr Protein Pept Sci 7:217–227
Xu X, Gerard AL, Huang BC, Anderson DC, Payan DG, Luo Y (1998) Detection of programmed cell death using fluorescence energy transfer. Nucleic Acids Res 26:2034–2035
Yang S-Y (2010) Pharmacophore modeling and applications in drug discovery: challenges and recent advances. Drug Discov Today 15:444–450
Yang S-T, Zhang X, Wen Y (2008) Microbioreactors for high-throughput cytotoxicity assays. Curr Opin Drug Discov Dev 11:111–127
Yang J, Copeland RA, Lai Z (2009) Defining balanced conditions for inhibitor screening assays that target bisubstrate enzymes. J Biomol Screen 14:111–120
Zang R, Li D, Tang I-C, Wang J, Yang S-T (2012) Cell-based assays in high-throughput screening for drug discovery. Int J Biotechnol Wellness Ind 1:31–51
Zhang JH, Chung TD, Oldenburg KR (1999) A simple statistical parameter for use in evaluation and validation of high throughput screening assays. J Biomol Screen 4:67–73
Zheng W, Spencer RH, Kiss L (2004) High throughput assay technologies for ion channel drug discovery. Assay Drug Dev Technol 2:543–552
Zheng CJ, Han LY, Yap CW, Ji ZL, Cao ZW, Chen YZ (2006) Therapeutic targets: progress of their exploration and investigation of their characteristics. Pharmacol Rev 58:259–279
Zlokarnik G, Negulescu PA, Knapp TE, Mere L, Burres N, Feng L, Whitney M, Roemer K, Tsien RY (1998) Quantitation of transcription and clonal selection of single living cells with beta-lactamase as reporter. Science 279:84–88

Web Articles

http://www.perkinelmer.com/Technologies/Assays-and-Reagents/Receptor-Ligand-Binding/default.xhtml
http://cen.acs.org/articles/92/web/2014/11/Tufts-Study-Finds-Big-Rise.html

Metagenomic Approaches for Novel Active Metabolites

12

Dharmaprakash Viszwapriya
and Shunmugiah Karutha Pandian

Abstract

Microorganisms are a proven source of bioactive metabolites with significant functionality and applications. Earlier studies on bioactive metabolite discovery relied solely on cultivable microbial species. However, advances in genomic techniques revealed that culturable microorganisms represent only a tiny fraction of the total microbial diversity, prompting the development of a new field for accessing the unculturable microbial community, termed metagenomics. Metagenomics overcomes the problems of cultivation techniques by exploring the genome of microorganisms as a whole (metagenome), rather than culturing it. Metagenomic-based functional analysis has identified a number of novel biomolecules at a pace which was not achievable using traditional culture-based screening techniques. The approach has also provided valuable insights into previously overlooked biochemistry of culturable organisms. Recent breakthroughs in next-generation sequencing technologies and bioinformatics have made large-scale sequencing and analyzing of metagenome feasible and cost-effective, which further accelerated research in this field. This chapter discusses different aspects of metagenomics and their application for identifying novel active metabolites, highlighting diverse array of strategies and recent discoveries.

Keywords

Metagenomics • Function driven • Metabolites • NGS • Novel • Sequence driven

D. Viszwapriya • S. Karutha Pandian (✉)
Department of Biotechnology, Alagappa University,
Science Campus, Karaikudi 630004, Tamil Nadu, India
e-mail: sk_pandian@rediffmail.com

S. Abdulhameed et al. (eds.), *Bioresources and Bioprocess in Biotechnology*,
DOI 10.1007/978-981-10-3573-9_12

12.1 Introduction

The world is teeming with plethora of microorganisms, which inhabit almost every ecological niche, ranging from deep sea to hydrothermal vent to human gut. They can be either free living or can form pathogenic or symbiotic relationships with eukaryotes or other prokaryotes. Despite their ubiquity, vast majority of them are not yet functionally characterized. This is largely due to the fact that nearly 99% of microorganisms in various environments are not amenable to culturing under laboratory conditions (Amann et al. 1995; Torsvik and Ovreas 2002). One of the preliminary observations which prompted the thought that cultured microorganisms did not represent the entire microbial community was the discrepancy between the size of microbial population estimated by microscopy and that was observed by dilution plating (popularly cited as 'great plate count anomaly') (Staley and Konopka 1985; Grimes et al. 1986; Torsvik et al. 1990). Further, the advent of cultivation-independent methods for studying microbial diversity like 16S rRNA-based phylogeny manifested the concept that microbial diversity is much more complex than previously thought and accessed by cultivation (Pace et al. 1986; Eden et al. 1991; Barns et al. 1994). The phenomenon of a hidden microbial world coupled with the understanding that they hold huge promise to unearth novel biomolecules led to the emergence of a new field of research called metagenomics.

Metagenomics is the study of metagenome, which represents collective DNA of all indigenous microorganisms present in the sample. It is also broadly referred as environmental genomics, population genomics and community genomics (Handelsman 2004). Metagenomics employs genomic technologies and bioinformatic tools to access the genetic content of microbiota present in a sample. Much of advances in microbial ecology, diversity, evolution and novel bioactive discovery over the past 10 years have been attributed to the field of metagenomics (Thomas et al. 2012). The term metagenomics was first used in 1998 by Handelsman and coworkers to denote the comprehensive analysis of a mixture of microbial genome, as in meta-analysis, which is the process of statistically combining separate analysis. This rapidly growing area has been proven to be a fruitful approach for accessing the uncultivable microbial community by circumventing the inherent and established limitation of traditional methods that rely on cultivation.

Metagenomics initially involved cloning of environmental DNA into a host that can be easily cultivated and hence was defined as 'the cloning and functional analysis of metagenome' (Handelsman et al. 1998). Recent advances in next-generation sequencing technologies have allowed metagenome to be sequenced and analysed directly without the need for cloning. Consequently, metagenomics has been redefined as 'the study of microbial communities directly from their natural environment using modern genomic techniques, bypassing the need for isolation and cultivation' (Chen and Pachter 2005).

Metagenomic technologies have enabled to delineate microbial diversity and access genetic information of unculturable microorganisms. It is also a powerful tool to reconstruct biochemical pathways and generate novel hypothesis of microbial function. For instance, proteorhodopsin-based photoheterotrophy was discovered through metagenomics (Beja et al. 2000). Another important application of metagenomics is its ability to mine metagenome for novel metabolites. Screening of metagenomic DNA for metabolites is generally based on two different approaches, namely, sequence-driven approach and function-driven approach. While the former screens the metagenome or libraries for particular DNA sequences, the latter primarily screens the library for an expressed trait. Both approaches have their own strengths and limitations; together, these approaches have provided insights into the uncultured world and the immense potential of novel metabolites they hold (Schloss and Handelsman 2003; Daniel 2005;). In this chapter, metagenomic tools are discussed as an excellent molecular technique with potent biotechnological and industrial applications. An overview of metagenomics is given, highlighting various metagenomic strategies employed for structural- and functional-based analysis of metagenomics, with particular emphasis on the discovery of novel active metabolites.

12.2 Metagenome: A Source of Novel Active Metabolites

A total of 10^{30} distinct microbial species are estimated to be present in the world, which represent a dynamic and extraordinary pool of global biodiversity (Turnbaugh and Gordon 2008). Owing to their enormous diversity, microbes harbour far greater genetic and metabolic diversity than plants and animals. The extreme variety of geographical diversity and habitat, in which they thrive, has enabled them to develop unique physiological and biochemical competence for survival. Consequently, they possess an incredible ability to synthesize novel metabolites, enzymes, drugs and variants of known biomolecules that are otherwise impossible or difficult to synthesize through chemical methods. For example, nearly 10,000 biologically active metabolites with therapeutic potential against HIV, inflammation, cancer and a number of human diseases are produced by the single genus *Streptomyces*. Hence, microbial diversity offers a precious source of bioactives which would benefit mankind and provide sustainable and innovative solutions to a broad range of problems of modern society (Lorenz and Eck 2005). However, majority of them have never been cultured, identified and classified, directly inferring that they go inaccessible for any kind of prospecting using traditional cultivation techniques. For instance, only 0.1% of the bacterial and 5% of the total fungal species are estimated to be reported so far (Felczykowska et al. 2012).

Metagenomic technology overcomes this bottleneck by exploring the genome of microorganisms as a whole (metagenome), rather than culturing it (Wilson and Piel 2013). Moreover, any culture-dependent methods for screening of active metabolites lack novelty as the possibility of mining out a new metabolite is a rare event. For over 70 years, culturable microorganisms, which constitute only 1% of the total microbial community, have been the prime reservoir of metabolite discovery. Only in the past two decades, the potential of unculturable microorganisms has gained prominence and henceforth explored for novel bioactive metabolites (Li and Qin 2005). The likelihood of finding novel metabolites from microbes depends primarily on the diversity of strains explored and their uniqueness, both of which can be readily met through metagenomics. Most of the biomolecules derived using metagenomic approaches are either entirely unrelated or meekly related to known genes proving its immense potential (Streit and Schmitz 2004).

12.3 Bioprospecting Metagenome

Diverse array of strategies has been developed to mine novel metabolites from metagenome. The process starts with the selection of ecological niche, metagenome extraction, direct sequencing or construction of metagenomic libraries, followed by screening for the target of interest using various methods. The overall scheme involved in the screening of novel active metabolites from metagenome is depicted in Fig. 12.1.

12.3.1 Sampling

Since microorganisms are omnipresent, metagenomic sampling can be done from virtually any habitat, both abiotic and biotic. Abiotic habitat includes freshwater, air, terrestrial soil, marine sediment, seawater and so on, while biotic habitat comprises sample collected from living organisms like plants, animals, insects, corals and so on. Even previously overlooked, biotic habitats like seaweed and seagrass are reported to host diverse bacterial phyla (Aravindraja et al. 2013; Viszwapriya et al. 2015) and have been explored using metagenomics recently (Martin et al. 2014; Singh et al. 2015). Besides these, microbiota of extreme environments like hydrothermal vents, dry deserts, volcanoes, deep ocean trenches, geographical poles, etc. can also be explored through metagenomics (Herbert 1992; Schiraldi and De Rosa 2002).

Based on the selection of environment, metagenomic studies fall into three major categories (Barone et al. 2014). The first category selects environment that is naturally enriched with the desired trait, such as xylanases from insect gut, amylases from bovine rumen, etc. However, it must be noted that selection of enriched environment does not guarantee greater detection frequency of the anticipated trait. The second category targets genomically diverse, naturally less extreme and more heterogeneous environment. To point a few, one gram of cultivated soil contains nearly 10^9 prokaryotic cells (Daniel 2005), and oceans constitute rich source of microorganisms with cell counts of 10^6–10^8 cells per millilitre (Fenical and Jensen 2006). Selecting heterogeneous environment would be a wise choice, since genomically diverse environment is ought to be metabolically diverse. For instance, culture-dependent mining of marine sources has revealed a number of novel antibiotics, antibiofilm compounds, anti-quorum-sensing agents and other pharmaceutically important metabolites, giving a cue about the immense metabolic profile hidden in their unculturable counterparts (Thenmozhi et al. 2009; Bakkiyaraj and Pandian 2010; Nithya et al. 2010; Gowrishankar et al. 2014; Padmavathi et al. 2014;). The

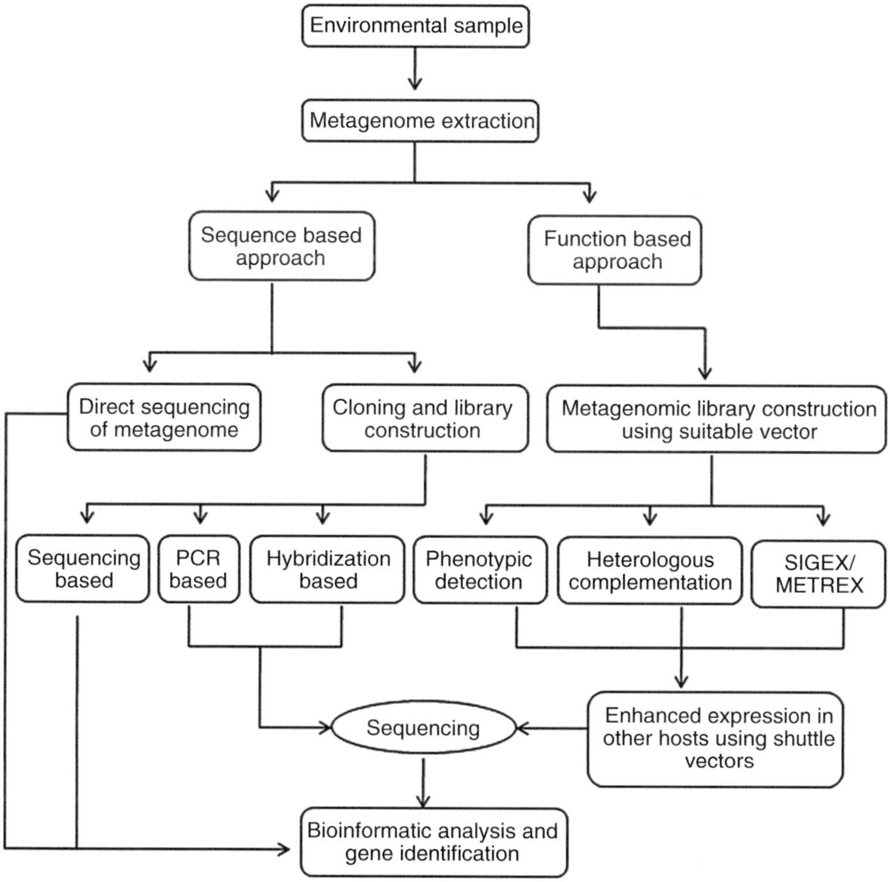

Fig. 12.1 Overview of various metagenomic strategies for identification of novel active metabolites

third category targets extreme environments to screen biomolecules that are stable and functional under such extreme conditions (Steele et al. 2008). As extremophiles produce metabolites that are functionally active in extreme environments, they find potential applications in industrial processes. The remarkable discovery of Taq DNA polymerase, a widely used thermostable enzyme in molecular biology, attests to this fact. In addition, the likelihood of identifying novel bioactives from the hostile environment is higher, since they are least explored and the indigenous microorganisms ought to have unique mechanisms to thrive in such extreme conditions (Schiraldi and De Rosa 2002; Simon and Daniel 2011).

12.3.2 Metagenome Extraction

Extraction of high-quality metagenomic DNA is one of the key factors that determine the success of metagenomic studies (Sharma et al. 2007; Purohit and Singh 2008). The quality of metagenome is determined by its yield, purity, fragment size and representativeness (Morgan et al. 2009; Ekkers et al. 2012). In practice, it is difficult to achieve all the parameters due to their negative reciprocal effect. For example, high yield of metagenome is often related to increased contamination and small DNA fragment size. Consequently, increased contamination interferes with enzymatic modifications of the isolated DNA, and small fragment size limits the analysis of large operons (Williamson et al. 2005). The quality of the metagenome is also affected by the physical and chemical structure of the sample. For example, unlike soil metagenome which is often contaminated with organic and inorganic compounds, metagenome extracted from water is usually of high quality, as the contaminating chemicals often remain in water after filtration. Nevertheless, numerous extraction protocols have been developed, specific to particular habitats, and some of them are even commercialized as DNA extraction kits (Rondon et al. 2000; Purohit and Singh 2008).

As microbiota from any habitat is typically diverse with unique cell wall characteristics and varying susceptibility to lysis, special methods are developed to lyse the heterogeneous population and extract their genome simultaneously. Each method has its own merits and demerits, and it is recommended to employ more than one extraction protocol for a single niche, to acquire maximum representation of the microbial community (Delmont et al. 2011). Metagenomic DNA extraction protocol is divided into two categories, namely, direct extraction and indirect extraction. Direct extraction involves direct lysis of cells contained in the sample followed by separation of DNA, whereas indirect extraction method separates cells from the sample followed by lysis and DNA extraction (Gabor et al. 2004; Daniel et al. 2005). The crude DNA extracted using both methods is then purified using standard procedures.

12.3.2.1 Direct Lysis Method

Metagenomic DNA isolated using direct lysis method is supposed to represent the microbial diversity in a better way, as this method does not include a cell separation step, and so microorganisms that are adhered to the sample particles also come into the picture (Daniel 2005). In addition, the method is more suitable when large quantity of DNA is required. However, the metagenomic DNA extracted using this method is often found to be sheared, contaminated and contains varying quantities of extracellular and eukaryotic DNA. Direct lysis method comprises soft and harsh lysis, both of which can be either used alone or in combination (Purohit and Singh 2008).

Soft lysis method involves disruption of microorganisms exclusively by enzymatic and chemical means. Lysozyme which cleaves the peptidoglycan layer of prokaryotic cell membrane is the most commonly used enzyme (Niemi et al. 2001). Depending on the need, other specific enzymes are also employed. For instance, achromopeptidase is used to improve the lysis of recalcitrant gram-positive bacteria like *Frankia*, and proteinase K, a broad-spectrum protein-digesting enzyme, is used to degrade membrane-associated proteins (Simonet et al. 1984; Rochelle et al. 1992; Zhou et al. 1996). Chemical method uses detergents like sodium dodecyl sulphate (SDS), cetyltrimethylammonium bromide (CTAB) and polyvinylpolypyrrolidone (PVPP) to dissolve the hydrophobic materials of cell membrane. Some protocol combines these detergents with heat treatment or chelating agents to facilitate lysis (Verma and Satyanarayana 2011).

On the other hand, harsh lysis protocols apply mechanical force in the form of bead beating, freeze thawing and sonication to disrupt cell wall (Kennedy et al. 2007). Bead beating is the most commonly used method wherein tiny beads made of glass or ceramic are mixed with an aqueous suspension of sample and subjected to agitation (Leff et al. 1995). Freeze-thaw method involves repeated cycles of freezing and thawing, which disrupt cells through ice crystal formation. Sonication uses ultrasonic waves in the range of 20–50 kHz and is more popularly used these days. The efficiency of harsh lysis method depends on duration and magnitude of the applied force. The major drawback of harsh lysis method is shearing of DNA even under optimized conditions (Siddhapura et al. 2010).

12.3.2.2 Indirect Lysis Method

Indirect lysis methods use mild mechanical force such as blending, rotating pestle homogenization or chemical means like cation-exchange resins followed by density-gradient or differential centrifugation to separate the cells from sample (Jacobsen and Rasmussen 1992; Lindahl and Bakken 1995). Although time-consuming and less efficient in terms of yield, the processes are less harsh than direct lysis methods (Holben et al. 1988). Compared to direct lysis approach, 10- to 100-fold reduction in DNA yield was observed (Gabor et al. 2004). Nevertheless, the DNA obtained using this method is almost entirely prokaryotic and free of sample contaminants like humic acids, polyphenols, etc. The DNA extracted using this approach is generally of high molecular weight and therefore is more preferred for large-insert libraries.

12.3.2.3 Contaminant Removal and Purification

Owing to the variant nature of ecological niche, DNA extraction protocols should be case specific and often need optimization. Standardization reduces lysis bias and ensures improved DNA quality. For example, polluted environment contains metal ions, chemical impurities and large number of organic and inorganic pollutants. These contaminants pose major problem, since they hinder molecular-biological manipulations like digestion, ligation and polymerase activity (Verma and Satyanarayana 2011). Basic purification steps like phenol-chloroform extraction and ethanol or polyethylene glycol precipitation are normally inadequate to remove such contaminants. Advanced techniques like gel elution, gel filtration, ion exchange and use of chemical agents such as activated charcoal and Amberlite resins are found successful in removing contaminants with varying levels of efficiency (More et al. 1994; Jackson et al. 1997; Sharma et al. 2007; Verma and Satyanarayana 2011).

12.3.3 Screening Metagenome for Novel Active Metabolites

Once an appreciable amount of high-quality DNA is extracted, it can be used to analyse microbial diversity or construct novel microbial genome or screen for biocatalyst, drugs, metabolites and other active biomolecules (Handelsman 2004). Since the current chapter is focused on the use of various metagenomic strategies to screen for novel active metabolites, we restrict ourselves to the same. Screening for metabolites can be done using two different approaches: sequence-driven approach and function-driven approach. While sequence-based approach relies on comparing the metagenome sequence with known sequences deposited in the database, function-based approach recognizes genes by their function rather than by their sequence. Sequence-driven approach can be either cloning dependent or independent, whereas function-driven approach is solely cloning dependent (Wilson and Piel 2013; Streit and Schmitz 2004). Representative list of metabolites identified using both the approaches is given in Table 12.1.

12.3.4 Function-Driven Approach

Function-based metagenomic approach is presumably the only approach that has the potential to identify entirely new classes of genes for both known and novel functions, because the approach is not dependent on the sequence information of the desired trait (Riesenfeld et al. 2004; Ferrer et al. 2009). The major advantage of this approach is that only full-length genes and functional gene products are identified. Therefore, the results are unambiguous (Simon et al. 2009). In addition, this technique facilitates purification of the desired biomolecules produced by a clone, since the extra- metabolite can be readily pinpointed using reference clone harbouring the vector alone (Lefevre et al. 2008). However, this approach requires analysis of large number of clones to recover few positive clones.

12.3.4.1 Library Size and Vector Choice

The process of function-driven approach starts with metagenomic library construction followed by functional screening of the desired trait and characterization of active clones via biochemical analysis and sequencing. Metagenomic library construction involves the same steps as genomic DNA library construction, except that metagenome extracted from the sample of interest is used as insert. Briefly, the metagenome is fragmented using restriction enzyme or randomly sheared by mechanical means (sonication, vigorous vortex, aspiration), ligated into an appropriate vector and used to transform suitable host. The success of function-driven approach depends on various parameters including the size of insert, vector system used, host, abundance of target gene in the source metagenome, efficiency of heterologous gene expression in surrogate host and assay method employed (Uchiyama and Miyazaki 2009).

Depending on the size of insert, metagenomic libraries are classified into two types, namely, small-insert libraries and large-insert libraries. Small-insert libraries

Table 12.1 Representative list of metabolites discovered using metagenomic strategies

Metabolite	Sampling site	Screening strategy	References
Borregomycin A and B (antiproliferative and antibiotic properties)	Soil	Sequence based	Chang and Brady (2013)
Violacein (antibiotic, pigment)	Soil	Function based	Feng et al. (2012)
Fasamycin (antibiotic)	Soil	Function based	Feng et al. (2012)
Zn-coproporphyrin III (anti-inflammatory drug)	*Discodermia calyx*, bacterial symbiont	Function based	He et al. (2012)
Minimide (antineoplastic compound)	*Didemnum molle*, microbiome	Sequence based	Donia et al. (2011)
Apratoxin A (cytotoxin)	*Lyngbya bouilloni*, cyanobacterial symbiont	Function based	Grindberg et al. (2011)
Indirubin (antibiotic)	Soil	Function based	Lim et al. (2005)
Patellamide (anticancer drug)	*Lissoclinum patella*, cyanobacterial symbiont	Function based	Long et al. (2005)
Onnamide A (antitumour drug)	*Theonella swinhoei*, bacterial symbiont	Sequence based	Piel et al. (2004)
Bryostatin (anticancer drug and memory enhancer)	*Bugula neritina*, bacterial symbiont	Sequence based	Hildebrand et al. (2004)
Pederin (toxic amide)	*Paederus* beetles, bacterial endosymbiont	Sequence based	Piel (2002)
Turbomycin A and B (antibiotic)	Soil	Function based	Gillespie et al. (2002)
Terragine (antibiotic)	Soil	Function based	Wang et al. (2000)

are constructed using plasmid vectors (pUC18, pHT01) which can accommodate insert size of only 0.5–8 kb. Large-insert libraries are generated using cosmid (pWE15, pJC8) and fosmid (pCC1FOS, pM0579), both of which can hold up to 40 kb insert, and bacterial artificial chromosome (BAC) vectors can hold ~150 kb insert. The choice of vector system depends on the quality of the extracted metagenome, desired vector copy number, required insert size and aim of the study. For instance, a highly sheared DNA is only suitable for the generation of small-insert libraries, which are appropriate for isolation of single gene encoding metabolic functions. On the other hand, large-insert libraries enable identification of complex pathways encoded by large gene clusters and operons. The presence of considerable flanking DNA on large inserts facilitates phylogenetic inference about the source of the fragment as well. The merits and demerits of both the libraries are detailed in Table 12.2. Two vectors, namely, pWE15 (cosmid vector) and pCC1FOS (fosmid vector), have been widely used for cloning large DNA fragments. The pCC1FOS has an added advantage that its copy number can be controlled by addition of

Table 12.2 Merits and demerits of small-insert and large-insert libraries

Advantages	Disadvantages
Small-insert libraries	
Simple to construct	Large number of clones must be screened to find positives
Suitable to clone sheared metagenome	Not suitable to screen metabolites encoded by large gene clusters
High copy number enables detection of poorly expressed genes	
Expression of heterologous genes from vector promoter is possible	
Large-insert libraries	
Screening small number of clones is sufficient to find positives	Technically complex
Suitable to screen metabolites encoded by large gene clusters	Low copy number hinders detection of poorly expressed genes
Suitable for partial genome characterization	Expression of heterologous genes from vector promoter is limited

arabinose, when used to transform appropriate host (e.g., *E. coli* EPI300). Antibiotics, antibiotic-resistance determinants, pigments, biocatalysts, microbial sensing signals and eukaryotic growth-modulating factors have been identified from metagenomic libraries constructed using pCC1FOS (Leveau and Gerards 2008; Riaz et al. 2008; Sabree et al. 2009; Donato et al. 2010; Cullingan et al. 2014).

12.3.4.2 Host Selection

E. coli is the most preferred host for the generation of metagenomic libraries. In fact, majority of the published studies have used *E. coli* for the initial construction and maintenance of metagenomic libraries. However, *E. coli* is not an ideal host, since it fails to express genes from more distantly related organisms. It has been reported that gene expression patterns differ significantly among different taxonomic groups of prokaryotes, and *E. coli* can efficiently express only 40% of the genes acquired by random cloning. This is due to the varied GC content of the promoter regions of the cloned genes, making it unrecognizable by *E. coli* transcriptional machinery. In some cases, the genes are expressed at relatively low levels, due to the differences in codon usage (Simon et al. 2009). Even in cases where transcription and translation are successful, problems are observed in post-translational modification and product-export phase. In many cases, the desired protein may not fold correctly due to the absence of required chaperons in *E. coli* host. To mitigate these limitations and to increase the detectable range of gene expression, additional hosts like *Streptomyces* spp., *Pseudomonas* spp., *Thermus thermophilus* and *Sulfolobus solfataricus* have been used (Wang et al. 2000; Li and Qin 2005; Albers et al. 2006; Angelov et al. 2009; Craig et al. 2010). Libraries constructed in *E. coli* can also be transferred to other hosts using shuttle cosmid or BAC vectors (Martinez et al. 2004).

Among the host candidates, *Streptomyces* spp. is considered to be the best for heterogeneous expression of biologically active secondary metabolites. The organism inherently produces many secondary metabolites, and their expression is controlled by intra- and intercellular signalling. These pre-existing biosynthetic pathways serve as adapted machinery for the synthesis and export of chemical entities expressed by the heterogeneous genes (Wang et al. 2000). Several secondary metabolites including non-ribosomal peptides, antibiotics and polyketides have been successfully expressed in streptomycetes (Salomon et al. 2004). In addition, there is high possibility of generating novel hybrid antibiotics, when genes of secondary metabolite pathway are introduced into this organism (Shen et al. 2001; Li and Qin 2005).

12.3.4.3 Metagenomic Library Storage

Storage conditions must be optimized, so as to preserve clone viability and also the original diversity of the library. Constructed metagenomic libraries are generally stored as pooled aliquots in liquid culture supplemented with corresponding vector-selective antibiotic and 15% glycerol under freezing conditions (−80 °C). It is critical to add appropriate antibiotic even during storage in order to prevent curing of recombinant vector. Glycerol acts as a cryoprotectant and ensures clone viability. It forms stable hydrogen bonds with water molecules and decreases the freezing point of the solution. By this way, it inhibits the efflux of intracellular water and prevents cell shrinking. It also decreases ice crystal formation, which may damage or rupture the cells. Stored aliquots can be revived in fresh media supplemented with antibiotic and incubated at optimum temperature for a brief period (1–2 h) under constant shaking condition. Extended incubation causes overgrowth of fast-growing clones which may hinder recovery of entire metagenomic library. The revived library is then used for function or sequence-based screening.

12.3.4.4 Functional Screening of Libraries

Metagenomic libraries can be functionally screened for active metabolites using various strategies, viz., phenotypic detection, heterologous complementation and customized intracellular functional screens like SIGEX, PIGEX and METREX. Each screening method holds intrinsic biases, and no single method can be employed to isolate wide variety of genes from metagenome. Appropriate screening method should be selected based on aim of the study and target of interest (Owen et al. 2012).

Phenotypic detection is based on visible phenotypes like colour, zone of clearance, etc. The screening process often uses chemical dyes or chromophore-bearing derivatives of enzyme substrates in the growth medium. The target gene or metabolite can be identified by visualizing the change in growth medium. Phenotypic detection is the most commonly used technique to screen metabolites, as it is simple and does not involve complex genetic manipulations. The sensitivity of this screening technique is very high, which offers an advantage to detect even rare clones. One of the examples for phenotypic detection is the screening of recombinant *E. coli* that confer carbonyl formation from polyols. The substrate (polyol) is supplemented in the indicator agar, which contains the indicator Schiff reagent (a mixture of sodium bisulphite and pararosaniline). Production of carbonyl from polyols converts the pale pink colour of Schiff base to dark red colour. Hence, carbonyl-forming clones are identified easily by the red-coloured zone surrounding the colonies (Daniel 2005). Another example is screening metagenomic library for quorum-sensing (QS) inhibitors in a lawn of bacterial indicator strain, *Chromobacterium violaceum.* The indicator strain produces a violet-coloured pigment called violacein, which is positively controlled by QS. Production of QS inhibitors by metaclones is identified simply by observing zone of violet-coloured inhibition around the clones (Riaz et al. 2008; Nithya et al. 2010). In addition to this, phenotypic screens have enabled discovery of numerous novel antibiotic candidates, by visually observing the zone of growth inhibition in bacterial and fungal lawns (Gillespie et al. 2002).

One of the limitations of phenotypic assays is the requirement of functional gene product secretion from the host strain for extracellular detection. Many metabolites of interests lack simple phenotypic assay making the screening process complicated. Besides, metagenomic libraries are usually large, which makes it cumbersome for single chemical analysis. Hence, alternate screening techniques that reduce time and target wide range of metabolites have been developed. One such technique is heterologous complementation-mediated functional screening. It is based on the complementation of host strains or their mutants, which require targeted gene for their growth under selective conditions, in which only recombinant clones that harbour and express the targeted gene can grow. To cite a few, complementation of Na^+/H^+ antiporter-deficient *E. coli* strain with soil-derived metagenome led to the identification of two novel genes that encode Na^+/H^+ antiporter (Majernik et al. 2001). Genes that confer resistance to antibiotics like tetracycline, kanamycin, tobramycin and amikacin have been identified by constructing metagenomic libraries in antibiotic-sensitive host, and the positive isolates are selected in medium supplemented with the corresponding antibiotic (Handelsman 2004). Another study targeted the bacterial β-D-glucuronidase enzyme, which has beneficial effects on human health. Screening of 4600 clones of a metagenomic library constructed using bacterial DNA extracted from human faeces using β-D-glucuronidase-deficient *E. coli* strain resulted in 19 positive clones (Gloux et al. 2011).

Substrate-induced gene expression (SIGEX) is a simple, semiautomated, less laborious and fast screening technique that selects clones harbouring catabolic genes induced by various substrates in concert with fluorescence-activated cell sorting (FACS). The technique was first proposed by Uchiyama and Watanabe in 2007 wherein they evaluated its utility for the screening of aromatic hydrocarbon-induced genes from a groundwater metagenomic library. The process is based on the concept that catabolic gene expression is usually induced by relevant compounds (commonly substrate) and often controlled by regulatory elements in its proximity. Hence, SIGEX screens clones harbouring desired catabolic genes that express only in the presence of substrate. The process uses an operon-trap vector designated p18GFP for constructing metagenomic libraries. The vector was designed with its cloning site sandwiched between the *gfp* structural gene and *lac* promoter.

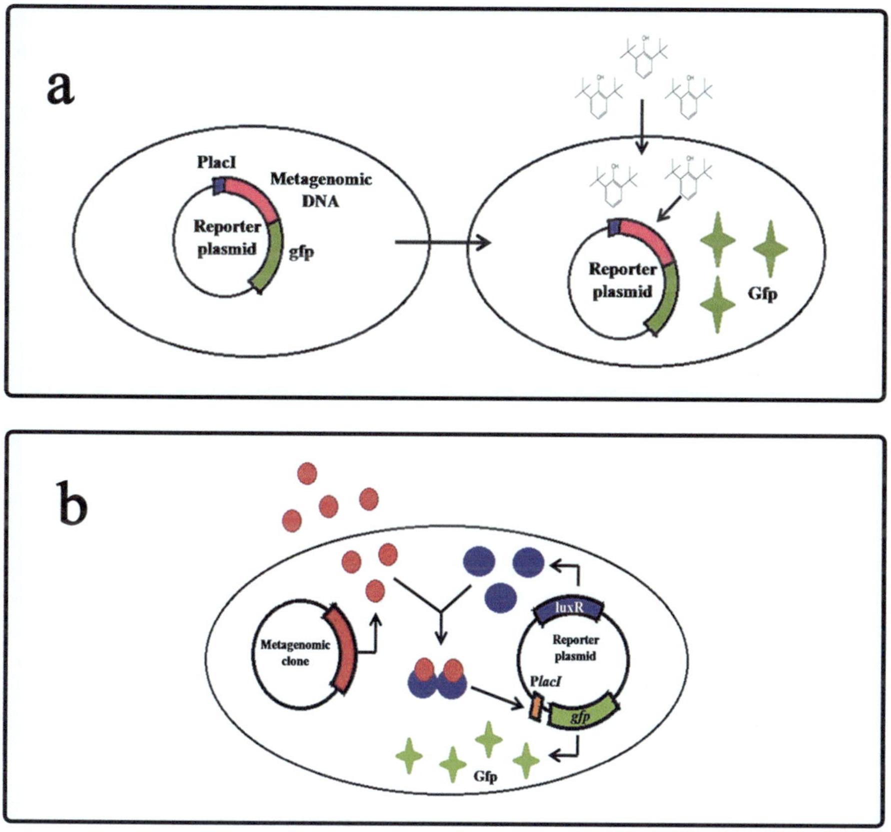

Fig. 12.2 Schematic representation of intracellular functional screens. (**a**) Substrate-induced gene expression (*SIGEX*): Metagenomic DNA is cloned into a promoter-trap vector upstream of GFP reporter gene. Promoters in metagenomic DNA which respond to specific substrates induce GFP expression (adapted and adopted from Sabree et al. 2009). (**b**) Metabolite-regulated expression (*METREX*): Diffusible metagenomic products that mimic QS signal forms complex with LuxR proteins, which bind to *lux*I promoter and induce GFP expression (Adapted and adopted from Williamson et al. 2005)

This allows cloning of metagenomic DNA upstream of *gfp* gene, thereby placing the expression of *gfp* under the control of promoters present in the metagenomic DNA (Fig. 12.2a). The constructed libraries were screened for catabolic genes in a three-step process. In the first step, self-ligated clones and clones which express *gfp* constitutively are removed by isopropyl β-D-thiogalactopyranoside (IPTG) induction in the absence of the substrate. This preliminary filtering greatly improves the efficiency of screening, as it almost completely eliminates false-positive clones. In the second phase, clones with desired catabolic genes are determined by *gfp* expression in the presence of substrate. Sorting and separation of green fluorescent protein (GFP)-expressing clones were done using FACS in both steps. Finally, in the third step, positive clones obtained in step two are characterized using biochemical assays and sequencing. Using this technique, 33 benzoate-degrading clones and 2 naphthalene-degrading clones from a total of 152,000 clones were isolated from groundwater metagenomic library (Uchiyama and Watanabe 2007).

SIGEX offers an efficient and economic way of high-throughput screening and finds importance in industrial applications for screening genes involved in antibiotic production or biodegradation induced by small molecules. It offers an alternate way to detect catabolic genes for which colorimetric and other plate-screening methods are not established. It enables identification of substrates for unknown enzymes and identification of function for hypothetical genes. Moreover, it does not require modified substrates as in colorimetric screenings that are generally more expensive and occasionally toxic. However, SIGEX has few limitations, the prime being its sensitivity to the orientation of genes. It cannot detect active clones in which the genes are cloned in the direction opposite to *gfp* (Yun and Ryu 2005). It fails to identify catabolic genes that express constitutively and the active clones that have a transcription terminator between the heterologous gene and *gfp*. Hence, it is not suitable for screening large-insert metagenomic libraries, which have abundant transcription terminators (Ermolaeva et al. 2000). This drawback should be critically considered, since the probability of finding a positive hit increases exponentially with the insert size (Gabor et al. 2004). Another important disadvantage is that substrates which do not migrate to the cytoplasm cannot be used in SIGEX. Alteration in FACS settings and media conditions can sometimes lead to false-positive and false-negative results. Despite these limitations, SIGEX is a very powerful high-throughput technique that can screen large number of clones in a relatively short time (Yun and Ryu 2005).

In SIGEX, activation of transcriptional regulators by other effectors rather than specific substrates leads to the recovery of false-positive clones. To circumvent this problem, Uchiyama and Miyazaki (2010) developed an altered version of SIGEX called product-induced gene expression (PIGEX), which uses a transcriptional activator, which is blind to the substrate but responds to the product and activates a promoter fused to reporter gene. Using this technique, the authors have identified three novel amidases which convert benzamide to benzoate using a benzoate-responsive transcriptional activator. Transcriptional activators can be searched in natural regulatory pathway or can be engineered to recognize particular product. For example, a study has engineered regulators that respond to 1,2,4-trichlorobenzene, a major product of γ-hexachlorocyclohexane (HCH) dehydrochlorination (Mohn et al. 2006).

Besides the techniques described above, which are used for screening broad range of metabolites, few other techniques are available that target a particular trait. One such technique is metabolite-regulated expression (METREX), an intracellular screen which detects molecules that mimic or inhibit quorum-sensing (QS) signals. METREX and SIGEX work on the same principle, but the difference is SIGEX captures promoters that are responsive to certain metabolites, whereas METREX detects metabolites which activate such promoters. METREX is designed in such a way that the heterologous DNA is introduced in the same cell, which harbours the biosensor plasmid. This allows identification of even poorly expressed metagenomic gene products (Williamson et al. 2005). QS signals (synthesized by *luxI* homologues) are small diffusible molecules, which, when bound to their receptor protein (synthesized by *luxR* homologues), act as a transcriptional factor and regulate the expression of various genes. The receptor-signal complex exhibits a positive feedback expression wherein it induces the expression of *luxI*. The biosensor plasmid used in METREX is pJBA132, which has *luxR* and the promoter of *luxI* fused to *gfp*. The host, *E. coli*, does not produce QS signals, and, hence, GFP is expressed only when QS signals are supplied exogenously or genes that synthesize QS signals are introduced into the cells. Metagenomic library is constructed in the *E. coli* strain harbouring the biosensor plasmid. The clones which express GFP are the ones that produce quorum-sensing signals (Fig. 12.2b). Conversely, clones that do not express GFP in the presence of exogenous QS signal are the ones that harbour gene(s), whose product inhibits QS signals (Guan et al. 2007).

In METREX, clones are sorted for GFP expression using FACS which can sort ~60,000 cells/min, and, hence, the technique is relatively fast. Using this technique, Williamson et al. (2005) identified 11 clones which synthesize QS signal analogue and 2 clones which inhibit QS signal. Most of the positive clones had no sequence similarity to known QS-related genes, attesting the ability of this technique to identify novel genes. The technique is so effective that it can detect clones that are overlooked in standard screening techniques for QS signals, in which the active molecule needs to be diffused out of the producing cell and reach the reporter strain. Typical QS signal identified by standard screening technique is acyl homoserine lactone. But METREX has revealed other chemically distinct classes of signal molecules, which can function similarly. The principle of METREX can be extrapolated for the discovery of other metabolites using specific promoters that gets activated by specific substrate.

12.3.4.5 Enrichment Strategies

One of the major problems in screening metagenomic libraries for metabolites is lower frequency of clones of a desired trait. It is quite likely that many rare microbes which produce novel metabolites are poorly represented in complex communities. The size of cloned DNA and, correspondingly, the number of clones that has to be screened to mine the gene of interest are determined by the frequency of organism containing desired gene in the sample. In order to increase this frequency, enrichment steps that specifically enrich the microorganisms harbouring the desired traits are used prior to library construction. This approach is also advantageous in generating metagenomic libraries, for the reason that it overcomes cloning difficulties caused by means of contaminants in the sample. Despite these advantages, enrichment technique is biased, as it causes loss of microbial diversity and predominantly selects fast-growing members of microbial community (Daniel 2005).

In general, enrichment strategies make use of specific carbon and nitrogen molecules as their sole energy source, which are selective for organisms containing the desired genes (Lefevre et al. 2008; Simon and Daniel 2011). For example, in one study, microorganisms that produce amidases, an enzyme which converts D-phenlglycine amide derivates into key intermediates for the production of semisynthetic β-lactam antibiotics, were selectively enriched by adding the soil sample to minimal medium containing a mixture of D-phenlglycine amide derivates as the sole nitrogen source. Metagenomic library was constructed in a leucine-auxotrophic *E. coli* strain using the DNA extracted from the enriched sample. Selection was done on medium containing phenylglycine-L-leucine, which would allow only the growth of clones capable of hydrolysing amide compounds. The study identified four amidase-positive clones, and all had low homology to known enzymes. One clone, designated pS2, was found to accumulate twofold-higher level of penicillin G compared to *E. coli* penicillin amidase and performed well in amoxicillin and ampicillin production. Subsequent diversity analysis using denaturing gradient gel electrophoresis (DGGE) showed 77% lower bacterial diversity in enriched sample than the original sample (Gabor et al. 2004).

One of the potential methods which not only enriches desired genomes but also offers straightforward strategy for selecting enriched DNA is stable-isotope probing (SIP). The technique uses ^{13}C-labelled substrate for enrichment. Microorganisms which use ^{13}C-labelled substrate will incorporate ^{13}C into their DNA. ^{13}C-labelled DNA is denser than normal DNA and hence can be readily separated using density-gradient centrifugation. The separated DNA is then used to construct metagenomic libraries and screened (Radajewski and Murrell 2002).

Another enrichment approach is bromodeoxyuridine (BrdU) enrichment, which is based on the principle that metabolically active cells will incorporate labelled nucleotide (BrdU) into their DNA. BrdU-incorporated DNA can then be selectively recovered by immunocapture. Addition of specific substrates along with BrdU enriches the microorganisms that can grow in the substrate and also enables to discriminate it from other members by immunocapturing. The captured DNA is subsequently used to construct metagenomic libraries and screened for the gene of interest (Urbach et al. 1999; Yin et al. 2000).

Both methods have limitations; the first issue is timing. Prolonged substrate feeding causes substrate recycling, thereby breaking the basis of enrichment. Additionally, immunocapture and density-gradient centrifugation may shear DNA making it difficult to screen complex pathways encoded on large segments of DNA. Comparatively, BrdU enrichment is cheaper and less laborious than SIP. Because, BrdU enrichment can be extrapolated to enrich the same sample for other target genes simply by providing BrdU in combination with different substrates, whereas SIP needs labelling of each and every substrate employed. However, there is a problem of cross-feeding in BrdU enrichment, since any of the metabolically active organism will take up BrdU, irrespective of substrate utilization (Schloss and Handelsman 2003).

Another proposed technique enriches consortia of clones which together accomplish a desired function. The principle underlying this technique is analogous to the isolation of bacterial species that collaborate to degrade (or synthesize) particular substrate. Each bacterium confers a particular gene product which is part of the mosaic degradation (or synthetic) pathway (Abraham et al. 2002). This suggests that it is not possible to capture all the genes of certain pathways on a contiguous piece of DNA and emphasizes the need to analyse multiple clones simultaneously.

Such mosaic pathway screening can also be applied in metagenomics, wherein the product of one clone acts as a substrate for another. Enrichment for this technique can be done by growing consortia of clones in liquid medium supplemented with the precursor substrate, which allows easy diffusion of the intermediate products, thereby enabling a cascade of reaction (Schloss and Handelsman 2003).

12.3.5 Sequence-Driven Approach

In sequence-based screening, identification of genes is based on homology search against known sequences. The approach aims for conserved DNA sequences of target genes. It is an efficient method to isolate metabolite-synthesizing genes with varying levels of similarity to known genes. Discovery of metabolite-encoding gene(s) by sequence-based screening is carried out either by analysing the metagenomic libraries or by directly exploring the metagenome. Metagenomic libraries for sequence-based screening are constructed similar to function-based screening, except that libraries can be constructed exclusively in *E. coli* without discrepancy. This is because the clones are screened by sequence rather than by functional expression. However, libraries tailor-made for functional-based screening can also be used for sequence-based screening (Banik and Brady 2010; Daniel 2005).

Library screening using sequence-based analysis typically employs hybridization probes or PCR primers designed to identify conserved DNA sequences in the target gene. These strategies can be applied to screen any kind of gene products whose sequence information is known. However, they have two major drawbacks. Foremost, the primers and probes are designed purely based on known sequences, which are in most cases purified from easily cultured laboratory isolate. Therefore, the designed primers and probes are biased towards genes that are similar to known genes and eliminate identification of novel genes. The second drawback is that the approach does not guarantee acquisition of full-length genes or gene clusters which are necessary for the production of desired product. For instance, PCR amplification using primers derived from conserved sequences often retrieves partial fragment of target genes, and additional steps are needed to recover whole gene sequence (Simon and Daniel 2009). Despite these limitations, sequence-based approach has identified several novel functional enzymes, pigments, antibiotics and other bioactive molecules. For instance, Banik and Brady (2008) isolated two novel glycopeptide-encoding gene clusters from a cosmid library comprising ten^6 clones derived from desert soil using PCR-based screening. Isolation of novel glycopeptide-synthesizing clusters is critical for the development of novel glycopeptide analogues which can substitute antibiotics. In another study, gene clusters responsible for production of theopederins, a group of polyketides with excellent antitumour activity, were isolated using PCR-based screening from a cosmid library derived from the complex metagenome of marine sponge *Teonella swinhoei* (Piel et al. 2004).

Meyer et al. (2007) introduced a sequence-based screening technique called subtractive hybridization magnetic bead capture. In this method, conserved regions of target genes are amplified using biotinylated degenerate primers, and the amplified fragments are immobilized on streptavidin-coated magnetic beads. The immobilized fragments are then used as probes to capture full-length genes from metagenomic DNA by hybridization. This approach is more specific than PCR-based approach and allows recovery of multiple gene targets in a single reaction.

Metagenomic libraries can also be screened using shotgun sequencing which provides large amount of data including phylogenetic relationship, millions of novel genes and deduced metabolic pathways. However, shotgun sequencing of metagenomic library is labour intensive particularly when genes of a desired trait are aimed. Furthermore, since the sequenced data is analysed based on sequence similarity, this method is not free from limitations. In few studies, microarray technology has been employed for screening metagenomic libraries (Pathak et al. 2009).

On the other hand, metagenome can be sequenced directly using shotgun approach and subsequently screened for active metabolites. Cloning independent screening techniques bypasses the most tedious step of generating metagenomic libraries. Moreover, it is less biased as it avoids arbitrary loss of genome sequences, which is an inherent problem in cloning and library construction. One of the pioneering large-scale metagenomic studies using shotgun sequencing is Sargasso seawater metagenome sequencing carried out by Venter et al. (2004). The study identified 1.2 million unknown genes including the discovery of rhodopsin-like photoreceptors in bacterial species. In addition, shotgun sequencing revealed the presence of more than 20 metabolite-synthesizing gene clusters in the industrially important microorganism, *Streptomyces avermitilis*. Before sequencing, it was known to produce only four metabolites (Omura et al. 2001). These studies indicate that whole-metagenome sequencing has the potential to identify homologues of gene encoding metabolites from uncultured microbial community, which can reveal novel metabolites or provide a means to improve the functionality of known metabolites.

Cloning independent strategy exploits the advantages of sequencing methods (Barone et al. 2014). The metagenome extracted from the sample is either sequenced as such or subjected to PCR amplification (targeting a particular gene using primers derived from conserved sequences) and then sequenced. In direct sequencing technique, metagenomic DNA is broken down randomly into numerous small fragments. The broken fragments are then size selected and sequenced using traditional Sanger-based sequencing or by advanced next-generation sequencing (NGS) techniques. The sequence reads obtained are clustered, and overlapping ends of different reads are assembled into a continuous sequence with the help of various bioinformatic tools and analysed.

Earlier metagenomic studies used Sanger-based capillary electrophoresis sequencing, which is tedious and costly. Development of high-throughput NGS technologies has made random shotgun sequencing of large sets of metagenome, simple and economically feasible. They have revolutionized the way metagenomic data are obtained (Tyler-Smith et al. 2015). NGS is a catch-all term used to describe a number of modern sequencing technologies like 454 pyrosequencing, Illumina sequencing, Ion torrent sequencing, SOLiD sequencing and single-molecule real-time (SMRT) sequencing (Morozova and Marra 2008). The principle of NGS

Table 12.3 Comparison of sequencing technologies (Mardis 2013; Metzker 2010)

Technique	Read length (bases)	Run time	Reads per run	Accuracy	Pros	Cons
Sanger sequencing	400–900	3 h	1	99.9%	Long individual reads	Not suitable for large-scale sequencing
454 sequencing	700	24 h	1 million	99%	Long read length, fast	Expensive, homopolymer errors
Illumina sequencing	50–300	11 days	6 billion	99%	High yield	Short reads
SOLiD sequencing	85	7–14 days	1.4 billion	99%	Low cost	Short reads, slow, palindrome sequence errors
Ion torrent sequencing	400	2 h	80 million	98%	Low cost, fast	Homopolymer errors
SMRT sequencing	3000 (up to 15,000)	20 min to 4 h	0.05 million	87%	No PCR, longest read length, fast	Low throughput, high error rate

technologies is similar to capillary electrophoresis sequencing. It involves DNA polymerase to catalyse the incorporation of fluorescently labelled nucleotides into a DNA strand during sequential cycles of DNA synthesis in which the incorporated nucleotides are identified at each cycle by fluorescent excitation. The difference is that unlike capillary electrophoresis sequencing, NGS processes million of fragments in a massively parallel fashion. Hence, NGS has the ability to produce huge data (up to ~6 billion reads in some platform) cheaply in a single run. Each NGS platform uses a unique combination of template preparation, sequencing chemistry and imaging and genome alignment/assembly methods. The technologies also differ in the quantum of data produced in terms of read length, number of reads per run and error frequency. The cost and time taken for a single run also differ among NGS platforms (Metzker 2010). A comparison of different NGS technology is provided in Table 12.3. Owing to the increasing demand for high-quality sequence information, NGS technologies are constantly improved, and advanced versions are released each year. Genomics and sequencing technologies have improved tremendously making it possible to sequence a single cell isolated from complex microbial community (Barone et al. 2014; Kvist et al. 2007). Furthermore, new platforms with improved sequencing chemistry, read length, read quality and data output are under development.

Among the NGS sequencing techniques, 454 pyrosequencing and Illumina sequencing are extensively used for metagenomic analysis and hence are briefly explained here. In pyrosequencing, metagenomic DNA is randomly fragmented and attached to microscopic beads. The attached fragments are then clonally amplified by emulsion PCR and deposited into the wells of a picotitre plate for parallel sequencing. Sequencing process involves sequential addition of all four dNTPs, which, if complementary to the template strand, get incorporated by DNA polymerase. The polymerization reaction releases pyrophosphate, which is converted via two enzymatic reactions to produce light. The light production is captured and converted into sequences.

Illumina sequencing workflow comprises three steps:

1. *Library preparation*: The sequencing library is prepared by random fragmentation of metagenomic DNA following ligation with adapters at the 5′ and 3′ ends. The adapter-ligated fragments are then PCR amplified (using primers complementary to the adapter) and gel purified.
2. *Cluster generation*: This is carried out on a planar, optically transparent surface similar to a microscope slide called flow cell. The flow cell has a lawn of surface-bound oligos, which are complementary to the adapter sequence. The prepared library is loaded into the flow cell wherein the fragments are captured by the lawn of oligos. The fragments are then amplified into distinct clonal clusters by bridge amplification.
3. *Sequencing*: Illumina uses proprietary four-coloured, reversible terminator nucleotides for sequencing. The use of four different fluorescent dyes enables application of all four nucleotides simultaneously rather than sequentially in order to interrogate a given nucleotide in the corresponding position in the template. In addition, natural competition which exists among the nucleotides minimizes incorporation bias and greatly reduces error rate. The 3′ end of all the four nucleotides is blocked by certain chemical modification to facilitate chain termination. Contrary to the dideoxynucleotides of Sanger sequencing, the termination can be reversed by removing the blocking groups and hence the term 'reversible terminator'. After each base extension, the fluorescent signal is recorded at each cluster. Subsequently, the reagents are washed away, fluorescent labels are cleaved, and the 3′ end of the incorporated base is unblocked for addition of next nucleotide. The cycle is repeated and the entire sequence of each cluster is sequenced. Illumina sequencing platform is able to provide hundreds of gigabases of data within hours of single sequencing run (Hodkinson and Grice 2015; Holt and Jones 2008).

12.3.5.1 NGS Data Analysis

The sheer volume of raw sequence reads generated by NGS makes data analyses a daunting task. Metagenomic NGS data analysis is even more challenging as the raw data is inherently noisy and contains reads of as many as 10,000 species. However, bioinformatics has also developed in parallel, and varieties of tools are available which can handle huge data and extract useful information. Data analysis involves a series of steps as represented in Fig. 12.3. The first step is pre-filtering, which removes low-quality and redundant sequences using bioinformatic tools like Eu-Detect, DeconSeq and CD-HIT-DUP (Schmeider and Edwards 2011). Once the data is filtered, the reads are assembled to obtain large genome contigs. Assembly can be performed by two means: reference-based assembly (co-assembly) and

Fig. 12.3 Flow diagram showing the steps involved in NGS data analysis for novel bioactive metabolite screening

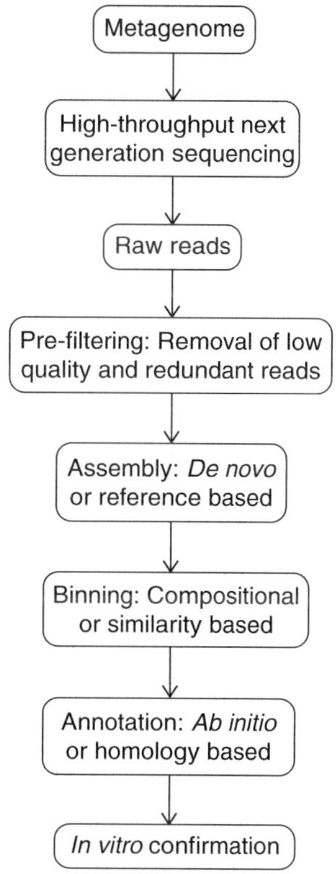

de novo assembly. Reference-based assembly works well, if the metagenomic data set contains sequences that are closely related to available reference genomes. *De novo* assembly requires large computational resources and is challenging for NGS platforms which generate short reads. Examples of assembly programmes include Phrap or Celera assembler, Velvet assembler, Newbler, MIRA and Meta-IDBA.

Assembly is followed by binning, which sorts DNA sequences into groups that represent a single genome or genome from closely related organisms. Two types of binning algorithms are available: compositional based and similarity based. The former looks for conserved nucleotide composition like GC content and distribution of k-mers, whereas the latter bins sequences based on the similarity of sequences with reference genome. Examples of compositional-based binning algorithms are PhyloPythia, S-SOM, PCAHIER and TACAO, while similarity-based binning softwares include IMG/M, MG-RAST, MEGAN and MetaPhyler. Binning softwares like PhymmBL and MetaCluster consider both composition and similarity (Thomas et al. 2012).

The next step is annotation wherein potential open reading frames are identified. Annotation methods are broadly classified into two types, namely, homology-based annotation and ab initio annotation. Homology-based annotation softwares like MEGAN4 use simple BLAST-like search to identify genes, whereas ab initio-based programmes like GeneMark and GLIMMER use intrinsic features of the sequence to predict coding regions. Certain annotation programmes like NPI SEARCH, anti-SMASH, ClusterMine360 and NaPDos are specifically designed to identify metabolite-synthesizing genes and pathways. Finally, the results obtained can be confirmed by wet laboratory experiments (Huson et al. 2011; Zhu et al. 2010).

12.4 Conclusions

Metagenomic strategy offers a powerful tool to discover novel bioactive metabolites from uncultivable microbial community. Plethora of novel biomolecules discovered using culture-independent approaches attests to this fact. The approach has not only facilitated functional characterization of unculturable bacteria but also provided valuable insights into previously unknown biochemistry of culturable organisms. To date, metagenomic studies have been used to explore only a few niches, indicating that the hitherto unexplored habitats are likely to be a rich source of novel metabolites. Though the whole metagenomic process has progressed tremendously from the time when the concept was introduced, improvements are still needed in order to maximize the potential of metagenomics for novel bioactive discovery. Development of standard protocols that can extract high-quality metagenome covering the entire biodiversity is needed. Challenges in heterologous gene expression should be addressed by refining host-vector system. In addition, screening strategies need to be revised, since each of the currently developed strategy has its own limitations. Intracellular functional screens like SIGEX, PIGEX and METREX are good alternatives, but they are limited and target only particular trait. Additional high-throughput, sensitive screening methods which can cover a broad range of bioactives should be developed. Advances in NGS technologies have made whole-metagenome sequencing looks like child's play. However, since most of the sequence-based data mining methods rely on homology-based screening, new bioinformatic genome mining strategies that employ pattern recognition-based algorithms designed to identify conserved yet unexplored metabolite-synthesizing genes and gene clusters are needed. With the growing rate of sequence information, it is expected that the number of novel genes identified through metagenomics will outcome the current ability of techniques which assign function to them. Collaborative efforts involving diverse fields like bacterial genetics, molecular biology, genomics, robotics, bioinformatics and synthetic biology are needed to address this potential problem.

References

Abraham WR, Nogales B, Golyshin PN, Pieper DH, Timmis KN (2002) Polychlorinated biphenyl-degrading microbial communities in soils and sediments. Curr Opin Microbiol 5:246–253

Albers SV, Jonuscheit M, Dinkelaker S, Urich T, Kletzin A, Tampe R et al (2006) Production of recombinant and tagged proteins in the hyperthermophilic archaeon *Sulfolobus solfataricus*. Appl Environ Microbiol 72:102–111

Amann RI, Ludwig W, Schleifer KH (1995) Phylogenetic identification and *in situ* detection of individual microbial cells without cultivation. Microbiol Rev 59:143–169
Angelov A, Mientus M, Liebl S, Liebl W (2009) A two-host fosmid system for functional screening of (meta) genomic libraries from extreme thermophiles. Syst Appl Microbiol 32:177–185
Aravindraja C, Viszwapriya D, Pandian SK (2013) Ultra deep 16S rRNA sequencing analysis of geographically similar but diverse unexplored marine samples reveal varied bacterial community composition. PLoS One 8(10):e76724
Bakkiyaraj D, Pandian SK (2010) *In vitro* and *in vivo* antibiofilm activity of a coral associated actinomycete against drug resistant *Staphylococcus aureus* biofilms. Biofouling 26:711–717
Banik JJ, Brady SF (2008) Cloning and characterization of new glycopeptide gene clusters found in an environmental DNA mega library. Proc Natl Acad Sci 105:17273–17277
Banik JJ, Brady SF (2010) Recent application of metagenomic approaches toward the discovery of antimicrobials and other bioactive small molecules. Curr Opin Microbiol 13:603–609
Barns SM, Fundyga RE, Jeffries MW, Pace NR (1994) Remarkable archaeal diversity detected in a Yellowstone National Park hot spring environment. Proc Natl Acad Sci 91:1609–1613
Barone R, De Santi C, Esposito FP, Tedesco P, Galati F, Visone M et al (2014) Marine metagenomics, a valuable tool for enzymes and bioactive compounds discovery. Front Mar Sci 1:1–6
Beja O, Aravind L, Koonin EV, Suzuki MT, Hadd A, Nguyen LP et al (2000) Bacterial rhodopsin: evidence for a new type of phototrophy in the sea. Science 289:1902–1906
Chang FY, Brady SF (2013) Discovery of indolotryptoline antiproliferative agents by homology-guided metagenomic screening. Proc Natl Acad Sci 110:2478–2483
Chen K, Pachter L (2005) Bioinformatics for whole-genome shotgun sequencing of microbial communities. PLoS Comput Biol 1:e24
Craig JW, Chang FY, Kim JH, Obiajulu SC, Brady SF (2010) Expanding small-molecule functional metagenomics through parallel screening of broad-host-range cosmid environmental DNA libraries in diverse proteobacteria. Appl Environ Microbiol 76:1633–1641
Culligan EP, Marchesi JR, Hill C, Sleator RD (2014) Combined metagenomic and phenomic approaches identify a novel salt tolerance gene from the human gut microbiome. Front Microbiol 5:189–197
Daniel R (2005) The metagenomics of soil. Nat Rev Microbiol 3:470–478
Delmont TO, Robe P, Cecillon S, Clark IM, Constancias F, Simonet P et al (2011) Accessing the soil metagenome for studies of microbial diversity. Appl Environ Microbiol 77:1315–1324
Donato JJ, Moe LA, Converse BJ, Smart KD, Berklein FC, Mc Manus PS et al (2010) Metagenomic analysis of apple orchard soil reveals antibiotic resistance genes encoding predicted bifunctional proteins. Appl Environ Microbiol 76:4396–4401
Donia MS, Ruffner DE, Cao S, Schmidt EW (2011) Accessing the hidden majority of marine natural products through metagenomics. ChemBioChem 12:1230–1236
Eden PA, Schmidt TM, Blakemore RP, Pace NR (1991) Phylogenetic analysis of *Aquaspirillum magnetotacticum* using polymerase chain reaction-amplified 16S rRNA-specific DNA. Int J Syst Bacteriol 41:324–325
Ekkers DM, Cretoiu MS, Kielak AM, van Elsas JD (2012) The great screen anomaly – a new frontier in product discovery through functional metagenomics. Appl Microbiol Biotechnol 93:1005–1020
Ermolaeva MD, Khalak HG, White O, Smith HO, Salzberg SL (2000) Prediction of transcription terminators in bacterial genomes. J Mol Biol 301:27–33
Felczykowska A, Bloch SK, Nejman-Falenczyk B, Baranska S (2012) Metagenomic approach in the investigation of new bioactive compounds in the marine environment. Acta Biochim Pol 59:501–505
Feng Z, Chakraborty D, Dewell SB, Reddy BVB, Brady SF (2012) Environmental DNA-encoded antibiotics fasamycins A and B inhibit FabF in type II fatty acid biosynthesis. J Am Chem Soc 134:2981–2987
Fenical W, Jensen PR (2006) Developing a new resource for drug discovery: marine actinomycete bacteria. Nat Chem Biol 2:666–673

Ferrer M, Beloqui A, Timmis KM, Golyshin PN (2009) Metagenomics for mining new genetic resources of microbial communities. J Mol Microbiol Biotechnol 16:109–123

Gabor EM, de Vries EJ, Janssen DB (2004) Construction, characterization and use of small-insert gene banks of DNA isolated from soil and enrichment cultures for recovery of novel amidases. Environ Microbiol 6:948–958

Gillespie DE, Brady SF, Bettermann AD, Cianciotto NP, Liles MR, Rondon MR et al (2002) Isolation of antibiotics turbomycin A and B from a metagenomic library of soil microbial DNA. Appl Environ Microbiol 68:4310–4306

Gloux K, Berteau O, Oumami HE, Beguet F, Leclerc M, Dore JA (2011) Metagenomic β-glucuronidase uncovers a core adaptive function of the human intestinal microbiome. Proc Natl Acad Sci 108:4539–4546

Gowrishankar S, Poornima B, Pandian SK (2014) Inhibitory efficacy of cyclo (l-leucyl-l-prolyl) from mangrove rhizosphere bacterium–*Bacillus amyloliquefaciens* (MMS-50) toward carcinogenic properties of *Streptococcus mutans*. Res Microbiol 165:278–289

Grimes DJ, Atwell RW, Brayton PR, Palmer LM, Rollins DM, Roszak DB et al (1986) The fate of enteric pathogenic bacteria in estuarine and marine environments. Microbiol Sci 3:324–329

Grindberg RV, Ishoey T, Brinza D, Esquenazi E, Coates RC, Liu WT et al (2011) Single cell genome amplification accelerates identification of the apratoxin biosynthetic pathway from a complex microbial assemblage. PLoS One 6:e18565

Guan C, Ju J, Borlee BR, Williamson LL, Shen B, Raffa KF et al (2007) Signal mimics derived from a metagenomic analysis of the gypsy moth gut microbiota. Appl Environ Microbiol 73:3669–3676

Handelsman J (2004) Metagenomics: application of genomics to uncultured microorganisms. Microbiol Mol Biol Rev 68:669–685

Handelsman J, Rondon MR, Brady SF, Clardy J, Goodman RM (1998) Molecular biological access to the chemistry of unknown soil microbes: a new frontier for natural products. Chem Biol 5:R245–R249

He R, Wakimoto T, Takeshige Y, Egami Y, Kenmoku H, Ito T et al (2012) Porphyrins from a metagenomic library of the marine sponge *Discodermia calyx*. Mol Biosyst 8:2334–2338

Herbert RA (1992) A perspective on the biotechnological potential of extremophiles. Trends Biotechnol 10:395–402

Hildebrand M, Waggoner LE, Liu H, Sudek S, Allen S, Anderson C et al (2004) *bryA*: an unusual modular polyketides synthase gene from the uncultivated bacterial symbiont of the marine bryozoan *Bugula neritina*. Chem Biol 11:1543–1552

Hodkinson BP, Grice EA (2015) Next-generation sequencing: a review of technologies and tools for wound microbiome research. Adv Wound Care 4:50–58

Holben WE, Jansson JK, Chelm BK, Tiedje JM (1988) DNA probe method for the detection of specific microorganisms in the soil bacterial community. Appl Environ Microbiol 54:703–711

Holt RA, Jones SJ (2008) The new paradigm of flow cell sequencing. Genome Res 18:839–846

Huson DH, Mitra S, Weber N, Ruscheweyh N, Schuster SC (2011) Integrative analysis of environmental sequences using MEGAN4. Genome Res 21:1552–1560

Jackson CR, Harper JP, Willoughby D, Roden EE, Churchill PF (1997) A simple, efficient method for the separation of humic substances and DNA from environmental samples. Appl Environ Microbiol 63:4993–4995

Jacobsen CS, Rasmussen OF (1992) Development and application of a new method to extract bacterial DNA from soil based on separation of bacteria from soil with cation exchange resin. Appl Environ Microbiol 58:2458–2462

Kennedy J, Marchesi JR, Dobson AD (2007) Metagenomic approach to exploit the biotechnological potential of the microbial consortia of marine sponges. Appl Microbiol Biotechnol 75:11–20

Kvist T, Ahring BK, Lasken RS, Westermann P (2007) Specific single cell isolation and genomic amplification of uncultured microorganisms. Appl Microbiol Biotechnol 74:926–935

Lefevre F, Robe P, Jarrin C, Ginolhac A, Zago C, Auriol D et al (2008) Drugs from hidden bugs: their discovery via untapped resources. Res Microbiol 159:153–161

Leff LG, Dana JR, McArthur JV, Shimkets LJ (1995) Comparison of methods of DNA extraction from stream sediments. Appl Environ Microbiol 61:1141–1143

Leveau JH, Gerards S (2008) Discovery of a bacterial gene cluster for catabolism of the plant hormone indole 3-acetic acid. FEMS Microbiol Ecol 65:238–250

Li X, Qin L (2005) Metagenomics based drug discovery and marine microbial diversity. Trends Biotechnol 23:539–543

Lim HK, Chung EJ, Kim JC, Choi GJ, Jang KS, Chung YR et al (2005) Characterization of a forest soil metagenome clone that confers indirubin and indigo production on *Escherichia coli*. Appl Environ Microbiol 71:7768–7777

Lindahl V, Bakken LR (1995) Evaluation of methods for extraction of bacteria from soil. FEMS Microbiol Ecol 16:135–142

Long PF, Dunlap WC, Battershill CN, Jaspars M (2005) Shotgun cloning and heterologous expression of the patellamide gene cluster as a strategy to achieving sustained metabolite production. ChemBioChem 6:1760–1765

Lorenz P, Eck J (2005) Metagenomics and industrial applications. Nat Rev Microbiol 3:510–516

Majernik A, Gottschalk G, Daniel R (2001) Screening of environmental DNA libraries for the presence of genes conferring Na+(Li+)/H+ antiporter activity on *Escherichia coli*: characterization of the recovered genes and the corresponding gene products. J Bacteriol 183:6645–6653

Mardis ER (2013) Next-generation sequencing platforms. Annu Rev Anal Chem 6:287–303

Martin M, Biver S, Steels S, Barbeyron T, Jam M, Portetelle D et al (2014) Functional screening of a metagenomic library of seaweed-associated microbiota: identification and characterization of a halotolerant, cold-active marine endo-ß-1, 4-endoglucanase. Appl Environ Microbiol 80:4958–4967

Martinez A, Kolvek SJ, Yip CLT, Hopke J, Brown KA, Mac Neil IA et al (2004) Genetically modified bacterial strains and novel bacterial artificial chromosome shuttle vectors for constructing environmental libraries and detecting heterologous natural products in multiple expression hosts. Appl Environ Microbiol 70:2452–2463

Metzker ML (2010) Sequencing technologies – the next generation. Nat Rev Genet 11:31–46

Meyer QC, Burton SG, Cowan DA (2007) Subtractive hybridization magnetic bead capture: a new technique for the recovery of full length ORFs from the metagenome. J Biotechnol 2:36–40

Mohn WW, Garmendia J, Galvao TC, De Lorenzo V (2006) Surveying biotransformations with a la carte genetic traps: translating dehydrochlorination of lindane (gammahexachlorocyclohexane) into lacZ-based phenotypes. Environ Microbiol 8:546–555

More MI, Herrick JB, Silva MC, Ghiorse WC, Madsen EL (1994) Quantitative cell lysis of indigenous microorganisms and rapid extraction of microbial DNA from sediment. Appl Environ Microbiol 60:1572–1580

Morgan JL, Darling AE, Eisen JA (2009) Metagenomic sequencing of an *in vitro*-simulated microbial community. PLoS One 5:e10209

Morozova O, Marra MA (2008) Applications of next-generation sequencing technologies in functional genomics. Genomics 92:255–264

Niemi RM, Heiskanen I, Wallenius K, Lindstrom K (2001) Extraction and purification of DNA in rhizosphere soil samples for PCR-DGGE analysis of bacterial consortia. J Microbiol Methods 45:155–165

Nithya C, Aravindraja C, Pandian SK (2010) *Bacillus pumilus* of Palk Bay origin inhibits quorum-sensing-mediated virulence factors in Gram-negative bacteria. Res Microbiol 161:293–304

Omura S, Ikeda H, Ishikawa J, Hanamoto A, Takahashi C, Shinose M et al (2001) Genome sequence of an industrial microorganism *Streptomyces avermitilis*: deducing the ability of producing secondary metabolites. Proc Natl Acad Sci 98:12215–12220

Owen JG, Robins KJ, Parachin NS, Ackerley DF (2012) A functional screen for recovery of 4′-phosphopantetheinyl transferase and associated natural product biosynthesis genes from metagenome libraries. Environ Microbiol 14:1198–1209

Pace NR, Stahl DA, Lane DJ, Olsen GJ (1986) The analysis of natural microbial populations by ribosomal RNA sequences. In: Marshall KC (ed) Advances in microbial ecology. Springer, Boston, pp 1–55

Padmavathi AR, Abinaya B, Pandian SK (2014) Phenol, 2, 4-bis (1, 1-dimethylethyl) of marine bacterial origin inhibits quorum sensing mediated biofilm formation in the uropathogen *Serratia marcescens*. Biofouling 30:1111–1122

Pathak GP, Ehrenreich A, Losi A, Streit WR, Gartner W (2009) Novel blue light-sensitive proteins from a metagenomic approach. Environ Microbiol 11:2388–2399

Piel J (2002) A polyketides synthase peptide synthetase gene cluster from an uncultured bacterial symbiont of *Paederus* beetles. Proc Natl Acad Sci 99:14002–14007

Piel J, Hui D, Wen G, Butzke D, Platzer M, Fusetani N et al (2004) Antitumor polyketides biosynthesis by an uncultivated bacterial symbiont of the marine sponge *Theonella swinhoei*. Proc Natl Acad Sci 101:16222–16227

Purohit MK, Singh SP (2008) Assessment of various methods for extraction of metagenomic DNA from saline habitats of coastal Gujarat (India) to explore molecular diversity. Lett Appl Microbiol 49:338–344

Radajewski S, Murrell JC (2002) Stable isotope probing for detection of methanotrophs after enrichment with $^{13}CH_4$. Methods Mol Biol 179:149–157

Riaz K, Elmerich C, Moreira D, Raffoux A, Dessaux Y, Faure D (2008) A metagenomic analysis of soil bacteria extends the diversity of quorum-quenching lactonases. Environ Microbiol 10:560–570

Riesenfeld CS, Goodman RM, Handelsman J (2004) Uncultured soil bacteria are a reservoir of new antibiotic resistance genes. Environ Microbiol 6:981–989

Rochelle PA, Fry JC, Parkes RJ, Weightman AJ (1992) DNA extraction for 16S rRNA gene analysis to determine genetic diversity in deep sediment communities. FEMS Microbiol Lett 100:59–65

Rondon M, August P, Bettermann A, Brady S, Grossman T, Liles M et al (2000) Cloning the soil metagenome: a strategy for accessing the genetic and functional diversity of uncultured microorganisms. Appl Environ Microbiol 66:2541–2547

Sabree ZL, Rondon MR, Handelsman J (2009) Metagenomics. In: Schaechter M (ed) Encyclopedia of microbiology. Elsevier, Amsterdam, pp 622–632

Salomon CE, Magarvey NA, Sherman DH (2004) Merging the potential of microbial genetics with biological and chemical diversity: an even brighter future for marine natural product drug discovery. Nat Prod Rep 21:105–121

Schiraldi C, De Rosa M (2002) The production of biocatalysts and biomolecules from extremophiles. Trends Biotechnol 20:515–521

Schloss PD, Handelsman J (2003) Biotechnological prospects from metagenomics. Curr Opin Biotechnol 14:303–310

Schmeider R, Edwards E (2011) Fast identification and removal of sequence contamination from genomic and metagenomic datasets. PLoS One 6:e17288

Sharma P, Capalash N, Kaur J (2007) An improved method for single step purification of metagenomic DNA. Mol Biotechnol 36:61–63

Shen B, Du L, Sanchez C, Edwards DJ, Chen M, Murrell JM (2001) The biosynthetic gene cluster for the anticancer drug bleomycin from *Streptomyces verticillus* ATCC15003 as a model for hybrid peptide – polyketide natural product biosynthesis. J Ind Microbiol Biotechnol 27:378–385

Siddhapura PK, Vanparia S, Purohit MK, Singh SP (2010) Comparative studies on the extraction of metagenomic DNA from the saline habitats of Coastal Gujarat and Sambhar Lake, Rajasthan (India) in prospect of molecular diversity and search for novel biocatalysts. Int J Biol Macromol 47:375–379

Simon C, Daniel R (2009) Achievements and new knowledge unraveled by metagenomic approaches. Appl Microbiol Biotechnol 85:265–276

Simon C, Daniel R (2011) Metagenomic analyses: past and future trends. Appl Environ Microbiol 77:1153–1161
Simon C, Herath J, Rockstroh S, Daniel R (2009) Rapid identification of genes encoding DNA polymerases by function-based screening of metagenomic libraries derived from glacial ice. Appl Environ Microbiol 75:2964–2968
Simonet P, Capellano A, Navarro E, Bardin R, Moiroud A (1984) An improved method for lysis of *Frankia* with achromopeptidase allows detection of new plasmids. Can J Microbiol 30:1292–1295
Singh RP, Kumari P, Reddy CRK (2015) Antimicrobial compounds from seaweeds-associated bacteria and fungi. Appl Microbiol Biotechnol 99:1571–1586
Staley JT, Konopka A (1985) Measurement of *in situ* activities of non-photosynthetic microorganisms in aquatic and terrestrial habitats. Annu Rev Microbiol 39:321–346
Steele HL, Jaeger KE, Daniel R, Streit WR (2008) Advances in recovery of novel biocatalysts from metagenomes. J Mol Microbiol Biotechnol 16:25–37
Streit WR, Schmitz RA (2004) Metagenomics–the key to the uncultured microbes. Curr Opin Microbiol 7:492–498
Thenmozhi R, Nithyanand P, Rathna J, Pandian SK (2009) Antibiofilm activity of coral-associated bacteria against different clinical M serotypes of *Streptococcus pyogenes*. FEMS Immunol Med Microbiol 57:284–294
Thomas T, Gilbert J, Meyer F (2012) Metagenomics-a guide from sampling to data analysis. Microb Inform Exp 2:1–12
Torsvik V, Ovreas L (2002) Microbial diversity and function in soil: from genes to ecosystems. Curr Opin Microbiol 5:240–245
Torsvik V, Goksoyr J, Daae FL (1990) High diversity in DNA of soil bacteria. Appl Environ Microbiol 56:782–787
Turnbaugh PJ, Gordon JL (2008) An invitation to the marriage of metagenomics and metabolomics. Cell 134:708–713
Tyler-Smith C, Yang H, Landweber LF, Dunham I, Knoppers BM, Donnelly P et al (2015) Where next for genetics and genomics? PLoS Biol 13:e1002216
Uchiyama T, Miyazaki K (2009) Functional metagenomics for enzyme discovery: challenges to efficient screening. Curr Opin Biotechnol 20:616–622
Uchiyama T, Miyazaki K (2010) Product-induced gene expression, a product-responsive reporter assay used to screen metagenomic libraries for enzyme-encoding genes. Appl Environ Microbiol 76:7029–7035
Uchiyama T, Watanabe K (2007) The SIGEX scheme: high throughput screening of environmental metagenomes for the isolation of novel catabolic genes. Biotechnol Genet Eng Rev 24:107–116
Urbach E, Vergin KL, Giovannoni SJ (1999) Immunochemical detection and isolation of DNA from metabolically active bacteria. Appl Environ Microbiol 65:1207–1213
Venter JC, Remington K, Heidelberg JF, Halpern AL, Rusch D, Eisen JA et al (2004) Environmental genome shotgun sequencing of the Sargasso Sea. Science 304:66–74
Verma D, Satyanarayana T (2011) An improved protocol for DNA extraction from alkaline soil and sediment samples for constructing metagenomic libraries. Appl Biochem Biotechnol 165:454–464
Viszwapriya D, Aravindraja C, Pandian SK (2015) Comparative assessment of bacterial diversity associated with co-occurring eukaryotic hosts of Palk Bay origin. Indian J Exp Biol 53:417–423
Wang GYS, Graziani E, Waters B, Pan W, Li X, McDermott J et al (2000) Novel natural products from soil DNA libraries in a streptomycete host. Org Lett 2:2401–2404
Williamson LL, Borlee BR, Schloss PD, Guan CH, Allen HK, Handelsman J (2005) Intracellular screen to identify metagenomic clones that induce or inhibit a quorum-sensing biosensor. Appl Environ Microbiol 71:6335–6344
Wilson MC, Piel J (2013) Metagenomic approaches for exploiting uncultivated bacteria as a resource for novel biosynthetic enzymology. Chem Biol 20:636–647

Yin B, Crowley D, Sparovek G, De Melo WJ, Borneman J (2000) Bacterial functional redundancy along a soil reclamation gradient. Appl Environ Microbiol 66:4361–4365

Yun J, Ryu S (2005) Screening for novel enzymes from metagenome and SIGEX, as a way to improve it. Microb Cell Factories 4:8–12

Zhou J, Bruns MA, Tiedje JM (1996) DNA recovery from soils of diverse composition. Appl Environ Microbiol 62:316–322

Zhu W, Lomsadze A, Borodovsky M (2010) *Ab initio* gene identification in metagenomic sequences. Nucleic Acids Res 38:e132

Ethnic Knowledge in Classification and Conservation: A Case Study of Wild Yam in Western Ghats

13

V. Balakrishnan, Nadesapanicker Anil Kumar, and M.K. Ratheesh Narayanan

Abstract

Among the wild food, yams especially the species belonging to the genus *Dioscorea* (lesser yam) occupy a unique place in the food basket of indigenous tribal communities and serve as a 'lifesaving' plant group to marginal farmers and forest-dwelling communities, during periods of food scarcity. We describe here the knowledge and practices of Kattunaikka with reference to conservation using ethnotaxonomical knowledge and sustainable utilization of *Dioscorea*. This is the result of a study conducted in Wayanad district – a hot speck in the Southern Western Ghats at different intervals covering all the important seasons and exploring different habitat types. The present study clearly shows that the ethnic communities use simple and convenient techniques for classifying the species that have utility value to them. They even understand the population of *Dioscorea* in each locality and do the harvesting in a sustainable manner. The method of classification by using the utilitarian character is part of their livelihood; hence, the classification methods originate spontaneously. However, in case of scientific system for delimiting the taxa, evidences from folk system of classification have not been recognized. It is true that understanding the folk system of classification is very difficult as the researcher has to think from the stakeholder (tribal) viewpoint. Folk system of classification will be a strong area of knowledge in the future in respect of the conservation and sustainable utilization of wild relatives of edible plants. The study revealed much valuable information not only pertaining to the edibility of *Dioscorea* but also its genetic diversity and thereby opening up the basis for correct identity of many of the taxonomically

V. Balakrishnan (✉) • N. Anil Kumar
Community Agro Biodiversity Centre, M.S. Swaminathan Research Foundation, Puthoorvayal, Kalpetta, Wayanad, Kerala, India
e-mail: balupattali@gmail.com; anilmaruthur@gmail.com

M.K. Ratheesh Narayanan
Department of Botany, Payyanur College, Payyanur, Kerala, India

S. Abdulhameed et al. (eds.), *Bioresources and Bioprocess in Biotechnology*, DOI 10.1007/978-981-10-3573-9_13

confusing taxa of this genus. The research findings open up the scope of rewarding Kattunaikka for their efforts in conserving diversity of *Dioscorea* from the available national fund for biodiversity conservation under the Biodiversity Act, 2002 or PPVFR Act, 2001.

Keywords
Ethnic knowledge • Classification • Conservation • Genetic diversity • Wild yam • *Dioscorea* • Western Ghats

13.1 Introduction

Since CBD, the role and responsibilities of ethnic communities and traditional knowledge holders have attracted wider attention for developing region-specific strategies and action plans in the area of conservation and sustainable utilization of biodiversity (IWGIA 2007, GEF (Global Environment Facility) 2006). Indigenous communities, because of their profound understanding about nature and the environment they are living in, are successful in biodiversity conservation and management of ecosystem services throughout the world. In India, for example, a country with diverse ethnic communities shows how important is traditional culture and habits for living in harmony with a variety of flora and fauna and using many of them along with the services of common properties like sea, rivers, lakes, mountains or any such naturally bestowed resource in a sustainable and equitable manner.

Historically, the biodiversity available in wild were much used for edible purpose, as evidenced by occurrence of over 30,000 plant species as food plants among the ethnic communities worldwide (FAO 2004). From India, there are 1532 edible wild food species reported, mostly from the Western Ghats and Himalayan regions of the country (Arora and Pandey 1996). Western Ghats of India is one of the richest emporia of ethnobotanical wealth (Henry et al. 1996; Pushpangadan and Atal 1984; Nair 1999). Tribal communities like *Kattunaika*, *Paniya*, *Cholanaikka*, *Kurichya*, *Kuruma*, *Thenkuruma*, *Mudukar* of Attappady area, *Kani tribe*, *Adiya* of Wayanad, *Maratti Naikas* of Kinningar, Kasaragod and South Canara are the major tribal communities depending upon various wild species as a famine food in the Western Ghats of India.

Among the wild food, yams especially the species belonging to the genus *Dioscorea* (lesser yam) occupy a unique place in the food basket of tribal communities across the tropical world. *Dioscorea* serve as a 'lifesaving' plant group to them, particularly the forest-dwelling communities of this region, during periods of food scarcity (Arora and Pandey 1996; Roy et al. 1988). *Dioscorea* is preferred even instead of the common staple like rice by ethnic community like *Kattunaikka* of Western Ghats. Various studies on ethnobotanical information on wild *Dioscorea* in India revealed its usage both as food and medicine (Abhyankar and Upadhyay 2011). Though this delicacy is not available for all seasons in wild, the ethnic communities adopt some innovative practices that help to preserve tubers for extended time after harvesting it from the forest. Planting the much preferred tubers in their backyards and mastery of the methods that convert inedible tubers into edible ones are some notable examples.

Knowledge and practices of Kattunaikka with reference to identification, conservation and sustainable utilization of *Dioscorea* are described here. This is the results of a study conducted at different intervals covering all the important seasons and exploring different habitat types in Wayanad district – a hot speck in the Southern Western Ghats. This long duration of study that specifically focused on the genetic diversity and usefulness of different taxa of *Dioscorea* helped to record the traditional knowledge of the Kattunaikka community from a variety of contexts like ethnotaxonomy, ethnobotany, community food gathering and farming activities as well as community rituals.

Wayanad District: A Hot Speck in a Global Biodiversity Hotspot Wayanad, a steep mountainous plateau in the Western Ghats – a global biodiversity hot spot with vast forest cover, extensive rice fields and high concentration of tribal communities – is an agrobiodiversity and a poverty hotspot (for the tribal communities like Paniya, Adiya and Kattunaikka) (Fig. 13.1). The district is an area, rich in biodiversity as evidenced by over 2100 flowering plants with 52 Red Data species and 650 endemic plant species of Western Ghats (Ratheesh Narayanan 2009). It is one of the 150 economically most backward districts of India with a total of 173,529 BPL (Below Poverty Line) cardholders. The tribal communities constitute 18.53 % of the total population of the district, which is the highest share of Adivasi population (about 31.2 %) of Kerala state of India. The major communities that inhabit the district are Paniya (45.63 %), Kurichya (16.68 %), Kattunaikka (11.26 %), Kuruma (13.86 %) and Adiya (7.39 %) (Census Commissioner 2011). Kurichya and Mullu Kuruma are traditionally agricultural communities, mostly involved in paddy cultivation. Paniya, Adiya, Kattunaikka, Kuruma and Kurichya are the low-income groups.

13.1.1 Study and Survey

The study carried out in five sites in the district and documented traditional knowledge, methods of genetic identification and way of description of *Dioscorea* by Kattunaikka men and women. The study was also focused on the usefulness, processing methods, management options and distribution and status in a gender specific manner. Fifty men and women from 5 hamlets (10 each) of Kattunaikka were partnered in this study.

The method used was confined to those techniques needed for the basic documentation of ethnobotanical knowledge in general and of ethnotaxonomical in particular. The local names and other key terms used to identity various species of this tuber group were accurately transcribed and cross-checked with several other people of the same community. Due care was also taken in documenting diverse aspects of natural habitats of yams like associated plant communities, soil types, geographical landmarks, climate zones and seasons. Management practices of this community

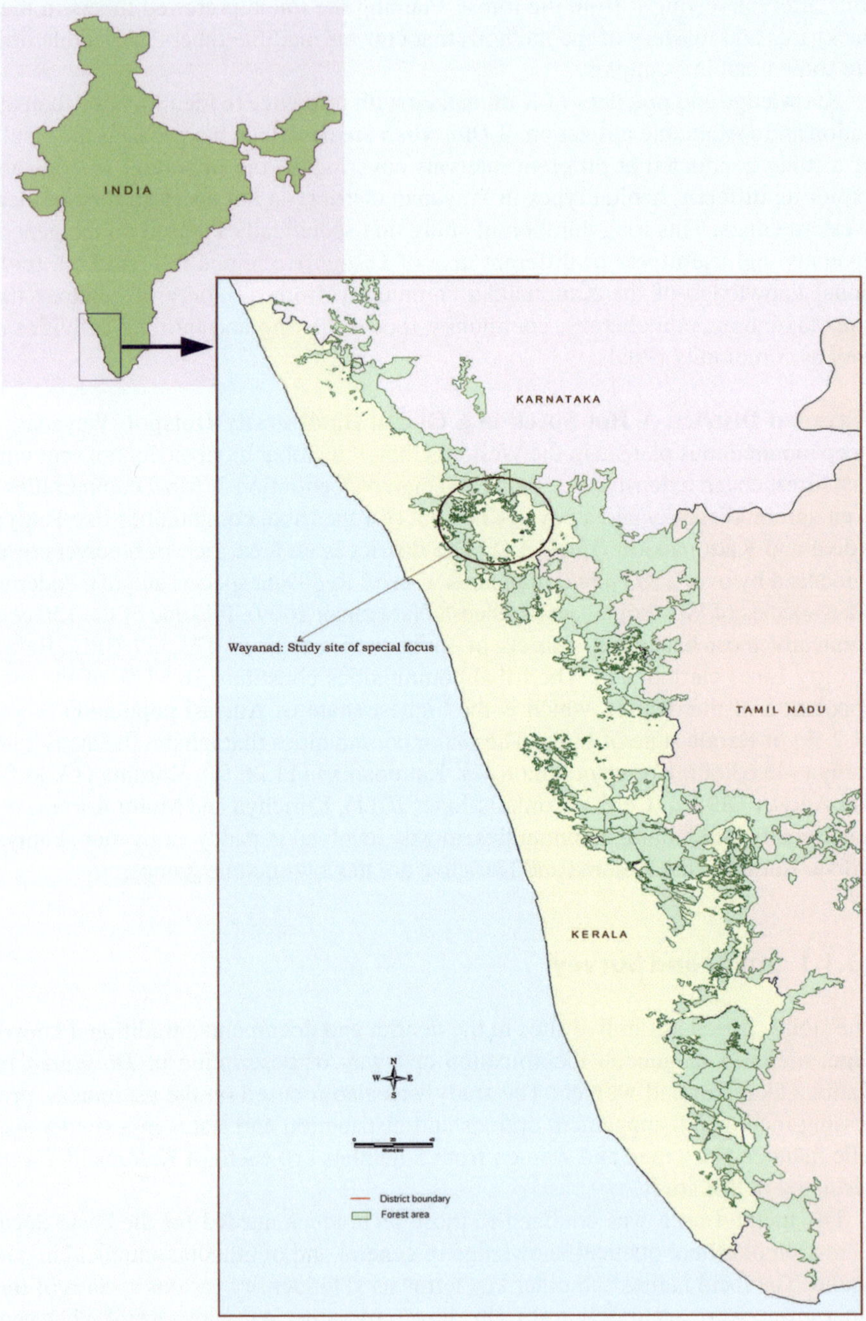

Fig. 13.1 Study area

in terms of the conservation of the rare or the highly preferred varieties for consumption were also recorded. Some of the methods explained in the publication by Rao, R.R. (1989) were adopted during the survey by making necessary changes depending on the tribal hamlets and forest types. Participatory rural appraisal (PRA) techniques (Chambers 1992, 1994) and tools such as resource mapping, resource flow mapping and seasonality mapping were used to understand the resource utilization, changes in management and issues related to conservation of wild yams. The targeted communities were consulted for their mode of use and its allied ethnobotanical information experienced during the entire field exploration trips. Care was given to document the knowledge in a gender-sensitized manner among different age groups.

13.1.2 Ethnotaxonomy of *Dioscorea*

Dioscorea, with about 630 scientifically described taxa belonging to the family *Dioscoreaceae*, is believed to have originated from Southeast Asia, West Africa and Pre-Columbian Tropical America (Wilkin 1998; Jayasurya 1984). Prain and Burkill (1936) reported occurrence of about 50 different *Dioscorea* taxa in India, largely in the West, East and Northeast regions. Taxonomy of quite a few species of this genus is considered to be problematic (Burkill and Prain 1936; Velayudhan et al. 1998), which is attributed to its high degree of continuous variability in morphological characters, especially of the aerial parts, such as leaves and bulbils. This continuous variation makes it difficult for the 'Linnaean taxonomists' to segregate distinctly the variations of this genus. For classifying *Dioscorea* species, the Linnaean system uses character classes pertaining to that of floral, fruit and seeds (reproductive characters) as well as the direction of twining of the stem. The size, shape and nature of tubers and bulbils are also used in certain cases as important taxonomic characters.

Whereas the folk system relies chiefly on the use classes of tubers such as edibility, taste, colour of the flesh, size, direction of growth, fibre content, cooking properties and occasionally its number. Classification of taxa based on use value and application by ordinary men and women is now called folk taxonomy (Peter Lister 2001). The 'folk taxonomists', unlike professional taxonomists depend upon a wide range of characters not only pertaining to the taxa in question but those associated with habitats and seasons. This classification system covers a wide range of categories, which could be referred to as 'use classes' that go along the usefulness of species and 'locality classes' which refer to the characters of locality (Peter Lister 2001). Xu Jianchu et al. (2001) have studied genetic diversity in taro (*Colocasia esculenta* Schott) of Araceae in China by using an ethnobotanical and genetic approach, and in this study, they have used the folk taxonomy tool by using the knowledge of 'Yunnan' tribal group to delimit the 20 traditional varieties of taros. The ethnotaxonomical approach seems to be useful in providing significant evidence and leads for delimiting precisely those taxa with continuous variation, particularly in case of species with use value. Therefore, to make the taxonomy of this genus less problematic, it requires a comprehensive approach in weighing characters for delimiting the taxa, perhaps by integrating the Linnaean, folk and modern taxonomic systems.

13.2 Kattunaikka of Western Ghats: *Dioscorea* Genome Saviours

As the name denotes, Kattunaikka ('Kadu' means forest and 'Naikka' means the leader) are the real heroes of forests, who by tradition gather all their major food from the wild. Kattunaikka – a community – still keep a nomadic tradition in living. Kattunaikka use more *Dioscorea* species compared to other ethnic communities that inhabit in Western Ghats. They in general live in interior forests and depend largely on forest biodiversity for their food and livelihood security. Their language is an admixture of all the Dravidian languages. There are about 212 Kattunaikka settlements in all forest regions of the Wayanad district (Anon. 1998) and constitute 9.93% of the total tribal population of Wayanad. Kattunaikka living along the border of Karnataka are also known by the name 'Then Kurumbas', since they collect 'then' (honey) from forests, and 'Cholanaikka' in Nilambur area of Malappuram district, as they mostly reside in 'chola' (shola) forests (Sukumaran Nair 1995). Their maximum concentration is seen in Pulpally, Sulthan Bathery and Muthanga regions, which are the dry zones of Wayanad district (Fig. 13.2). Shrinking forest resources, deterioration of traditional knowledge and dwindling opportunities in the agriculture sector have substantially affected the community, making them one of the most vulnerable sections of the tribal communities of the study area.

But, Kattunaikka men and women can easily identify any *Dioscorea* species even in dry form without much ambiguity. Women could identify it even by observing the tuber alone. For identifying, first they smell it or remove the outer skin of the tuber; if that is also not enough, they break the tuber and smell it and in some case taste it as fresh. Identification by considering the tuber character alone is a unique ability acquired by women of Kattunaikka through years of experience. This skill was investigated further in identifying not only the wild species but the different varieties also in a systematic manner. The perception and years of experience of Kattunaikka community in utilizing wild yams appeared to be sustainable in many ways compared to other communities of this region (Hemal et al. 2001; Balakrishnan et al. 2003).

Among the various tubers, *D. hamiltonii*, *D. belophylla*, *D. oppositifolia* and *D. pentaphylla* var. *pentaphylla* are the much-consumed varieties, but except some varieties of *D. pentaphylla*, all others are now rare in distribution in Wayanad. Kattunaikka attribute many reasons for its rarity such as the management practices of the Forest Department like replacing ground cover with other species and fire line clearing, invasion of exotic species, absence of forest fire, etc. Forest fires, they say contribute high nutrition to the soil, which favours luxuriant growth of many of the tubers and also helps in the quick and successful germination of bulbils and tubers. It is studied that overexploitation of these tubers is very minimum from the part of

Fig. 13.2 Field trip with tribal people

Kattunaikka as they always leave a piece of tuber in the cavity from which the collection is made to ensure its availability for the next season too.

It was reported by Kattunaikka members that there is a growing trend among other communities; the tubers are harvested from wilderness in an unsustainable manner. Among the wild *Dioscorea*, majority are without bulbils. Regeneration in many of the taxa which is not producing bulbils needs much conservation priority since its regeneration is mainly through the tuber. So unsustainable harvesting of the same from the wilderness creates a major loss of the gene pool. Even though many of the tribal communities are aware of the practices adopted for sustainable harvest, they give less attention to go for it, particularly the youngsters among them. They are aware of the threat status of such species in their immediate locality but do not have much idea about the rarity of them in the neighbouring area. As some very senior members of the tribal community say, many of the taxa such as *D. belophylla*, *D. hamiltonii*, some varieties of *D. pentaphylla*, *D. pubera* and *D. oppositifolia* are once common in the immediate locality of their settlements but are now very rare and not even located in the recent past.

Here, Kattunaikka are an exception. Kattunaikka women in many of the sites surveyed have attempted in domesticating some of the *Dioscorea* such as *D. hamiltonii*, *D. pentaphylla* var., 'Nalla noora' (*Dioscorea pentaphylla* var. *linnaei*), *D. oppositifolia* except *D. belophylla* because of the difficulty in harvesting and the necessity for a full-length tuber, for replanting to domesticate. Among the *pentaphylla* complex, they prefer 'Nalla Noora' (*D. pentaphylla* var. *linnaei*) and 'Noora korana' (*D. pentaphylla* var. *jacquemontii*) to domesticate, as they are tastier and nutritious than the related ones. The species selection for domestication chiefly depends on the availability of the variety and cooking quality. Women prefer the ones that are good for making curry (Vedavally and Anil Kumar 1998). They also do the steaming and roasting of the tuber depending on the presence or absence of both hard and soft fibres. It is observed that the species common to the nearest locality were not selected for domestication, even if they are rare. Among the cultivated species, they distinguish varieties of *D. alata* from the domesticated species *D. hamiltonii*, though both look similar.

13.3 Kattunaikka's Classification System

Kattunaikka calls *Dioscorea* as 'Kalasu', which is very diverse in this district. The study revealed that Kattunaikka community is aware about 22 different types of *Dioscorea* from this district. Among the varieties known to them, **Venni kalasu** (*D. hamiltonii* Hk. f.), **Hekku kalasu** (*D. belophylla* Voight) and **Kavala kalasu** (*D. oppositifolia* L.) are predominantly seen in interior evergreen and moist deciduous forests. **Shoddi kalasu** (*Dioscorea* sp.) is seen in dry deciduous forests and **Ere kalasu** (*Dioscorea* sp.) in rocky grasslands. The varieties such as **Noora kalasu** (*D. pentaphylla* L.), **Nara kalasu** (*D. wallichii* Hk. f) and **Hendhiride** (*D. bulbifera* L) are seen on forest openings and **Booji kavala** (*D. pubera* Blume) on marshy areas. Kattunaikka collect *Dioscorea* from almost all these landscapes but

Table 13.1 Ecotype/habitat of various class of *Dioscorea*

Sl. no.	Local name	Scientific name	Ecotype/habitat
1	'Venni kalasu'	*D. hamiltonii*	Interior evergreen and moist deciduous forest
2	'Hekku kalasu'	*D. belophylla*	Moist deciduous forest
3	'Kavala kalasu'	*D. oppositifolia*	Interior evergreen, semievergreen and moist deciduous forest
4	'Shoddi kalasu'	*Dioscorea* sp.	Dry deciduous forest
5	'Ere kalasu'	*Dioscorea* sp.	Rocky grassland and laterite soil in dry deciduous forest
6	'Noora kalasu'	*D. pentaphylla*	Forest openings and disturbed habitats
7	'Nara kalasu'	*D. wallichii*	Bamboo forest and disturbed habitats.
8	'Hendhiride'	*D. bulbifera*	Forest openings and stream/riverside vegetations
9	'Booji kavala'	*D. pubera*	Marshy area in the dry deciduous forest

frequently from the forests and other such unmanaged habitats, as these are the ones they frequently access (Table 13.1).

Kattunaikka are experts in identifying various *Dioscorea* by relying on cooking qualities, texture, colour, growth habit and smell of tubers. They also take into consideration the size, shape and the number of tubers. They use 10 different character classes and 44 character states of tubers for classification. They also use characters of leaf, stem, bulbils, habitat, soil and harvest season in identifying the taxa. There are 27 such character states that are used. Thus, in all, 71 characters are used (Table 13.2). By this process, at least 22 distinct taxa can be identified each based on the abovementioned characters. According to Kattunaikka, these characters are reliable and diagnostic, despite the similarity in the morphology of aerial parts. The type of characters used by them could be classified into **biological**, which are related to the vital functions of the tuber and **cryptic** – those characters that are hidden. This is evidenced, for instance, in the case of *D. pentaphylla*, *D. belophylla* and *D. hamiltonii* where many varieties look alike except in the characteristics of tubers. This approach in general is different from the typical Linnaean system followed in classification of the genus where morphology, anatomy, palynology, embryology, cytology, genetics, reproductive features and a whole set of such evidences are used.

The classification system used by this community reveals more number of taxa when compared to those reported by taxonomists from the whole of Southern Western Ghats. Kattunaikka know, for instance, six different forms such as 'Nalla Noora', 'Chena Korana', 'Hendhi korana', 'Korana', 'Noora Korana' and 'Manal korana', from the scientifically described species, *D. pentaphylla*, which is a complex known with four varieties in the Western Ghats region such as *linnaei*, *communis*, *rheedei* and the *pentaphylla* proper (Fischer and Gamble 1928; Velayudhan et al. 1998 and Jayasurya 1984) to the Linnaean taxonomists. Prain and Burkill (1938) reported the occurrence of a variety named *jacquemontii* in this complex from Bombay, which is very similar to *linnaei*, but differs by having long and

Table 13.2 Character and character states used by 'Kattunaikka'

Sl. no	Character class	Character	Character state
1	Tuber	1. Number	1,2, >3, <3
		2. Shape	Globose, Oblong, Linear, Irregular
		3. Direction of growth	Vertical, horizontal, tuber starts just below stem base, tuber seen at the end of long thread, stem up to 1 m deep, stem up to soil surface, deep, shallow, starts bending just below the ground, starts bending deep below the ground
		4. Taste	Sweet, sour, itching
		5. Size	Small (not enough for one family), large (enough for one family)
		6. Edibility	Eat fresh, simple processing, thorough processing, inedible
		7. Cooking quality	Curry, boiling, frying
		8. Fibre	Hard fibre, soft fibre, no fibre, more, less
		9. Mucilage	More, less
		10. Colour of flesh	Yellow, white, reddish yellow
2	Leaf	1. Number of leaflets	Simple, complex
		2. Structure	Presence of hairs, absence of hairs
3	Stem	1. Spine	Whole, base only
		2. Shape	Angled, terrate, flat
		3. Colour	Green, black, red
4	Bulbils	1. Globose	Present, absent
		2. Elongate	
5	Habitat	1. Dry area	Rocky grassland, forest opening, dry deciduous forest, moist deciduous forests, evergreen forest
		2. Wet area	Streamside, forest swamps
6	Soil	1. Laterite	Rich, poor
		2. Black	
		3. Red	
7	Harvesting season	1. September–Nov	
		2. March–April	

elongated bulbils. But the existence of this variety is known only from that location. Taxonomists reported about 16 varieties in the *pentaphylla* complex alone throughout its range of distribution (Prain and Burkill 1938). In India, reports of all these varieties are known from different eco-regions of the country (Hooker 1894; Gamble 1915–1936; Prain and Burkill 1938).

Kattunaikka classify *Dioscorea* into six groups, namely, 1. **Noora Korana**, 2. **Venni**, 3. **Hekku-Heruman**, 4. **Naara**, 5. **Kavala-shoddi** and 6. **Hendhiride**, and

Fig. 13.3 Classification tree

two with single species ('Nara' and 'Salu kalasu') (Fig. 13.3). Each group is examined in terms of their characteristics, quality of the tuber, habitat and distribution within the district.

The **Noora Korana** complex divided into two basic groups viz. 'Noora' and 'Korana' based on two major characters such as tuber, which gives 'nooru' (powdery flesh) and with glabrous leaves and tuber with 'korana' (fibrous flesh) and

Fig. 13.4 Tubers in *D. pentaphylla* complex (Noora, korana, Chena korana, Hendhi korana, Noora korana, Manalkorana)

pubescent leaves (see key given below). The 'Noora' again is classified into 'Nalla noora' and 'Kottu Noora', and 'Korana' into 'Korana' typical, 'Chena korana', 'Noora korana', and 'Hendhi korana' and 'Manal Korana'. Thus, there are seven entities that exist in this species of which six belong to *D. pentaphylla* complex (Fig. 13.4) and one ('kottu noora') belongs to yet another species *D. hispida*. The 'Nalla noora' match with variety *linnaei*, 'Korana' can be identified as variety *rheedei* , the 'Noora korana' to variety *jacquemontii* and 'Hendhi korana' as variety *communis*. 'Chena korana' and 'Manal Korana' do not match with any of the hitherto described variety in this complex. Since the characters of bulbils and tubers of none of the described varieties of *pentaphylla* complex fall under this variety, it can be concluded that the variety 'Chena korana' and 'Manal Korana' is left with no scientific name.

Key to Noora Korana Complex

1. Tubers of nooru category (flesh turns into powder after cooking). Leaves glabrous.... Noora.

   1. Tubers of korana category (flesh turns fibrous after cooking). Leaves pubescent. Korana.

Key to Noora Group

1. Tubers cylindric, fibrous, edible without processing;......... Nalla noora (D. pentaphylla var. linnaei)

   1. Tubers globose, nonfibrous, edible only after processing ... Kottu Noora (D. hispida)

Key to *Korana* Group

1. Season for Harvest March–April. 2

1. Season for Harvest September–November. 3

2. Tubers spherical, flesh pale yellow in colour...........*Chena korana.* (*D. pentaphylla* var.)

2. Tuber resembles the shape of a wild boar, flesh white in colour.......... *Hendhi korana* (*D.pentaphylla*, var. *communis*)

3. Tuber very fibrous, difficult to break...... *Korana* (*D. pentaphylla* var. *rheedei*)

3. Tuber less fibrous, easy to break..4

4. Tubershort, *grows in laterite soil*............................... *Noora korana* (*D. pentaphylla* var. *jacquemontii*)

4. Tuber very long, grow in sandy soil...................... Manal Korana (*D. pentaphylla*)

The 'Nalla noora', among these varieties, is the very commonly consumed tuber. As the name refers, 'nalla' means safe or ideal to eat. The tuber is single, elongates up to 1 m. in length is less fibrous and becomes powdery when cooked and delicious to taste. This variety is common in forest fringes of deciduous vegetation types. The 'Kottu noora' is referred to *D. hispida*, which is not as common as the *pentaphylla* complex in the district. This tuber requires thorough processing before consumption. The chopped tubers would be kept in fast-flowing running water for over 24 h. No other community in the district consumes this variety, which shows its dangerous quality. Many other communities including Kattunaikka use it as a fish-stupefying agent.

The 'Korana' is commonly used for making curry, occasionally as stewed cake, because of its high fibre content. The tuber is difficult to break by hand and has got an itching character, which can be evaded by thorough washing. This is rare in distribution in the district and is restricted to the riparian habitats. September to November is the harvest season of this tuber (Fig. 13.5). 'Chena korana' is also rare and occur usually in disturbed evergreen forests. It has got the shape of a *chena* (= elephant foot yam) and very (hard) fibrous in nature. This is used only for the purpose of steaming and roasting and harvested during March–April. 'Noora korana' (Fig. 13.6) is with spines all over the stem and more like noora in appearance of the tuber but with rusty pubescent leaves. It is consumed in the form of only steamed dishes. 'Hendhi korana' tuber has got the shape of 'hendhi', meaning wild boar in their dialect and with thick root hairs all over the tuber (Fig. 13.5). March to April are the months for its harvest. The tuber is consumed in the form of roasted dishes. It is seen along the streamside thickets and moist deciduous forests but rare in distribution. 'Manal Korana' grows in sandy soil along the riverside. 'Manal' means sand and the tuber that grows in sandy soil got this name 'Manal Korana'. The plant is with three to five leaflets, pubescent, stem with more prickles and bulbils very small. The tuber is consumed in the form of curry as well as roasted dish. Harvesting season of 'Manal Korana' is September to November.

Fig. 13.5 Noora korana

Fig. 13.6 Hendhi korana

Another group of tubers, which the tribal people consume, is the 'Venni' collection. There are two different types of 'Venni', namely, 'Kalu venni' and 'Mooda venni'. 'Kalu venni' has got very long 'kalu' (leg), which grows vertically much deeper into the soil up to 2 m in length. It does not bear bulbils in any season. This is consumed in the form of curry and eaten afresh because of its less fibrous content. 'Kalu venni' is identified as *D. hamiltonii* Hook. f. In 'Mooda venni', the tuber-bearing stalk is small and the stem base is spinous. This tuber is also less fibrous and thus eaten afresh or in the form of liquid dishes. The aerial part of this plant bears bulbils, which are very similar to that of *D. alata.* This is very rare in distribution and seen along riparian bushes and swampy forests only. The months of March to April are ideal for its harvest.

Hekku is yet another group of tubers with three distinct folk varieties, 'Hekku typical', 'Hekku heruman' and 'Herman' (see key to taxa). The 'Hekku' is commonly seen in moist deciduous forest and grows very deep in the soil as long thread to bear the tuber, which is highly delicious among all the tubers they consume. This is usually consumed in the form of liquid dishes. March to April are the months for its harvest but rare in distribution. 'Hekku heruman', is also equally delicious as the 'Hekku' but is restricted in distribution. The tuber is irregular in shape and borne on a long thread. It is characterized by its black-coloured stem and spines at the base as well as the nodal regions. This is also very rare in distribution and seen only in moist deciduous forests. In 'heruman', the number of tubers is usually three and is irregularly shaped. Kattunaikka consume this usually in the form of curry. Like the 'Hekku heruman', this also is rare in the district and seen only in moist deciduous forests.

Key to the Hekku-Heruman Group

1. Tubers 3...*Heruman*

1. Tubers single in number...2

2. Tuber regular in shape, on a long thread; stem base and nodes not spinous .... *Hekku.*

2. Tuber irregular in shape, on a short thread; stem base and nodes spinous... *Hekku heruman.*

The *Naara* set consists of two varieties, 'Naara' typical and 'Naara mooyan', which are distinguished between the number of the tubers and its orientation of growth in the soil. Both of these have got horizontally growing tubers, the former grows near the surface of the subsoil, whereas the latter starts its growth much deeper in the soil. 'Naara' is more fibrous in nature and consumed in the form of roasted dishes, particularly curing stomach disorders and for constipation. 'Naara mooyan' resembles the morphology of aerial parts with that of 'Hekku' (*D. belophylla*) but differs in the growth habit, where it grows horizontally against the vertically growing character of the 'Hekku'. The 'Naara' is harvested during September–October and 'Naara mooyan' during March–April and are seen in moist deciduous to evergreen forests.

1. Harvesting season September to October.................................... **Naara**

1. Harvesting season March to April .... **Naara mooyan**

The **Kavala-shoddi** group consists of four taxa, *Kavala kalasu*, *Ere kalasu*, *Shoddi kalasu* and *Booji kavala* (see the key for identification). *Kavala* except *Booji kavala* is the most popular tuber among all the tribes of Wayanad, as it is very tasty and commonly found in moist forests on which normally the community depends more (Fig. 13.7).

'Booji kavala', the tubers and bulbils are eaten fresh or in the form of liquid dishes. It is very restricted in distribution and seen normally in the swampy forests of the interior wet and dry deciduous forests. The Kattunaikka families domesticated this species because of its high delicious character and its rare occurrence in interior forests. March to April is the ideal season for harvesting. Because of the absence of fibres, they are using it for making curry.

Fig. 13.7 Kavala shoddi tubers

'Ere kalasu' is rare in distribution and seen in deciduous forests on forest fringes. The tuber would be sweet, if they harvested at the time of flowering of the plant, which is normally during September–November. 'Shoddi' is also one of the much-consumed tubers, but the quantity of the tuber is too small for an entire family and therefore not much preferred. This is also rare in distribution and seen only in rocky grasslands.

1. Harvest time March to April.......... .... *Booji kavala*
1. Harvest time September to November....... ....**2**
2. Tuber >1, with stalk.. *Shoddi kalasu*
2. Tuber 1, without stalk...**3**
3. Tuber grows horizontal to the soil .... *Ere kalasu*
3. Tuber grows vertical to the soil. .... *Kavala kalasu*

The sixth group distinguished is 'Hendhiride'. According to Kattunaikka, the 'Hendhiride' collection has got two types, the '*bolle* Hendhiride' with pale green stem and 'choppu Hendhiride' with red stem. Both are inedible except for wild boar and *Paniya* – another tribal community in the district. 'Hendhi' in their language refers to wild boar. The 'Hendhiride' could be identified as *D. bulbifera*, in which the 'bolle' is variety *bulbifera* and chopu is variety *vera*. The '*bolle* Hendhiride' is with comparatively larger tuber but smaller bulbils. It is rarer than the other.

1. Stem green colour.... *Hendhiride bolle* (*D. bulbifera* var. *bulbifera*)
1. Stem red colour................ Hendhiride chopu (D. bulbifera var. vera)

Two other varieties standing singly are *Salu Kalasu* and *Nara*, which are very rarely consumed. No other community in the district are aware about the edible nature of these two species, which itself shows its usefulness in terms of the edibility at the time of scarcity (Table 13.3).

'Salu kalasu' is consumed during acute famine season. This taxon is identified as *D. tomentosa* characterized by the presence of peculiar kind of fibres and mucilage, which cause some irritating sensation especially to children. It is common and seen usually in disturbed forests.

'Nara' is identified as *D. kalkapershadii*, which is a little-known species. The characteristics are cylindrical tuber with spines even on the tuber with rich and hard fibres, spinous stem and petiole and compound leaves with five to seven pubescent leaflets. This variety is extremely rare in distribution and seen in highly disturbed habitats with much sunshine. Moreover, this taxon is seen along with rich population of *D. pentaphylla* and *D. tomentosa*. Burkill and Prain (1936) described this as a coarse variety of *D. pentaphylla* and *D. tomentosa*.

13.4 Discussion

Importance of Ethnic Knowledge in Conservation Wild *Dioscorea* is grown in different agroclimatic zones and in diverse landscapes of Southern Western Ghats. It can also be observed that tribal hamlets are seen more in the area where there is rich availability of wild food diversity. It can be presumed that the settlement of

Table 13.3 Method of cooking and harvest season of various *Dioscorea* sp.

Sl. no	Local name	Scientific name	Method of cooking	Harvesting season
1	'Noora' (Nalla Noora)	*D. pentaphylla* var. *linnaei*	Curry	September–November
2	'Korana'	*D. pentaphylla* var. *rheedei*	Curry, steaming	September–November
3	'Chena korana'	*D. pentaphylla* var.	Steaming	March–April
4	'Noora korana'	*D. pentaphylla* var. *jacquemontii*	Steaming	September–November
5	'Hendhi korana'	*D. pentaphylla* var. *communis*	Roasting	March–April
6	'Manal korana'	*D. pentaphylla* var.	Roasting, curry	September–November
7	'Kottu noora'	*D. hispida*	Curry	March–April
8	'Mooda venni'	*Dioscorea* sp.	Curry	March–April
9	'Kalu venni'	*D. hamiltonii*	Curry	March–April
10	'Hekku'	*D. belophylla*	Curry	March–April
11	'Hekku heruman'	*Dioscorea* sp.	Curry	September–November
12	'Heruman'	*Dioscorea* sp.	Curry	September–November
13	'Naara'	*D. wallichii*	Roasting	September–November
14	'Naara mooyan'	*Dioscorea* sp.	Curry, steaming	March–April
15	'Kavala kalasu'	*D. oppositifolia*	Curry	March–April
16	'Ere kalasu'	*Dioscorea* sp.	Curry, steaming	September–November
17	'Shoddi kalasu'	*Dioscorea* sp.	Curry, steaming	March–April
18	'Hendhiride bolle'	*D. bulbifera* var. *vera* Prain and Burkill	–	March–April
19	'Hendhiride chopu'	*D. bulbifera* var. *bulbifera* Wight	–	March–April
20	'Salu'	*D. tomentosa*	Roasting	September–November
21	'Nara'	*D. kalkapershadii* Prain and Burkill	Roasting	March–April
22	'Booji kavala'	*D. pubera*	Curry	March–April

tribal hamlet in a particular locality is mainly based on the easy access of wild food especially the tuberous one.

There are many ethnic names for *Dioscorea*, in which Kattunaikka holds 22 names, followed by Cholanaikka 19 and Paniya 14 (Table. 13.4, Fig. 13.8). By using all the utilitarian characters and folk morphological units, Kattunaikka

Table 13.4 Various taxa of *Dioscorea* identified by different tribes from study area

Sl. no	Name of tribe	Taxa (folk name)		Total no.
1	Kattunaika	1. Noora	2. Heruman	22
		3. Korana	4. Naara	
		5. Chena korana	6. Naara mooyan	
		7. Noora korana	8. Kavala kalasu	
		9. Hendhi korana	10. Ere kalasu	
		11. Manal korana	12. Shoddi kalasu	
		13. Kottu noora	14. Hendhiride bolle	
		15. Mooda venni	16. Hendhiride chopu	
		17. Kalu venni	18. Salu	
		19. Hekku	20. Nara	
		21. Hekku heruman	22. Booji kavala	
2	Cholanaika	1. Venni	2. Narna mayil	19
		3. Chola venni	4. Noora	
		5. Hekku venni	6. Erna	
		7. Kavala	8. Nooramanche	
		9. Korana	10. Kavala manche	
		11. Nara	12. Eyyan	
		13. Nukappan	14. Chaval	
		15. Chornu	16. Paluvenni	
		17. Mayilu	18. Kattu gundan	
			19. Nalla gundan	
3	Paniya	1. Panni Noora	10. Nokappan	14
		2. Nara	11. Chapa noora	
		3. Kavala	12. Venni	
		4. Noora	13. Noolan	
		5. Neinoora	14. Bolle	
		6. Vellan		
		7. Chelapu		
		8. Naravayan		
		9. Kalakkizhangu		
4	Muduga	1. Jalkilangu	2. Noorekilangu	12
		3. Narankilangu	4. Kuranekilangu	
		5. Chavalkilangu	6. Venni	
		7. Perkukilangu	8. Erekilangu	
		9. Cholakavalakilangu	10. Majavukilangu	
		11. Kavalakilangu	12. Karaperakukilangu	
5	Kani	1. Neduvan	2. Karinkizhangu	8
		3. Pinnan parakizhangu	4. Kattuvallikilangu	
		5. Nooran	6. Thavalaikilangu	
		7. Kavalan	8. Siruvallikilangu	

(continued)

Table 13.4 (continued)

Sl. no	Name of tribe	Taxa (folk name)		Total no.
6	Kuruma	1. Vennagu	2. Njarakangu	6
		3. Kuruddakangu	4. Neelangu	
		5. Perungalkondakangu	6. Adankangu	
7	Kurichya	1. Naran	2. Pullu noora	8
		3. Noora	4. Eruma noora	
		5. Kariyali noora	6. Panni noora	
		7. Kazhuthu noora	8. Chalippa	
		1.	9. Mothirakudam	
8	Adiya	1. Adde ganasu	2. Nereganasu	5
		3. Chooreganasu	4. Kurudaganasu	
		5. Mudikeganasu	6.	
9	Maratinaika	1. Suna kirangu	2. Nare	7
		3. Adda nali	4. Poo nare	
		5. Marlu ghande	6. Bol ghande	
		1.	7. Airol	
10	Irular	1. Sheelikilangu		2
		2. Vellakodikizhangu		

identify correctly these 22 taxa, which are grown in diverse habitats like evergreen forest, dry deciduous forest, streamside, disturbed forest, tea estate, etc. Among these, *Dioscorea tomentosa*, *Dioscorea hispida* and *Dioscorea bulbifera* were consumed in the acute famine season by only two communities such as Cholanaikka and Paniya.

Though Kattunaikka are knowledgeable enough to identify all taxa of *Dioscorea* distributed in Wayanad, they consume only 18 of them in normal case, but during famine season, about 20 taxa are consumed. Paniya community can identify 14 but consume in normal case only 10 and during famine season 11 taxa. In case of Kuruma community, a comparatively advanced group in cultivation and less dependency over forest can identify only 6 taxa and they consume only 4 even during famine season. Kurichya community can identify 8 different taxa and they consume only 5 taxa. (Fig. 13.9).

Kattunaikka normally uses 7 character classes, 24 characters and 63 character states for identifying *Dioscorea* in all seasons, whereas morphotaxonomic system used 4 character classes, 15 characters and 44 character states (Fig. 13.10). For correct identity for its immediate food value, they mainly depended only on two character states such as character of tuber and the harvesting season of tuber. Among all the above characteristic features, they mainly depends on qualitative characters such as cooking quality, season of harvest, quality and growth character of tuber, etc. for knowingly or unknowingly to delimit the genetic diversity in *Dioscorea*. The targeted group calls each taxon in separate folk name for locating and familiarizing

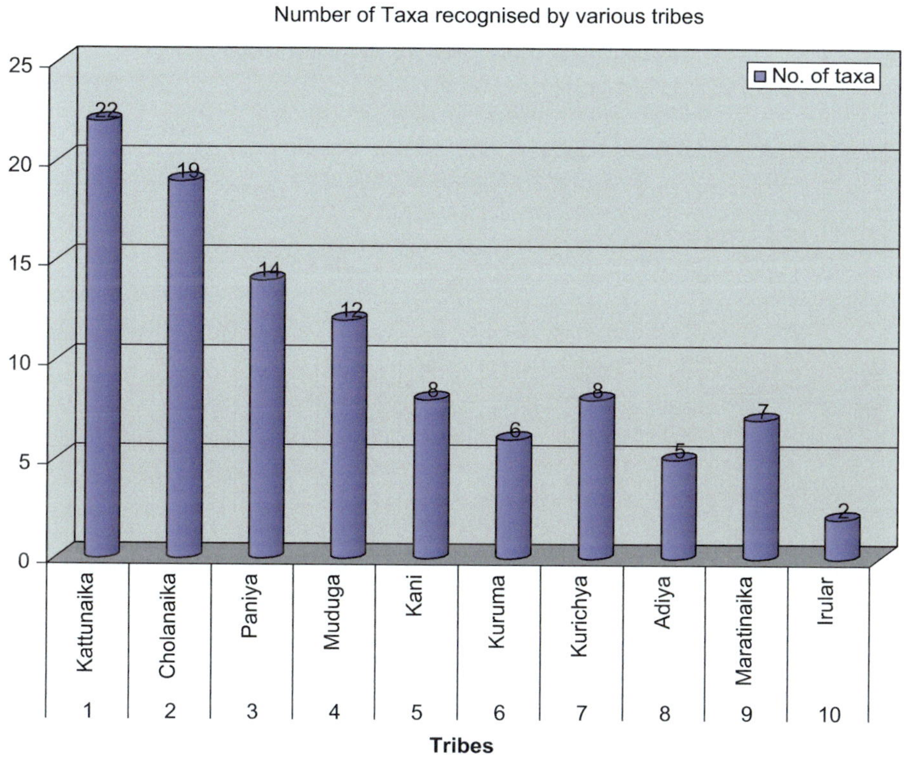

Fig. 13.8 Different taxa of *Dioscorea* identified by tribes from study area

this famine food to all age groups in the tribal area. The meaning of folk names itself explains the morphological characters of the tuber and habitat of the plant. These characters perceived by the tribal community for years of experience which in turn will be helpful for the younger generation in the tribal family for easily locating the food *Dioscorea* (Anil Kumar and Vedavalli 2016).

Much preferred taxa were identified for the years of trial and error method and also through the traditionally transferred indigenous knowledge. Such high-quality tubers can be brought to domestic level for wider cultivation and easy accessibility. Tubers of *Dioscorea hamiltonii*, *D. belophylla*, *D. pubera*, *D. pentaphylla* var. *pentaphylla* and *D. oppositifolia* are some of the highly tasteful much sought and preferred by both men and women of the tribal families. Such taxa can be domesticated or can be used as breeding material for improvement of modern-cultivated varieties or for the improvement of the same wild taxa by using molecular technologies.

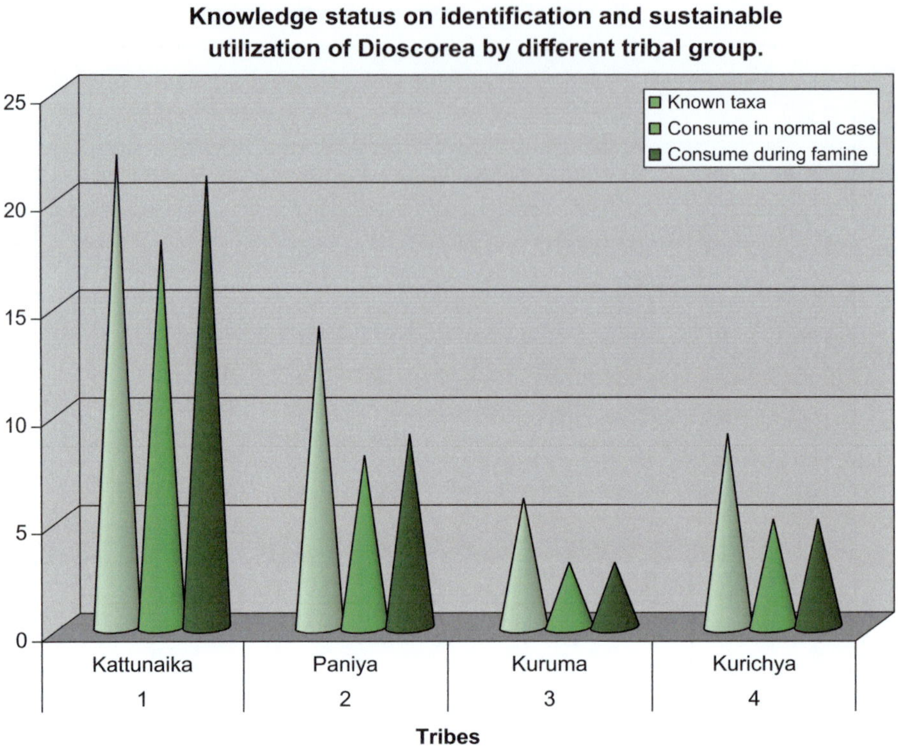

Fig. 13.9 The knowledge status in identification and sustainable utilization of *Dioscorea* by different tribal groups

13.5 Importance of Folk Language in Conservation

The persistence of knowledge on biological resources mainly lay on its usage and identity. Naming in local language a genetic resource that is useful or harmful to humans is essential and such language should be easily transmittable across generations. The Kattunaikka in Wayanad area speaks a dialect of *Kannada*, but the younger generation converse in Malayalam – the language of Kerala state. As mentioned earlier, they named almost all *Dioscorea* taxa found in Wayanad district in the admixture of Kannada-Malayalam languages. It is very rare, a plant of *Dioscorea* is known to different communities in the same ethnic name as originally called by a particular user community of a particular locality. The major reasons could be attributed for this are the geographical break in the distribution of different varieties of *Dioscorea*, absence of proper interaction between different tribal groups using same language and variations in the processing techniques for edibility of the tuber adopted by different communities. So the distribution and edibility of some of the *Dioscorea* varieties could only be identified by knowing its correct ethnic name. Taxa like *D. pentaphylla*, as it is widely distributed, is known commonly by the

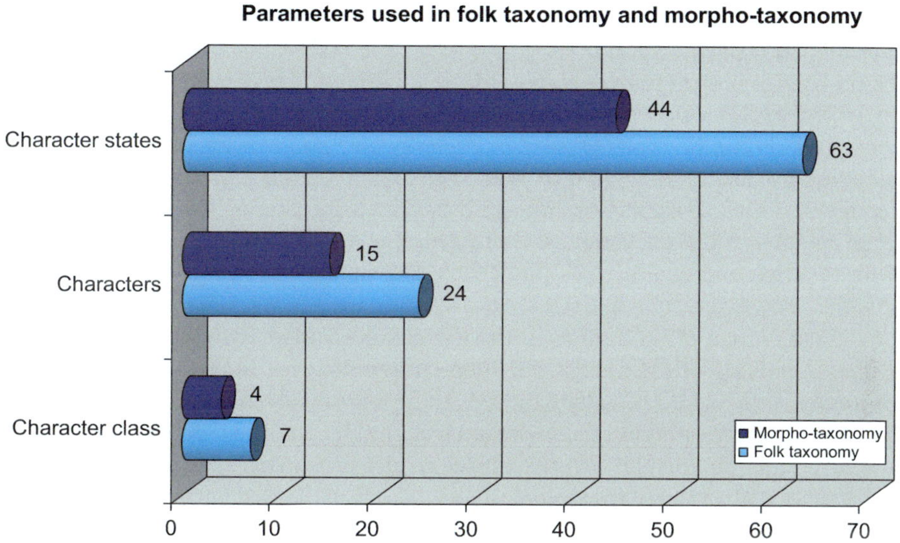

Fig. 13.10 Parameters used in folk and morphotaxonomy of *Dioscorea*

ethnic name 'Noora' throughout its range of distribution, whereas *D. bulbifera*, a variety, which is edible for some community but nonedible for others, is known by different names in tribal habitations across a wider geographic area. The oral tradition of passing knowledge in the community between different age groups also has not being practiced these years. The major reason is that younger generation is not much interested in possessing this knowledge or using the tribal language, which results in lessening the accessibility to this wild resource. This is can be illustrated by the example of *D. pentaphylla* and its ethnotaxonomic knowledge among different age group of people among Kattunaikka. 'Noora' and 'Korana' are the two taxa of the species *D. pentaphylla* known to the younger generation, whereas the elders of them are familiar with seven taxa in the same ethnic group called 'Noora Korana' group. This indicates for an effective conservation strategy for wild plant genetic resources; there is a need for comprehensive approach in preserving the ethnic languages followed by different communities.

13.6 Ethnotaxonomy and India's PPVFR Act 2001

The present study clearly shows that the ethnic communities use simple and convenient techniques for classifying the species that have utility value to them. They could even locate the population of *Dioscorea* in each locality and harvest them in a sustainable manner. The method of classification by using the utilitarian character is part of their livelihood; hence, the classification methods originate spontaneously. However, in case of scientific system for delimiting the taxa, evidences from folk system of classification have not been recognized. It is true that understanding the folk system of classification is very difficult, as the researcher has to think from the stakeholder tribal viewpoint. The folk system of classification will be a strong area of knowledge in the future in respect of the conservation and sustainable utilization of wild relatives of edible plants. Therefore, ethnobotanists and other scientists, who work on contributions of tribal community men and women, have to give more attention to folk classification of wild species. This is very important in order to recognize farmers' rights – a concept established for sharing the benefits arising from breeding of plant varieties. Since *Dioscorea* is an important food crop, the wild relatives and landraces of this genus gain much significance in breeding, so that the knowledge of people who conserve such species is taken into account. One of the definitions of the farmer in the Protection of Plant Varieties and Farmers' Rights Act, 2001 of India reads thus: 'one who conserves and preserves, severally or jointly, with any person any wild species or traditional varieties or adds value to such wild species or traditional varieties through selection and identification of their useful properties'. So this clearly tells, a farmer is not only a person who is just engaged in cultivation of those domesticated crops in the past but one who continuously adds value to such crops and finds new crops from wild relatives. In this sense, Kattunaikka are 'farmers of forests' who add value to conservation of an important plant genetic resource. The scope of rewarding Kattunaikka for their efforts in conserving diversity of *Dioscorea* from the available national fund for Biodiversity conservation under the Biodiversity Act, 2002 or PPVFR Act, 2001 has to be explored. The present system developed in India for the registration of farmers varieties should also apply in the case of tribes (farmer of forest) who conserve wild edible plants through selection and identification of their useful properties.

13.7 Conclusions

Among the wild edible plants, *Dioscorea* is the much-harvested major source of food for forest-dependent communities of the Western Ghats region, especially by the Kattunaikka live in Wayanad district. A detailed ethnobotanical study on a single genus *Dioscorea* is the first holistic effort to know this genus from Southern Western Ghats from a genetic diversity and food security perspective. The study brought out in detail the folk classification methods that help the communities to better understand and utilize this genetic resource. The results of this study indicate that the ethnotaxonomical system of *Dioscorea* gives an insight for classical taxonomists in weighing different character groups for classifying plant groups with high plasticity of characters. Further, the study highlights various means and measures adopted by this community for conserving and managing the rare/threatened and much preferred *Dioscorea* tubers.

The study revealed much valuable information not only pertaining to the edibility of *Dioscorea* but also its genetic diversity and thereby opening up the basis for correct identity of many of the taxonomically confusing taxa of this genus. The men and women of Kattunaikka community are well versed in the utilization aspects of *Dioscorea* in terms of its identification, habitat, associated plants, maturity and sweetness of the tuber, its availability, and in identifying the matured tuber ideal for consumption. Effective conservation and sustainable utilization of wild genetic resources depend mainly on the associated ethnic knowledge especially the use value. In general, a species without any knowledge on its utility character is not considered for its care or protection by the communities. Apart from such knowledge on the utility aspects, the skills on classification and identification of taxa are also essential for conserving a species. Getting information on indigenous uses and management of a large number of taxa of wild *Dioscorea* gathered with the active participation of tribal knowledge holders are worthwhile formulating strategies for conservation and sustainable use of this important PGR of the tropical world.

References

Abhyankar RK, Upadhyay R (2011) Ethnomedicinal studies of tubers of Hoshangabad, M.P. Bulletin of Environment, Pharmacology and Life Sciences 1(1):57–59

Anil Kumar N, Vedavalli L (2016) Conservation of family farming heritage. In: Family farming meeting the zero hunger challenge-food and agriculture organization of the United Nations and M.S. Swaminathan Research Foundation (MSSRF), Published by Academic Foundation, pp 245–268

Anonymous (1998) Survey report, District Tribal Development Project, Wayanad, Kerala

Arora RK, Pandey A (1996) Wild edible plants of India, diversity, conservation and use. National Bureau of Plant Genetic Resources, New Delhi, pp 15–16

Balakrishnan V et al (2003) Ethnotaxonomy of *Dioscorea* among the Kattunaikka people of Wayanad District, Kerala, India. Plant Genet Resour News Lett 135:24–32

Burkill IH, Prain D (1936) An account of the Genus Dioscorea in the East. Ann Roy Bot Gard Culcutta, Longman/London

Chambers R (1994) The origins and practices of participatory rural appraisal. World Development 22(2):953–969

Fischer CEC, Gamble JS (1928) Flora of the Presidency of Madras..Superintendent, Govt. Printing, Calcutta. III, pp 1507–1513

Food and Agriculture Organization of the United Nations (FAO) (2004) Annual report: the state of food insecurity in the world, monitoring the progress towards the world food summit and millennium development goals, Rome

Gamble JS (1915–1936) The flora of the Presidency of Madras. Adlard & Son Ltd., London

Gamble JS, Fischer CEC (1915–1936) Flora of the Presidency of Madras, 11 parts. Adlard & sons Ltd., London

GEF (Global Environment Facility) (2006) The role of local benefits in global environment programs, Evaluation report no. 30. GEF Evaluation Office, Washington, DC

Hemal SK, Anil Kumar N, Rasheed PA (2001) Wild food management in Wayanad, Kerala. An Exploratory study. Community Agrobiodiversity Centre, M.S.Swaminathan Research Foundation, Puthoorvayal, Kalpetta, Wayanad

Henry AN, Hosagouder VB, Ravikumar K (1996) Ethno-medico-botany of the southern Western Ghats of India. In: Jain SK (ed) Ethnobiology in human welfare. Deep Publications, New Delhi, pp 173–180

Hooker JD (1894) Flora of British India, vol VI

IWGIA (2007) The indigenous world 2007. By Sille Stidsen and regional editors. International World Group for Indigenous Affairs, Copenhagen, 594p

Jayasurya AHM (1984) Systematic arrangement of the genus *Dioscorea (Dioscoreaceae)* in Indian Sub-continent. Revised hand book to the Flora of Ceylon IX

Nair KKN (1999) Ethnobotany and its application in biodiversity conservation. The Botanica 49:66–68

Office of the Registrar General and Census Commissioner Census of India (2011) Office of the Registrar General and Census Commissioner, Government of India New Delhi, viewed 15 February 2016

Peter Lister (2001) http://www.uws.edu.au/vip/listerp/ethtax.htm

Prain D, Burkill IH (1936) An account of the Genus *Dioscorea* in the East. Part I – the species which twines to the left. Superintendent, Government Printing, Bengal Government Press, Calcutta, pp 1–210

Prain D, Burkill IH (1938) An account of the Genus *Dioscorea* in the East. Part II – the species which twines to the right. Superintendent, Government Printing, Bengal Government Press, Calcutta, pp 211–528

Pushpangadan P, Atal CK (1984) Ethno-medico-botanical investigations in Kerala 1. Some primitive tribals of Western Ghats and their herbal medicine. J Ethnopharmacol 11:59–77

Rao RR (1989) Methods and techniques in ethnobotanical study and research: some basic considerations. In: Jain SK (ed) Methods and approaches in ethnobotany. Society of Ethnobotanists, Luckow, 192 p

Ratheesh Narayanan MK (2009) PhD thesis, Flora of Wayanad, India

Roy B, Holder AC, Pal DC (1988) Plants for human consumption in India, Flora of India series 4. Botanical Survey of India, Calcutta, pp 63–65

Sukumaran Nair (1995) Wayanad and her tribal Sons, Discover Wayanad the Green paradise, pp 41–43

Vedavally L, Anil Kumar N (1998) Gender dimensions in biodiversity management case study Wayanad, Kerala. In: Swaminathan MS (ed) Gender dimensions in biodiversity management. Konark Publishers.4, Delhi, pp 96–106

Velayudhan KC, Muralidharan VK, Amalraj VA, Asha KI (1998) Genetic resource of yams of Western Ghats. Indian Journal of plant genetic resources, of Western Ghats. National Bureau of Plant genetic Resources, Regional Station, Vellanikkara, Thrissur 11(1):69–80

Wilkin P (1998) A morphometric study of *Dioscorea quartiniana* A.Rich. *(Dioscoreaceae)*. Paper accepted for publication August 1998.Herbarium, Royal Botanic Gardens, Kew, Richmond, Surrey, TW9 3AB, UK

Xu Jianchu, Yang Yongping, Pu Yingdong, Ayad WG, Eyzaguirre PB (2001) Genetic diversity in Taro (*Colocasia esculenta* Schott, *Araceae*) in China: an ethnobotanical and genetic approach. Econ Bot 55(1):14–31

Part III

Bioprospecting

Prospecting Extreme Niches 14

Indu C. Nair

Abstract

Origin of life in the universe is an ever-challenging concept for researchers in all branches of science, and it often places many unanswerable questions before the entire scientific community. However life through extremophiles is emerging as one of the acceptable hypothesis in this respect. Exploitation of the extremophiles and investigation into the conditions in which they survive often bring new information regarding the mechanism by which these microbes metabolize and survive hard surroundings, many of which resemble the extreme conditions at the beginning of the life on earth. The wide diversity, immense adaptation, and the efficient catabolic potential of the microorganisms add huge support to their chance of pre-existing before the realization of present life forms on earth and add the possibility of carrying life across planets in the universe. Information regarding the extremophiles, their structural and chemical adaptations, their genome content and its significance, and the enzymes and other products from extremophiles contribute endless scope for exploring prospects of these systems for future benefits.

Keywords

Bioprospecting • Extremophiles • Metagenomics • Bioprocess

I.C. Nair (✉)
Department of Biotechnology, SASSNDP Yogam College,
Konni, Pathanamthitta, Kerala 689699, India
e-mail: inducnair73@gmail.com

S. Abdulhameed et al. (eds.), *Bioresources and Bioprocess in Biotechnology*,
DOI 10.1007/978-981-10-3573-9_14

14.1 Introduction

Science is the knowledge of facts, phenomenon, laws, and proximate causes verified by exact observation and organized experiments built on logic. Need-based developments are highly essential for the proper exploitation of the services of science. Designing new strategies for bioprospecting the extreme niches becomes significant in this viewpoint. The multifarious ecosystem of nature has been scrutinized by humans for useful products. In this respect, nature never fails human and has been serving as a source of novel entities and substances. The modern science has discovered lot of molecules from living cells that have specific applications as food, drug, and industrial ingredients. As time passes, the need to replace the old ones by more effective substances increases as in the case of antibiotics. Inventory and screening have been always there to explore possibilities of new areas and novel compounds, and the attempts to commercialize the products are also progressing. Exploring the biodiversity for identification and production of new compounds and making them available on an industrial scale comprise the realms of bioprospecting. A constructive approach toward searching and development of new resources for chemical compounds, genes, microbes, other organisms, etc. is termed bioprospecting. It enables to explore new habitats for screening of substances of significance.

Plants are considered to be the main reservoir of useful chemicals/metabolites. Thousands of plant-derived compounds are familiar to us, which find vast areas of administration. Animal products are also of high market value. Man makes use of such compounds as therapeutic agents, food, biocontrol agents, and other compounds of commercial value. Indigenous information is very helpful in primary screening of compounds. Indigenous information has originated from long-time observations and is comparatively reliable. Microorganisms form the next group, which has been screened for countless number of uses. Today microbial technology has emerged as a reliable means to investigate and develop strategies for searching new compounds. Though the microbial diversity is yet to be explored even in common resources, it enthralls scientists, when it is extended to the extreme environments in search of microbial diversity and unique products. It opens an arena of research, which has immense value in the scientific and commercial aspects. The molecular techniques as well as classical culture techniques can act as tools for improving the status of bioprospecting to a more reliable state in the case of extremophiles (Banos et al. 2015).

It is observed that the knowledge about the life forms of the extreme physical parameters has expanded the boundaries of knowledge about the evolution, history of life, origin, and possibilities of this unique phenomenon in other parts of the universe. The present chapter introduces a raft of metabolic capabilities inherent in the extremophiles and the scope of their utilization. The different approaches toward screening is also discussed which paves fundamentals for bioprospecting.

14.2 Extreme Niches

Mesophilic organisms including humans live in a narrow range of temperature and pressure. They respire oxygen and cannot withstand very high or very low pH ranges. They are protected from damaging radiations. Once, it was assumed that life existed only in such physical parameters referred to as standard conditions. While a small area of earth is occupied by such conditions, most of the portions are denoted as extreme states, which cannot be inhabited by mesophilic organisms. These external extremes include temperature, pressure, salinity, pH, radiation, etc. and are seen on the earth and other extraterrestrial regions. The physical parameters in such regions show extremes in one or in combination. On the earth there are organisms which can thrive over such extreme conditions of habitat and are known as extremophiles. Microbial communities dominate the extreme niches, and they attract the scientific world through their specialized molecular mechanisms of stress tolerance.

Studies indicated the presence of microorganisms, which shows tolerance for any extreme conditions. It is understood that for many microbes isolated from such areas, extreme conditions are essential for their existence, and they find it difficult to survive in standard physical state. The microbes living in the extreme niches are often primitive Archaebacteria with many peculiar adaptations. They have special structural features, and their functional components like proteins are well suited to an extreme lifestyle. They have evolved a range of modified metabolic pathways. The microbes in the geothermal vents are capable of utilizing SO_2, the otherwise poisonous gas. In the vents the temperature ranges from 60 to 460 °C where water exists as a supercritical fluid. Chemolithotrophic and chemo-organotrophic modes of life exist. Many barophilic organisms widely occupying the ocean floors live under high pressure and in darkness and have special adaptations to bear with high pressure. Different types of organisms and their preferred extreme niches are given in Table 14.1.

Table 14.1 Distribution of extremophiles

Niches	Types of organisms
Hyperpressure	Barophiles occupy the high-pressure areas like ocean bottoms
Hypersaline	Regions where salt concentration is higher than that of seawater. Halophiles survive here
Extreme cold	Psychrophiles live at a temperature of 5 °C or below persistently
Extreme hot	Thermophiles are seen in areas with elevated temperature, usually above 40°. The niches include black smokers, hydrothermal vents, etc.
Areas with radiation	Areas where high degree of radiation occurs due to nuclear fallout
Extreme pH	Basophiles dwell in the habitat where pH is above 9 (sometimes with regular intervals)
Acidic	The pH is below 5. Acidophiles survive under these conditions
Anhydrous	Habitats where there is no free water
Anaerobic	No free oxygen is available here
Altered by human	Such habitats are intensely changed due to human activities

14.3 Bioprospecting the Extremophiles

The bioprospecting of the extreme environment opens up a host of possibilities to exploit the microbial properties for bio-industry. They can survive because of the "extreme" adaptations in the cellular machinery like enzymes which are able to function at very stress conditions, the presence of protective substances like osmoprotectants, change in the composition of macromolecules to prevent denaturation, the presence of other bioactive compounds, etc.

14.3.1 Barophiles

Barophiles or piezophiles are microorganisms which survive under high pressure. They are defined as organisms growing at pressures greater than atmospheric pressure (0.1 MPa). They exhibit a universal distribution on the ocean floors where pressure is higher than 380 atm. Obligate and facultative barophiles are there and since they live in the cold dark conditions, they are poor in DNA repair mechanism and are very sensitive to UV radiations. Increased binding capacity of enzymes is observed in these organisms, and pressure-controlled gene expression is a key factor in the survival of these organisms. The product of the gene cyd D is essential for growth under high pressure (Kato and Querishi 1999). The presence of eicosapentaenoic acid in the membrane is another adaptation, which maintains membrane fluidity. *Moritella* sp., *Colwellia* sp., etc. are common examples. *Xenophyophore* is a eukaryotic sponge-like organism dwelling in the deep-sea bottoms.

High-pressure and cold conditions decrease membrane fluidity. As an adaptation to prevent this, highly unsaturated fatty acids are incorporated to the cell membrane. PUFA has generally very low melting point and stabilizes the membrane systems. EPA and DHA present in these extremophiles are having long chains, usually rare in other bacteria and are synthesized through special anaerobic pathways (Yano et al. 1998). A stable membrane system that integrates unusual fatty acids shows the homeoviscous adaptation by altering the fatty acid composition according to environmental changes. The polyunsaturated fatty acids act as a protecting agent in the deep marine environment. Furan fatty acid in *Shewanella putrefaciens* is the scavenger of hydroxyl and peroxyl radicals (Carla et al. 2012).

A pressure ranging from 0.1 Mpa to 1 GPa, (i.e., 1 bar *to* 10 K bar) affects only intermolecular distances and conformations. More than 30 K bar pressure can change bond angles and covalent bond distances. Primary structure changes are effected above the pressure change of 20 K bar (Raid 2005). Eicosapentaenoic acids present in the membrane system are helpful in maintaining the fluidity under cold conditions and high pressure (Kato et al. 1988). Barophiles and their enzymatic machinery can be exploited for pressure-controlled reactions, bioremediation, bioengineering, etc. High pressure affects the constitution of cell membrane and may increase the rate of monosaturation and polyunsaturation in the membrane. Under high pressure at sea bottom, membrane fluidity was kept increased by elevated unsaturation. A moderately barophilic *Shewanella* sp. with different DNA-binding proteins at different pressures was shown to possess a pressure-regulated operon, which was cloned and sequenced (Sathyanarayana et al. 2005). It was reported that in some bacteria, porin expression is influenced by pressure differences and fluidity of the membrane. A pressure-regulated operon was identified in a barophilic bacterium DB6705, and an ORF encodes CydD protein that is required for cytochrome complex in the aerobic respiratory chain, suggesting the apparent importance of membrane components in high-pressure adaptation (Sathyanarayana et al. 2005).

14.3.2 Halophiles

Halophiles are seen in high salt concentrations usually above 36% w/v salt. Though salted surroundings induce osmosis in normal cells, halophiles are well adapted to prevent this. The phenomenon of halotolerance is mainly due to the presence of osmoprotectants. Halophiles accumulate osmoprotectants in their cells to prevent the osmotic loss of water, and this increases the internal osmolarity. The mesophilic organisms dry out in such conditions due to loss of water from the cytoplasm. Many compounds like amino acids, polyols, betaines, sugars, ectoines, etc. or their derivatives act as osmoprotectants. These are synthesized in the cell or collected from the external environment.

Another judicious mechanism practiced by halophiles is to allow the influx of potassium ions, which is found in the archaeal group called Haloarchaea. The entire cellular conditions cope up with this adaptation. The halophiles are chemotrophs and can survive in anaerobic conditions. The halophiles have acidic proteins with most of the amino acids being negatively charged and capable of functioning in a salty environment (Madern et al. 2000). The red color of saltern lakes is due to the presence of carotenoid pigments in the halophiles. Different ranges of salinity occur when evaporation takes place in hypersaline environments. Various microbial populations flourish according to the salt gradient showing diverse adaptations in salinity. In solar salterns, algal blooms are also observed. Protozoa, yeast, and fungi are among the microbial populations reported. Microbial mats containing predominantly photosynthetic unicellular Cyanobacteria and purple sulfur bacteria are present in the anoxic zones. At highest salinities, pleomorphic Haloarchaea grows with rod, cocci, triangular, and square-shaped morphologies. A metagenomic study reported the presence of minute amino acid utilizing nanobacteria belonging to this group (Sarma and Sarma 2001). The presence of similar bacteria is observed in human intestinal tract also. The evolutionary position and physiological capability make them significant. There are many saline lakes containing Cyanobacteria, Aphanothece halophytica, and other phototrophic bacteria which form the top layer of the hypersaline lakes. Anaerobic forms are also observed in saline lakes. The eukaryotic halophiles include algae like *Dunaliella* sp. and protozoan like *Fabrea salina* growing in moderately concentrated regions.

The halophiles find traditional use in food and nutraceutical industry. They are used to ferment food and are involved in the manufacture of solar salt. The specialized molecules in the cellular system, which enable them to survive under high salt concentrations, find a variety of uses in basic research and applied fields. Many of them are thermophiles also. The chromoprotein, containing retinal, can be used as a neurological probe, as in biocomputing. Halophytic cells act as a source of many enzymes, which are stable under hot conditions. Polyamines are used in cosmetics for improving hydration properties (Thorsten et al. 2008). They act as a source of PHB. The presence of gas vesicles makes many halophiles buoyant, and gas vesicles are used for bioengineering of floating molecules. Production of stabilizing osmolytes and stress protein are coming into commercialization (Sarma and Sarma 2001).

A modified Entner-Doudoroff pathway and many other deviations from the central pathways are exhibited by halophiles. Some can produce PHB under excess carbon conditions. Halophiles have surface negative charges, which are essential for the solvation of halophilic protein and for the prevention of protein denaturation and aggregation. In bacteria, a specialized portion of membrane called purple membrane carries light-dependent transmembrane protein pump consisting of rhodopsin (Sarma and Sarma 2001).

Carotenoids are the second largest group of natural pigment that has a significant role in the health-related aspects. They are effective antioxidants, which can fight against oxidative stress. The free radical scavenging activity is related to their molecular structure. Carotenoids form the precursor of vitamin A and functions in the proper maintenance of child health and survival helping in development of antibodies, vision, and general resistance.

According to Oren (2010), two processes in which halophiles are profitably employed are carotenoid production and production of an organic compound ectoine. Carotenoids have a variety of food and pharmaceutical applications. Ectoine is used as a stabilizer for enzymes and as an ingredient in cosmetics. Bacteriorhodopsin is used as a photosensor (Oren 2010). Carotenoid production from halophiles is well studied and even attempted at industrial level because of high rate of production, inexpensive and fast downstream processing, and provision for adjusting physical parameters. The trial for increasing the production is promising, and NaCl concentration below 16% w/v induces high levels of production. Downstream processing becomes easy because lower salt concentration bursts the cells.

Enzymes at high ionic strength can be isolated from halophiles. Bacterioruberin is another product of interest in research. This is a protectant against high intensities of light. This is said to influence photoreactivation also. It has a role in controlling membrane fluidity (Banos et al. 2015).

14.3.3 Psychrophiles

Psychrophiles show effective mechanism to thrive over cold temperatures. They are capable of surviving under a temperature range of −20 °C to +10 °C. During low temperature, normally microorganisms show reduced type of biochemical reactions. Bacteria are isolated from regions of troposphere and stratosphere where temperature is about −40 °C.

Many of the bacteria isolated from a cold environment are psychrotrophs like *Bacillus*, *Arthrobacter*, *Corynebacterium*, etc. Different lineages of psychrophiles and psychrotrophs are identified through phylogenetic analysis. They include Archaebacteria and Eubacteria (Sathyanarayana et al. 2005). Microalgae and yeast are also reported from the cold soil sediments. Here, the microbial activity is restricted to the unfrozen water inside the soil. Many picoplanktons in the polar regions belong to Crenarchaeota group. The presence of antifreeze proteins, cold-adapted enzymes, cryoprotectants, increased antioxidant activity, etc. is observed in psychrophiles.

During low temperature, normally microorganisms show reduced type of biochemical reactions. It is reported that the psychrophiles could overcome the increased viscosity of water at low temperatures. Cold acclimatization proteins (CAps) exist in them, which is an important adaptation to cope with the low temperature.

14.3.4 Thermophiles

Thermophiles are detected from geothermal vents universally. They are quite able to survive in boiling temperatures where life would be in danger for other organisms. Some thermophiles show tolerance to multiple extremes like acidic or alkaline environment in addition to high temperatures.

Thermophiles are mostly chemo-organotrophic or chemolithotrophic. Their metabolism is diverse, and they include methanogens, sulfate reducers, nitrate reducers, etc. *Aquifex* sp. is a strict chemolithotroph which can utilize molecular hydrogen and thiosulfate as electron donors. *Crenarchaeota* exhibit hyperthermophilic character with ether-linked lipid in the cell membrane, which consists of isoprenoids instead of fatty acids. The cells are pleomorphic with wide size ranges and are able to synthesize heat-stable molecules, including enzymes. Mesophilic enzymes are denatured at high temperatures, but the enzymes from thermophiles shows stability. They are suitable models for studying thermostability of proteins and are even able to adjust with acidic or alkaline pH extremes.

Methanococcus jannaschii produces a thermostable protease, which shows maximum activity at 116 °C, and the enzyme is active till 130 °C. Thermophiles are also characterized by special membrane lipids. *Thermotoga maritima* has 15,16-dimethyl-30-glyceryloxy-triacontanedioic acid which confers protection against hydrolysis at high temperatures (Carla et al. 2012).

Thermophiles occupy a very significant place among the extreme microbes when their applications are considered. Thermostable enzymes from these organisms form a target, which attract industrial processes. Thermostable enzymes are exploited for research purposes, too. There has been an intense search for thermophilic enzymes, which exhibit superior properties at elevated temperatures.

An enzyme or protein is said to be thermophilic when the transition temperature, Tm is very high (unfolding temperature), or it has long half-life at a given high temperature. Most of the proteins in thermophile are thermostable, but not all. Inside the cell there are protective measures so that the intracellular enzymes may be less thermostable, but all the extracellular enzymes are essentially thermostable.

It is well known that *Thermus aquaticus* and *Thermococcus litoralis* are the sources providing thermostable enzymes for PCR. In the repeated cycles of PCR, a temperature of more than 90°C is required. Ordinary polymerases denature at this temperature while taq polymerase and similar enzymes perform the copying with fidelity. The polymerase chain reaction was revolutionized by the employment of *Thermus aquaticus* DNA polymerase. Thermostable enzymes are useful in the production of highly pure chiral compounds. These are important in pharmaceutical production. Principles of green chemistry prefer thermophilic enzymes as biocatalyst. L-amino acylase from *Thermococcus litoralis* is used in the resolution of amino acids and amino acid analogs. Y Lactam is a building block for antiviral carbocyclic nucleotides. The enzyme amidase/Y lactamase is used in the synthesis of this compound, and *Sulfolobus solfataricus* produces the enzyme. An alcohol dehydrogenase similar to Zn containing alcohol dehydrogenase is produced from *Aeropyrum pernix*. The haloacid dehalogenase is another thermostable enzyme with striking applications and can be used in biotransformation and bioremediation (Little child 2011). *Thermotoga maritima* is found to be a source of endoxylanases Xyn A and xyn B. Many industrial processes are based on enzymes, and the thermoactive enzymes are able to promote this field. Environmental friendly methods can be developed using these enzymes.

Glycosyl hydrolases, proteases, lipases, etc. are produced from thermophiles and were characterized. Some are cloned and expressed in *E. coli*. Since this is not scalable, the expression is studied in fungal cell factories (Bergquist et al. 2014). Thermophilic enzymes find a vital role in biorefineries, which form a recent trend in industrial field. An initial step in biorefining is the fractionation of crops, which needs high temperature to produce cellulose, hemicelluloses, starch, etc. *Thermostable amylomaltase* is produced from *Thermus* and *Thermococcus* sp. (Turner et al. 2007). Even though there are numerous reports on the enzymes, as per the academic and patent literature there are only limited uses at industrial level. More studies and research are needed for the judicious exploitation of the thermophilic cellular machinery.

The mechanism, which enables the survival of these microorganisms in such extreme situations, is the presence of the enzymes called extremozymes. They have special mechanisms to keep their functionally active structure with folding and motifs at elevated temperatures. Microbial methanogenesis is a key process in the carbon cycle of earth, and hyperthermophilic and hydrogenotrophic thermophiles play a major role in the primary production of ecosystems in deep seas (Takai et al. 2005). Takai group has developed a system for cultivation and incubation of microorganisms under high hydrostatic pressure corresponding to deep seafloor characteristics.

High level of potassium and magnesium present in the intracellular environment protects the bacterial DNA from denaturation. Reverse gyrases are observed in DNA of thermophiles, which keep the DNA in the positively supercoiled situation. Polyamines are also abundant in these cells. About 16 polyamines are reported from

Thermus thermophilus, which are long and help in maintaining the cellular functions at high temperatures. In mesophilic cells polyamines are very low, and the observed ones belong to short-chain groups. Long polyamines prevent the denaturation of DNA than shorter polyamines. Intrinsic and extrinsic factors and battery of enzymes enable the thermophiles to protect their DNA. Linear polyamines containing propyl groups lead to increase in melting temperature which are proportional to the nitrogen atoms in the individual polyamines suggesting that charge-charge interactions contribute to stabilizing effect.

14.3.5 Radiation-Surviving Microorganisms

The most efficient strains of bacteria, which can resist the effects of radiation, belong to the group *Deinococcus*. It survives continuous and prolonged ionizing radiations as well as gamma radiation. The *Deinococcus* strains are not pathogenic. The genome of all organisms is affected by even weak radiations, and the existing repairing mechanisms cannot overcome the damage caused to the genetic material. Studies reveal that several different biological mechanisms operate in radiation-surviving bacteria to protect the genome and intracellular machinery from radiations. Extraordinary repair capabilities are attributed to this group of organisms. The genome is enriched with repetitive sequences including transposons like insertion sequences (IS). Desiccation-resistant proteins are also isolated from *Deinococcus* cells. *Deinococcus radiodurans* is capable of accumulating large amount of Mn+. This is helpful in fenton oxidation, and Mn phosphate superoxide radicals scavenge the oxidizing species (Stone 2009). *Deinococcus* is considered to be a multiple extremophile, which can survive dehydration and acid conditions. Four cells normally stick together forming a tetrad. Smooth colonies are observed to be pink. *Thermococcus gammatolerans*, *Chroococcidiopsis*, and *Rubrobacter* are other radiation-resistant microorganisms. Decomposition of organic compounds heavy metal removal could be performed by the modified bacterium. The main application of the organism is bioremediation of irradiated sites. *Deinococcus radiodurans* can survive high degree of radioactivity, and it is reported that they can scavenge heavy metals like mercury. In Deinococci, genome is not a single copy but exists as multiple copies. Homologues recombination is possible among these, and repair mechanisms operate rapidly. The cell wall has got thick peptidoglycan layer, which is perforated. The plasma and outer membranes are separated by an uncharacterized layer, and the fourth layer is the outer membrane; fifth is a vivid electroluscent zone followed by the sixth layer of regularly packaged hexagonal protein subunits (Makarova et al. 2001). Some strains exhibit a carbohydrate coat over this. This is known as the toughest bacterium. One of the greatest advantages is that these organisms are not pathogenic. The facts about Deinococci point out the possibilities of microbial life in the extraterrestrial regions.

14.3.6 Alkaliphiles

Alkaliphiles are divided into alkaliphiles and haloalkaliphiles. Haloalkaliphiles need high salt concentrations, but alkaliphiles can survive under less alkaline conditions. Both aerobic and anaerobic alkaliphiles are being reported. Haloalkaliphiles isolated from Lake Magadi of Kenya live under a pH 11. The GC content of DNA was about 62.5 mol%. They have large gas vacuoles and carry a new osmoprotectant, 2 sulfotrehalose (Horikoshi 1999).

The alkaliphiles are rich sources of alkaline proteases. They find application in hide dehairing process performed under alkaline conditions. As detergent additives they are remarkable choices, and about 30% of the commercial enzyme produced is used in the field of detergents. Alkaliphiles can be used to decompose gelatinous coating of x ray films and recovery of silver. Alkaline amylases are also being investigated. Pullulanases, cellulases, glucanotransferases, etc. are promising enzymes in the industrial field. Alkaline lipases are also studied. Alkaline xylanases are important in biobleaching. Many are thermostable also. Organic acids, siderophores, cholic acid derivatives, carotenoids, antibiotics, and enzyme inhibitors are also obtained from alkaliphiles.

In basic research, the special adaptations of alkaliphiles draw the attention of scientists. Slightly alkaline cytoplasm is maintained by alkaliphiles. The intracellular pH regulation is dependent on the removal of sodium ions from cytoplasm into the medium by H^+ or Na^+ antiporters which are efficient at transporting H^+ into the cell at the expense of Na^+ export from it (Sathyanarayana et al. 2005). In alkaliphiles internal cellular balance is maintained by these antiporters. Flagellar motion also derives energy from sodium gradient. At elevated pH levels, the peptidoglycan of the cell wall shows higher degree of cross-linking. This increases the rigidity of cell wall. Sodium-dependent ATP synthesis has not been identified in alkaliphiles, and ATP synthesis in alkalophilic *Bacillus* species have been demonstrated to be exclusively protein translocating (Sathyanarayana et al. 2005).

14.3.7 Acidophiles

Acidophiles thrive over acidic pH and are mostly capable of bioleaching and biooxidation. The metal sulfides are solubilized by these organisms, releasing metals and acids. The archaeon *Ferroplasma* is an example for this. These organisms can also be used in ore floatation concentrates for the removal of metal from ore. Some of these organisms are the causes of harm to nature due to their acid releasing capacity.

14.3.8 Extraterrestrial

Modern theories on the origin of life suggest that the molecular seeds of life appeared on earth just after the big bang. These life molecules traveled through space and flourished wherever they had suitable conditions to evolve. This theory of panspermia is gaining much attention now. In the extreme environment of space, they wait for an apt situation to evolve. The early periods of earth were similar to these extreme environments. The condition was anaerobic, very hot with striking, lightning, and volcanic eruptions. Studies reveal that microorganisms have inhabited the earth during these extreme conditions since 3.5–4 billion years.

14.4 Tools for Bioprospecting Extremophiles

Isolating the extremophiles and keeping and growing them in the lab are absolutely difficult and challenging. Hence the metagenomic approach is extensively practiced in exploiting extremophiles. However under controlled situations, extremophiles from certain conditions can be isolated, can be grown, and can be kept in the laboratory conditions.

Metagenomics opens up venues for the study and implementation of microbial biodiversity. Recently vast number of bioactive compounds is screened from microorganisms through metagenomic approaches. It provides information about the microbial groups, which cannot be otherwise grown in culture conditions. Sequence-based and bioinformatics analysis provides information about the relationship among groups and their taxonomical positions.

Isolation of microorganisms from extreme environment is not an easy task as in most of the cases as they may exist as viable but not cultivable. It will be very difficult to redesign the exact environment in which they have been surviving. As per the great plate anomaly (Staley & Konopka 1985), only a small percentage of bacteria form colonies in plating procedure. Under these circumstances, the most reliable alternative is to depend upon the isolation of nucleic acids from environmental samples. This novel strategy helps the researcher to avoid the vigorous culturing strategies and therefore getting an added advantage of precisely tracing the molecular identity of the isolate. The isolated DNA may be representative of microbial genome, of the structural diversity of the samples, and of certain functional genes.

PCR amplification and screening is also very useful in scrutinizing for new genes. Such genes are expressed in suitable host, and optimization is achieved. The complexity of metagenomics is reduced by the single-cell genomics, where a single cell can be isolated from a sample, and the genome can be amplified. Library construction is an effective method for molecular level prospecting. Fosmid libraries are constructed for the screening of compounds from marine sponges.

Information regarding the metabolic activity and gene expression profiling can be collected through the analysis on the RNA isolated from the same environmental samples. Once the nucleic acids are isolated, nucleic acid hybridization, Polymerase chain reaction, 16S or 18S r RNA analysis, phylogenetic analysis, molecular typing, sequencing, and cloning, expression of functional gene and its characterization using the tools of bioinformatics, etc. can be done to explore the functional uniqueness of the organisms under extreme niches which will ultimately help in better prospecting of the extremophiles.

The functional heterogeneity of the extremophiles can be traced through quantitative PCR where the genes are amplified in real time facilitating more investigations into the functional roles and productivity of these organisms in the environment. Fluorescence in situ hybridization (FISH) is another useful technique using which more information on the morphology, localization, and the functional role of the organisms in a selective environment can be studied. Genomics involving functional and structural genomics and metagenomics (Jo et al. 1998) involving the application of structural genomics to the environment have been extensively used to characterize the extremophiles.

Structural and comparative genomics help us to understand how these extremophiles have adapted and evolved in peculiar conditions. But to understand how these unique organisms react to the surroundings, extensive analysis in terms of functional genomics is essential. It can be very well attained through transciptomics, proteomics, and also through metabolomics. Extremophiles including acidophiles, psychrophiles, thermophiles, and other categories survive and function under these extreme conditions mainly due to the selective and timely expression of various functional genes. As many of these organisms are uncultivable under laboratory conditions, it is not easy to get into the level of functional genes. The most reliable way is to depend upon metagenomics, which concentrates only on the pool of nucleic acid samples in a given extreme environment. Metagenomic approach definitely brings valuable information regarding gene diversity and gene functionality in a given situation.

Metagenomic approach to trace the potential of extremophiles include shotgun cloning of the selected gene followed by its sequencing and library construction which can later be utilized for tracing the gene for novel product or pathway. Many such attempts have been successful. Metagenomic approach has been extensively used in the biological production of many industrially important products including novel biocatalysts, new antibiotics, new medically important products, biosurfactants, and also for bioremediation (Yasir et al. 2014). The metagenomic approach has been instrumental in determining the crystal structure of RNase H1 in characterizing an esterase from a extremophile with a different and unique active site (Olalla et al. 2014) and novel polyhydroxy alkonate synthase encoding gene (Marcus et al. 2011).

However the identity of the isolate and function of the gene traced cannot be explored in all the cases of metagenomics approach. If DNA sample collected do not carry 16S or 18S r DNA component, it will be challenging to assign the species to which the DNA sample belongs. Similarly as in most of the cases of functional genes from extremophiles, assigning a function to the isolated novel gene is also a difficult task. Metagenomic approach has revolutionized the field of microbial ecology and industrial biotechnology particularly in the bioprospecting of microorganisms under extreme niches (Barone et al. 2014).

These extreme isolates can be screened for novel products or functions and can be exploited through careful formulation of the bioprocess. Before attempting to isolate organisms from the extreme niches, a careful and deep understanding of the factors leading to extreme conditions is necessary. The isolates collected must be

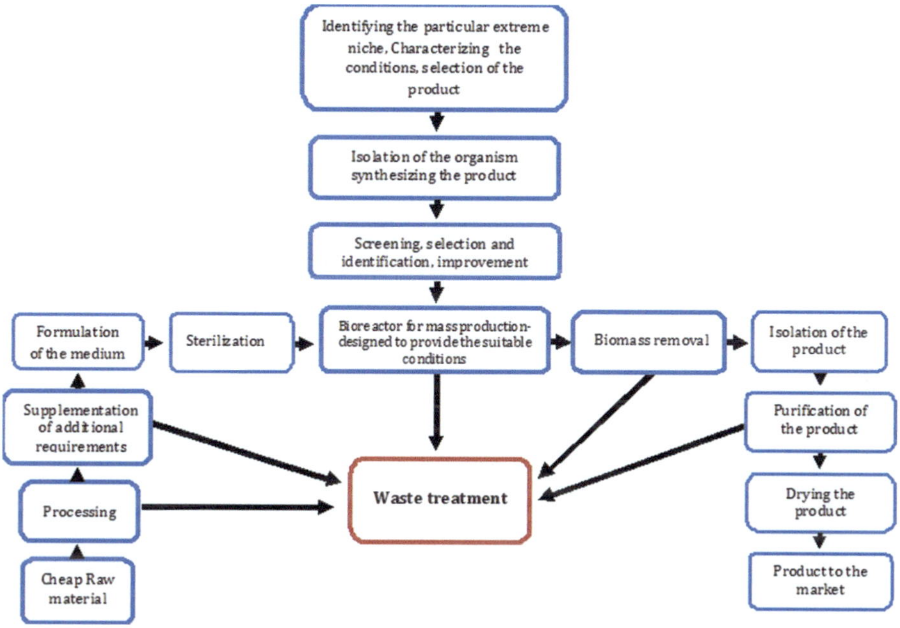

Fig. 14.1 Prospecting the extreme niches: a typical flow chart for the bioprocess for extremophiles

purified, identified, and subjected to screening for a desired product. Once the organism is selected, the medium and the conditions for getting the maximum product should be optimized. The production of the metabolite can be carried out in a typical bioreactor specifically designed to provide the required conditions of growth and metabolite production. A fine control of various parameters, like pH, temperature, dissolved oxygen concentration, etc., contributing to the production of the metabolite should be carefully monitored. Once the maximum product is attained, the extraction of the product and its purification can be done by suitable techniques like centrifugation and chromatography. The concentrated product may be processed and packed to bring into the market (Fig. 14.1.).

14.5 Conclusions

The research on extreme life has great influence in developing new technologies. The use of thermostable nucleases is a familiar example. It throws light on the possibilities of existence of life beyond earth, and it explores the chances of colonizing new areas in space. With the advent of modern tools in molecular biology and refined approaches in bioprocess, bioprospecting of extremophiles could be made more fruitful and ever increasing. The new products, process, and information generated through these exercises will definitely benefit mankind to make a better quality life progressing and prospering on a green chemistry approach.

References

Banos RM, Garbayo I, Vilchez C, Benete MJ, Martinez-Espionasa RM (2015) Carotenoids from halobacteria and their potential in biotechnology. Mar Drugs 13(9):5508–5532

Barone R, Santi CD, Esposito FP, Tedesco P, Galati F, Visone M, De Scala AD, De Pascaale D (2014) Marine metagenomics, a valuable tool for enzymes and bioactive compounds discovery. Front Mar Sci 1:38

Bergquist PL, Morgan HW, Saul D (2014) Selected enzymes from extreme thermophiles with application in biotechnology. Curr Biotechnol 3(1):45–49

Carla CC, Carvato R, Caramigo MJ (2012) Lipids of prokaryotic origin at the base of marine food webs. Mar Drugs 10(12):2698–2714

Horikoshi K (1999) Alkaliphiles: some applications of their products for biotechnology. Microbiol Mol Biol Rev 63(4):735–750

Jo H, Rondon M, Brady SF, Clardy J, Goodman RM (1998) Molecular biological access to the chemistry of unknown soil microbes. Chem Biol 5(10):R245–R249

Kato C, Querishi MH (1999) Pressure response in deep sea piezophilic bacteria. J Mol Biol Biotechnol 1:87–92

Kato C, Li L, Nogi Y, Nakamura Y, Tamaoka J, Horikoshi K (1988) Extremely barophilic bacteria isolated from the Mariana trench, challenger deep at a depth of 11000 meters. Appl Environ Microbiol 64(4):1510–1513

Littlechild JA (2011) Thermophilic archaeal enzymes and application in biocatalysts. Biochem Soc Trans 39(1):155–158

Madern D, Ebel C, Zaccai G (2000) Halophilic adaptation of enzymes. Extremophiles 4:91–98

Makarova KS, Arvind L, Wolfi YI, Tatusov RL, Minton KW, Koonin EV, Daly MJ (2001) Genome of the extremely radiation resistant bacterium *Deinococcus radiodurans* viewed from the perspective of comparative genomics. Microbiol Mol Biol Rev 65(1):44–79

Marcus S, Wang C, Voget GM, Streit WR, Driscoll B, Charles TC (2011) Harvesting of novel polyhydroxyalkanoate (PHA) synthase encoding genes from a soil metagenomic library using phenotype screening. FEMS Microbiol Lett 321(2):150–156

Olalla LL, Cerdan ME, Sisco MG (2014) New extremophilic lipases and esterases from metagenomics. Curr Protein Pept Sci 15:445–455

Oren A (2010) Industrial and environmental application of halophilic microorganisms. Environ Technol 31(8–9):825–834

Raid MM (2005) Chemistry in extreme conditions. Elsevier, Amsterdam

Sarma SD, Sarma PD (2001) Halophiles. in Encyclopedia of Life Sciences, Nature Publishing Group

Sathyanarayana T, Kumar CR, Shivaji S (2005) Extremophilic microbes: diversity and perspectives. Curr Sci 89(1):78–90

Staley JT, Konopka A (1985) Measurement of in situ activities of non-photosynthetic microorganisms in aquatic and terrestrial habitats. Annu Rev Microbiol 39:321–346

Stone M (2009) Metal balance helps explain survival of microbial 'superhero. Microbe 4(4):164–165

Takai K, Moyer CL, Miyazaki M, Nogi Y, Hirayama H, Nealson KH, Horikoshi K (2005) *Marinobacter alkaliphilus* sp. nov., a novel akaliphilic bacterium isolated from sub sea floor alkaline serpentine mud from ocean drilling program site 1200 at south Chamorro seamount, Mariana forearc. Extremophiles 9:17–27

Thorsten B, Louis P, Galinski EA (2008) Heterologous ectoine production in *E. coli* by passing the bottle neck. Saline Syst 4:12

Turner P, Mamo G, Karlson EN (2007) Potential and utilization of thermophiles and thermostable enzyme in biorefining. Microb Cell Factories 6:9. doi:10.1186-1475-2859-6-9
Yano Y, Nakayama A, Ishihara K, Saito H (1998) Adaptive changes in membrane lipids of basophilic bacteria in response to changes in growth pressure. Appl Environ Microbiol 64(2):479–485
Yasir B, Singh SP, Konwar BK (2014) Metagenomics: an application based perspective. Chin J Biol, Article ID-146030, p 7

Gene Banks and Bioprospecting 15

R. Aswati Nair, K.S. Jayakumar, and P. Padmesh Pillai

Abstract

Gene banks are biorepositories, which preserve the genetic material of the organism for long-term storage and subsequent utilization of the bioresource in a sustainable manner. Unlike biodiversity hotspots, these are manmade repositories, which conserve the biological material under *in situ* or *ex situ* conditions, by simulating the ecological niche. The status and nature of the taxa actually define the conservation strategy required for its preservation, and accordingly various technology-driven methods have been developed across the globe. However, conservation measures need to be adequately compensated not only to make them self-sustained but also to reward the communities concerned, as they are the custodians of this biological wealth and the associated traditional knowledge. The signatories to the Convention on Biological Diversity (CBD), armed with the provisions available in its article for access to genetic resources and benefit sharing (ABS), have formed a legal framework to protect their biodiversity and interest. A judicious and balancing strategy, on a case-to-case basis, formulated

R.A. Nair
Department of Biochemistry and Molecular Biology, Central University of Kerala, Kasaragod, Kerala 671316, India

School of Biotechnology, National Institute of Technology Calicut (NITC), Calicut, Kerala, India

K.S. Jayakumar
Biotechnology and Bioinformatics Division, Jawaharlal Nehru Tropical Botanic Garden and Research Institute, Thiruvananthapuram, Palode – 695562, Kerala, India

P.P. Pillai (✉)
Department of Genomic Science, Central University of Kerala, Kasaragod, Kerala 671316, India
e-mail: padmeshpillai@gmail.com

S. Abdulhameed et al. (eds.), *Bioresources and Bioprocess in Biotechnology*,
DOI 10.1007/978-981-10-3573-9_15

by the member countries linking their preserved germplasm with bioprospecting efforts, will certainly pay dividend in the future.

Keywords
Plant genetic resources (PGRs) • Gene banks • Botanic garden • Seed bank

15.1 Introduction

Plants form a vital part of the world's biological diversity by maintaining the Earth's environmental balance and ecosystem stability. Plant genetic resources (PGRs) constitute the most valuable and essential material to meet the current and future needs of crop improvement programmes (Hammer and Teklu 2008). Besides ensuring sustainable food security and protecting environment, PGRs have also fuelled the development of new industrial products as well as major research breakthroughs (Beattie et al. 2005). Being finite resources of economic and ethical value, it becomes the primary responsibility of the region containing this potential wealth to make systematic collection and conservation of PGR, for posterity (FAO 1996a, b). In addition to the perspective of conservation, currently there is a greater need for conserving as many crop variety options as possible for future breeding programmes to develop biotic/abiotic resistance/tolerance (Plucknett et al. 1987). The agriculture sector is confronted with multiple problems such as emergence of new pests and diseases due to the global phenomenon of climate change. Since the nature of these problems is unpredictable, genetically diverse PGRs have to be collected and conserved for future use before any irrecoverable loss occurs to genetic diversity. Studies indicate that many plant species are threatened by a combination of environmental and social factors that include loss of habitat due to deforestation and urbanization, pollution and climate change, invasive alien species, over-exploitation, changes in consumer/farmer preferences and predominance of preferred crop varieties (St Amand et al. 2000; Teklu and Hammer 2006), all of which adversely affects the biodiversity. Estimates indicate that of the 80,000 edible plant species, only 30 species account for >99% of those under human consumption (Saad and Rao 2001). FAO estimates suggest that since 1900, there has been a loss of approximately 75% of genetic diversity in agricultural crops (Saad and Rao 2001; FAO 2010). The present number of globally threatened plant species is listed as 11,233 in the International Union for Conservation of Nature (IUCN) Red List (IUCN Red list 2015). Statistics from IUCN Red List indicate that the disappearance of such vital and large amount of biodiversity presents one of the greatest challenges for the world community [Secretariat of the Convention on Biological Diversity (CBD) 2009]. Genetic variation being the raw material for evolutionary change within populations (Frankel and Soulé 1981), PGR conservation is a fundamental concern to revert the unabated gene erosion. Erosion of these genetic resources may decrease the evolutionary potential of species to persist in the face of abiotic and biotic challenges and will pose a severe threat to the world's food security. The predicament and challenges posed to global biodiversity were addressed by CBD in 2002 by adoption of the updated Global Strategy for Plant Conservation (GSPC) 2011–2020 to check this continuing loss of plant diversity and to ensure sustainable development (Secretariat of the CBD 2009). In this context, concerted effort to conserve PGRs assumes greater significance and holds promise to the posterity.

15.2 Gene Bank Concept

To preserve the valuable agro-biodiversity for future generations, local, regional and international efforts have been setting up collections of genetic resources, called gene banks, for many decades. Importance of the establishment of gene bank for preserving tropical and sub-tropical germplasm was spearheaded by FAO (Wilkes 1983) in 1960 by recommending establishment of a global network of crop gene banks (Frankel and Bennett 1970; Frankel and Hawkes 1975). Major milestone in the field of PGR conservation was made by Nikolai Vavilov (1920–1940) following the acceptance of the concept of "centres of origin of plants and animals" (Vavilov 1926). The systematically assembled germplasm collections of Vavilov and colleagues resulted in establishing the world's first gene bank in 1940. These collections harbour traits that include drought and heat tolerance, nutritional quality, disease resistance and other characteristics that can contribute to more sustainable and productive agriculture.

Unequivocally, botanic gardens and gene banks are repositories where plant material is collected, stored, documented and made available for redistribution. While botanic gardens display their live collections in the form of field gene banks, preserved genetic diversity of the plant is maintained either as propagules or *in vitro* cultures, in gene banks. Gene banks are sometimes also referred to as an *ex situ* conservation facility, as biological materials are conserved outside their natural habitat. Many of the gene banks that constitute priceless repositories aim at conserving wild germplasm, in addition to preserving genetic diversity, as core collections of PGR from centres of origin and elsewhere.

Therefore, the purpose of gene banks is primarily to conserve the available genetic diversity in the taxa and subsequently make this material, together with associated information, available for future use in research and plant breeding. As evident, conservation programmes, either *in situ* or *ex situ*, demand tireless effort, time and money, and hence mere preservation will not sustain the programme in the long run (Engels 2004). However, translating this rich repository of genetic and chemical diversity in plants for identification of novel genes and molecules of therapeutic and biological significance, through bioprospecting programmes, will be not only rewarding but also render conservation as a sustainable programme.

15.3 Components of Gene Bank

PGRs can be conserved either within or outside their natural habitat, and accordingly they are called as *in situ* and *ex situ* conservation strategies, respectively. PGR conservation in reserves/protected areas and on-farm and home gardens falls under *in situ* conservation approaches that conserve biodiversity at species and ecosystem level in their natural environments (McNeely 1996; Watson and Eyzaguirre 2002). *In situ* conservation approach is dynamic and maintains the evolutionary potential of species (Frankel 1970; Frankel and Soulé 1981; Ledig 1988, 1992). For successful *in situ* conservation in reserves/protected areas, regular inventorying of species, observations on general ecological conditions, habitat alterations and monitoring of genetic diversity by studying morphological and molecular variations need to be carried out (Berg 1996). Sustainable *in situ* conservation requires community participation and must benefit local communities (McNeely 1994, 1996; Brush 1995). Participatory conservation approaches will help in bridging traditional knowledge (TK) with scientific theories on conservation, sustainable utilization and benefit sharing (Arunachalam 2014). The realization that traditional 'farming' system is an important reservoir of unique genetic diversity (Altieri and Merrick 1987) led to adoption of conservation strategies for preserving crop plants and their wild relatives in the field and is referred to as on-farm conservation (Jarvis and Hodgkin 2000). Such an *in situ* conservation helps in maintaining traditional crop cultivars called landraces in addition to conserving the traditional farming systems (Hodgkin et al. 1993; Ramanatha Rao et al. 1997). Home garden conservation akin to on-farm conservation and consisting of multiple units involving a community plays an important role in genetic diversity conservation and development of a wide spectrum of vegetables, fruits, medicinal plants and spices (Evenson 1986; Michon et al. 1986). Both on-farm and home garden conservation approaches help in conserving neglected and underutilized species. The two approaches are dynamic in that they allow both natural and human selection processes to continue (Sthapit et al. 2009). *In situ* conservation approaches are vulnerable to natural and human-made disasters and changes in management practices.

As mentioned earlier, gene banks can operate through *in situ* or *ex situ* mode or, as more often practised, a combination of both by suitable balancing strategy (Plucknett and Smith 2014). The preservation of species *in situ* offers the advantage of allowing natural selection to act, which cannot be recreated *ex situ*. Although maintenance of viable and self-sustainable populations of wild species in their natural state represents the ultimate goal, habitat destruction is inevitable, and endangered species need to be preserved before they become extinct.

Therefore, means of *ex situ* conservation is more explored and developed for maintenance of diversity, as it also provides avenues to study the various aspects of threat, endangerment and rarity encountered by the taxa, in its natural habitat. Besides the *ex situ* approaches also serve to conserve PGR that would otherwise not survive due to destruction of their natural habitat, replacement with landraces, modern varieties and changes in consumer/farmer preference owing to improved performance of modern varieties (Saad and Rao 2001; Jarvis et al. 2008; Ogwu et al. 2014). Furthermore, it also has the advantage of preserving plant material and making it available for research purposes, without damaging the natural populations.

The *ex situ* conservation practices globally adopted for gene bank activities are briefly discussed below.

15.4 Botanic Gardens

Conventional *ex situ* approaches include conservation of PGR in botanic gardens, field gene bank, pollen bank, DNA bank, seed bank, *in vitro* storage and cryopreservation (Hawkes et al. 2000; Tao 2013). PGR conservation by way of establishing botanic gardens served as ideal grounds for conserving and introducing economic plants to new regions besides conducting breeding experiments for new and improved varieties (Wyse Jackson and Sutherland 2000; Heywood 2011). The first comprehensive attempt to involve botanic gardens in PGR conservation was initiated in 1953 at an International Colloquium of the Sub-committee of Botanic Gardens of the Union of Biological Sciences (IUBS). Using information from world botanic gardens, the International Red Data Book for angiosperms was prepared (Melville 1970) to document the threatened PGR, funded by IUCN (International Union for Conservation of Nature). The plan for conservation of threatened plant taxa is entrusted with IUCN Botanic Gardens Conservation Secretariat [BGCS, presently Botanic Gardens Conservation International (BGCI)]. Globally botanic gardens contain at least 100,000 plant species (Heywood 2011) that constitute approximately 30% of known plant species. All the botanic gardens are listed in the International Botanic Gardens Directory (Heywood et al. 1990) compiled by BGCI. The Plant Search Database developed by BGCI currently holds records for over 1,328,115 collections representing 483,310 taxa in botanic gardens (www.bgci.org/plant_search). Of the more than 3314 botanic gardens on a global scale (www.bgci.org/garden_search; BGCS 1989; Heywood et al. 1990), around 230 are in tropical countries (Heywood 1987). The statistics indicate that there is considerable imbalance in global distribution of botanic gardens (Fig. 15.1) as evident from the existence of few botanic gardens focusing on PGR conservation in tropical countries (Groombridge 1992) despite the fact that tropics account for more than 90% of world diversity. In India, there are over 33 government-managed botanic gardens.

Botanic gardens can also be considered as a field gene bank, only difference being that the latter are traditionally used for conserving perennial plants or species of agriculture importance that produce non-orthodox or recalcitrant seeds, vegetatively propagated plants and plants that produce no/less seeds (SGRP-CGIAR). Botanic gardens on the other hand focus on conservation of wild, ornamental, rare and endangered species.

15.5 Seed Bank

Technological developments for long-term preservation of biological materials by drying them to low moisture content (MC) (Probert 2003) followed by storing at sub-zero temperatures saw the emergence of dramatic changes in PGR conservation strategies. This led to the development of seed, cryo- and pollen bank facilities. However, the major consideration for long-term conservation of germplasm collections is to preserve the seeds as propagules. For this, the determination of the seed

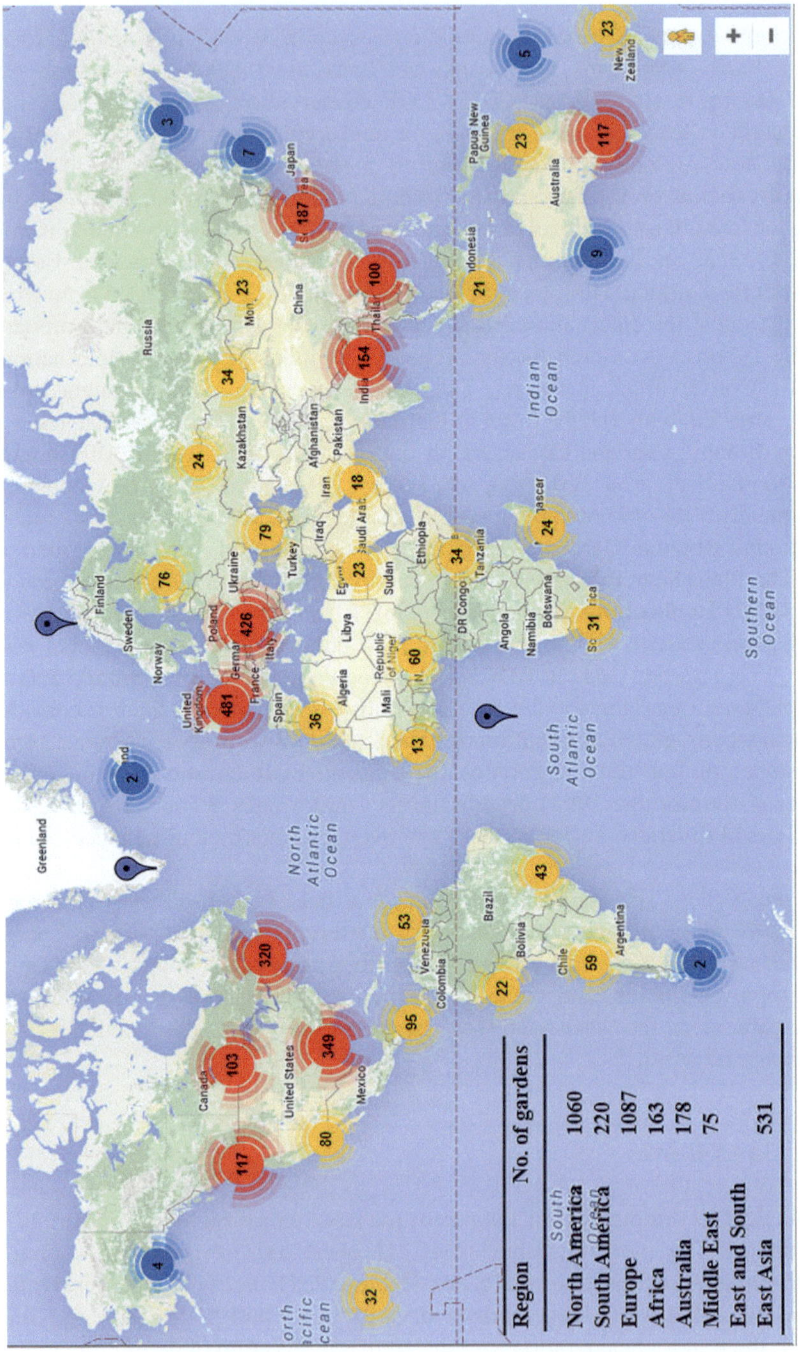

Region	No. of gardens
North America	1060
South America	220
Europe	1087
Africa	163
Australia	178
Middle East	75
East and South East Asia	531

Fig. 15.1 Global distribution of botanic gardens (Source: www.bgci.org/map.php)

behaviour of each individual species to be preserved during storage under dry and cold condition needs to be identified and standardized. If the seeds can be dried to low-percentage humidity such as below 8%, in the majority of cases, they are found to withstand very cold temperatures of below 20 °C. When the seeds tolerate these extreme conditions and remain viable for years together in storage, they are classified as orthodox (desiccation tolerant) as opposed to recalcitrant (desiccation intolerant) when they do not. Undeniably, the most cost-effective method of preserving PGR for long-term *ex situ* conservation is through the storage of seeds under very specific conditions, following techniques well developed for various plant species by research institutes across the globe. The main advantage of seed banking, as seen through success stories, is that it allows large populations to be preserved and genetic erosion to be minimized, by providing optimum conditions and reducing the need for regeneration. The success or efficiency of a seed bank is dependent more on continuous viability monitoring and regeneration or re-collection, when the viability of the sample drops below a minimum level.

As *ex situ* PGR conservation by seed banks is efficient, reproducible, of low maintenance cost, conserves a wide diversity of each target taxon and has ease of characterization and utilization. In view of this, there has been an increase in the number of seed storage facilities. Many of the botanic gardens have developed adjuncts for seed banks in addition to *in vitro* banks and cryo-banks. Globally the botanic gardens are reported to maintain 250,000 seed bank accessions (Wyse Jackson 1999). BGCI figures estimate that currently 400 botanic gardens have seed banks maintaining seeds in long-term and medium-term storage (www.bgci.org/resources/seedbanks). In view of the fact that PGR conservation as seeds is safe, cheap and convenient, more than 80% of PGR is conserved as seeds in various gene banks (FAO 1996a, b; Laliberte 1997). Estimates suggest that among the global collection of accessions in gene banks, approximately 6.1 million plant accessions are stored as seed banks in 1320 national, international and regional gene banks [World Information and Early Warning System-Food and Agriculture Organization (WIEWS-FAO) databases] around the world.

It is now pertinent to mention the establishment of Svalbard Global Seed Vault (SGSV) in the permafrost in North Pole in 2008, which provided a safe seed storage facility to safeguard the accessions of various global gene banks. Various gene banks across the world have deposited duplicates or backup copies of accessions in SGSV (Table 15.1). Currently SGSV has holdings of over 8,60,000 samples originating from different parts of the world.

Technically, for storage of seeds in gene bank, the healthy material after cataloguing is multiplied, if required, and is sorted for either short-term (ambient temperature or air-conditioned) or medium-term storage (0 °C to –5 °C) in hermetically sealed bottles/laminated aluminium packets or long-term storage at temperatures below –20 °C after drying to low moisture content (MC) (2–7%) as per standard specifications (Royal Botanic Garden Seed Information Database) and sealing under vacuum (Rao et al. 2006). As specified in Genebank Standards (FAO/IPGRI 1994), the lower the MC and lower the storage temperature, the longer is the seed storage life. Only plant species that produce orthodox seeds can withstand the drying and low-temperature conditions specified in Genebank Standards (FAO/IPGRI

Table 15.1 Germplasm holdings in Svalbard Global Seed Vault (SGSV) of selected global gene banks

	Number of duplicates deposited in SGSV			
Depositor institute	Accessions	Taxa	Species	Countries
Centre for Genetic Resources (Netherlands)	19,713	388	260	152
Department of Agriculture, Food and Rural Development (Ireland)	100	4	4	4
N.I. Vavilov All-Russian Scientific Research Institute of Plant Industry (Russian Federation)	5278	411	112	96
National Plant Germplasm System (United States of America)	88,647	2061	1580	18
Nordic Genetic Resource Center (Norway)	17,571	492	322	79
AVRDC – The World Vegetable Centre (Taiwan)	14,411	163	120	117
Centro Internacional de Agricultura Tropical (CIAT), Brazil	50,272	647	640	137
The International Maize and Wheat Improvement Center (CIMMYT – Centro Internacional de Mejoramiento de Maíz y Trigo), Mexico	1,30,291	20	11	101
International Center for Agricultural Research in the Dry Area (ICARDA), Syria	1,16,484	780	375	129
ICRISAT, India	1,10,014	14	12	134
International Rice Research Institute (IRRI), Philippines	1,22,060	67	50	129
NBPGR, India	25	1	1	1

Source: www.nordgen.org/sgsv; Accessed January 2016

1994). When going for PGR conservation by seed banks, maintenance of viability and genetic integrity of the seeds is of prime significance [Seed Information Database (SID)-Royal Botanic Gardens) and hence is assessed periodically according to standard protocols as outlined by the International Seed Testing Association (ISTA, 2008). Even though majority of plant species produce orthodox seeds, there are many economically important plant species that produce recalcitrant seeds; a list detailing this has been published in the *Handbooks for Genebanks No. 1* (Cromarty et al. 1990). Recalcitrant seeds cannot be dried below a critical MC and are sensitive to storage at low temperature, and hence PGR conservation for plant species producing recalcitrant seeds, for vegetatively propagated plants and for plants that produce little/no seeds and alternative methods have been developed with their respective merits, such as field gene banks, *in vitro* germplasm collections and pollen and DNA banks. This has been an alarming problem with botanic gardens as many species in their germplasm collection belong to the category of recalcitrance. Therefore, the other complementary techniques developed for *ex situ* conservation are as follows.

15.6 Field Gene Banks

Field gene banks are living collections of plant taxa, adopted as the main conservation strategy for long-lived perennials, recalcitrant species and vegetatively propagated species, by simulating the ecological niche prevalent in their natural habitat. However, the strategy is limited by the large land area required and also the magnitude of effort needed to maintain and protect them from natural disasters. Furthermore, outbreeders require controlled pollination for regeneration from seed. Despite these besetments, in many circumstances, they are the only available option for the conservation of important germplasm.

15.7 In Vitro Storage of Germplasm

Cryopreservation and plant tissue culture methods are the two widely practised techniques for *in vitro* mode of germplasm preservation. *In vitro* cultures involve development of sterile cultures and are often used for species wherein seed conservation is not feasible (Engelmann and Rao 1996). Cryopreservation is another promising *ex situ* conservation strategy for stable long-term conservation of PGR (Kartha and Engelmann 1994) wherein tissues are stored at ultra-low temperature, commonly using liquid nitrogen at −196 °C that will suspend/reduce metabolic activity of the tissues (Reed 2008; Lync et al. 2011). The storage of germplasm in laboratory conditions (*in vitro*) is specially suited for the long-term conservation of recalcitrant species and vegetatively propagated species. The cryopreservation method has so far been successful with only a relatively few species, as it requires more standardization with a wide range of reagents, but is a very promising technique for long-term storage. Though the technique requires dedicated equipments and skilled manpower, the potential is enormous for long-term preservation of planting materials without losing their regeneration ability. However, more research is necessary to define the mechanisms of desiccation and chilling injury and to investigate methods of alleviating it (Eberhart et al. 1991).

Although micropropagation method is relatively inexpensive, it is labour intensive and time consuming, as it needs careful standardization of media components and culture conditions for whole plant regeneration. The situation is more complicated with tree species compared to herbs and shrubs owing to inherent biochemical and physiological intricacies associated with tree forms.

15.8 Pollen Banks

Depending on the taxonomic status of the plant species, PGR conservation by preserving pollen grains is possible either by storage under appropriate physical conditions (0–10 °C, 10–30% relative humidity) or by cryopreservation techniques that ensures pollen viability (Barnabas and Kovacs 1997). Pollen is a rich source of diverse alleles within a gene pool and hence can be an effective propagule for gene banks. The advantages with easiness of pollen storage and shipment provide researchers with increased options when designing their breeding programmes. As with other *ex situ* methods, there are different steps in pollen banking that needs optimization, which again differ within species. This includes such steps like pollen collection, desiccation, viability testing and longevity assessment, which fortunately have been developed for many species of interest. However, pollen banks are beset with such technical problems like limited pollen production in some species, labour-intensive steps in collection and further processing, non-availability of standardized processing or viability testing protocols for a wide range of taxa and poor regeneration potential of aged pollen. Despite these challenges, pollen is still a valuable genetic resource for conservation as it provides breeders and researchers with an additional propagule that may be immediately useful in their programmes. Furthermore, pollen preservation may be useful for base collections of species that do not produce orthodox seeds.

15.9 DNA Banks

Another *ex situ* conservation strategy is by development of a DNA bank (Callow et al. 1997) that aims at conserving DNA from the target plant species by constructing genomic libraries. DNA collection in the form of banking has become an important resource in the global effort to address biodiversity crisis and hence a choice of preservation. *Ex situ* conservation by DNA bank is generally not followed for PGR conservation in view of the many disadvantages associated with the technique such as being time consuming, conserving only a single sample in one library and impossibility of regenerating entire plant from DNA. However, the potential remains promising and challenging as it certainly complements the ongoing activities undertaken in *ex situ* conservation programmes.

As widely accepted, DNA sequence analysis is useful in the identification and delimitation of species and higher taxa and is also increasingly important in DNA taxonomy and DNA barcoding (Kristiansen et al. 2005; Chase et al. 2005). It is a powerful resource for molecular phylogenetics that include both biogeographical and phylogeographical studies. Furthermore, DNA sequence information can easily be made available via the Internet and made globally accessible within existing initiatives such as the Global Biodiversity Information Facility (Bisby 2000; http://www.gbif.net), Species 2000 (http://www.species2000.org) and regional databases. Therefore, the creation of a network of DNA banks (DNA Bank-Net) is nevertheless a complementary effort in gene bank activity.

As observed, the use of a single conservation method will not be appropriate to conserve genetic diversity of all plant species. Therefore, combination of different methods is adopted for maximizing effective PGR conservation programmes (Engels and Visser 2003). The Consultative Group on International Agricultural Research (CGIAR) centres established 11 gene banks for conservation and utilization of PGR. Estimates indicate that there are approximately 1750 gene bank collections around the world (FAO 2010). Among the global collection of accessions in gene banks, 60% are in long-term or medium-term facilities, 8% in short-term facilities and remainder conserved in field gene banks or by cryopreservation techniques (Saad and Rao 2001).

15.10 Documentation of Gene Bank

The conserved germplasm in any of the earlier discussed *ex situ* techniques is not of much significance unless it is properly documented and characterized at different levels, to understand the existing variability. Thus analysis of variability is the prerequisite to initiate programmes towards translating the richness in diversity to products/processes of utility by bioprospecting efforts. Towards enhancing usability of conserved PGR collections and for identification of distinct accessions, documentation of acquisition process, field management, regeneration, characterization, evaluation and distribution, data is important (Painting et al. 1993). Standards for documenting data such as the Biodiversity International (Biodiversity International 2013) crop descriptors as well as genetic marker descriptors (De Vicente et al. 2004; Alercia et al. 2012) should be used. Global effort to collect and conserve germplasm by setting up gene banks worldwide was stimulated with the realization that the world was rapidly losing much of its agro-biodiversity. Success of this collection effort opened up new challenges on maintaining the collections and characterizing those using molecular tools in addition to morphotaxonomic keys (Bretting and Widrlechner 1995). Wide varieties of molecular technologies are available for germplasm characterization and are increasingly being used to complement traditional germplasm management approaches (Spooner et al. 2005). Molecular approaches will solve taxonomic ambiguities by providing information on genetic variation within and between populations thereby eliminating redundancies and guiding the assembly of core collections of manageable size (Frankel and Brown 1984; Hamon et al. 1995) for germplasm conservation. Many research programmes are already underway in which entire collections are screened with molecular markers. Examples of these include (1) the EU-funded project, "Molecular Markers for Genebanks" (http://www.cgn.wur.nl/pgr/research/lettuce/) for evaluating lettuce collection of the Centre for Genetic Resources, Netherlands, (2) TEGERM (http://www.biocenter.helsinki.fi/bi/tegerm/) and (3) CGIAR Challenge Program, "Unlocking Genetic Resources in Crops for the Resource-Poor" (http://www.cgiar.org/research/res_cppilot.html). Many bioinformatic tools have been developed for analysis of the enormous data generated from molecular characterization experiments such as EU-funded "GENE-MINE" (http://www.gene-mine.org/) and BBSRC-funded project "GERMINATE" (http://bioinf.scri.sari.ac.uk/germinate/). For many of the endangered medicinal plant species, very little of the material is available in the gene banks; one reason that could be attributed is that the best means for conserving many of the medicinal plants is by *in situ* methods in their natural habitats in reserves/national parks (Heywood and Dulloo 2005). In the face of an increasing public interest in the use of plants as medicines and fast disappearance of natural vegetation, molecular methods are increasingly being used for characterizing germplasm conserved by *in situ* approaches, including endemic and threatened plants for which genetic interrelationships and place of origin can be clarified. As an example, in one of the author's Department of Biotechnology [Government of India (GoI)]-funded National Gene Bank Programme on Medicinal Plants, two DNA markers, viz. Randomly Amplified Polymorphic DNA (RAPD) and Inter-simple Sequence Repeat (ISSR), were used for genetic diversity assessment of *Decalepis arayalpathra* [(Joseph & V. Chandras) Venter], a critically endangered endemic medicinal plant with narrow and restricted distribution. The accessions were collected from populations spread across three different locations in the Bonacaud region, a forest reserve area in the Western Ghats of India. In the light of the extent of genetic variation in the three populations analysed (Fig. 15.2), the study could help in making appropriate recommendations for *in situ* conservation.

For comprehensive storage of information, increased attention is being paid for computerized documentation and management of resources (Carrascal et al. 1995; Quek and Ramanatha Rao 1995; Quek and Zhang 1995; Hunt and Sherill 1998). Several germplasm information management systems have been developed specifically for information management of gene banks such as GRIN-Global (http://www.ars-grin.gov), GENESYS (http://www.genesys-pgr.org), Mansfield Database (IPK) (http://mansfeld.ipk-gatersleben.de/pls/htmldb_pgrc), SESTO (NordGen) (http://www.nordgen.org/sesto) and International Crop Information System (ICIS) platform (http//irri.org/knowledge/tools/international-crop-information-system).

15.11 Restoration Activities of Gene Bank

Apparently, the purpose of gene banks is not to simply safeguard genetic diversity for the future but also to encourage its use. They provide direct help to farming communities, for example, by reintroducing 'extinct' landraces or by providing material for crop breeding programmes. Plenty of illustrations can be cited wherein the gene banks have made significant difference to farmers' livelihoods and agricultural output (Table 15.2). In Kenya, napier (or elephant) grass – a valuable fodder crop for thousands of smallholder dairy farmers – has suffered from attacks by a devastating smut fungal disease. The Kenyan Agricultural Research Institute was able to develop disease-resistant varieties of this tropical grass after accessing napier grass germplasm housed by the International Livestock Research Institute's (ILRI) fodder crop gene bank. Meanwhile, the International Rice Research Institute (IRRI) analysed 4317 rice varieties and found that 100% of the varieties released by IRRI and 90% of non-IRRI varieties had at least one gene bank accession in their pedigree. These examples are proof of the vital role of gene banks in agriculture and for ensuring food security. In view of this, concerted efforts are underway at national and international levels to streamline and regulate PGR accessibility at global level.

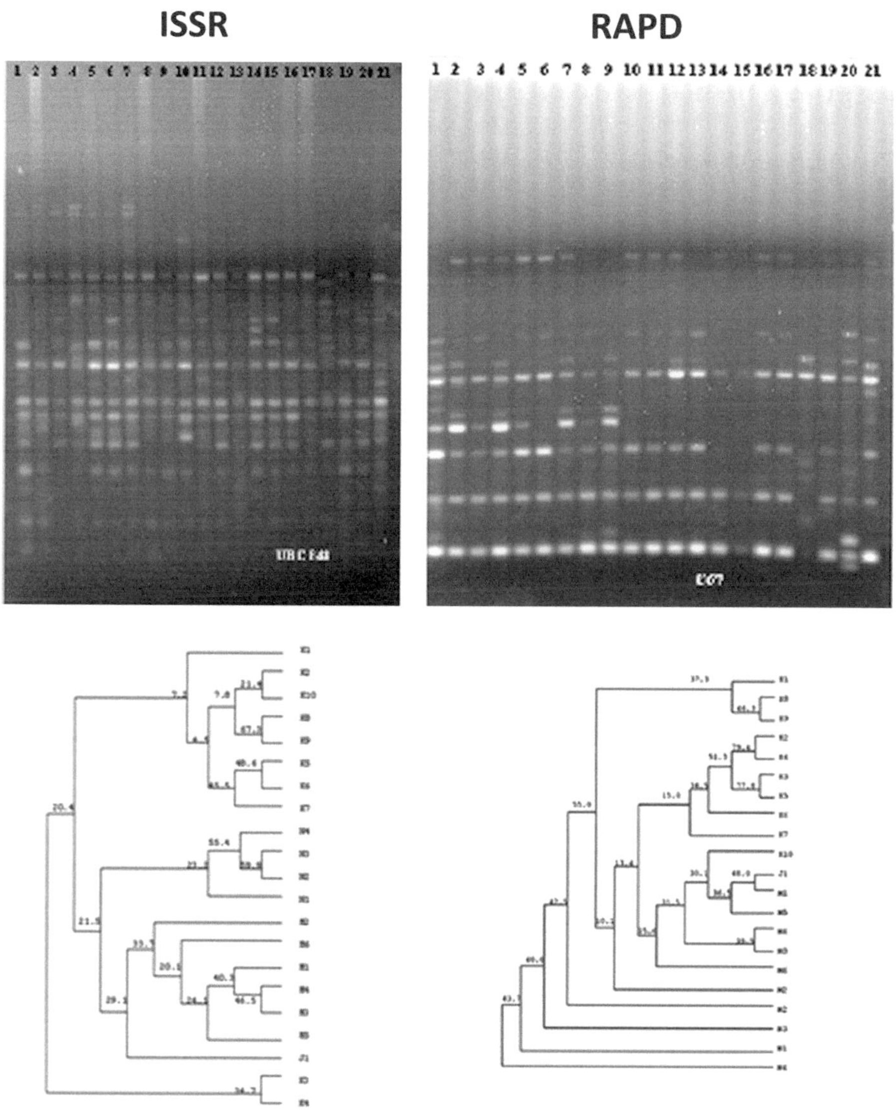

Fig. 15.2 (**a**) ISSR and RAPD profile of *Decalepis arayalpathra* using respective primers (**b**) ISSR and RAPD phenogram based on UPGMA analysis of *Decalepis arayalpathra.* By and large, in both the phenograms, grouping of accessions was in accordance to their geographical distribution

15.12 Gene Bank Activities: Indian Scenario

In terms of plant diversity, India is ranked tenth in the world with approximately about 46,042 species of flowering and non-flowering plants that account for more than 11% of the world's floral diversity (Goyal and Arora 2009). The wide diversity

Table 15.2 Examples showing the role of gene bank in PGR restoration

S. no.	Species	Contributing gene bank	Examples
1	Rice (*Oryza perennis*)	IRRI	Resistant to ragged stunt virus; collections deposited in IRRI reintroduced in Taiwan
2	Maize	CIMMYT	84 maize accessions lost following severe earthquake (1971) in Nicaragua were replenished
3	Maize	CIMMYT	Interruptions in electricity supply (1975) caused extensive damage to germplasm collections in Nicaragua that were replenished
4	Rice	IRRI	Political strife (1970) disrupted agricultural production in Kampuchea, and unique rice varieties lost were restored
5	Wild barley	Vavilov Institute	Wild species that became extinct in Ethiopia were reintroduced

Source: Plucknett and Smith (2014)

of PGR in the Indian subcontinent can be attributed to its varied geography, diverse ecosystems, rich agricultural heritage and ethnic diversity (Jacob et al. 2015). India is one of the 12 centres/regions of diversity of crop plants in the world (Zeven and de Wet 1982) with rich genetic diversity occurring in several crop plants and their wild progenitors and approximately 33% of its botanic wealth being endemic (Nayar 1980). These are concentrated in the 4 global biodiversity hotspots (Northeast India, the Western Ghats, Northwest Himalayas and the Andaman and Nicobar Islands) (Myers et al. 2000) and 15 biogeographic zones (Western Himalayan Zone, Eastern Himalayan Zone, Western Dry Region, Western Ghats, Lower Gangetic Plain, Middle Gangetic Plain, Upper Gangetic Plain, Trans Ganga Plain, Eastern Plateau and Hills, Central Plateau and Hills, Western Plateau and Hills, Southern Plateau and Hills, Eastern Coastal Plains and Hills, Gujarat Plain and Hills, the Island Region) (Khanna 1989). The traditional agricultural practices in India have also contributed to enhance the diversity in PGR by a continuous stream of introductions of new crops and their cultivars by man since ancient times. Current estimates indicate that the rich floristic wealth found in India numbers around 166 species (Zeven and de Wet 1982) with wild gene pools of direct or indirect utility accounting for around 320 species, distributed in 15 phytogeographical/agro-ecological zones (Arora and Nayar 1984). Due to this rich diversity of cultivated and the wild crop gene pools in phytogeographical/agro-ecological zones, India is the primary centre of diversity for crops and a secondary centre of diversity for many African crops like finger millet, sorghum, cowpea, cluster bean, okra, etc. besides high regional diversity in certain crops like mung bean, sword bean, tomato, citrus, cardamom, ginger, turmeric, tuber crops, etc. due to geographical contiguity with the Far East and/or the Indo-Malayan region.

India's rich crop genetic diversity is evident in the occurrence of primitive landraces and wild types of various cereals (e.g. rice), legumes (e.g. mung bean, cowpea, pigeon pea) and other horticultural crops (sugarcane, banana, mango, citrus, eggplant, okra, cucumber, sesame, jute, black pepper, ginger and yams) (Smith and Blumberg 1990). Due to the threat of habitat degradation, their collection and conservation have become a national priority. Even though the need to conserve genetic resources was taught and decreed in China and India as early as 700 BC, a systematic and scientific attempt to conserve gene pool in India by *ex situ* approaches is a relatively young venture. The National Gene Bank (NGB) established by the Indian Council of Agricultural Research (ICAR) as part of the National Bureau of Plant Genetic Resources (NBPGR) in 1985–1986 (NBPGR 2011) was a start-up initiative in this direction. Besides housing the seed bank, the NGB also contains a cryoconservation bank, *in vitro* tissue culture bank and field gene bank. The seed bank contains a total of 392,163 accessions of 1761 spp. (NBPGR 2014), and the *in vitro* tissue culture bank contains approximately 1969 accessions of 158 species (Goyal and Arora 2009). Around 8493 accessions of 720 species are conserved in cryobanks (Goyal and Arora 2009), and for vegetatively propagated plants, about 40,000 germplasm accessions are being maintained by a network of field gene banks (MoEF 2010). NGB in India is one among the three largest national gene banks in the world, the other two being in China and the USA (Zhang et al. 2014). The NGB at NBPGR operates in a network mode to have an effective linkage with other national and international organizations engaged in PGR management and is entrusted to maintain the base collection. The major partners in the network include ten regional stations/base centres/quarantine centres, spread over different phytogeographical zones in the country, and 59 national active germplasm sites located in various Indian Council of Agricultural Research (ICAR) institutes, National Research Centres (NRC) and All India Coordinated Crop Improvement Projects and State Agriculture Universities (SAUs) (Sharma and Brahmi 2011; Patra et al. 2015). Information of germplasm maintenance at various institutes across India is summarized in Table 15.3.

The country also has one of the world's largest germplasm repositories of food crops grown in the semiarid tropics at International Crop Research Institute for Semi-arid Tropics (ICRISAT) located in Patancheru, Andhra Pradesh. The gene bank facility at ICRISAT that contains hundreds of wild species as well as many farmer-developed varieties or landraces, which are fast disappearing in nature, has been a major source of diversity for crop improvement programmes in India. The facility that maintains more than 120,000 accessions serves as a world repository for five major crops (sorghum, pearl millet, chickpea, pigeon pea and groundnut) and six lesser-known millet species (finger millet, foxtail millet, proso millet, kodo millet, little millet and barnyard millet). It contains around 555 wild pigeon pea accessions, 37,949 sorghum accessions and 720,000 dry land cereal germplasm (Reddy et al. 2006). The ICRISAT gene bank was designed to withstand natural disasters and has numerous backup systems to keep the collection secure. Towards providing additional safeguards and for promoting national and regional research programmes, three smaller facilities were also established in Africa. The facility has also ensured that nearly 100,000 samples have been duplicated and are stored at the Svalbard Global Seed Vault in Norway under the auspices of the Global Diversity Trust and the Nordic Gene Bank.

Table 15.3 Information of germplasm maintenance by *ex situ* approaches at selected Indian institutes

Crop	Institute	Approximate number of accessions conserved	Conservation method
Rice	Central Rice Research Institute (CRRI) Cuttack and	12,000	Field bank
	CRRI, Hyderabad	24,000	Seed bank
Wheat and barley	Directorate of Wheat Research (DWR), Karnal	7000	Seed bank
Maize	Directorate of Maize Research (DMR), New Delhi	2500	Seed bank
Sorghum	NRC on Sorghum, Hyderabad	7366	Seed bank
Pearl millet	AICRP on Pearl Millet	3100	Seed bank
Small millets	AIC Small Millets Improvement Project, Bangalore	13,290	Seed bank
Pulses	Indian Institute of Pulses Research (IIPR), Kanpur	6395	Seed bank
Groundnut	NRC on Groundnut, Junagadh	8963[a] and 84[b]	[a]: Seed bank; [b]: Field bank
Oilseeds (sunflower, safflower, castor)	Directorate of Oil Seeds Research (DOR), Hyderabad	10550[a] and 1329[b]	[a]: Seed bank; [b]: Field bank
Rapeseed and mustard	NRC on Rapeseed and Mustard, Bharatpur	8082	Seed bank
Soybean	NRC on Soybean, Indore	2500	Seed bank
Oil palm	NRC on Oil Palm, Pedavegi, A.P	103	Field bank
Cotton	Central Institute for Cotton Research (CICR)	8879	Seed bank
Jute	Central Research Institute for Jute and Allied Fibres (CRIJ&AF), Barrackpore	3226[a] and 1427[b]	[a]: Seed bank; [b]: Field bank
Fodder crops	Indian Grassland & Fodder Research Institute (IGFRI), Jhansi	6267	Seed bank
Sugarcane	Sugarcane Breeding Institute (SBI), Coimbatore	5861	Seed bank
Banana, plantain	NRC on Banana, Tiruchirappalli	907	Field bank
Potato	Central Potato Research Institute (CPRI), Shimla	457[a]; 22342[b]; 1471[c]	[a]: Seed bank; [b]: Field bank;[c]: *in vitro*/cryo-bank
Cashew	NRC for Cashew, Puttur	519	Field bank
Citrus species	NRC on Citrus, Nagpur	150	Field bank
Grapes	NRC for Grapes, Pune	600	Field bank

(continued)

Table 15.3 (continued)

Crop	Institute	Approximate number of accessions conserved	Conservation method
Leechi, bael, aonla and jackfruit	NRC on Litchi, Muzaffarpur	2426	Field bank
Medicinal and aromatic plants	NRC on M & AP, Anand	190	Field bank
Mango, Guava, litchi	Central Institute for Subtropical Horticulture (CISH), Lucknow	848	Field bank
Mulberry	Silkworm and Mulberry, Hosur	806	Field bank
Onion and garlic	NRC for Onion and Garlic, Pune	1066	Field bank
Orchids	NRC for Orchids, Pakyong, Sikkim	1500	Field bank
Plantation crops	Central Plantation Crops Research Institute(CPCRI), Kasaragod	522	Field bank
Spices	Indian Institute of Spices Research (IISR), Kozhikode	5695	Field bank
Tea	UPASI Tea Research Foundation (TRF), Valparai	400	Field bank
Tobacco	Central Tobacco Research Institute (CTRI), Rajahmundry	2359	Field bank
Tropical fruits	Indian Institute of Horticulture Research (IIHR), Bangalore	1983[a]; 1754[b]	[a]: Seed bank; [b]: Field bank
Tuber crops	Central Tuber Crops Research Institute (CTCRI), Thiruvananthapuram	3871	Field bank
Vegetables	Indian Institute of Vegetable Research (IIVR), Varanasi	16,139	Seed bank

Source: Sharma and Brahmi (2011)
[a]Seed bank; [b]Field bank and [c]*In vitro*/cryobank

Among the major groups of crops conserved in global gene banks, 40% are cereals and 15% are food legumes, while vegetables, roots and tubers, fruits and forages each account for less than 10% (WIEWS-FAO database). Medicinal, spice, aromatic, aquatic and ornamental species of relevance for food and agriculture are rarely found in long-term global public collections. For conservation of medicinal plants, forest departments of states of Andhra Pradesh, Karnataka, Kerala and Tamil Nadu along with the Foundation for Revitalisation of Local Health Traditions (FRLHTs) have set aside 54 forest patches as medicinal plant conservation areas (MPCAs) measuring 200–500 ha each. These selected MPCAs are representative of all forest types and climatic zones of the peninsular region and harbour many species of medicinal plants threatened with extinction (Bhatt 2004). The Himalayan Forest Research Institute, University of Horticulture and Forestry in Solan, Himachal

Pradesh, India and Ayurveda Centre in Himachal Pradesh are also involved in *ex situ* conservation of medicinal plants of higher-altitude and cold desert areas.

India being a signatory to CBD (http://www.cbd.int/convention/convention.shtml), three NGBs were set up by DBT, (GoI) under G-15 countries, a Gene Bank for Medicinal and Aromatic Plants (GBMAP) programme that envisages survey, collection and characterization of wild genetic resources and strengthening of infrastructural facilities for conservation and development of a national database. The three NGBs were set up each at NBPGR in New Delhi; Central Institute for Medicinal and Aromatic Plants (CIMAP) in Lucknow, for conserving MAPs of northern region; and Jawaharlal Nehru Tropical Botanic Garden and Research Institute (JNTBGRI) in Kerala for MAP conservation in peninsular India.

In the NGB for MAP located at CIMAP, Lucknow, there are around 1282 germplasm accessions of 161 plant species in field gene bank at CIMAP, Lucknow, and its 4 research centres situated across the country. In addition to being maintained in field gene bank, CIMAP, Lucknow, also has around 533 accessions of 42 plant species in seed gene bank and 85 accessions of 30 plant species in *in vitro* bank. The NGB at JNTBGRI was set up in 1993 wherein MAP germplasms of peninsular India are conserved through maintenance of field gene banks, seed bank, *in vitro* facility and cryo-bank. Records indicate that the facility conserves approximately 2200 accessions in field gene banks and over 177 accessions of 20 species in seed gene bank.

The country is also credited with the world's largest collection of spice germplasm that includes ginger, pepper, turmeric, cardamom, nutmeg, clove, cinnamon and allspice germplasm in the National Repository for Spice Germplasm at the Indian Institute of Spices Research (IISR), Calicut (Tyagi et al. 2004). The spice germplasm collection at IISR is maintained in *ex situ* and *in situ* conservatories with the Spice Germplasm Database holding documentary information on 2317 black pepper accessions, 436 cardamom accessions, 570 ginger accessions and 780 turmeric accessions (www.spices.res.in/spicegenes).

For conservation of medicinal plants of Western Himalayan Region, DBT established a gene bank at Indian Institute of Integrative Medicine, Jammu [formerly Regional Research Laboratory (RRL)]. For exclusively conserving Ayurvedic plants, a gene bank comprising of field, in vitro and seed bank was established by DBT at the Arya Vaidya Sala Kottakkal, Kerala, India.

15.13 *In Situ* Conservation Approaches in India

PGR diversity is concentrated in centres of diversity (Vavilov 1926), which are often the centres of crop domestication (Harlan 1971). Such centres are thus the prime targets for *in situ* conservation. *In situ* conservation is carried out in field/natural environments for wild relatives of crop plants or wild species. These strategies aim at maintaining the evolutionary processes that continue to shape genetic diversity. Several species-specific projects are being implemented by the Ministry of Environment and Forests (MoEF, GoI), for conservation of flagship plant species like citrus, orchid and rhododendron (MoEF 2010) in their natural ecosystems. For *in situ* conservation of wild and cultivated species/varieties of *Citrus*, the National Citrus Gene Sanctuary in West Garo Hills District, Meghalaya, was established. The sanctuary is spread over an area of 47 km^2. This *in situ* conservation is complemented by *ex situ* collections of *Citrus* species at eight sites in different parts of the country. The rich orchid diversity in the north-east region prompted establishment of Orchid Research and Development Centre (ORDC) at Tipi and Sessa Orchid Sanctuary in Arunachal Pradesh for *in situ* conservation of orchids. For conservation of rhododendrons that are endemic to Sikkim Himalayas, the Shingba rhododendron and the Barsey rhododendron sanctuaries were established.

15.14 Community Seed Banks

Agriculture has tremendously benefited from unstructured marketing system involving bartering. Towards strengthening this bargaining power of farming communities and for protecting traditional knowledge in addition to deriving equitable share from the commercial benefits that may arise from utilizing these resources, community gene banks were established for conservation of local landraces. These comprise medium-term storage facilities wherein PGR conservation is pursued actively by farming communities and non-government organizations (NGOs). One such example is the *Beej Bachao Andolan*, a movement in Garhwal Himalayas to save indigenous seeds from being lost with the introduction of hybrid varieties. The movement was successful in *in situ* conservation of several hundred indigenous varieties of seeds, including those of 40 different crops, oilseeds, medicinal plants and vegetables (Bhatt 2004). The Biological Diversity Act formulated in 2002 in India addresses the challenges of documenting and conserving traditional knowledge. The setting up of People's Biodiversity Registers (PBRs) is a step in this direction, wherein trained scientists along with local community members chronicle the indigenous local flora and fauna and associated traditional knowledge (PBR 2013).

15.15 Bioprospecting

Bioprospecting or biodiversity prospecting in brief is the exploration, extraction and screening of biological diversity and indigenous knowledge for commercially valuable genetic and biochemical resources. As gene banks are repositories of biodiversity richness in the form of live germplasm collections or parts thereof, they also provide instant and accessible resource for various screening procedures meant for isolation of useful traits of agronomic, therapeutic or biological utility. Interestingly, bioprospecting programmes not only complement the gene bank activity but also strengthen their purpose of existence. In the present global scenario, no country is self-sufficient in terms of biodiversity, and even the most biodiversity-rich nations have to depend on other parts of the globe for meeting their bioresource-based needs. Apparently, it has become increasingly significant that bioprospecting programmes hold the key for future development. Incidentally, bioprospecting covers a wide range of commercial activities that have applications in diverse industrial sectors including pharmaceuticals, food and beverages, biotechnology, seed, crop protection, horticulture, botanic medicines, cosmetics and personal care. It is estimated that the current size of the bioprospecting industry ranges between US$ 17.5 million and US$ 30 million and expected to touch about US$500 million by 2050. The commercial significance of the industry, however, is potentially far higher since it provides valuable leads for new product development in several important industrial sectors, as a result of which the industry is driven to look for new species or new products, to expand and diversify in its commercial activity through bioprospecting. As such, they enter into collaborative programmes with collectors in different countries to procure their needed supply of bioresources.

15.16 Linking Gene Banks with Bioprospecting

The botanic gardens and organized gene banks, harbouring the available and existing variability in taxa, are repositories of biodiversity that could cater the needs of both public sector, non-profit R&D institutions and commercially oriented private sector companies. However, several of the biodiversity-rich nations have already developed legal framework with stringent regulations in place, to access and benefit sharing of their biological richness for commercial activity. The legal framework available for member countries to regulate ABS (access to genetic resources and benefit sharing) under the provisions of CBD is shown in Table 15.4.

The major goal of this exercise is to emphasize that bioprospecting programmes in any dimension should provide incentives in terms of monetary benefit to the member country and it should reach the grassroot level so that the community which is the real custodian of biological wealth or traditional knowledge (TK) is benefited. The Panama ICBG programme and the development of Jeevani, an Ayurvedic drug formulation based on the TK of local community in India, are illustrious paradigms linking bioprospecting with bioresource and TK.

In Panama, the project termed International Cooperative Biodiversity Groups (ICBG) was funded by the National Institutes of Health, the US National Science Foundation and the US Department of Agriculture from 1998 to 2008 (Kursar et al. 2007). The programme consists of collections, assay of samples for activity against specific disease targets and purification of bioactivity-guided fractions for active compounds. Through this programme, the scientific community of Panama was greatly benefited from manpower training, infrastructure development and incentives for their conservation programmes. The other notable example is the commercialization and benefit sharing of Jeevani, an Ayurvedic drug developed based on the TK of Kani tribe of Southern India on the medicinal plant, *Trichopus zeylanicus* var. *travancoricus* (Pushpangadan et al. 1988). This benefit-sharing system resulted in implementing Article 8(j) of the Convention on Biological Diversity (CBD), and Dr. Pushpangadan who led this multidisciplinary scientific team was awarded the

Table 15.4 The legal options available for member countries under CBD to regulate access and benefit sharing of their biodiversity

ABS legislative options	Selected countries pursuing these options
Environmental Framework Laws – enabling in nature; entrust competent national authority with responsibility of providing more specific ABS legislation	Gambia, Kenya, Malawi, Republic of Korea, Uganda
Sustainable development, nature conservation or biodiversity laws – implement a number of CBD provisions, take up conservation and sustainable use of biodiversity, establish principles of PIC (Prior Informed Consent) and MAT (Mutually Agreed Terms)	Costa Rica, Eritrea, Fiji, India, Mexico, Peru
Dedicated or stand-alone national laws and decrees on access to genetic resources – specifically design frame works for regulating access	Philippines, draft laws in Brazil
Modification of existing laws and regulations – such as those governing wildlife, national parks, forestry, land and fisheries – to include ABS provisions	Nigeria at national level; State of Sarawak, Malaysia and Western Australia at sub-national level
Regional Measures – multilateral agreements establishing common principles and procedures for regulating access	The countries of the Andean Pact (Bolivia, Colombia, Ecuador, Peru and Venezuela). Regional initiatives under discussion by South East Asian countries and proposed by members of OAU

Source: Glowka (1998); cited in ten Kate (1999) and Afreen and Abraham (2009)

UNEP Borlaug Award in 1998 and the UN-Equator Initiative Award in 2002, at the World Summit held at Johannesburg in 2002.

15.17 Challenges and Constraints

Despite holding an impressive number of accessions and their importance in ensuring food security future, gene banks suffer myriad challenges that include the following:

1. Considerable cost of characterizing, evaluating and cataloguing genetic resources.
2. Lack of secure funding.
3. Due to inadequate characterization of accessions in many gene banks, only a small proportion of PGR is actually used by plant breeders.
4. Due to political reasons, farmers do not have easy access to materials in many gene banks.
5. In developing countries that contain many of the most valuable collections, electricity is expensive, and state-of-the-art facilities are lacking. For example, NGB at NBPGR relies on electricity-based seed storage at −64°F (−53°C). Considering

the expense, other storage methods that cost less are being explored such as the use of liquid nitrogen or development of solar energy to power the gene banks.
6. Infrastructure and human resource for conducting research and development especially in the emerging areas of biodiversity conservation need to be augmented.
7. Strengthening documentation systems.
8. Strengthening in-service training and orientation courses for personnel engaged in conservation programmes.

15.18 Conclusions

The organized gene bank activity, which started as early as 1940, continues to be a driving force in the collection and preservation of plant genetic resources all over the world. Provisions embedded in the articles of CBD actually provided impetus to biodiversity-rich nations, to be proactive in conserving their innumerable and valuable species. However, the major lacuna that still remains is the translation of the preserved diversity of PGRs at the genetic and chemical levels to products of utility. Access to genetic resources and benefit sharing should not be an impediment but rather a judicious negotiation tool for the biologically rich yet economically poor nations, to safeguard and compensate their biological wealth and associated traditional knowledge. The incentives/compensations received would certainly sustain their conservation activities and wellbeing of the communities as well.

Acknowledgements We gratefully acknowledge the Directors of NIT, Calicut; JNTBGRI, Thiruvananthapuram and Vice Chancellor, Central University of Kerala (CUK) for extending all logistic and infrastructure support, in the preparation of the manuscript. We also thankfully acknowledge the Department of Biotechnology, Government of India, for supporting the National Gene Bank Programme at JNTBGRI, as some of the data from the cited programme was reproduced in the text.

References

Afreen S, Abraham BP (2009) Bioprospecting: promoting and regulating access to genetic resources and benefit sharing. Decision 36(3):121–158

Alercia A, Diulghero S, Mackay M (2012) FAO/bioversity multi-crop passport descriptors (MCPD V.2). Rome, FAO and Bioversity International. Available at: http://www.bioversityinternational.org/uploads/tx_news/1526.pdf

Altieri MA, Merrick LC (1987) *In situ* conservation of crop genetic resources through maintenance of traditional farming systems. Econ Bot 41(1):86–96

Arora RK, Nayar ER (1984) Wild relatives of crop plants in India. NBPGR Sci. Monogr. No. 9. 90 p

Arunachalam V (2014) Participatory conservation: a means of encouraging community biodiversity. Plant Genet Resour Newsl 122:1–6 Available: http://nbaindia.org/uploaded/Biodiversityindia/4th_report.pdf. Accessed 20 Oct 2014

Barnabas B, Kovacs G (1997) Storage of pollen. In: Shivanna KR, Sawhney VK (eds) Pollen biotechnology for crop production and improvement. Cambridge University Press, Cambridge, pp 293–314

Beattie AJ, Barthlott W, Elisabetsky E, Farrel R, Kheng CT, Prance I, Rosenthal J, Simpson D, Leakey R, Wolfson M, ten Kate K (2005) New products and industries from biodiversity. In: Millennium ecosystem assessment. ecosystems and human well-being: current states and trends. Island Press, Washington, DC, pp 271–295
Berg T (1996) Dynamic management of plant genetic resources: potentials of emerging grass-roots movements. Rep. No. Study No. 1. FAO
Botanic Gardens Conservation International (BGCI) https://www.bgci.org
Bhatt S (2004) Biodiversity. In: Chhokar K, Pandya M, Raghunathan M (eds) Understanding environment. Sage, New Delhi
Biodiversity International (2013) Crop descriptors list. Rome. Available at: http://www.bioversity-international.org/index.php?id=168
Bisby FA (2000) The quiet revolution: biodiverity informatics and the internet. Science 289:2309–2312
Bretting PK, Widrlechner MP (1995) Genetic markers and plant genetic resource management. Plant Breed Rev 13:11–86
Brush SB (1995) *In situ* conservation of landraces in centers of crop diversity. Crop Sci 35:346–354
Callow JA, Ford-Lloyd BV, Newbury HJ (1997) Biotechnology and plant genetic resources: conservation and use; No. 19 in Biotechnology in Agriculture Series. Cab International, Wallingford
Carrascal MJ, Pau LF, Reiner L (1995) Knowledge and information transfer in agriculture using hypermedia: a system review. Comput Electron Agric 12:83–119
Chase MW, Salamin N, Wilkinson M, Dunwell JM, Kesanakurthi RP, Haidar N, Savolainen V (2005) Land plants and DNA barcodes: short term and long term goals. Philos Trans R Soc Lond B Biol Sci 360(1462):1889–1895
Cromarty AS, Ellis RH, Roberts EH (1990) Handbooks for Genebank No. 1. The design of seed storage facility for genetic conservation. IBPGR, Rome
De Vicente MC, Metz T, Alercia A (2004) Descriptors for genetic markers technologies. IPGRI, Rome
Eberhart SA, Roos EE, Towill LE (1991) Strategies for long-term management of germplasm collections. In: Falk DA, Holsinger KE (eds) Genetics and conservation of rare plants. Oxford University Press, New York. pp 135–145 (ISBN 0-19-506429-1)
Engelmann F, Ramanatha Rao V (1996) In vitro conservation of plant genetic resources: an overview of activities at the International Plant Genetic Resources Institute (IPGRI). In: Normah MN, Narimah MK, Clyde MM (eds) Proceedings of the International Workshop on In Vitro Conservation of Plant Genetic Resources July 4–6, 1995, Kuala Lumpur. UKM, Bangi, Malaysia, pp 241–252
Engels JMM (2004) Plant genetic resources management and conservation strategies: problems and progress. In: Davidson CG, Trehane P (eds) Proceedings XXVI IHC—IVth international symposium taxonomy of cultivated plants. Acta Horticulturae 634, ISHS 2004 Publication
Engels JMM, Visser, L (2003) A guide to effective management of germplasm collections. Handbooks for Genebanks No. 6. IPGRI, Rome
Evenson S (1986) Farmer practice and production study: characterization of home gardens in Aur Jaya (Sitiung Vc). Rep. No. No. 32. Tropsoils
FAO (1996a) Food and agriculture organization of the united nations. Report on the state of the world's plant genetic resources for food and agriculture. FAO, Rome
FAO (1996b) Global plan of action for the conservation and sustainable utilization of plant genetic resources for food and agriculture. FAO, Rome
FAO (2010) The second report on the state of the world's plant genetic resources for food and agriculture. FAO, Rome
FAO/IPGRI (1994) Genebank standards. FAO/IPGRI, Rome, 13 p
Frankel OH (1970) Genetic conservation in perspective. In: Frankel OH, Bennett E (eds) Genetic resources in plants. Blackwell, Oxford, pp 469–489

Frankel OH, Bennett E (1970) Genetic resources in plants – their exploration and conservation, International biological programme handbook 11. Blackwell, Oxford

Frankel OH, Brown AHH (1984) Current plant genetic resources-a critical appraisal. In: Genetics: new frontiers, vol 4. Oxford/IBH Publishing Co., New Delhi, pp 1–11

Frankel OH, Hawkes JG (1975) Crop genetic resources for today and tomorrow. Cambridge University Press, Cambridge. http://nbaindia.org/uploaded/Biodiversityindia/4th_report.pdf

Frankel OH, Soulé ME (1981) Conservation and evolution. Cambridge University Press, Cambridge

Goyal AK, Arora S (2009) India's 4th National report to CBD. Ministry of Environment and Forests, Government of India (http://nbaindia.org/uploaded/Biodiversityindia/4th_report.pdf)

Groombridge B (ed) (1992) Global biodiversity: status of the Earth's living resources. Compiled by the World Conservation Monitoring Centre. Chapman and Hall, London, p 594

Hammer K, Teklu Y (2008) Plant genetic resources: selected issues from genetic erosion to genetic engineering. J Agric Rural Dev Trop Subtrop 109(1):15–50

Hamon S, Dussert S, Noirot M, Anthony F, Hodgkin T (1995) Core collections—accomplishments and challenges. Plant Breed Abstr 65:1125–1133

Harlan JR (1971) Agricultural origins: centers and non-centers. Science 174:468–474

Hawkes JG, Maxted N, Ford-Lloyd BV (2000) The ex situ conservation of plant genetic resources. Kluwer, Dordrecht, pp 1–250

Heywood VH (1987) The changing role of the botanic gardens. In: Bramwell, David et al (eds) Botanic gardens and the world conservation strategy. Academic, London, pp 3–18

Heywood VH (2011) The role of botanic gardens as resource and introduction centres in the face of global change. Biodivers Conserv 20:221–239

Heywood VH, Dulloo ME (2005) In situ conservation of wild plant species: a critical global review of best practices. IPGRI Technical Bulletin 11. IPGRI, Rome, Italy

Heywood VH, Heywood CA, Jackson PW, World Wildlife Fund, International Association of Botanical Gardens, Botanic Gardens Conservation Secretariat. (1990) International directory of botanical gardens, V. Koenigstein, Germany; Champaign, Ill: Published by Koeltz Scientific on behalf of WWF--World Wide Fund for Nature, Botanic Garden Conservation Secretariat, International Association of Botanical Gardens

Hodgkin TH, Ramanatha Rao V, Riley K (1993) Current issues in conserving crop landraces. Presented at the FAO-IBPGR On-Farm Conservation Workshop, 6–8 December 1993, Bogor, Indonesia

Hunt ED, Sherrill D (1998) Data management systems for in situ genetic conservation: developing a standard environmental model. In: Zencirci N, Kaya Z, Anikster Y, Adams WT (eds) Proceedings of international symposium on in situ conservation of plant genetic diversity, Antalaya, Turkey, 4–8 November 1996. Central Research Institute for Field Crops, Ankara, pp 301–308

ISTA (International Seed Testing Association) (2008) International rules for seed testing. ISTA, Bassersdorf

IUCN (2015) The IUCN red list of threatened species. Version 2015.4 (http://www.iucnredlist.org) Accessed 1 Feb 2006

Jacob SR, Tyagi V, Agrawal A, Chakrabarty SK, Tyagi RK (2015) Indian plant germplasm on the global platter: an analysis. PLoS One 10(5):e0126634

Jarvis DI, Hodgkin T (2000) Farmer decision making and genetic diversity: linking multidisciplinary research to implementation on-farm. In: Brush SB (ed) Genes in the field: on-farm conservation of crop diversity. Lewis Publishers, Boca Raton, pp 261–278

Jarvis A, Upadhyaya H, Gowda CL, Agrawal PK, Fujisaka S (2008) Climate change and its effect on conservation and use of plant genetic resources for food and agriculture and associated biodiversity for food security. Thematic study for the SoW Report on PGRFA. FAO, Rome

Kartha KK, Engelmann F (1994) Cryopreservation and germplasm storage. In: Vasil IK, Thorpe TA (eds) Plant Cell and Tissue Culture. Kluwer, Dordrecht, pp 195–230

Khanna SS (1989) The agro-climatic approach. In: Survey of Indian agriculture. The Hindu, Madras, pp 28–35

Kristiansen K, Cilieborg M, Drabkova L, Jorgensen T, Petersen G, Seberg O (2005) DNA taxonomy: the riddle of Oxychloe. Syst Bot 30:284–289
Kursar TA, Caballero-George CC, Capson TL, Cubilla-Rios L, Gerwick WH, Heller MV, Ibáñez A, Linington RG, McPhail KL, Ortega-Barría E, Romero LI, Coley PD (2007) Linking bioprospecting with sustainable development and conservation: the Panama case. Biodivers Conserv 16:2789–2800
Laliberté B (1997) Botanic garden seed banks/genebanks worldwide, their facilities, collections and networks. BGC News 2(9):18–23
Ledig FT (1988) The conservation of diversity in forest trees. Bioscience 38:471–479
Ledig FT (1992) Human impacts on genetic diversity in forest ecosystems. Oikos 63:87–108
Lync P, Souch G, Trigwell S, Keller J, Harding K (2011) Plant cryopreservation: from laboratory to genebank. Asia-Pac J Mol Biol Biotechnol 18(1):239–242
McNeely JA (1994) Biodiversity: conservation and development in Asia: how protected areas can contribute to local communities. Chin Biodiver 2:75–57
McNeely JA (1996) The role of protected areas for conservation and sustainable use of plant genetic resources for food and agriculture. In: JMM Engels (ed) Situ conservation and sustainable use of plant genetic resources for food and agriculture in developing countries. Report of a DSE/ATSAF/IPGRI Workshop, 2–4 May 1995, Bobb-Rottgen, Germany. A joint publication of IPGRI, Rome, Italy and DSE, Feldafing, Germany, pp 27–41
Melville R (1970) Red data book, Vol 5-Angiospermae. IUCN, Morges
Michon G, Mary F, Bompard J (1986) Multistoried agroforestry garden system in West Sumatra, Indonesia. Agrofor Syst 4:315–338
MoEF (Ministry of Environment and Forest), GoI, New Delhi, India. http://www.indiaenvironmentportal.org.in/files/BiodiversityTarget2010.pdf
Myers N, Mittermeier RA, Mittermeier CG, da Fonseca GAB, Kent J (2000) Biodiversity hotspots for conservation priorities. Nature 403:853–858 pmid: 10706275 doi:10.1038/35002501
Nayar MP (1980) Endemism and patterns of distribution of endemic genera (angiosperms). J Econ Taxon Bot 1:99–110
NBPGR (2011) NBPGR: a compendium of achievements. National Bureau of Plant Genetic Resources, New Delhi, 58 p
NBPGR (2014) Available: www.nbpgr.ernet.in.
Ogwu MC, Osawaru ME, Ahana CM (2014) Challenges in conserving and utilizing plant genetic resources (PGR). Int J Genet Mol Biol 6(2):16–22
Painting KA, Perry MC, Denning RA, Ayad WG (1993) Guidebook for genetic resources documentation. IPGRI, Rome
Patra BC, Ray S, Ngangkham U, Mohapatra T (2015) Rice. In: Singh M, Upadhyaya HD (eds) Genetic and genomic resources for grain cereals improvement. Academic, London, pp 1–80
People's Biodiversity Register (PBR) (2013) Revised PBR guidelines. National Biodiversity Authority, Chennai, Tamil Nadu. www.nbaindia.org/content/105/30/1/pbr.html. Accessed Jan 2016
Plucknett DL, Smith NJH (2014) Gene banks and the world's food. Princeton University Press, Princeton, 264 p
Plucknett DL, Smith NJH, Williams JT, Anishetty NM (1987) Gene banks and the world's food. Princeton University Press, Princeton, 233 p
Probert RJ (2003) Seed viability under ambient conditions and the importance of drying. In: Smith RD, Dickie JB, Linington SH, Pritchard HW, Probert RJ (eds) Seed conservation: turning science into practice. Royal Botanic Gardens, Kew, pp 337–365
Pushpangadan P, Rajasekhran S, Ratheesh Kumar PK, Jawahar CR, Velayudhan Nair V, Lakshmi N, Sarad Amma L (1988) Arogyapacha (*Trichopus zeylanicus* Gaertn.). The Ginseng of Kani Tribes of Agasthyar Hills (Kerala) for Evergreen Health and Vitality. Anc Sci Life 7:13–16
Quek P, Ramanatha Rao V (1995) Plant genetic resources and information networks. In: Arora RK, Ramanatha Rao V (eds) Proceedings of the South Asia national coordinators meeting on plant genetic resources, 10–12 January 1995. IPGRI Office for South Asia, BARC, Bangladesh, 147–154

Quek P, Zhang Z (1995) New concepts in information management for plant genetic resources in APO region. Presented at the XVIII Pacific science congress on population, resources and environment: prospects and initiatives, 5–12 June, Beijing, China
Ramanatha Rao V, Riley KW, Zhou M, Valmayor R, Iwanaga M (1997) Biodiversity in plant genetic resources of Asia. In: Park YG, Sakamoto S (eds) Biodiversity conservation of plant genetic resources in Asia. Japan Scientific Societies Press/IPGRI, Tokyo/Serdang, pp 1–19
Rao NK, Hanson J, Dulloo ME, Ghosh K, Nowell D, Larinde M (2006) Manual of seed handling in genebanks. Handbooks for Genebanks No. 8. Biodiversity International, Rome
Reddy VG, Upadhyay HD, Gowda CLL (2006) Current status of sorghum genetic resources at ICRISAT: their sharing and impacts. J SAT Agric Res 2(1) 5 pages
Reed BM (ed) (2008) Plant cryopreservation. A practical guide. New York, USA, Springer
Royal Botanic Gardens, Kew. Seed Information Database (SID). Predict seed viability module (available at: http://data.kew.org/sid/viability/percent1.jsp). Convert RH to water content (available at: http://data.kew.org/sid/viability/mc1.jsp). Convert water content to RH (available at: http://data.kew.org/sid/viability/rh.jsp). Kew, UK
Saad MS, Rao VR (eds) (2001) Establishment and management of field Genebank, a training manual. IPGRI-APO, Serdang
Secretariat of the Convention on Biological Diversity (2009) The convention on biological diversity plant conservation report: a review of progress in implementing the Global Strategy of Plant Conservation (GSPC), 48 p
SGRP-CGIAR Crop Genebank Knowledge Base. Field genebanks. Available at: http://cropgenebank.sgrp.cgiar.org/index.php?option=com_content&view=article&id=97&Itemid=203&lang=english
Sharma SK, Brahmi P (2011) Gene bank curators: towards implementation of the international treaty on plant genetic resources for food and agriculture by the Indian National Gene Bank. In: Frison C, López F, Esquinas-Alcazar JT (eds) Plant genetic resources and food security: stakeholder perspectives on the international treaty on plant genetic resources for food and agriculture. FAO Biodiversity International/Earth scan, London, pp 183–196
Smith MA, Blumberg AY (1990) Conservation of India's plant genetic resources: USAID's largest biodiversity activity. Diversity 6(2):7–9
Spooner D, van Treuren R, de Vicente MC (2005) Molecular markers for genebank management. IPGRI Technical Bulletin No. 10
St Amand PC, Skinner DZ, Peaden RN (2000) Risk of alfalfa transgene dissemination and scale-dependent effects. Theor Appl Genet 101:107–114
Sthapit B, Padulosi S, Mal B (2009) Role of On-farm/In situ Conservation and Underutilized Crops in the Wake of Climate Change. National Symposium on Recent Global Developments in the Management of Plant Genetic Resources, NBPGR, New Delhi, India, pp 17–18
Tao KL (2013) Complementary conservation strategies for plant genetic resources. In: Mandal BB, Chaudhury R, Engelmann F, Mal B, Tao KL, Dhillon BS (eds) Conservation biotechnology of plant germplasm. NBPGR/IPGRI, New Delhi/Rome, 51
Teklu Y, Hammer K (2006) Farmers perception and genetic erosion of Ethiopian tetraploid wheat landraces. Genet Resour Crop Evol 53:1099–1113
ten Kate K, Laird SA (1999) The commercial use of biodiversity: access to genetic resources and benefit-sharing. Earthscan, London
Tyagi RK, Abraham Z, Latha M, Velayudhan KC, Ravindran PN, Nirmal Babu K, George JK, Agrawal A, Dhillon BS (2004) Conservation of spices germplasm in India. Indian J Plant Genet Resour 17(3):163–174
Vavilov NI (1926) Centers of origin of cultivated plants. pp 22–135. In: Dorofeyev VF (ed) Origin and geography of cultivated plants, English translation of Vavilov's works, 1992 ed. Cambridge University Press, Cambridge
Watson JW, Eyzaguirre PB (eds) (2002) Proceedings of the Second International Home Gardens Workshop: Contribution of home gardens to in situ conservation of plant genetic resources in farming systems, 17–19 July 2001, Witzenhausen, Federal Republic of Germany. International Plant Genetic Resources Institute, Rome

Wilkes G (1983) Current status of crop germplasm. Crit Rev Plant Sci 1:133–181
Wyse Jackson P (1999) Experimentation on a large scale – an analysis of the holdings and resources of botanic gardens. BGC News 3(3):27–30
Wyse Jackson PS, Sutherland LA (2000) International agenda for botanic gardens in conservation. Botanic Gardens Conservation International, Kew
Zeven AC, de Wet JMJ (1982) Dictionary of cultivated plants and their regions of diversity. Centre for Agricultural Pub. and Documentation, Wageningen 259 p
Zhang J, Xin X, Yin G, Lu X, Chen X (2014) In vitro conservation and cryopreservation in national genebank of China. In: Reed BM (ed) Proceeding of second international symposium on plant cryopreservation. Acta Horticulture 1039: 309–317

Bioprospecting Traditional Medicine 16

Chethala N. Vishnuprasad and C.P. Unnikannan

Abstract

Traditional medicines use herbal formulations as the predominant strategy for managing various disease conditions. The formulations are designed based on the concepts of holism and are expected to exert a poly-pharmacological effect. The unique philosophy of disease management in traditional medicines and the potential poly-pharmacological effects of the formulations can open up new horizons in drug discovery research. It can set new road maps for medical pluralism as well as holistic therapeutic approaches. However, the conventional bioprospecting strategies for TMFs are largely confined to the discovery of a single, target-specific bioactive molecule from their constituent plants. Such kind of a reductionist approach could not successfully address the basic therapeutic and pharmacological concepts of traditional medicines. Here, the chapter discusses the significance of bioprospecting TMFs, emphasizing the necessity of a novel integrative and transdisciplinary approach for bioprospecting. The chapter also discusses how a better bioprospecting strategy can be adopted for traditional medicines using *Ayurveda* and traditional Chinese medicines as examples.

Keywords

Bioprospecting • Traditional medicines • *Ayurveda* • Traditional Chinese medicine • Network pharmacology • Poly-pharmacology

---

C.N. Vishnuprasad (✉)
School of Life Sciences, Institute of Transdisciplinary Health Science and Technology (TDU), #74/2, Jarakabande Kaval, Yelahanka, Bangalore 560064, India
e-mail: drcnvp@gmail.com

C.P. Unnikannan
Weizmann Institute of Science, Rehovot 76100, Israel

S. Abdulhameed et al. (eds.), *Bioresources and Bioprocess in Biotechnology*,
DOI 10.1007/978-981-10-3573-9_16

16.1 Introduction

The evolutionary history of healthcare practices and medicines in the world are concurrent with the evolution of ethnicities and their cultures. The healthcare practices and products that evolved in different geographically separated ethnic groups, referred as traditional medicines (TMs) which are solely based on the indigenous materials and are designed to meet the needs of the specific ethnic group, remained as the mainstream medical practice for years. However, the later development of the modern systems of medicine provided a more unanimous global outlook in healthcare management and gradually took over the position as mainstream or primary healthcare across the globe, leaving ethnomedicines and TM as complementary and alternative medical (CAM) practices. However, the dynamic relationships between people and environment as well as the holistic approaches in healing illness make CAMs an important component of current therapeutic schemes. One of the characteristic features of TM is the use of herbal and non-herbal natural resources, individually or in combination, as the primary means of drug formulation with plants and plant products as the base for any formulation. This concept of herbalism, followed in TMs, is not strictly based on "evidences" gathered through "modern scientific methods," and thus the modern medicine recognizes traditional medicines and ethnomedicines as alternate medicines.

The World Health Organization (WHO) defines traditional medicine as "the sum total of the knowledge, skills and practices based on the theories, beliefs, and experiences indigenous to different cultures, whether explicable or not, used in the maintenance of health as well as in the prevention, diagnosis, improvement or treatment of physical and mental illness". Although the revolutionary changes in modern medicine caused a greater setback for traditional medicines, huge population in several Asian and African countries still rely on traditional medicine as their primary healthcare system. Figure 16.1 depicts the major forms of TM practiced in different parts of the world.

In Asia, the widely accepted traditional medicines include *Ayurveda*, *Siddha*, *Unani*, traditional Chinese medicine (TCM), traditional Korean medicine (TKM), Kampo (Japan), etc. Besides these, innumerable numbers of folk medicines, home remedies, and other regional or ethnicity-specific healing practices further expand the list of global traditional medicines. Generally, the traditional medical knowledge is transmitted through orally from individual to individual or family to family, and in many of such cases, a written document of the knowledge system was lacking. The lack of documented, evidence-based details on the efficacy and safety limits the global acceptance of traditional medicines. However, globally, there is an emerging trend of opting for traditional medicines (CAMs) in conjunction with modern medicine to combat various disease conditions, mainly because of their nutraceutical approaches, lesser side effects, and lower economic burden to the patients (Medagama and Bandara 2014). While accepting the importance of traditional medicines, the WHO makes a note of the potential negative or dangerous effects that arise from the inappropriate use of traditional medicines. In light of these facts, the WHO calls for the need of proper research to ascertain the efficacy

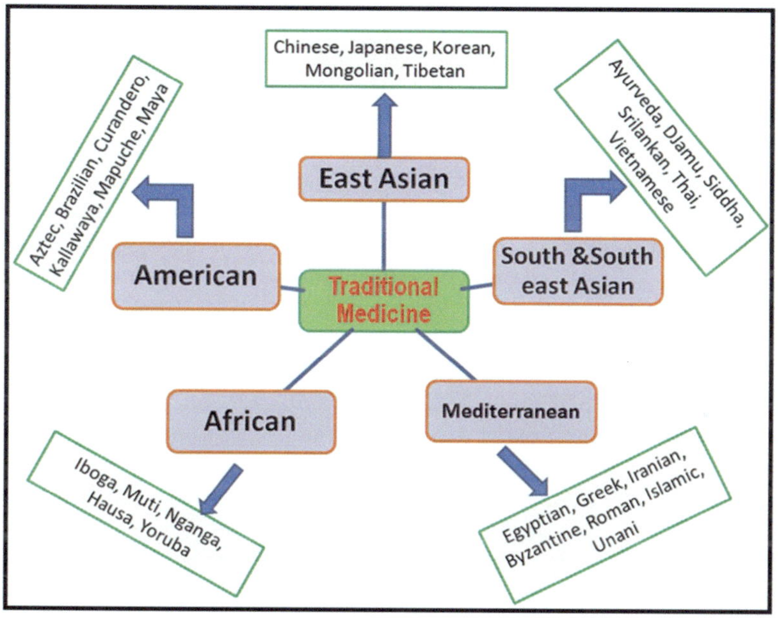

Fig. 16.1 *Traditional medicine diversity*: Figure portrays the different types of TMs and the regions where they are prevalent. The figure includes the most common or well-known traditional medical practices. Innumerable numbers of folklore medicines and tribal medicines are also deemed as TM

and safety of the TMs. Considering the significance and future needs, a proper scientific exploration and bioprospecting of TM is a necessary requirement for the present-day healthcare sector. The chapter here highlights the importance of bioprospecting traditional medicines and discusses the various transdisciplinary approaches in bioprospecting TM using *Ayurveda* and TCM, the two most successfully practicing TM in the world, as classical examples. An innovative bioprospecting strategy should have an epistemologically sensitive methods and rigorous experimentation using modern science.

16.2 Bioprospecting: A Value Addition to TM

Bioprospecting in its broad sense indicate a process of discovering the biological values of substances of natural origin. It became an integral component of modern drug research industry. In its early periods, bioprospecting was largely centered only on plant diversity, whereas in the present days, its spectrum had expanded to other biodiversity forms, viz., insects, microorganisms, and other natural resources as well. Historically plants and their bioactive phytochemicals have made outstanding contributions in the field of therapeutics. The history of botanical therapeutics can be traced back to thousands of years with plants and plant extracts being the primary source for medicines in all traditions across the world (Ngo et al. 2013).

The practice had gradually evolved and transformed into different forms of traditional or complementary and alternative medicines (CAMs) over the years (Bhasin 2007). Even now, these CAMs remain as one of the important sources of new medicines and therapeutics, and they hold the promises of managing diseases through a holistic approach. The personalized holistic therapeutic approaches as well as the high success rate in clinical practices made TM an integral part of global healthcare, and it also opens up new horizons in the drug discovery. However, inadequate evidences and proof of concepts in a contemporary scientific perspective limited the global acceptance of TM. Many of them have disappeared with the passage of time (Phondani 2011). Even then, properly documented, codified forms of TM survived all sorts of impediments and continue to serve as effective complementary and alternate medical systems to meet the requirements of human health management. Some of the examples include *Ayurveda*, TCM, *Unani*, *Siddha*, and various other Oriental medicines. In India, particularly, *Ayurveda* is the first option as a complementary medical practice. It has an inimitable diagnostic and therapeutic approach based on the multifactorial aspects of the disease. The unique philosophy of disease management in TM as well as its potential poly-pharmacological effects can open up new horizons in drug discovery research and can set new road maps for medical pluralism and holistic therapy. However, the conventional bioprospecting strategies for TMFs are largely confined to the discovery of a single target-specific bioactive molecule from their constituent plants. Such kind of a reductionist approach could not successfully address the basic therapeutic and pharmacological concepts of TM. An intelligent, integrative, and transdisciplinary approach of bioprospecting is needed to revitalize the traditional medical practices, which essentially should result in translating the traditional medical knowledge into a globally accepted scientific language, to cope with the increasing needs of medical pluralism.

16.3 Deciphering Information from Traditional Medical Literature

All over the world, traditional medical knowledge are documented in the form of manuscripts written in languages and scripts that were indigenous to that particular tradition. Before the invention of printing technology, the medical inventions and knowledge were handwritten and documented in different forms such as wood blocks, papyrus, etc. In India, palm leaf manuscripts were used as the means of documenting all knowledge systems from ancient times. For a complete and meaningful bioprospecting, of any traditional medical knowledge, the primary step is the preparation of transcripts from the available manuscripts. Most of the time, the language barriers and unavailability of quality manuscript materials limit the process. Moreover, the systematic study of traditional medical manuscripts requires a well-planned transdisciplinary approach with experts from manuscriptology, languages, medical science, and other basic science areas. Deciphering the knowledge content in manuscripts could broaden the scope of bioprospecting the TM. However, unfortunately, this vital part of traditional knowledge repository remains under noticed.

In India alone, there are thousands of manuscripts in the field of medicine with only a small fraction made available in the form of edited texts. The fundamental principles of *Ayurveda* were recorded in the form of three classical texts, viz., *Charaka Samhita*, *Sushruta Samhita*, and *Ashtanga Hridaya*, collectively known as *Bruhatrayis* (the Great Trilogy). In addition, there are several regional commentaries to these classical texts; plus some independent works based on the fundamental principles are also available on different aspects of *Ayurveda*, viz., drug formulations, diagnosis, pharmacognosy, etc. (Narayanaswamy 1981). Several of them are currently available in the form of printed editions, transcripts, as well as manuscripts. However, the situation is that only a few hundred manuscripts (out of thousands available) are repeatedly reprinted and studied, giving an impression that these are the sole and whole of the contributions from traditional medicines. This situation basically confines the researcher to the available texts and thus restricts the horizontal and vertical spaces of bioprospecting.

16.4 Epistemological Bridging: The Real Challenge

The epistemology of TM is essentially based on the geographical and cultural features of the ethnic groups where it is originated and developed. This is characteristically different from the epistemology of modern medicine where it has unanimous and globally accepted descriptions for disease and therapy. The concepts of health and disease in TMs are different from the modern allopathic concepts but are more similar to the Hippocratic theory of "humorism." According to Hippocrates' theory, the health is a harmonious balance of four humors. To quote Hippocrates:

> *The body of man has in itself blood, phlegm, yellow bile, and black bile; these make up the nature of the body, and through these he feels pain or enjoys health. Now, he enjoys the most perfect health when these elements are duly proportioned to one another in respect to compounding, power and bulk, and when they are perfectly mingled. Pain is felt when one of these elements is in defect or excess, or is isolated in the body without being compounded with all the others.* (http://www.greekmedicine.net)

Indian TMs, with *Ayurveda* being the most refined and systematic traditional medical practice, also follow a concept more similar to the Hippocratic theory of "humorism." Being a holistic medical science, *Ayurveda* and other Indian TMs primarily focus on maintaining an optimal health and improving body's natural resistance to protect the body from various diseases. Globally, TMs achieve the "holism" through an integrated regulation of day-to-day activities of body, mind, spirit, and the emotions. Particularly in *Ayurveda*, good health is defined as a harmonious balance of three basic bodily humors, viz., *Vata* (wind), *Pitta* (bile), and *Kapha* (phlegm). This is known as the *Tridosha* concept. Body, mind, and consciousness synergistically play a vital role in maintaining this balance to make the human physique healthy. According to *Ayurveda* human body is a framework of five elements (*Pancha-mahabhuta*) and seven tissues (*Saptadhatu*), which are the fabric components of the physique (Fig. 16.2). The appropriate functioning of *Pancha-mahabhuta*

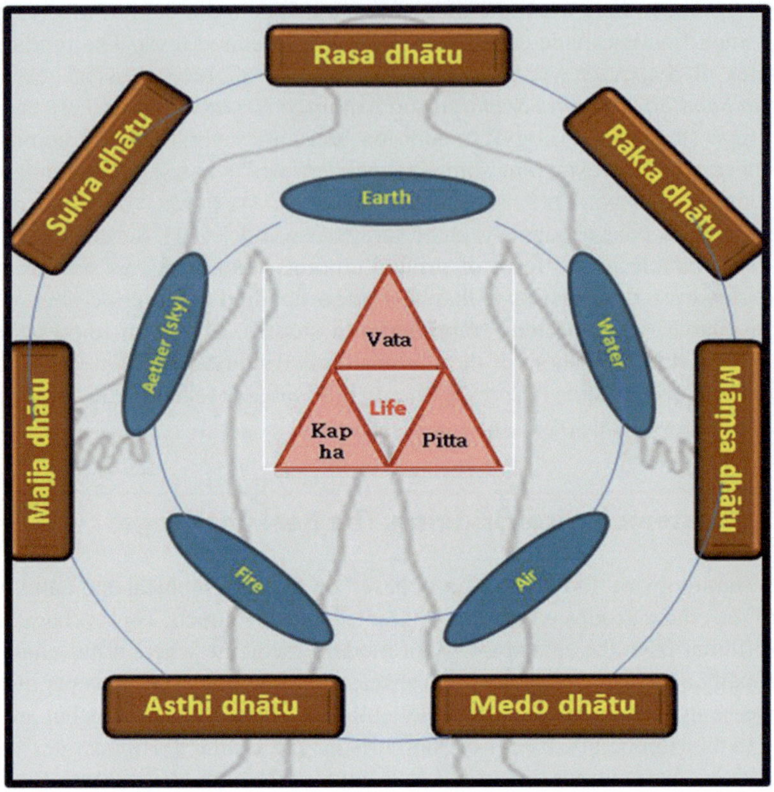

Fig. 16.2 Pictorial representation of the fundamental concepts of *Ayurveda*: *Pancha-mahabhuta* and *Saptadhatu*, the fabric components of the physique, under the influence of *Tridosha* modulate health and disease in human

and *Saptadhatu* is extremely regulated by the balance of the aforementioned humors, viz., *Vata*, *Pitta*, and *Kapha*.

Each individual has a unique combination of *Vata*, *Pitta*, and *Kapha*, and the vitiation of humors is regarded as the major reason for the onset of a disease. As disease and its treatment in *Ayurveda* are conceptualized based on the *Tridosha* theory and the combinations of three humors are characteristic to a person, the treatment strategies of *Ayurveda* are more personalized. Not only from a therapeutic perspective, conceptualization of health and its maintenance are also based on the same *Tridosha* theory. Although the terminologies and languages vary regionally, the fundamental *Pancha-mahabhuta*, *Saptadhatu*, and *Tridosha* concepts remain more or less the same in all the traditional medical systems in India (Fig. 16.3).

The situation is the same with other traditional medicines across the world but with regional and ethnicity-specific differences in epistemology. TCM also emphasizes in building resistance to illness and points out if the body is rich in *qi* (the life energy or life force) it will not get susceptible to external pathogens. In TCM, the basic structure is composed of three parts in which the first is consistent with

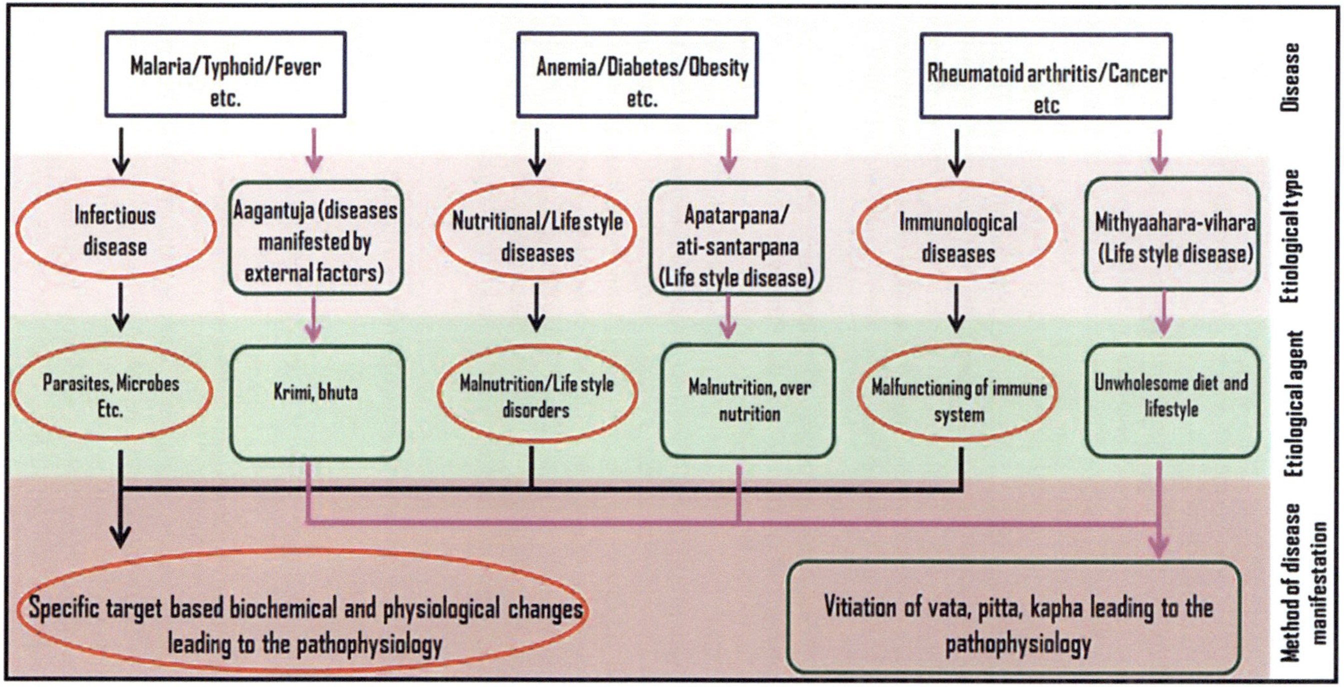

Fig. 16.3 Illustration of epistemological differences in etiology and mode of disease manifestation in TM vs modern medicine using *Ayurveda* as an example: The *red circles* represent the modern medicine arm and the green squares represent the *Ayurveda* arm. The picture indicates the qualitative parameters of *Ayurveda* vs the quantitative parameters of modern medicine

modern medicine, the second is involuntarily beyond modern medicine, and the third needs further evaluation. The knowledge and practices in agreement with modern medicine would find it relatively easy to have an epistemological bridging, whereas the knowledge and practices that are not recognized in modern medicine that may be valuable in the future practice of modern medicine would require logical bioprospecting methods to establish an epistemologically sound bridge between the two concepts of disease management.

As western medicines follow a reductionist approach (through understanding the intricate components and processes involved in health and disease), the TM like *Ayurveda* follows a holistic approach (understanding the system as a whole) in defining health and disease. Integrating these two epistemologically different concepts is difficult but is the need of the hour. Some of the recent attempts in this direction found to have good results and could set the tone of research for the future. A path-breaking study reported recently from India had attempted to correlate the genomic variations with the classification of *prakriti* (*Vata*, *Pitta*, *Kapha* constitutional type of body). A genome-wide SNP (single nucleotide polymorphism) analysis of 262 male individuals belonging to three *prakritis* concludes that *prakriti* classification, the foundation stone for the practice of *Ayurveda*, has a genetic basis and resonates with personalized medicines (Govindaraj et al. 2015). The study reported an association of PGM1 gene with the *Pitta* type that correlates the characteristics of *Pitta* in *Ayurveda* (digestion, metabolism, and energy production) with the functions of PGM1 gene (key functions in many metabolic pathways). Similarly an earlier study from India showed individuals from contrasting constitutional types (*prakriti*) exhibit striking differences in biochemical and hematological parameters and genome-wide expression levels. The study concludes by saying classification of constitutional types helps to uncover genes that contribute system-level differences in individuals which could lead to differential disease predisposition (Prasher et al. 2008). Few other genomic studies also validate the concepts of *prakriti* and personalized medicines in *Ayurveda* (Juya et al. 2012; Rotti et al. 2015).

16.5 Potentials of Bioprospecting TM: *Ayurveda* as an Example

One of the fundamental slogans of TM is to preserve the health of a person in a good condition and treat the person in case of any disease. This is achieved through maintaining a physiological and psychological balance with proper food, lifestyle, and medications. Although the three aforementioned arms are equally important, the medicine part had gained more importance. The concepts of food and lifestyle are, in fact, included in the medication part as the complete effect of a medicine can be achieved only when the patient follows appropriate food and lifestyle. This is well aligning with renowned concept "food is medicine," proposed by great philosophers of western and eastern medicine such as *Hippocrates* and *Charaka*. Owing to its rich biodiversity, the Indian subcontinents have recorded around 15,000 medicinal plants among which more than 700 plants have been reported in *Ayurveda* (Meena et al. 2009). Although Ayurvedic medicines have components of herbal, mineral, and animal origin (which are used separately or in combination), herbal formulations are the common therapeutic/preventive substances in *Ayurveda*. This is primarily because of their abundance and commonness in the nature. The whole plant or parts of the plants (leaves, flowers, seeds roots, barks, stems, etc.) are used as medicines and are prepared in the form of decoction (*Kwatha*), hot and cold infusions (*Phanta*, *Hima*), powders (*Churna*), medicated oils (*Taila*), and so on (Frawley and Ranade 2000). These medicines, whether they are single or poly-herbal, exert their biological effects through the synergistic combination of various bioactive phytochemicals present in them (Kim et al. 2015). Albeit the enormous amount of studies resulted in successfully isolating and identifying bioactive compounds from individual constituent plants of the formulations, most of the follow-up attempts to develop them as an effective therapeutic molecule (drug) was not successful (McClatchey 2005). The failure is almost certainly due to an inadequate or inefficient strategy to address the characteristic pharmacological and biological concepts of *Ayurveda*. The emergence of synthetic chemistry with modern pharmacology caused a greater setback in natural product-based drug applications. Nevertheless, in recent years, natural product chemistry is reemerging. This is fundamentally due to their superior chemical diversity and safety over synthetic pure chemical compounds. But the lack of scientific validation for majority of the herbal formulations remains a major impediment in the integration of herbal medicines to the mainstream medical practices. It is also important that the bioprospecting approach needs to focus more on the formulation, because studies on individual plants need not necessarily address all the possible pharmacodynamic and pharmacokinetic properties of a formulation as the combinatorial effects of the components in a formulation and their post-digestive modifications can greatly alter the pharmacodynamic and pharmacokinetic properties. The combinatorial effect of diverse bioactive compounds in the formulation could produce a network pharmacological effect, and the phenomenon of synergistic herb-herb or compound-compound interactions requires novel strategies to understand it better.

16.6 Concept of Poly-pharmacology

In classical pharmacology, drugs are defined as chemical substances, which are able to bring out a desired change in the biological system through their interactions with the system. Conventionally these interactions are believed to have high levels of complementarity where the drug interacts with its target site through a lock-and-key hypothesis. Historically, this common analogy of drug action has been the strategy of pharmacological industry for new drug discovery. Chemists attempt to develop the compounds by designing the "druggable" molecules more specific to its receptor. However, this leading paradigm of drug discovery is gradually changing to a more complex understanding of drug action, wherein the drug, the key, may interact with various targets as well as a single target may get activated by multiple drugs. This, in principle, indicates that the drug can have an effect on whole set of molecular interactions (interactome) in the human system as well as in the human disease networks (diseasome) (Barabasi et al. 2011). The conventional one-drug, one-target philosophy of drug design had transformed to one-drug, multiple-target concept making it a new paradigm of drug design and drug discovery. This new paradigm of poly-pharmacology also discusses the probability of multiple drugs (or bioactive molecules in the formulations) acting on multiple targets resulting in a network of various physiological effects. Although the modern molecular drug therapies are rooted in the above mentioned one-drug, one-target philosophy, several of those drugs are reported for their multi-targeting activities. Reddy and Zhang (2013) in their recent article on poly-pharmacology discusses the multi-target activity of molecular drugs and the unintended side effects that causes concluding this as one of the main reasons by which drugs are frequently withdrawn from the market.

Both single and poly-herbal formulations used in TMs stand as classical examples for poly-pharmacology and network pharmacology, by potentially modulating multiple therapeutic targets to get a holistic pharmacological effect. The synergism of compounds in the formulation confers several beneficial effects that a typical single molecular drug cannot produce. A recent study from India demonstrated how pharmacological networks of medicinal plants and their formulations can be constructed for understanding their putative actions and mechanisms using *Triphala*, one of the most studied formulations of *Ayurveda*, as an example. *Triphala*, a mixture of fruits of *Emblica officinalis/Phyllanthus emblica* (Amalaki/Indian gooseberry), *Terminalia bellerica* (Vibhitaki), and *Terminalia chebula* (Haritaki), is considered to be an effective formulation for the management of various diseases. The study reports that *Triphala*, as a composite formulation, is involved in 78 diseases through 44 targets and 7 out of the several bioactives present in *Triphala* are shared among its three constituent plants (Chandran et al. 2015). The study was basically attempted to demonstrate how pharmacology network of medicinal plants and their formulation could be constructed and used to understand putative actions, indications, and mechanisms. Through this article some of the important perceptions the authors put forward are:

1. A multi-target, multi-ingredient formulation may be a better approach than the one-gene, one-target, one-drug approach.
2. Drug discovery need not to be confined to the discovery of a single target-specific bioactive molecule.
3. Bioprospecting and standardizing synergistic, safe, and effective formulations, with proper scientific evidences, can offer better and economical alternatives.

The individual constituent plants of *Triphala* formulation are very well studied, and several bioactives were isolated and identified with the help of modern research (Baliga et al. 2012). However, such studies focusing on bioassay-guided isolation and identification of active compounds through a reductionist approach failed to hit the success line of drug discovery process. The approach remains the same for most of the studies using herbal formulations. Table 16.1 lists out some of the bioprospecting studies attempted with various kinds of traditional medicines.

Table 16.1 TMFs in modern scientific research: The table lists out the common TMFs studied for their biological properties along with the disease/biological marker studied and model system used

S. no	Type of formulation	Disease/biological marker studied	Model system	References
1	*Triphala*	Apoptosis of cancer cells, colitis, preventing dental caries, arthritic and nephrotoxic protection, epithelial to mesenchymal cell transition, immunomodulatory effect	Cell models, rats	Baliga et al. (2012)
				Baliga (2010)
				Chandran et al. (2015)
				http://www.ncbi.nlm.nih.gov/pubmed/?term=VaddeR%5BAuthor%5D&cauthor=true&cauthor_uid=26167492
				Kalaiselvan and Rasool (2015)
				Baskaran et al. (2015)
				Sivasankar et al. (2015)
				Belapurkar et al. (2014)
				Bhattacharjee et al. (2015)
				Bag et al. (2013)
				http://www.ncbi.nlm.nih.gov/pubmed/?term=SrinageshJ%5BAuthor%5D&cauthor=true&cauthor_uid=23422630 http://www.ncbi.nlm.nih.gov/pubmed/?term=DeepG%5BAuthor%5D&cauthor=true&cauthor_uid=16471318
2	*Amalaki Rasayana*	Apoptosis, aging and fecundity, neurodegeneration	*Drosophila*	Dwivedi et al. (2015)
				Dwivedi et al. (2012)

(continued)

Table 16.1 (continued)

S. no	Type of formulation	Disease/biological marker studied	Model system	References
				Dwivedi et al. (2013)
3	*Rasa-sindoor*	Apoptosis, aging and fecundity, neurodegeneration	*Drosophila*	Dwivedi et al. (2015)
				Dwivedi et al. (2012)
				Dwivedi et al. (2013)
4	*Laghupanchamoola*	Analgesic and hypnotic	Rat	Ghildiyal et al. (2014)
5	*Hridayarnava rasa*	Genotoxicity	Swiss albino mice	Jagtap et al. (2014)
6	*Swarna makshika bhasma*	Genotoxicity	Rats	Pavan et al. (2012)
7	*Vasanta kusumakara rasa*	Neurobehavioral activity	Rats	Kumar and Gupta (2012)
8	*Nagabhasma*	Antidiabetic		Dwivedi et al. (2013)
				Rajput et al. (2013)
9	*Rasayana avaleha*	Cancer therapy	Human studies	Vyas et al. (2010)
10	*Rasagenthi lehyam*	Prostate cancer	Human studies	Ranga et al. (2004)
11	*Iron containing preparations*	Anemia and iron deficiency	Clinical study	Sharma et al. (2007)
12	*Nishamalaki*	Diabetes	Clinical study, animal studies	Dawane et al. (2014)
				Rao et al. (2013)
13	*Chyawanprash*	Hepatoprotective activity	Mice and rat models	Menon and Krishnan Nair (2013)
				Jose and Kuttan (2000)

14	*Brahma rasayana*	Hepatoprotective activity, chromosomal aberrations	Mice and rat models	Menon and Krishnan Nair 2013
				Guruprasad et al. (2012)
15	*Kushmanda ghrita*	Depressive illness	Clinical study	Chandre et al. (2011)
16	*Vardhamana pippali rasayana*	Rheumatoid arthritis	Clinical study	Soni et al. (2011)
17	*Dhanwantharam thailam*	Allergy and dermatitis	Case study	Lakshmi (2014)
18	*Panchagavya ghrita*	Neuropharmacological action	Rats and mice	Joshi et al. (2015)
19	*Various Chinese medicines*	Cancer	Various cancer cells and animal models	Nie et al. (2016)
20	*Various Unani medicines*	Orofacial diseases	Clinical and preclinical models	Hongal et al. (2014)

The list indicates that most of them were either clinical studies or *in vivo* model studies looking at specific clinical parameters related to the disease of choice. As an example, the study on *Brahma rasayana* and *Chyawanprash* shows their ability to alleviate cisplatin-induced acute nephrotoxicity using mice model (Menon and Krishnan Nair 2013). Although the study presents important observations, they are not correlated to the pharmacodynamics and pharmacokinetics of the formulation. Particularly in the case of Ayurvedic formulations, though revalidating their clinical effects using modern parameters is essential, it is not mandatory. However, generating proof of concept and understanding their mode of action from a pharmacodynamic, pharmacokinetic, and basic science point of view are necessary for the integration of *Ayurveda* and modern medicine. Besides drug discovery research, such integration is the present-day requirement in disease biology as it can shed light into several hitherto unknown aspects of disease etiology and progression.

16.7 Methods of Bioprospecting TM

Traditional medical systems had evolved in conjunction with the cultural and philosophical aspects of its land of origin. As an example, the TMs in India had evolved in the lights of Vedic and non-Vedic traditions of Indian philosophy (*Astika* and *Nastika Darsana*). This close association makes TM unique in its epistemology, and it is drastically different from the modern concepts of pharmacology and medicines. This is one of the major hurdles in translating TM knowledge into the contemporary scientific language.

The epistemological differences start with the fundamental definitions of health and disease. Taking *Ayurveda* as an example, from the etiological perspective, it defines a disease principally as a condition that causes an imbalance of *Tridoshas*. Those agents or attributes that vitiate the *dosha* balance are the etiological agents as per *Ayurveda*, while in modern medicine the etiological agents are defined as more tangible entities and have qualitative and quantitative properties. Similarly, in TCM, the theories based on *qi* and *meridians* appear to be ambiguous with the concepts of modern medicine, and the description on internal "organs" like the kidney and spleen is far different from the modern anatomical and physiological theories. Also, the TCM therapy based on the *Yin-Yang* principle and the model of five elements, similar to the Ayurvedic concepts, are not comparable with the modern evidence-based medicines.

Secondly, from a pharmacological perspective, TMs use formulations containing multiple bioactive molecules. Unlike the target-specific molecular drugs used in modern medicine, TM formulations exert a poly-pharmacological effect. In *Ayurveda*, the pharmacological properties of a formulation are specific terms as *Rasapanchaka* properties (taste, *rasa*; potency, *veerya*; post-digestive effect, *vipaka*; other unique effects, *prabhava*; and pharmacological action, *karma*), whereas in case of TCM as *five flavors* (acrid, sweet, bitter, sour, and salty) and *four natures* (hot, warm, cool, and cold). These terminologies and concepts are far different from the pharmacodynamic and pharmacokinetic concepts of modern medicine. Third is the pathophysiology and biochemistry of disease progression in TM. Although it is possible to find an approximate correlation of disease terminologies in TM with that of modern medicine (e.g., *Pandu* as anemia), an exact one-to-one comparison is almost impossible. Modern medicine has a structural approach that qualitatively and quantitatively measures different biochemical parameters and understands the disease progression, while TM uses a systemic approach that focuses more on qualitative measurements to understand the disease progression. Many of the modern biochemical parameters like blood glucose, lipid profile, specific antibiotics, hormones, etc. were not even mentioned in TM. Thus for bioprospecting TM formulations, the strategies should design judiciously emphasizing the fundamental concepts of TM (holism) and how it can compare with concepts of modern medicine (reductionism). The path-breaking study described in the previous section (Govindaraj et al. 2015) is a classic example of correlating the concepts of health and disease in *Ayurveda*.

16.8 Methods to Understand the Pharmacological and Phytochemical Concepts

Medicinal plants are the hot targets for drug discovery because the modern drug industry is looking for better bioactive molecules that can effectively interact with specific targets without non-specificity. Conversely the *Rasapanchaka* and *Vata-Pitta-Kapha* concepts of *Ayurveda* or the *four nature* and *five flavors* concepts of TCM look for strengthening and restoring the body's natural resistance to prevent and cure diseases. Nevertheless, it is the proven fact that for any biological effect a drug, whether it is single molecular drug or mixture of bioactives, needs to exert several biochemical interactions in the body. Most of the time, molecular drugs are believed to have a target-specific action, whereas in poly-drug formulation the effect is obviously multi-targeted. As TM uses crude herbal mixtures, there is no appropriate relationship between selective reactive moiety (bioactives) and the traditionally explained pharmacological property such as *Rasapanchaka*. The traditional pharmacological properties are the gross effect of chemical, biochemical, and pharmacological properties of the formulation. There are quite a few numbers of studies attempted to correlate the Ayurvedic pharmacology concepts with modern science. One of the recent studies has used chromatographic techniques to determine the *Kashaya* (astringent) rasa (taste) of two important *Kashaya* dominant drugs *Dolichos biflorus* Linn. and *Bauhinia variegata* Linn. (Kolhe et al. 2014). The study highlighted that correlating *Rasapanchaka* with pharmacological activity has significance as it can help in lead optimization and thus reduces the time and cost in a drug discovery process (Joshi et al. 2007; Rath et al. 2014). Another study on plant-taste activity of 183 Ayurvedic medicinal plants compared their traditional descriptions on taste and ethno-pharmacological actions. The study reports a statistically significant association between the plant-taste and ethno-pharmacological activities confirming the taste-activity relationship and demonstrates how statistical models can be used to predict ethno-pharmacological actions of medicinal plants depending on their taste (Gilca and Barbulescu 2015). Although the results are encouraging, there are no many such attempts using poly-herbal formulations. In one of the studies, a poly-herbal Ayurvedic formulation called *pathyadi varti* has been studied for pharmacognostical and physicochemical properties using phytochemical tools. Even though the study describes some of the physicochemical properties, it did not attempt to correlate it with the Ayurvedic concepts of pharmacology (Gupta et al. 2012).

Further, the modern tools of chemistry and phytochemistry have successfully used to validate other pharmacological concepts of TMs. As an example from *Ayurveda*, a recent study had provided justifications for the use of substitutes for rare plants (*Abhava Pratinidhi Dravya*) in the *Ayurveda*, using modern scientific tools. The rationale for selecting substitutes in *Ayurveda* is based on the *Rasapanchaka* property. The study based on *Aconitum heterophyllum* and *Cyperus rotundus* showed that despite of their taxonomic and morphologic dissimilarities, there are similarities in their Ayurvedic profiles. Interestingly these similarities are reflected in the phytochemical and HPLC profiles of these plant drugs. In the article, the authors raise the point of focusing on broad groups of chemicals rather than individual compounds mentioning the previous unsuccessful attempts to correlate chemistry and action of these plants. A possible reason for this could be that most of such studies were based on selected group of phytochemicals like alkaloids, terpenes, etc. and did not have a holistic approach with the plant (Venkatasubramanian et al. 2010; Adams et al. 2013). Similarly, another study done in brine shrimp bioassay model compared the bioactivity of a recommended method of *Ayurvedic* drug preparation with those of non-recommended methods. The study showed that the bioactivity (LC50) of the recommended preparation of milk decoction (*ksheerapaaka*) of *Piper longum* was 25 times greater than the preparations that were not used in *Ayurveda*, indicating the importance of traditional methods of preparations and the importance of studying the formulations in its original form (Sudha et al. 2004).

The scope of bioprospecting TM formulation can also extend to its unique therapeutic procedures as well. The use of medicinal fumes is an example in this. Medicinal fume is one of the drug administration methods described in various TMs, including *Ayurveda*, and it has promising research opportunities from phytochemical and ethno-pharmacological viewpoints. The formulations used in fumigation are characteristic and cannot be used in any other form to get the therapeutic effect (Vishnuprasad et al. 2013). Similarly the Ayurvedic procedures like *Basti* (medicated enema), *Abhyanga* (oil massage), and *Panchakarma* (biopurificatory methods) also have unique formulations which restore the body's *Tridosha* balance through multiple pharmacological effects (Thatte et al. 2015; Gujarathi et al. 2013; Auti et al. 2013).

16.9 Understanding the Biological Activities of TM: Lessons from *Ayurveda* Research

The systemic approach (of TM) and the structural approach (of modern medicine) are the fundamental principles that explain the biological activities of a drug/formulation. However, these are very unique, and it is almost impossible to find a proper meeting point for these concepts. Consider the example of diabetes mellitus – the disease described in traditional Indian medicine as *Prameha* or *Madhumeha* is generally correlated to "diabetes" in the modern medicine and used as a synonym for diabetes mellitus. Diabetes in modern medicine is described around the insulin-blood glucose concept, whereas in Indian TM, both these fundamental concepts insulin and blood glucose are absent. In *Ayurveda*, *Prameha* is defined as excess and turbid urination. Among the 20 types of *Prameha*, *Madhumeha* has a characteristic sweetness in the urine. The disease *Madhumeha*, from an etiological and clinical perspective, appears to have some similarities with diabetes mellitus. Nevertheless, from a disease management perspective, the therapeutic strategies prescribed for *Madhumeha* are quite applicable to diabetes mellitus and practiced successfully. This epistemological difference in disease correlation is true for almost all disease conditions, whether it is metabolic or infectious, which makes it difficult to integrate the biological effects studied from Ayurvedic and modern science perspectives.

Some of the recent studies on *Rasayana* could be sited as good examples for how to resolve this issue and how good bioprospecting can be done for Ayurvedic formulation for bridging the concepts of *Ayurveda* and modern science. *Rasayana* is one of the eight disciplines described in *Ayurveda*. It deals with methods such as dietary recipes and regimen, herbal and mineral supplements, and health-promoting lifestyle that are said to enhance quality of life and delay aging (Balasubramani et al. 2011). A study reported on iron deficiency anemia has attempted to understand the Ayurvedic *Rasayana* concept using modern biological experiments. While iron deficiency is the etiological agent of anemia in modern medicine, the corresponding disease entity in *Ayurveda*, called *Pandu*, is a disease caused by derangement in *jatharagni*, the digestive and absorptive "fire" in the body that leads to malabsorption of nutrients. The relatively less success rate of medicinal supplementation of iron and folic acid for treating anemia is mainly due to the failure of the system to assimilate the iron in the body. For the management of *Pandu*, *Ayurveda* uses *Phyllanthus emblica* L., which is a *Rasayana* drug having *agni vardhaka* (*enhance agni*) property. The study using cell models showed that amla juice increases the dialysability and uptake of iron into Caco2 and HepG2 cells. This to a certain extent correlates and validates the Ayurvedic concept of managing anemia through enhancing the *Agni*. Additionally the study also showed that the amla juice significantly increased the iron bioavailability compared to an equivalent amount of ascorbic acid as control, indicating the possible synergistic effects in the formulation. Ascorbic acid is a known iron bioavailability enhancer (Venkatasubramanian et al. 2014). Similarly another *Rasayana* plant, pomegranate, is studied for the enhancement of life-span, reproductive ability, stress tolerance, and resistance to infection using

Drosophila melanogaster as the bioassay model. The transdisciplinary research model showed an increase in life-span and fecundity with an improvement in resistance to oxidative stress and infections (Balasubramani et al. 2014). However, it would be difficult at this stage to find an appropriate quantitative biomarker, say like modulation of "x" or "y," to measure the Ayurvedic concepts of *Agni* or *Rasayana*.

As a proof of concept to the pharmacological networking that addresses the potential holistic therapeutic approaches of Ayurvedic formulations, study of *Withania somnifera* on Alzheimer's disease transgenic mice is a very good example. The study, using Alzheimer's disease transgenic mice, showed that oral administration of a semi-purified extract of the root of *Withania somnifera*, consisting predominantly of withanolides and withanosides, enhanced expression of low-density lipoprotein receptor-related protein (LRP) in brain microvessels and the Aβ-degrading protease neprilysin (NEP) in 14 d–21 d after a substantial decrease in brain Aβ levels, whereas a significant increase in liver LRP and NEP occurred much earlier, at 7 d, and was accompanied by a rise in plasma sLRP. The study shows a remarkable therapeutic effect of *W. somnifera* mediating through upregulation of liver LRP indicating that targeting the periphery offers a unique mechanism for Aβ clearance and reverses the behavioral deficits and pathology seen in Alzheimer's disease models (Sehgal et al. 2012).

Although these studies are in an infant stage, they have successfully marked a milestone in transdisciplinary research for deciphering the concepts of TM using modern scientific tools and opened up new horizons for bioprospecting Ayurvedic formulation.

16.10 Conclusions

Bioprospecting TMs are the need of the hour because a multi-targeted, multi-ingredient formulation strategy could have a better therapeutic effect than conventional one-gene, one-target, one-drug approach. However, most of the bioprospecting attempts are confined to the discovery of a single target-specific bioactive molecule approach, from individual plants and their active fractions. The major drawback of such attempts is their inability to correlate between concepts of TM and the modern medicine. A proper bioprospecting need to focus on the overall biological and pharmacological activity of the herb/formulation with all possible integrations to the fundamental principles of TM. Attempts in this direction have shown encouraging success, and such studies have set the tone for prospective scientific research and bioprospecting of traditional medicines for the future.

References

Adams SJ, Kuruvilla GR, Krishnamurthy KV, Nagarajan M, Padma V (2013) Pharmacognostic and phytochemical studies on Ayurvedic drugs Ativisha and Musta. Rev Bras 23:398–409

Auti SS, Thakar AB, Shukla VJ, Ravishankar B (2013) Assessment of Lekhana Basti in the management of hyperlipidemia. Ayu 34:339–345

Bag A, Bhattacharyya SK, Pal NK (2013) Antibacterial potential of hydroalcoholic extracts of triphala components against multidrug-resistant uropathogenic bacteria--a preliminary report. Indian J Exp Biol 51:709–714

Balasubramani SP, Venkatasubramanian P, Kukkupuni SK, Patwardhan B (2011) Plant-based *Rasayana* drugs from *Ayurveda*. Chin J Integr Med 17:88–94

Balasubramani SP, Mohan J, Chatterjee A, Patnaik E, Kukkupuni SK, Nongthomba U, Venkatasubramanian P (2014) Pomegranate juice enhances healthy lifespan in *Drosophila melanogaster*: an exploratory study. Front Public Health 2:245

Baliga MS (2010) *Triphala*, Ayurvedic formulation for treating and preventing cancer: a review. J Altern Complement Med 16:1301–1138

Baliga MS, Meera S, Mathai B, Rai MP, Pawar V, Palatty PL (2012) Scientific validation of the ethnomedicinal properties of the Ayurvedic drug *Triphala*: a review. Chin J Integr Med 18:946–954

Barabási AL, Gulbahce N, Loscalzo J (2011) Network medicine: a network-based approach to human disease. Nat Rev Genet 12:56–68

Baskaran UL, Martin SJ, Mahaboobkhan R, Prince SE (2015) Protective role of *Triphala*, an Indian traditional herbal formulation, against the nephrotoxic effects of bromobenzene in Wistar albino rats. J Integr Med 13:115–121

Belapurkar P, Goyal P, Tiwari-Barua P (2014) Immunomodulatory effects of *Triphala* and its individual constituents: a review. Indian J Pharm Sci 76:467–475

Bhasin V (2007) Medical anthropology: a review. Stud Ethno-Med 1:1–20

Bhattacharjee R, Nekkanti S, Kumar NG, Kapuria K, Acharya S, Pentapati KC (2015) Efficacy of *Triphala* mouth rinse (aqueous extracts) on dental plaque and gingivitis in children. J Investig Clin Dent 6:206–210

Chandran U, Mehendale N, Tillu G, Patwardhan B (2015) Network pharmacology of *Ayurveda* formulation *Triphala* with special reference to anti-cancer property. Comb Chem High Throughput Screen 18:846–854

Chandre R, Upadhyay BN, Murthy KH (2011) Clinical evaluation of *Kushmanda Ghrita* in the management of depressive illness. Ayu 32:230–233

Dawane J, Pandit VA, Deshpande SS, Kuvalekar AA, Mandpe A, Wele A, Dalvi M (2014) Evaluation of antidiabetic activity of Nishamalaki on streptozotocin induced type II diabetes. Int J Phytomed 6:595–600

Dwivedi V, Anandan EM, Mony RS, Muraleedharan TS, Valiathan MS, Mutsuddi M, Lakhotia SC (2012) *In vivo* effects of traditional *Ayurvedic* formulations in *Drosophila melanogaster* model relate with therapeutic applications. PLoS One 7:e37113

Dwivedi V, Tripathi BK, Mutsuddi M, Lakhotia SC (2013) Ayurvedic *Amalaki Rasayana* and *Rasa-Sindoor* suppress neurodegeneration in fly models of Huntington's and Alzheimer's diseases. Curr Sci 105:1711–1723

Dwivedi V, Tiwary S, Lakhotia SC (2015) Suppression of induced but not developmental apoptosis in *Drosophila* by Ayurvedic *Amalaki Rasayana* and *Rasa-Sindoor*. J Biosci 40:281–297

Frawley D, Ranade S (2000) *Ayurveda*, nature's medicine. Lotus Press, Wisconsin, pp 125–165

Ghildiyal S, Gautam MK, Joshi VK, Goel RK (2014) Analgesic and hypnotic activities of *Laghupanchamula*: a preclinical study. Ayu 35:79–84

Gilca M, Barbulescu A (2015) Taste of medicinal plants: a potential tool in predicting ethnopharmacological activities? J Ethnopharmacol 174:464–473

Govindaraj P, Nizamuddin S, Sharath A, Jyothi V, Rotti H, Raval R, Nayak J, Bhat BK, Prasanna BV, Shintre P, Sule M, Joshi KS, Dedge AP, Bharadwaj R, Gangadharan GG, Nair S, Gopinath PM, Patwardhan B, Kondaiah P, Satyamoorthy K, Valiathan MV, Thangaraj K (2015) Genome-wide analysis correlates *Ayurveda Prakriti*. Sci Rep 5:15786

Gujarathi RH, Gokhale VM, Tongaonkar JN (2013) Effect of *Basti* (oil enema) therapy for the management of cough in pertussis. Ayu 34:397–400

Gupta VB, Manjusha R, Ravishankar B, Harisha CR, Shukla VJ, Khant DB (2012) Pharmacognostical and physicochemical analysis of *pathyadi varti* – a polyherbal *Ayurvedic* formulation. Int J Pharm Life Sci 3:1643–1648

Guruprasad KP, Subramanian A, Singh VJ, Sharma RS, Gopinath PM, Sewram V, Varier PM, Satyamoorthy K (2012) *Brahmarasayana* protects against Ethyl methanesulfonate or Methyl methanesulfonate induced chromosomal aberrations in mouse bone marrow cells. BMC Complement Altern Med 12:113

Hongal S, Torwane NA, Pankaj G, Chandrashekhar BR, Gouraha A (2014) Role of *Unani* system of medicine in management of orofacial diseases: a review. J Clin Diagn Res 8:12–15

http://www.greekmedicine.net/whos_who/Hippocrates.html

Jagtap CY, Chaudhari SY, Thakkar JH, Galib R, Prajapati PK (2014) Assessment of genotoxic potential of *hridayarnava rasa* (a herbo-mineralo-metallic ayurvedic formulation) using chromosomal aberration and sperm abnormality assays. Toxicol Int 21:242–247

Jose JK, Kuttan R (2000) Hepatoprotective activity of *Emblica officinalis* and *Chyavanaprash*. J Ethnopharmacol 72:135–140

Joshi K, Hankey A, Patwardhan B (2007) Traditional phytochemistry: identification of drug by 'taste'. Evid Based Complement Alternat Med 4:145–148

Joshi R, Reeta KH, Sharma SK, Tripathi M, Gupta YK (2015) *Panchagavya Ghrita*, an *Ayurvedic* formulation attenuates seizures, cognitive impairment and oxidative stress in pentylenetetrazole induced seizures in rats. Indian J Exp Biol 53:446–451

Juya RC, Negi S, Wakhode P, Bhat S, Bhat B, Thelma BK (2012) Potential of ayurgenomics approach in complex trait research: leads from a pilot study on rheumatoid arthritis. PLoS One 7:e45752

Kalaiselvan S, Rasool M (2015) *Triphala* exhibits anti-arthritic effect by ameliorating bone and cartilage degradation in adjuvant-induced arthritic rats. Immunol Investig 44:411–426

Kim HU, Ryu JY, Lee JO, Lee SY (2015) A systems approach to traditional oriental medicine. Nat Biotechnol 33:264–268

Kolhe RH, Acharya R, Shukla VJ (2014) Role of thin-layer chromatography in ascertaining *Kashaya Rasa* (astringent taste) in medicinal plants on the concept of *Samana* and *Vichitra Pratyayarabdha* principles of *Ayurveda*. Ayu 35:179–183

Kumar G, Gupta YK (2012) Evidence for safety of Ayurvedic herbal, herbo-metallic and *Bhasma* preparations on neurobehavioral activity and oxidative stress in rats. Ayu 33:569–575

Lakshmi C (2014) Allergic contact dermatitis (type iv hypersensitivity) and type I hypersensitivity following aromatherapy with *Ayurvedic* oils (*Dhanwantharam thailam*, *Eladi* coconut oil) presenting as generalized erythema and pruritus with flexural eczema. Indian J Dermatol 59:283–286

McClatchey WC (2005) Medicinal bioprospecting and ethnobotany research. Ethnobot Res Appl 3:189–190

Medagama AB, Bandara R (2014) The use of complementary and alternative medicines (CAMs) in the treatment of diabetes mellitus: is continued use safe and effective? Nutr J 13:102

Meena AK, Bansal P, Kumar S (2009) Plants-herbal wealth as a potential source of ayurvedic drugs. Asian J Tradit Med 4:152–170

Menon A, Krishnan Nair CK (2013) Ayurvedic formulations ameliorate cisplatin-induced nephrotoxicity: preclinical studies on *Brahma Rasayana* and *Chyavanaprash*. J Cancer Res Ther 9:230–234

Narayanaswamy V (1981) Origin and development of *Ayurveda*: (a brief history). Anc Sci Life 1:1–7

Ngo LT, Okogun JI, Folk WR (2013) 21st century natural product research and drug development and traditional medicines. Nat Prod Rep 30:584–592

Nie J, Zhao C, Deng LI, Chen J, Yu B, Wu X, Pang P, Chen X (2016) Efficacy of traditional Chinese medicine in treating cancer. Biomedical Rep 4:3–14

Pavan BS, Biswajyoti P, Jalaram TH, Ravishankar B, Gupta VB (2012) Evaluation of subchronic genotoxic potential of *Swarna Makshika Bhasma*. Ayu 33:418–422

Phondani PC (2011) Worth of traditional herbal system of medicine for curing ailments prevalent across the mountain region of Uttarakhand, India. J Appl Pharm Sci 1:81–86

Prasher B, Negi S, Aggarwal S, Mandal AK, Sethi TP, Deshmukh SR, Purohit SG, Sengupta S, Khanna S, Mohammad F, Garg G, Brahmachari SK, Indian Genome Variation Consortium, Mukerji M (2008) Whole genome expression and biochemical correlates of extreme constitutional types defined in *Ayurveda*. J Transl Med 6:48

Rajput D, Patgiri BJ, Galib R, Prajapati PK (2013) Anti-diabetic formulations of *Nāga bhasma* (lead calx): a brief review. Anc Sci Life 33:52–59

Ranga RS, Girija R, Nur-e-Alam M, Sathishkumar S, Akbarsha MA, Thirugnanam S, Rohr J, Ahmed MM, Chendil D (2004) *Rasagenthi lehyam* (RL) a novel complementary and alternative medicine for prostate cancer. Cancer Chemother Pharmacol 54:7–15

Rao G, Bhat S, Rao SG, Bhat PG (2013) Effect of treatment with '*Nishamalaki'* powder on glycemic control and markers of erythrocyte oxidative stress in diabetic rats compared to troglitazone. Int Pharm Sci Rev Res 19:127–134

Rath SK, Panja AK, Nagar L, Shinde A (2014) The scientific basis of rasa (taste) of a substance as a tool to explore its pharmacological behavior. Anc Sci Life 33:198–202

Reddy AS, Zhang S (2013) Polypharmacology: drug discovery for the future. Expert Rev Clin Pharmacol 6:41–47

Rotti H, Mallya S, Kabekkodu SP, Chakrabarty S, Bhale S, Bharadwaj R, Bhat BK, Dedge AP, Dhumal VR, Gangadharan GG, Gopinath PM, Govindaraj P, Joshi KS, Kondaiah P, Nair S, Nair SN, Nayak J, Prasanna BV, Shintre P, Sule M, Thangaraj K, Patwardhan B, Valiathan MV, Satyamoorthy K (2015) DNA methylation analysis of phenotype specific stratified Indian population. J Transl Med 13:151

Sehgal N, Gupta A, Valli RK, Joshi SD, Mills JT, Hamel E, Khanna P, Jain SC, Thakur SS, Ravindranath V (2012) *Withania somnifera* reverses Alzheimer's disease pathology by enhancing low density lipoprotein receptor-related protein in liver. Proc Natl Acad Sci 109:3510–3515

Sharma DC, Chandiramani D, Riyat M, Sharma P (2007) Scientific evaluation of some ayurvedic preparations for correction of iron deficiency and anemia. Indian J Clin Biochem 22:123–128

Sivasankar S, Lavanya R, Brindha P, Angayarkanni N (2015) Aqueous and alcoholic extracts of *Triphala* and their active compounds chebulagic acid and chebulinic acid prevented epithelial to mesenchymal transition in retinal pigment epithelial cells, by inhibiting SMAD-3 phosphorylation. PLoS One 10:e0120512

Soni A, Patel K, Gupta SN (2011) Clinical evaluation of *Vardhamana Pippali Rasayana* in the management of *Amavata* (Rheumatoid Arthritis). Ayu 32:177–180

Sudha VBP, Geeta UG, Venkatasubramanian P (2004) Bioactivity of traditional preparation of *Piper longum* L. (*Piperaceae*). J Trop Med Plants 5:179–182

Thatte U, Chiplunkar S, Bhalerao S, Kulkarni A, Ghungralkar R, Panchal F, Vetale S, Teli P, Kumbhar D, Munshi R (2015) Immunological and metabolic responses to a therapeutic course of *Basti* in obesity. Indian J Med Res 142:53–62

Venkatasubramanian P, Kumar SK, Nair VS (2010) *Cyperus rotundus*, a substitute for *Aconitum heterophyllum*: Studies on the *Ayurvedic* concept of *Abhava Pratinidhi Dravya* (drug substitution). J Ayurveda Integr Med 1:33–39

Venkatasubramanian P, Koul IB, Varghese RK, Koyyala S, Shivakumar A (2014) *Amla* (*Phyllanthus emblica* L.) enhances iron dialysability and uptake in *in vitro* models. Curr Sci 107:1859–1866

Vishnuprasad CN, Pradeep NS, Cho YW, Gangadharan GG, Han SS (2013) Fumigation in *Ayurveda*: potential strategy for drug discovery and drug delivery. J Ethnopharmacol 149:409–415

Vyas P, Thakar AB, Baghel MS, Sisodia A, Deole Y (2010) Efficacy of *Rasayana Avaleha* as adjuvant to radiotherapy and chemotherapy in reducing adverse effects. Ayu 31:417–423

Microbial Repositories in Bioprospecting 17

Gayathri Valsala and Shiburaj Sugathan

Abstract

Microbial diversity is an integral part of ecosystem and is a ticket to each country's prosperity because it enables the development of a framework for long-term conservation and facilitates targeted bioprospecting initiatives that can potentially catalyse industrial innovation. Microbial culture banks, also called culture collections, are the key repositories of biodiversity. Microbial cultures from the collections are used by scientists to make sure that the materials they use are authentic. This is particularly important in research leading to peer-reviewed publications and patent. Reference strains are also of paramount importance for clinical diagnostic testing, food, water and environmental microbiology testing and validation studies. There are 708 microbial culture collections registered under World Federation of Culture Collections (WFCC). This chapter deals with the current international status of microbial culture collections, their functions and services they provide to meet the current and future demands relevant to biodiversity conservation and biotechnology.

Keywords

Microbial diversity • Bioprospecting • Microbial Repository • Culture Collection • WFCC

G. Valsala • S. Sugathan (✉)
Division of Microbiology, Jawaharlal Nehru Tropical Botanic Garden and Research Institute, Palode, Thiruvananthapuram, Kerala 695562, India
e-mail: gayathri.v.biotech@gmail.com; drshiburaj@gmail.com

S. Abdulhameed et al. (eds.), *Bioresources and Bioprocess in Biotechnology*,
DOI 10.1007/978-981-10-3573-9_17

17.1 Introduction

Microbes form an important component of ecosystems and carry out various functions essential for the biosphere including nutrient cycles of nitrogen, phosphate, oxygen and carbon, in bioremediation of pollutants, soil building and even in the life of plants and animals. They are an essential part of food chains and food webs and are responsible for the decomposition of waste materials and dead organic matter of plants and animals. The heterogeneity of biosphere has assisted in the creation of a wide array of microorganisms with potential to produce novel pharmaceuticals, enzymes and fine chemicals. Proper management and utilisation of microbial diversity has a supreme role in development of industrial and commercial applications. Even the pathogenic microbial stocks have to be maintained for ready availability in academic and industrial research. Advancements in the field of biotechnology have paved way for the isolation, culturing and characterisation of new microorganisms (Smith et al. 2008).

Microbial culture collections or repositories are crucial resource centres for preserving and supplying microbial specimens. These microbial resource centres (MRCs) are living libraries of microorganisms which are indispensable for scientific research in life science. They serve as repositories of living microbial cells, genomes and information relating to heredity and the functions of biological systems. They are also responsible for the supply of microorganisms in primary and applied scientific research, including those for commercial application in fields such as pharmaceuticals, cosmetics and industrial enzymes. Microorganisms are distributed ubiquitously on earth, and this large number of microbes has to be identified, classified and preserved. Microbial repositories contain collections of cultivable organisms of algae, bacteria, fungi (including yeasts), protozoa and viruses and their replicable molecules like genomes, plasmids, viruses, cDNAs and also uncultivable microorganisms (as metagenome libraries).

The demand for bioprospecting of microbial resources has escalated in recent time which has contributed to an increase in the number and quality of microbial repositories. World's microbial diversity represents 50% of total biodiversity in terms of species numbers. The microbial world is the biggest unexplored biodiversity reservoir in the world with only <1% of the bacterial species and <5% of fungal species currently known. A large fraction of the world's microbial wealth remains untapped because of difficulties in sample collection from extreme locations and difficulties in laboratory culturing (Staley et al. 1997). A large numbers of microbial species are uncultivable and require novel techniques of isolation and identification. The available microbial resources need to be conserved and used in a sustainable way along with due protection for intellectual property, which can be done through ex situ safeguarding through microbial culture collections. In addition to their role in scientific, industrial, agricultural, environmental and medical research, microbial repositories also serve as an interface between governments, industry and the public, assisting policymakers and public in understanding the value of microbial conservation.

Despite the ecological and economic value of microbial diversity, its conservation and management has been largely ignored. One of the reasons for this is the unawareness regarding the role of microbial diversity to the economy and in improving quality of life. Hence there is a compelling need to convince policymakers to undertake microbial conservation and management.

17.2 Microbial Diversity

Microorganisms were discovered by Antonie van Leeuwenhoek in 1677. Louis Pasteur in 1857 described their properties and uses in his study on lactic acid fermentation. The role of microbes as causative agents of infections was observed by Robert Koch in 1876 as part of his study of anthrax. Microorganisms are distributed in the kingdoms of Monera, Protista, Fungi and a part of plants of the Whittaker's five-kingdom system. Recent evolutionary studies based on ribosomal RNA sequences show that the evolutionary distance between bacteria to animals and plants is farther than the distance between plants and animals from each other. On the basis of this, Woese et al. (1990) proposed a new superior concept of domains over the kingdom classification, with three domains, Archaea, Bacteria and Eucarya. Microorganisms are now regarded as a collection of evolutionarily different organisms with bacteria, actinomycetes, cyanobacteria, etc. under domain Bacteria, while methanogens, extremely thermophilic organisms, extremely halophilic organisms, etc., are placed in the domain Archaea and moulds, yeasts, basidiomycetes, algae and protozoans, etc. are under Eucarya.

The diversity of microorganisms on Earth is truly astounding. About 1 billion bacteria are present in just a gram of soil of which only 1% can be cultured (Yandigeri et al. 2010). It is predicted that Earth is inhabited by as many as 1 trillion (10^{12}) microbial species (Locey and Lennon 2016). This figure indicates that 99.999% of microbial taxa still remain undiscovered and stresses the need for continued investigation. The total number of species of virus, bacteria, algae, fungi and lichens recorded from India and the world are summed up in Table 17.1.

Table 17.1 Total number of microbial species of in India and World

		Number of known species		Percentage of occurrence
SI. No.	Type	World	India	in India
1.	Virus and Bacteria	11,813	986	8.77%
2.	Algae	40,000	7284	18.21%
3.	Fungi	98,998	14,883	15.09%
4.	Lichens	17,000	2401	14.12%

Source: Chapman (2009) and Singh and Dash (2014)

17.3 Conservation of Microbial Diversity

Biodiversity surveys rarely attempt at systematic study of microorganisms in different habitats. The reason for this is the labour intensiveness of sampling, culturing and identification of enormous numbers of microorganisms. Microbes make up about 50% of the living protoplasm on this planet, and hence microbial resources deserve greater attention (Yandigeri et al. 2010).

Microbial diversity includes the range of variability among all types of microorganisms in the natural world. Microbes are sources of new genes and organisms that are of value to biotechnology, in monitoring and predicting environmental changes, conservation and restoration of higher organisms and in understanding biological interactions and evolutionary history. In order to fully harness the benefits of microbial resources, a better understanding of their diversity is required. In addition to their ecological role, many microorganisms are also of immense scientific and economic benefits. Pathogenic microbes cause thousands of mortality every day. Medical research of epidemics requires pure culture of the causative organism.

Microbial diversity offers numerous benefits. In agriculture, asymbiotic nitrogen fixers, mycorrhizal microbes and plant growth-promoting rhizobacteria (PGPR) like *Pseudomonas fluorescens*, *Bacillus subtilis*, *Bradyrhizobium japonicum*, etc. can be used as biofertilizers for improving crop nutrition. Some microbes assist in solubilisation of insoluble nutrient sources. Entomopathogenic fungus can be used as biocontrol agents against insect pests, and antagonistic microbes are involved in the suppression of growth and activity of plant pathogens. Microbial alkaloids, toxins and antibiotics can control plant pests. Some microbes like *Trichurus spiralis, Paecilomyces fusisporus* and *Trichoderma viridae* can act as compost accelerators at low C: P and C: N ratios, which is achieved through supplementation with rock phosphate and urea. Microbes are essential for degradation of complex plant and animal wastes and in nutrient cycles. Cellulose and lignin degrading microorganisms like *Trichoderma viridae*, *Scytalidium thermophilum*, *Fusarium moniliformae* and *Phaenerochaete chrysosporium* can be used for the management of agro-waste materials and bioconversion of distillery wastes and dyes (Singh and Nain 2014). They can also be used in bioremediation of oil spills, industrial waste and pesticides by oil/hydrocarbon degrading microbes and in bio-mineralisation of heavy metals and reclamation of degraded soils.

In industries, microbes are used in brewing of wine and beer, in the preparation of dairy products like cheese, curd and yoghurt. They also serve as sources of antibiotics, clinically important molecules and industrial enzymes. For humankind to benefit from biodiversity, we must understand and utilise its potential through proper identification and maintenance of biological resources (Yandigeri et al. 2010).

The importance of microbial culture banks has been recognised by industrial countries, which is the reason why majority of public collections are present in the Northern Hemisphere. Europe and America hold 56% of collections and 71% of the identified microorganisms. In contrast, the number of microbial culture collections in biodiversity rich regions, where approximately 20–30% of total novel isolations occur, is very low. The large numbers of novel bacterial and fungal species in the diversity hot spots are untapped reservoirs that hold bioeconomic potential (Komagata 1998).

India is one among the 12 countries of megadiversity with 1,27,000 species of listed plants, animals and microbes. Even though we are expected to have a rich microbial diversity, lack of skilled experts and adequate support restricts their discovery. India has 14,833 species of fungi, 2401 lichens, 17,000 flowering plants with an estimated 6 fungi per plant; therefore, it is anticipated that there are approximately 102,000 fungi (Arora et al. 2005).

Human interferences and natural calamities have resulted in the extinction of many organisms including plants, animals and microbes. All organisms serve as a reservoir for numerous other organisms, majority of which are microbes. The root microbiome, a dynamic community of microorganisms associated with plant roots, is unique for each plant species. The microbial communities inside the root and in the rhizosphere are distinct from each other and differ from the microbial communities of bulk soil (Gottel et al. 2011). Thus any loss of higher plants might result in the extinction of associated microorganisms and cause irreversible damage to biodiversity. While plants can be conserved as such in nurseries, botanical gardens or in/as cryobanks, seed banks and gene banks, microbes have to be preserved as pure cultures under laboratory conditions, followed by validation and preservation. Thus microbial culture collections serve not only in the preservation of microbes but also help in retrieving if they are completely lost from natural system. In situ microbial conservation method is inadequate because the organisms are prone to rapid evolution and are sensitive to environmental factors that can lead to their elimination.

The Convention on Biological Diversity (CBD) has recognised the importance of conservation of microorganisms and the significance of ex situ collections and benefit sharing. Molecular biology techniques and non-culture tools can be employed for the identification and characterisation of a vast diversity of microorganisms (Davison et al. 1998). A large fraction of microbes are unculturable, and hence unconventional isolation and identification strategies need to be developed (Smith et al. 2008). Metagenomic analysis of microbial communities does not require prior culturing and can be used for the identification of a vast majority of microbial biodiversity that can be missed by cultivation-based methods. It involves either shotgun or PCR directed sequencing of all genes from all the members of the sampled communities (Fig. 17.1).

The conservation and exploitation of microbial biodiversity still remains a challenge. *Ex situ* conservation strategies should be developed to conserve critical biodiversity elements, and the approaches to discover and understand microbial communities should be coordinated. The economic applications of microbial biodiversity should be identified, and the income received should be used to fund research, collection maintenance and biosystematics (Smith et al. 2008).

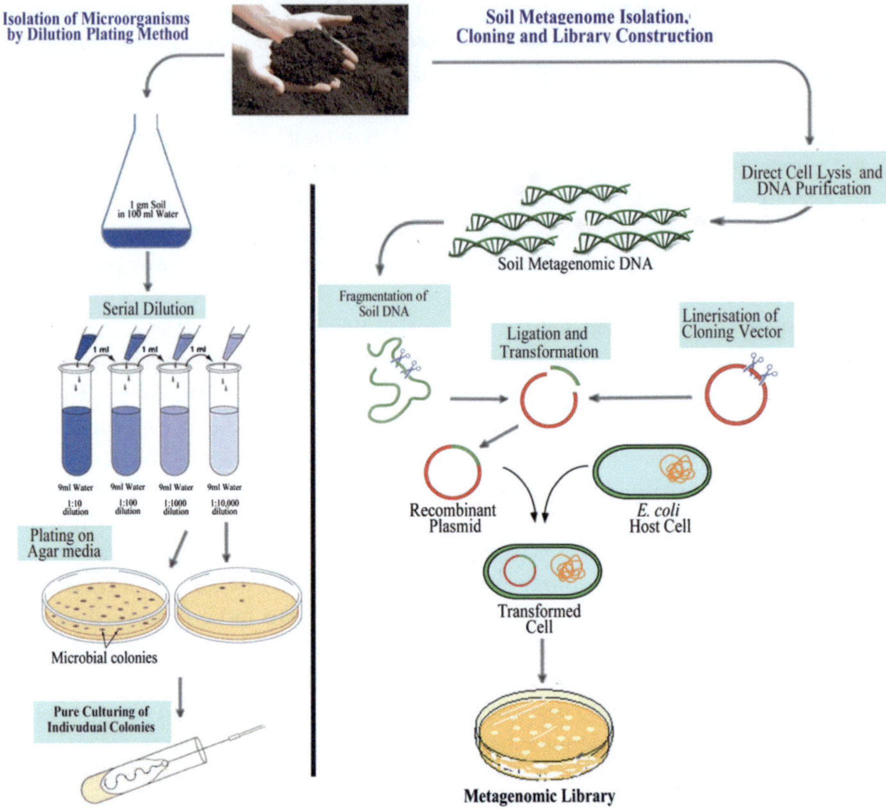

Fig. 17.1 Schematic representation of isolation of cultivable microorganisms by conventional dilution plating method and direct isolation of DNA from soil for construction of metagenome library

17.4 History of Microbial Repositories

Prof. Frantisek Král (1846–1911) established the first microbial culture collection in 1890 at the German University of Prague, Czech Republic, where fungal cultures were made commercially available (Smith et al. 2014). He had 30 years' experience in manufacturing laboratory glass products, after which he worked as a technician at the Institute of Hygiene of the German University of Prague. His expertise in glass manufacturing was the reason for his appointment as the director of bacterial collection by Prof. Soyka. Later he held the post of associate Professor of Bacteriology. Král published the first catalogue of microorganisms from a culture collection in 1900. After his death in 1911, his culture collection was acquired by Professor Ernst Pribham, who transferred it to the University of Vienna and issued several catalogues listing the microbes in the collection. A portion of this collection was taken to Loyola University in Chicago by Prof. Pribham in the 1930s. Some of the cultures in the collection still remains at Loyola University while others were transferred to the American Type Culture Collection (ATCC, USA) after Pribham's death (Uruburu 2003; Çaktu and Turkoglu 2011). After Král's collection, many other culture collections were established; however only few collections established in the early era of microbiology survived the politically and scientifically turbulent twentieth century.

The oldest culture collections that are still in use are the Mycothèque de l'Université catholique de Louvain (MUCL) in Louvain-la-Neuve, Belgium, established in 1892, and the Collection of the Centraal bureau voor Schimmel cultures (CBS) founded in 1906, in Utrecht, the Netherlands (Uruburu 2003). MUCL is one among the earliest fungal collections which currently holds over 25,000 strains of filamentous and yeast-like fungi, representing over 3300 species of Ascomycetes, Basidiomycetes, Hyphomycetes and Zygomycetes. The CBS Fungal Biodiversity Centre, founded as an institute of the Royal Netherlands Academy of Arts and Sciences, was the first independent microbial centre to undertake the preservation and supply of a wide range of fungi. Now it holds a world-renowned collection of living filamentous fungi, yeasts and bacteria (Hawksworth 1985; Smith et al. 2014). The well-known American Type Culture Collection (ATCC) was established in 1925 at Washington. Now ATCC is located in Manassas, Virginia.

There is an international connection between recognised microbial repositories lead by World Federation of Culture Collections (WFCC). It was in 1946, Prof. P. Houduroy established a centralised information facility at the University of Lausanne, Switzerland. Information about the strains maintained in different

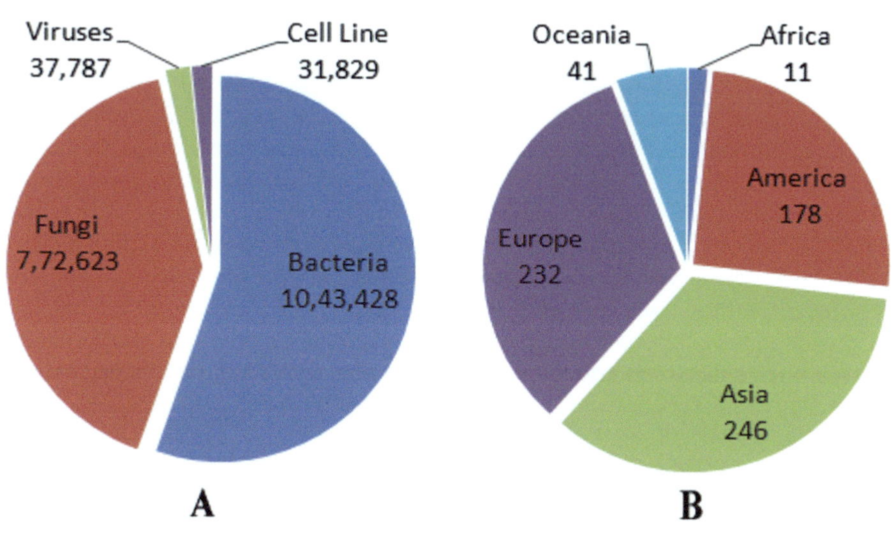

Fig.17.2 The charts showing the number of holdings of different type of microbial cultures in the culture collections (**a**) and region-wise distribution of Culture Collections (**b**), prepared based on database available with Culture Collections Information Worldwide (CCINFO). CCINFO is a world directory of all registered culture collections maintained by World Data Centre for Microorganisms (WDCM) (http://www.wfcc.info/ccinfo/statistics)

Table 17.2 Some important microbial culture collections in the world

Acronym	Name	Country	Type of agency
ATCC	American Type Culture Collection	USA	Government
BIOCEN (BioCC)	Centro Nacional de Biopreparados	Cuba	Industry
DSMZ	Deutsche Sammlung	Germany	Government
ACAM	The Australian Collection of Antarctic Microorganisms	Australia	University
CCFC	Canadian Collection of Fungal Cultures	Canada	Government
ACCC	Agricultural Culture Collection of China	China	Government
CIP	Collection de L'Institut Pasteur	France	Private
HNCMB	Hungarian National Collection of Medical Bacteria	Hungary	Government
ICMP	International Collection of Microorganisms from Plants	New Zealand	Semi-government
PPPPB	South African Plant Pathogenic and Plant Protecting Bacteria	South Africa	Semi-government
CCOS	Culture Collection of Switzerland	Switzerland	Private, University
KCCC	Culture Collection of Soybean Rhizobia and Green Microalgae	Thailand	Private, Industry
VTCC	Vietnam Type Culture Collection	Vietnam	Government
CBS	Centraal bureau voor Schimmel cultures	Netherlands	Semi-government
MUM	Micoteca da Universidade do Minho	Portugal	University

collections was provided by this facility. This centre became associated with International Association of Microbiological Societies (IAMS) and is now renamed as International Union of Microbiological Societies (IUMS). At a conference on culture collections held in Canada in 1962, the IAMS was asked to form a section on Culture Collections under the chairmanship of Prof. Skerman. This was reorganised in 1970 and the section became the World Federation of Culture Collections (WFCC). Through the World Data Centre of Microorganisms (WDCM) established in Mishima, Japan, WFCC has collected and stored information on several hundreds of collections throughout the world (Uruburu 2003). There are at present 708 culture collections in 72 countries registered in World Data Centre for Microorganisms (WDCM) over the world which together they hold almost 2.5 million strains of a wide range of microorganisms (http://www.wfcc.info/ccinfo/, accessed August 25, 2016) (Fig. 17.2 and Table 17.2).

17.5 Types of Microbial Culture Collections

Culture collections are classified into four categories, namely, Private or In-house Culture Collections, Specialised Collections, Research Collections and Public Collection or Service Collections (Mahilum-Tapay 2009).

17.5.1 Private or In-House Culture Collections

These collections are established and maintained by particular organizations, institutions, laboratories, hospitals or individual companies to meet their requirements. Collections of this type may hold cultures ranging from small to large, but catalogues are not generally available. The cultures are supplied on a discretionary basis and are not usually open to the public.

17.5.2 Specialised Collection

They are collections of specific group of microorganisms in specialised fields such as brewing, antibiotic production, plant or animal pathology, etc. in which a large number of strains have been isolated and characterised and detailed taxonomic study carried out. Examples include UTEX Culture Collection of Algae at the University of Texas, USA, and Culture Collection of Algae and Protozoa (CCAP), UK.

17.5.3 Research Collections

Research collections are created by individual researchers or teams as a part of their research programmes. These collections often include novel and peculiar strains of scientific importance, but they lack long-term storage facilities, and hence the cultures could be lost when researchers change positions, retire or pursue different lines of research.

17.5.4 Public Collection or Service Collections

These collections are directly or indirectly supported by government funds and are established for the purpose of public service. All details of the cultures contained are generally listed in publicly available catalogues, and authenticated microbial cultures can be obtained from them on request at less than full economic cost. They are large collections and the number of strains contained varies from collection to collection. Microbial Type Culture Collection (MTCC), Chandigarh, India, is a public culture collection funded jointly by the Department of Biotechnology (DBT) and the Council of Scientific and Industrial Research (CSIR), Government of India.

17.6 Services Offered by Culture Collections

Microbial culture collections provide authentic biological material for high-quality research and teaching in the form of reference strains and also play a vital role in the conservation, prospecting and sustainable use of microbial resources (Sharma and Shouche 2014). The primary functions of microbial repositories involve collection, accession maintenance, preservation, identification, documentation, patent deposition, cataloguing, safe keeping, screening and distribution of microbial cultures on request, preparation of printed or online catalogues of collect culture data and making them accessible to the researchers. Thus researchers and taxonomists should be able to easily identify a strain of particular interest from the database provided by culture collection centres. In addition, microbial repositories also provide training courses and workshops related to the identification and maintenance of microorganisms to personnel from medical, environmental, industry or government laboratories who are involved in isolating and identifying microorganisms, diagnosing disease, quality control, fermentation, culture management, etc. Research initiatives related to taxonomic studies and microbiological preservation are also carried at microbial culture collections. The services offered by a culture collection depend upon the size and interests of the collection.

17.6.1 Culture Deposition

In a culture collection, new depositions or accessions can be made by researchers. They will be made available for the depositors and other researchers on request. On deposition of a culture, a unique accession number will be allotted by the centre. According to International Code of Bacteriological Nomenclature, for any valid publication of new species, its type strain should be designated and deposited in one or more established culture collections. Microbial repositories will have curators for each specific group of microorganisms, who conduct research and are responsible for the development and maintenance of their collections and new acquisitions. Most journals today require deposition of nucleic acid or protein sequences in public repositories along with the deposition of cultures from which the sequences are derived. All specific information like the source of isolation, taxonomic status, published data, special properties, reason for deposit, growth medium, growth temperature, pH and optimum procedure for long-term maintenance of a strain have to be provided while making a deposition (Malik and Claus 1987).

Microbial cultures that are usually considered for deposition are:

(a) Published strains of newly named taxa
(b) Type, neotype, selected reference strains, specific and unique biotypes
(c) Strains with special properties and applications (bioassay, quality control, etc.)
(d) Strains with applications in agriculture, biotechnology, medicine, education, etc.
(e) Strains mentioned in patent applications
(f) Genetically manipulated strains, plasmid carriers, special mutants, etc.

17.6.2 Distribution of Cultures

An important service of culture collections is to supply of authentic and viable strains for research, teaching or industrial applications. Supply of culture will require a fee, but the supply of cultures between Culture Collections is on an exchange basis.

Microbial repositories usually provide guarantee for viability, for purity and, to a certain extent, to the properties cited in publications or their catalogues. There is a chance for alteration of characteristics of certain mutated strains, phage hosts and strains with plasmids, and hence Culture Collections often insist on reporting of discrepancies about a strain received from them. Microbial repositories can import cultures from abroad, but licences or import permits are required for this. The details on export and import restrictions of microorganisms are published in the News Letters of WFCC. The distribution of patent cultures is usually restricted to only authorized persons, subject to the approval of the patentee (Crespi 1985). Patent cultures are accepted on payment of a fee for their maintenance of the cultures.

17.6.3 Microbial Identification

Bacteria and fungi have been conventionally identified by isolation in culture followed by enzymatic reactions and morphological analyses. The identification of microorganisms, however, remains a challenge because of tedious and days-long biochemical and staining protocols and are prone to errors (Braga et al. 2013). Molecular techniques based on DNA amplification and sequencing provide more secure molecular identification of microorganisms. The 16S rRNA gene homology analysis has provided a tool for estimating bacterial phylogeny, which has led to rapid changes in bacterial taxonomy. Sequencing of the internal transcribed spacer (ITS) and large subunit (LSU) regions of rRNA, followed by comparative sequence analysis, has been the 'gold standard' for molecular identification of most fungi (Tsui et al. 2011). This molecular technique is fast and accurate but is dependent on sequence quality in existing reference databases. Recently, identification based on mass spectrometry (MS), especially MALDI-TOF-MS, has been shown to be an alternative accurate and fast method to identify unknown bacteria on the genus, species and even subspecies level, based on profiles of proteins and peptides derived from whole bacterial cells (Braga et al. 2013).

17.6.4 Patent Cultures

Microbial repositories accept and maintain cultures which are subject to patent application, mostly at the payment of specific fee. Such cultures will be catalogued and given an accession number, but their distribution will be subjected to the approval of the patentee while the patent is pending. In 1977, an International Treaty was signed by 78 countries at Budapest and came into force in 1980 and is administered by World Intellectual Property Organization (WIPO), Geneva. As per provisions of the treaty, cultures deposited at any of the International Depositary Authority (IDA) are recognised by the member states for the patent purpose. A patent becomes valid only if the microorganisms cited in the patent application are deposited with an IDA recognised culture collection (Malik and Claus 1987). As on June, 2016, there are 45 culture collections recognised by IDA. In India there are two depositories, namely, Microbial Culture Collection (MCC), Pune and Microbial Type Culture Collection and Gene Bank (MTCC), Chandigarh, that have acquired the status of International Depositary Authority under the Budapest Treaty.

17.6.5 Data Documentation and Catalogue Preparation

All major microbial repositories collect and document data regarding the specimens in their collection. Detailed record on strain characteristics, nutritional requirements (growth medium, temperature, pH, biochemical characters) along with literature references, application and history are documented. These data are published in the form of catalogues and catalogue supplements, which serve as a source of information for microbiologists. An International Centre for Information on Culture Collections has been established at Lausanne, Switzerland, for regular exchange of information between different collections.

17.6.6 Research and Training

All major culture collections offer training and consultation services in the fields of microbial cultivation, enrichment, isolation, identification and modern preservation techniques.

17.6.7 Safety and Security

It is mandatory to carry out safety operations in the collections including biosafety, chemical and physical safety, etc. and also to ensure good laboratory practice. Risk assessments must be carried out before cultures are brought into the collection, and specific procedures are applied. Adequate controls should be implemented to manage risk, to all who may come into contact with cultures, products and services provided. Particular attention needs to be given to the containment and bio-security aspects of strains which are potentially harmful to man, animals or crops.

17.7 Preservation of Microbial Cultures

Culture collections have a need to maintain healthy, viable microbial cultures for many months or years. Microorganisms occur as communities of mixed populations in nature. Once a strain or species is isolated and cultured in laboratory conditions, they become unstable, and their characteristics change on continuous maintenance in artificial media and conditions. Genetically engineered microbes and those containing plasmids may undergo changes with time when cultured. Isolation, selection, research and development of microorganisms with industrial applications required huge investment and effort. The authenticity and production efficiency of such cultures are also essential for patent application, and the loss of stability in such cultures may cause serious disruption to industrial processes (Malik and Claus 1987). Once a strain is selected and preserved, a distribution and seed stock should be maintained and stored separately. An original culture should also be preserved without subculturing. A preserved culture should be checked for its viability, purity and identity before making it available outside the collection.

Different methods are used for short-term and long-term microbial preservation of bacteria and fungi (Prakash et al. 2013). These include repeated subculturing, preservation on agar beads, slant-grown cultures overlaid with oil, using silica gel and other sterile supports, cryopreservation and lyophilisation. The techniques used in preservation of microorganisms can be divided in two groups. The first group aims at minimum detention of the cell vitality through hypobiosis, a state known as 'cells at rest'. The second group of techniques that include drying, freezing at low temperatures and lyophilisation involves maintenance of microorganisms in an anabiotic state.

Anabiosis, the state or ability of an organism to reversibly decrease or stop its vital activities, is a form of adaptation to survive and preserve under adverse conditions. Many microorganisms are able to survive under varying environmental stresses like warmth, cold, freezing and thawing, drying and moisturising. Sporulating bacteria and fungi are the most resistant to drying, chilling and heating. Many microbes that do not form spores also remain viable even after prolonged refrigeration. The properties of the state of anabiosis are:

- Absence or drastic reduction of metabolism
- Preservation of cellular structure for a long time
- Survival in the absence of free water
- Increased resistance to extreme conditions
- Ability to recover the vital activity under optimal conditions

These properties can be applied for long-term preservation of cultures (Uzunova-Doneva and Donev 2005).

The technique applied should cause only minimal damage to cells and must be able to stabilise genotypic, morphological and physiological features as well as viability until it is required for future use. A single universal preservation technique cannot be used for all microorganisms due to variation in susceptibility among them to different preservation techniques (Malik and Claus 1987). Among the different methods used, cryopreservation and lyophilisation are the most widely used in culture collections. Cryopreservation is the storage of cultures at temperatures that retard chemical reactions of around −70 °C and below. Freeze drying or lyophilisation involves the sublimation of ice from frozen material at reduced pressure directly from the solid phase to the gas phase. It requires storage either under vacuum or at atmospheric pressure in an inert gas.

17.7.1 Pre-preservation Conditions

The factors to consider prior to preservation of a microbiological culture are cultivation method, temperature, composition and pH of medium, aeration, age, physiological condition and concentration of the culture at the time of conservation (Uzunova-Doneva and Donev 2005).

17.7.1.1 Nutrient Media

Microorganisms can be cultured either in liquid broth or solid agar nutrient media for cell culture preparation before conservation. The composition of nutrient medium affects the cell resistance. Researchers have two opposing views regarding the choice of growth media. Culturing on rich nutrient media is recommended by some for obtaining a high percentage of viable cells. The opposing view is the use of nutrient poor media which induces cell metabolism reorganisation towards energy storage. Any media that increases the accumulation of proteins, carbohydrates and lipids in the cells increases its resistance lyophilisation. Tween 80 and oleic acid are usually added to medium to increase cell viability during preservation.

The acidity of medium used is another factor that influences cell resistance during conservation. Each organism has an optimal range of pH, under which it shows maximum growth. The media used to grow cells prior to preservation should be at its optimal pH for maximum survival. Similarly, optimal incubation temperature and aeration should be provided for growth.

In yeasts, medium level of aeration provides oxygen necessary for yeast metabolism, mediates proper mixing of nutrients and removes carbon dioxide released. *Saccharomyces cerevisiae* cultivated in aerobic conditions shows more resistant to hypo- and hypertonic shock compared to anaerobically grown cultures.

17.7.1.2 Culture Age

Choice of culture age is another important factor that influences percentage vitality of preserved microbial cultures. The optimum duration of culture varies with different organisms. Microorganisms are more resistant to freezing and dehydration at the end of the logarithmic growth phase or at the beginning of the stationary phase. For example, 48 hours culture of *S. cerevisiae* is most appropriate for conservation.

17.7.1.3 Cell Concentration

Studies on the correlation between population density and survival after lyophilisation show that the percentage of damaged cells increases at extreme conditions with decrease in cell concentration, but the increase in concentration leads to higher intercellular contacts, toxic metabolite production, etc. that reduces the vitality. Most culture collections recommend an initial concentration 10^8–10^{10} cells/ml.

17.7.1.4 Equilibration

It involves preparation of microbial cells to stabilise the cellular structure before conservation. The physical parameters are optimised osmotically using a cryoprotectant before cooling in a controlled way to protect cells from freezing injury. Various biochemical changes occur during equilibration causing either positive effect or negative effect on the metabolic processes before preservation. The cells undergo a transfer stage that prepares them to withstand the extreme conditions like osmotic and temperature shock used of conservation. Capacity for resistance varies among different microbes. Spore forming organisms retain high viability under most conservation methods. This is because spores are naturally resistant with low water content. Non-spore forming microorganisms show relatively low resistance to cryogenic and lyophilic treatment. Generally speaking, prokaryotes are more resistant than the eukaryotes, of which Gram-positive bacteria have higher resistance than Gram-negative ones.

17.7.2 Cryoprotectants

Preservation at ultra-lower temperatures causes injury to cells in two ways. Direct damage to cells from ice crystals formed and secondary damage caused by the increase in concentration of solutes resulting in osmotic imbalance during ice formation. Cryoprotectants are used to protect cells from cryo-injuries during cryopreservation (Prakash et al. 2013). The cryoprotectants may be of endocellular or extracellular type. The important properties of cryoprotectants are:

- Ability to maintain viability, morphological, biochemical, taxonomical and genetic properties of the microorganism during preservation
- It should be non-toxic and have high water solubility
- Should show colligative properties, which is the collective properties that a solution attains in the presence of the compound
- Low eutectic temperature, i.e. the temperature at which a eutectic mixture freezes or melts, preventing any structure forming after the solvent has been removed. A eutectic mixture is a mixture of two or more components that do not interact chemically but, which at certain ratios, inhibit the crystallization process of one another resulting in a system having a lower melting point than either of the components
- Prevent hyper concentration of salts in the suspension
- Stabilisation of hydrogen bonds in the crystal lattice to prevent formation of large crystals
- Good penetration into cells in the case of endocellular protectants
- Non-reactive and should not precipitate at high concentrations

17.7.2.1 Endocellular Cryoprotectants

These compounds penetrate into cells through cell membrane and create conditions for reducing water content inside cells at low temperatures to reduce the damaging effect of the concentrated solutes. They lower the freezing point of water, promote hydrogen bond formation and vitrification of solvents and prevent ice crystal formation inside the cells. The overcooling of the suspension before freezing results in small crystal formation, which reduces mechanical damage to cells during the cryogenic treatment. Endocellular cryoprotectants are considered ideal for microbial preservation. The most commonly used endocellular protectants are glycerol and dimethyl sulphoxide (DMSO) (Fuller 2004).

17.7.2.1.1 Glycerol

Addition of glycerol increases the stability of microorganisms during dehydration, lowers active water (aw) and preserves the cell vitality during freezing. Usually 10–15% glycerol is used prior to cryopreservation. An increase in synthetic processes before the anabiotic transition is observed in some microorganisms after the addition of extra glycerol to nutrient media. Even though glycerol shows good penetration at physiological temperatures, its penetration capability lowers at low temperatures and consequently provides less protection.

17.7.2.1.2 Dimethyl Sulphoxide (DMSO)

The cryoprotective activity of DMSO is concentration dependent. Ten percent concentration is used by most of the world culture collections. DMSO has a better penetrating ability than glycerol but it is toxic at higher concentrations.

17.7.2.2 Extracellular Cryoprotectants

Extracellular cryoprotectants do not penetrate the membrane of a eukaryotic cell or the bacterial cell wall. For this reason, they are incapable of preventing cell membrane rupture that occurs as a result of intracellular water-crystal formation. Examples of these non-penetrating molecules are dextran, hydroxyethyl starch (HES) and polyvinylpyrrolidone (PVP). These molecules function to reduce hyperosmotic gradients that form as the samples transition through the freezing process. As less water is available to function as a solute outside the cell, salt concentrations drastically increase. The increased osmotic pressure on cells cause alterations in cell shape, cellular interactions, subcellular distribution of molecules and ultimately cell viability.

17.7.2.2.1 Dextran

Being chemically inert, it is either used alone or in combination with other cryoprotectants. It has been successfully used for the preservation of viruses and microorganisms.

17.7.2.2.2 Hydroxyethyl Starch (HES)

It is a modified natural polymer of amylopectin that acts as a protectant that is non-toxic and biologically inert.

17.7.2.2.3 Polyvinylpyrrolidone (PVP)

Many hypotheses have been put forward regarding the cryoprotection mechanism of PVP. One is that PVP protect cells with its ability of cell penetration by pinocytosis. According to another hypothesis, the protective mechanism of PVP is by its ability to bind with the cell membrane and form an envelope around it.

Combined media for microbial preservation containing different sugar concentrations (sucrose, glucose, trehalose), colloids (gelatin, agar, peptone, milk and sera) and salts (sodium glutamate) are also used in cryopreservation. They result in a higher percentage of cell viability when compared to single component protective media.

17.7.3 Preservation Techniques

17.7.3.1 Continuous Subcultivation

This method involves frequent subculture of an organism from a depleted to fresh agar nutrient medium. It is one of the oldest and simplest preservation techniques and is used for short-term maintenance and preservation in laboratories and industries. The initial inoculum used should be from multiple colonies because use of a single colony will increase the probability of unwanted selection. The main demerit of this technique is the possibility of loss of innate strain properties and activity.

17.7.3.2 Cryopreservation

Cryopreservation refers to the preservation of biological materials at cryogenic temperatures, using deep freezers (−80 to −150 °C), dry ice (−80 °C) or liquid nitrogen (−196 °C). Storage at ultra-low temperatures at or below liquid nitrogen is the least damaging preservation technique for bacteria. Low temperature halts metabolism, protects proteins and DNA from denaturation and slows the mobility of cellular water and helps protect cells for long periods of time. Preservation at −196 °C is ideal because there is no chance of DNA mutation at this temperature. Microbial cells can also be stored at −80 °C for short-term preservation. During cryopreservation, the microbial culture is mixed with a cryoprotectant like glycerol or DMSO in cryovials and stored either immersed in liquid nitrogen (−196 °C) or in its vapour phase (−135 to −150 °C). Storage in liquid phase can result in entry of liquid nitrogen into the cryovials and its bursting during rapid thawing, and hence storage in vapour phase is safer (Prakash et al. 2013). Much research has been carried out on the optimum rate of cooling. Controlled cooling at the rate of −1 to −5 °C min^{-1} and rapid thawing at 37 °C water bath have proved most successful in maintaining cell viability (Smith et al. 2008). Slow warming can result in recrystallization of ice causing damage to cells. Dumont et al. reported high cell recovery at low and high cooling rates, while intermediate cooling rates were found to be detrimental to cell viability. They also found that the cellular response to cooling is not only dependent on cooling rate but also on the cell size, water permeability and presence of a cell wall (Dumont et al. 2004).

17.7.3.3 Lyophilisation

Lyophilisation or freeze drying is a technique in which water molecules are directly removed from frozen material by sublimation under vacuum-reduced pressure. It has been successfully used in the preservation of bacteria, yeasts and the spores of filamentous fungi. Freeze-drying process prevents cellular shrinkage or other structural changes and helps retain viability. Most microbial repositories prefer this technique for long-term preservation because of low cost of maintenance and ease of transportation of lyophilized cultures. Since the ampoules are completely sealed, they are protected from infection and infestation and hence stay viability for many years. Storage should be done out of direct sunlight; chilled storage reduces the rate of deterioration and extends shelf life.

It has been observed that most non-sporulating fungi, some species of yeast and certain bacteria lose viability following lyophilisation. Lyo-injury can occur to cells during the cooling or drying stages. During the phase change during freeze drying, the liquid crystalline structure of cell membranes can degenerate to gel phase, disrupting its fluid-mosaic structure and causing leakage of the membrane, ultimately resulting cellular damage. The optimisation of lyoprotectants and suspension media is essential for successful lyophilisation of certain microorganisms. A mixture of lyoprotectants and matrix materials or excipients should be used along with the cells for lyophilisation. Maximum cell viability and stability of lyophilised cultures can be obtained with stationary-phase cultures preserved in borosilicate ampules with a 1–2% final moisture content stored at 4 °C in the dark (Morgan et al. 2006; Smith et al. 2008). Skimmed milk as such or in combination with inositol is a suitable protectant for fungi. Another protectant is trehalose, which protect membranes by attaching to the phospholipids, replacing water and lowering the transition temperature. After lyophilisation, the ampoules or vials must be sealed, which can be done either using heat sealed glass or butyl rubber bungs in glass vials. The former is preferred as or butyl rubber bungs may leak over long-term storage.

17.7.3.4 Silica Gel Storage

Vegetative cells, propagules or dormant stages like spores, cysts or sclerotia of an organism can be preserved by inoculating onto cold silica gel and dehydrating by air drying to enable storage without growth or metabolism. It has been found to be successful with sporulating fungi, with cultures maintaining morphological stability for 7–20 years. The technique is relatively simple and cheap and does not require expensive apparatus. The main disadvantage is that it is limited to sporulating fungi and is unsuitable for non-sporulating fungi or fungi with delicate or complex spores (Smith et al. 2008). The recovery rate of cells is often low, and there is a possibility of introducing contaminants during repeated retrievals.

17.7.3.5 Liquid Drying

Liquid drying (L-drying) is used as an alternative to lyophilisation for preserving bacteria that are sensitive to freezing like *Spirilla* and *Azomonas insignis*. In this procedure, drying occurs directly from the liquid phase.

17.7.3.6 Gelatine Discs

Bacterial cultures are suspended in melted nutrient gelatine, poured as drops into Petri dishes and allowed to solidify. These drops are then freeze dried, or dried over a desiccant, and the resultant flat discs are stored over silica gel. Revival of the culture is done by placing a single disc in warmed broth, and the resulting suspension is plated onto a suitable growth medium. This technique is most suitable for short-term storage of limited number of cultures. The main advantages are ease of handling and storage.

17.8 Quality Control

Microbial repositories are closely associated with academic and industrial research laboratories as providers of reference strains for microbiology and molecular biology studies, and hence the quality controls to be maintained are high and demanding. They are responsible for more than 60% of descriptions of prokaryotic species, which indicates the level of professional expertise offered by such facilities. Quality control checks include accurate inventory with backup facility, validation of storage temperature and quality, proper specimen handling, minimum subculturing, availability of sufficient distribution stocks and preservation. Characterisation of morphological, anatomical, physiological, immunological and molecular properties of strains is carried out before and after preservation (Broughton et al. 2012; Prakash et al. 2013). Techniques derived from various biological disciplines like biochemistry, bacteriology, mycology and ecology to provide anatomical, physiological and biochemical data are used to characterise and authenticate microorganisms. In addition to long-term preservation of microbial cultures, microbial culture collections should maintain the genotypic and phenotypic stability of its preserved cultures, which is essential for the authentication of previous findings. Molecular techniques like amplified fragment length polymorphism, multilocus sequence typing, pulsed-field gel electrophoresis, PCR binary typing (PBIT), real-time PCR and whole genome comparison can be used to monitor the genetic drift in bacteria, fungi and microalgae (Muller et al. 2007; Ragimbeau et al. 2008).

It has been made mandatory for all culture collections to implementation the international criteria in quality management to provide high-quality microbial cultures and authentic information. They are now required to provide access to genes and genetic elements of deposited strains with associated information. The use of bioinformatics tools has improved data handling, cross referencing and in the transformation of collection catalogues into digital format.

The use of next-generation sequencing techniques for elucidation of whole genome sequences of microbes and matrix-assisted laser desorption/ionisation time-of-flight mass spectrometry in microbial typing have reduced the cost and time for comparison of genotypic and phenotypic integrity of microbes before and after preservation.

17.9 Importance of Microbial Repositories in Bioprospecting

Bioprospecting is the exploration and sustainable utilisation of biological resources for commercially valuable genetic and biochemical resources. Majority of the drugs in use today are derived from microorganisms. Of these, 45% are produced by actinomycetes, 38% by fungi and 17% by unicellular bacteria (Bérdy 2005). Since the beginning of industrial microbiology, enormous numbers of microorganisms have been isolated from various environments with applications in scientific research and industrial fermentation. A large numbers of these valuable isolates have become lost and are no longer available due to lack of facilities for preservation or because of change in field of study by the researcher. Once lost, it is difficult to isolate a species with exactly the same properties even if samples are collected at the same location. Microbial repositories can play an important role to prevent such loss by proper maintenance of cultures and their supply on demand. Microbial prospecting being a popular field of research, a large number of new microbial strains and species are expected to be isolated in near future, which will require the help of microbial repositories. Establishment of new culture collections and exchange of information among them are crucial for advancements in microbiology, microbial industry and biotechnology.

17.10 Culture Collections in India

India, being one among the 12 countries of megadiversity, is rich in floral, faunal and microbial diversity. Many novel species of fungi and bacteria, including actinomycetes, viruses and cyanobacteria, have been isolated by Indian researchers, and they form an invaluable gene pool. Indian Council of Agricultural Research (ICAR) pioneered microbial diversity studies in India with initial focus on crop improvement for food and economy. India's rich microbial diversity has yet been effectively identified and catalogued owing to the lack of adequate support and expertise (Arora et al. 2005).

India endorsed the Biological Diversity Act (BDA) in 2002 after signing the Convention on Biological Diversity (CBD). The Biological Diversity Act gives effect to the provisions made in the CBD. It addresses access to biological resources and associated traditional knowledge to ensure equitable benefit sharing arising out of their use to the country and its people, thereby contributing to achieving objectives of the CBD. India is one of the first few countries to have enacted such legislation. There are 30 microbial repositories in India according to WFCC statistics. According to WDCM, India ranks third in holding of microbial cultures (194,174 accessions), after USA with 261,637 and Japan with 246,343 microbial accessions. The list of all WDCM registered culture collections in India are listed in Table 17.3.

Table 17.3 Indian culture collections registered with World Data Centre for Microorganisms (WDCM)

Acronym	Name and address
ABRC	**Anaerobic Bacterial Resource Centre**, University of Hyderabad, Andhra Pradesh, 500046
AYL	**WHYLABS RESOURCE CENTRE FOR MICROORGANISMS**, Hyderabad, Andhra Pradesh, 500001
BAB	**Bank A Bug**, Gandhinagar, Gujarat, 382011
BDU	**National Facility for Marine Cyanobacteria**, Bharathidasan University, Tiruchirappalli, Tamil Nadu, 620024
BT	***Bacillus thuringiensis***, M.S.Swaminathan Research Foundation, Taramani, Chennai, Tamil Nadu, 600113
CCDMBI	**Culture Collection**, Department of Microbiology, Bose Institute, Calcutta, West Bengal, 700009
CIPDE	**Collection of Insect Pathogens, Dept. of Entomology**, Marathwada Agricultural University, Maharastra
CM	***Chroococcus minor***, JJT University, Jhunjhunu, Rajasthan, 331403
DBV	**Division of Standardisation**, IVRI, Izatnagar, Uttar Padesh
DMSRDE	**DMSRDE Culture Collection**, DRDO, Kanpur, Uttar Pradesh, 208013
DUM	**Delhi University Mycological Herbarium**, Department of Botany, University of Delhi, Delhi, 110007
EntoPatho	**Entomopathogen**, M.S. University of Baroda,Vadpdara, gujarat, 390002
GFCC	**Goa University Fungus Culture Collection and Research Unit**, Goa University, Taleigao, Goa, 403206
ITCC	**Indian Type Culture Collection**, Division of Plant Pathology, Indian Agricultural Research Institute, New Delhi, 110012
MCC	**Microbial Culture Collection**, National Centre for Cell Science, Pune University Campus, Ganeshkhind, Pune, Maharashtra, 411007
MCM	**MACS Collection of Microorganisms**, MACS-Agharkar Research Institute, G.G.Agarkar Road, Pune, 411004
MPKV	**Biological Nitrogen Fixation Project College of Agriculture**, Mahatma Phule Agricultural University, Pune, Maharashtra, 411005
MRCJ	**Col. Sir R. N. Chopra, Microbial Resource Center Jammu**, CSIR-Indian Institute of Integrative Medicine, Canal Road, Jammu & Kashmir, 180001
MTCC	**Microbial Type Culture Collection & Gene Bank**, IMTECH, Sector 39-A, Chandigarh, 160036
NAIMCC	**National Agriculturally Important Microbial Culture Collection**, National Bureau of Agriculturally Important Microorganisms, Kushmaur, Mau, Uttar Pradesh, 275101
NCCPF	**National Culture Collection of Pathogenic fungi**, Postgraduate Institute of Medical Education and Research (PGIMER), Sector-12, Chandigarh, Chandigarh, 160012
NCDC	**National Collection of Dairy Cultures**, National Dairy Research Institute, Karnal, Haryana, 132001

(continued)

Table 17.3 (continued)

Acronym	Name and address
NCIM	**National Collection of Industrial Microorganisms**, National Chemical Laboratory (CSIR), Dr. Homi Bhabha Road, Pashan, Pune, Maharashtra, 411008
NFCCI	**National Fungal Culture Collection of India**, MACS Agarkar Research Institute, G.G. Agarkar Road, Pune, Maharashtra, 411004
NIICC	**NII Microbial Culture Collection**, National Institute for Interdisciplinary Science and Technology (CSIR), Pappanamcode, Trivandrum, Kerala, 695019
NMCC	**North Maharashtra Microbial Culture Collection Centre**, North Maharashtra University, Umavi nagar, Jalgaon, Maharashtra, 425001
NTCCI	**Culture Collection**, Microbiology and Cell Biology Laboratory, Indian Institute of Science, IISc, Bangalore, Karnataka, 560012
UMFFTD	**Food and Fermentation Technology Division**, University of Mumbai, Nathalal Parekh Marg, Mumbai, Maharashtra, 400019
VBCCA	**Visva-Bharati Culture Collection of Algae**, Department of Botany, Visva-Bharati University, Santiniketan, West Bengal, 731235
VPCI	**Fungal Culture Collection**, Vallabhbhai Patel Chest Institute, University of Delhi, Delhi, 110007

Source: WFCC–MIRCEN World Data Centre for Microorganisms

17.11 Conclusions

Conservation and provision of microbial resources for research is now recognised as being an essential component in the advancement of the life sciences. The extent of diversity of the microbial world is largely unknown, and of that which is known, the genotypic and phenotypic diversity is enormous. Convention on Biological Diversity (CBD) is the most significant international agreement that ensures the need for the conservation of microbial diversity, sharing of microbial resources and their utilisation. In situ conservation strategies are practically not applicable to the conservation of microbial diversity, possible only through maintaining as pure cultures under laboratory conditions, followed by preservation. The main objectives of culture collections are to act as a depository, to supply authentic microbial cultures and to provide related services to the scientific community working in research institutions, universities and industries. Biotechnology has a pivotal role in pharmaceutical, industry and academia sectors. There is an increasing demand on culture collections for authenticated, reliable biological material and associated information that have paralleled with the growth of biotechnology. The full potential of microbial diversity is yet to be harnessed. A coordinated approach to resource provision is an urge to accelerate innovation and discovery. Further development and investment is needed in the microbial domain at global level.

References

Arora DK, Saikia R, Dwievdi R, Smith D (2005) Current status, strategy and future prospects of microbial resource collections. Curr Sci 89(3):488–495

Bérdy J (2005) Bioactive microbial metabolites. J Antibiot 58(1):1–26

Braga PAC, Tata A, dos Santos VG, Barreiro JR, Schwab NV, dos Santos MV, Eberlin MN, Ferreira CR (2013) Bacterial identification: from the agar plate to the mass spectrometer. RSC Adv 3:994–1008

Broughton R, Buddie AG, Smith D, Ryan MJ (2012) The effect of cryopreservation on genomic stability in strains of the fungus *Trichoderma*. CryoLetters 33:299–306

Çaktu K, Turkoglu EA (2011) Microbial culture collections: The essential resources for life. Gazi Univ J Sci 24(2):175–180

Chapman AD (2009) Numbers of living species in Australia and the world, 2nd edn. Australian Government, Department of the Environment, Water, Heritage and the Arts, Canberra

Crespi RS (1985) Microbiological inventions and the patent law-the international dimension. Biotechnol Genet Eng Rev 3:1–37

Davison A, Brabandere J, Smith D (1998) Microbes, collections and the MOSAICC approach. Microbiol Aust 19(1):36–37

Dumont F, Marechal PA, Gervais P (2004) Cell size and water permeability as determining factors for cell viability after freezing at different cooling rates. Appl Environ Microbiol 70:268–272

Fuller BJ (2004) Cryoprotectants: the essential antifreezes to protect life in the frozen state. CryoLetters 25:375–388

Gottel NR, Castro HF, Kerley M, Yang Z, Pelletier DA, Podar M, Karpinets T, Uberbacher E, Tuskan GA, Vilgalys R, Doktycz MJ, Schadt CW (2011) Distinct microbial communities within the endosphere and rhizosphere of *Populus deltoides* roots across contrasting soil types. Appl Environ Microbiol 77(17):5934–5944

Hawksworth DL (1985) Fungus culture collections as a biotechnological resource. Biotechnol Genet Eng Rev 3:417–453

Komagata K (1998) Microbial diversity and the role of culture collections. Pure Appl Chem 70(11):2110–2118

Locey KJ, Lennon JT (2016) Scaling laws predict global microbial diversity. Proc Natl Acad Sci 113(21):5970–5975

Mahilum-Tapay LM (2009) The importance of microbial culture collection and Gene Banks in biotechnology. In: Doelle HW, Rokem JS, Berovic M (eds) Biotechnology. Encyclopaedia of Life Support Systems (EOLSS). Eolss Publishers, Oxford

Malik KA, Claus D (1987) Bacterial culture collections: their importance to biotechnology and microbiology. Biotechnol Genet Eng Rev 5:137–197

Morgan CA, Herman N, White PA, Vesey G (2006) Preservation of micro-organisms by drying; a review. J Microbiol Methods 66:183–193

Muller J, Day JG, Harding K, Hepperle D, Lorenz M, Friedl T (2007) Assessing genetic stability of a range of terrestrial microalgae after cryopreservation using amplified fragment length polymorphism (AFLP). Am J Bot 94:799–808

Prakash O, Nimonkar Y, Shouche YS (2013) Practice and prospects of microbial preservation. FEMS Microbiol Lett 339:1–9

Ragimbeau C, Schneider F, Losch S, Even J, Mossong J (2008) Multilocus sequence typing, pulsed-field gel electrophoresis, and fla short variable region typing of clonal complexes of *Campylobacter jejuni* strains of human, bovine, and poultry origins in Luxembourg. Appl Environ Microbiol 74:7715–7722

Sharma A, Shouche Y (2014) Microbial Culture Collection (MCC) and International Depositary Authority (IDA) at National Centre for Cell Science, Pune. Indian J Microbiol 54(2):129–133

Singh P, Dash SS (2014) Plant discoveries 2013 – New genera, species and new records. Botanical Survey of India, Kolkata

Singh S, Nain L (2014) Microorganisms in the Conversion of Agricultural Wastes to Compost. Proc Indian Natl Sci Acad (Special Section) 80(2):473–481

Smith D, Ryan MJ, Stackebrandt E (2008) The *ex situ* conservation of microorganisms: aiming at a certified quality management. In: Doelle HW, EJ DS (eds) Biotechnology. Encyclopaedia of life support systems. Eolss Publishers, Oxford

Smith D, McCluskey K, Stackebrandt E (2014) Investment into the future of microbial resources: culture collection funding models and BRC business plans for biological resource centres. Springer Plus 3:81

Staley JT, Castenholz RW, Colwell RR, Holt JG, Kane MD, Pace NR, Saylers AA, Tiedje JM (1997) The microbial world: foundation of the biosphere. American Academy of Microbiology, Washington, DC, p 32

Tsui CKM, Woodhall J, Chen W, Chen W, Levesque CA, Lau A, Schoen CD, Baschien C, Najafzadeh MJ, de Hoog GS (2011) Molecular techniques for pathogen identification and fungus detection in the environment. IMA Fungus 2(2):177–189

Uruburu F (2003) History and services of culture collections. Int Microbiol 6:101–103

Uzunova-Doneva T, Donev T (2005) Anabiosis and conservation of microorganisms. J Cult Collect 4(1):17–28

Woese CR, Kandler O, Wheelis ML (1990) Towards a natural system of organisms: proposal for the domains Archaea, Bacteria, and Eucarya. Proc Natl Acad Sci U S A 87:4576–4579

Yandigeri M, Mesapogu S, Yadav A, Arora DK (2010) Microbial culture banks: custodian of real natural wealth. Paper presented at national conference on biodiversity, development and poverty alleviation, Uttar Pradesh State Biodiversity Board, 22 May 2010

Medicinal Plants of India: Implications in Modern Medicine

18

Nimisha Vijayan, T.S. Swapna, G. Seghal Kiran, and Sabu Abdulhameed

Abstract

The linkage between biodiversity and human health is well established. Conserving forest biodiversity by valuing and harnessing it, as medicine, is consistent with poverty reduction and local public health prevention efforts. Global demand for herbal medicines is accompanied by reducing supply of medicinal plants due to overharvesting, habitat loss, and agricultural encroachment. Plants are always considered as a primary source of drugs in traditional and alternative system of medicine in various forms such as crude form, juice, decoction, and crude extracts. About 80% people of the world, particularly in the rural areas of developing countries, continue using traditional resources in healthcare. This chapter deals with the assessment of the richness of ethnomedicinal plant species used by the local tribes in forest areas and a basic information on the ethnomedicinal plants, its conservation strategies, and correlation between the ethnomedical usage of medicinal plants and modern medicines.

Keywords

Medicinal plants • Biodiversity • Ethnobotany

N. Vijayan • S. Abdulhameed (✉)
Inter University Centre for Bioscience, Department of Biotechnology and Microbiology, Kannur University, Thalassery Campus, Kannur, Kerala 670661, India
e-mail: nimivjn@gmail.com; drsabu@gmail.com

T.S. Swapna
Department of Botany, University of Kerala, Karyavattom, Thiruvananthapuram, Kerala 695581, India
e-mail: swapnats@yahoo.com

G. Seghal Kiran
Department of Food Science and Technology, Pondicherry University, Puducherry 605014, India
e-mail: seeshegal@yahoo.co.in

S. Abdulhameed et al. (eds.), *Bioresources and Bioprocess in Biotechnology*, DOI 10.1007/978-981-10-3573-9_18

18.1 Introduction

18.1.1 Ethnobotanical Base of India

The scientific investigation of indigenous knowledge of different ethnic group or communities is highly relevant for a utilization of the same in applying along with modern practices. Traditional ethnobotanical knowledge of plant resources forms the basis for the use of plants and plant products to treat variety of diseases (Singh 1993). Medico-ethnobotany is emerging these days as it deals with the interrelation of botany with the indigenous knowledge about medicinal properties of plants, which finds their application coupled with modern medicine. This in turn will help in protecting the cultural ethnicity and biodiversity of the region. Traditional knowledge about the medicinal plants and the method of application are kept as a secret by the indigenous healers (Fisseha 2007). The lack of recorded documents increases the risk of losing the immensely valuable data (Fisseha 2007). This fact has triggered a lot of interest in initiating studies on scientifically unearthing the ethnomedicinal practices kept secret for a long time by different ethnic groups. Documenting the traditional knowledge is significant for the conservation of biodiversity and sustainable utilization of natural resources (Muth et al. 2006).

India is known for its rich resources of medicinal plants, perfumes, spices, etc., and India owes the existing climatic condition for this richness in biodiversity. Traditional knowledge is maintained through the successive herbal medicinal practitioners all over India who are practicing Ayurveda – the Indian system of medicine – Unani, and Siddha systems of healthcare. In fact, knowledge about Ayurvedic treatment practices and medicinal plants are scientifically well documented in authentic ancient texts, including Rigveda, Atharvaveda, Charaka Samhita, Sushruta Samhita, and Dhanvantari-Nighantu, but information regarding the medicinal plants and practices involved in Siddha system is still very much restricted to traditional healers, and there is a need for more emphasis to document the same. Abasta (1986) reported that Indian traditional system of medicine uses the medicinal plant resources in nearly 10,000 formulations. Further, 74% of plant-derived prescription drugs are chemically isolated from active compounds of traditional medicinal plants, and these days there is an enhanced interest in research in this direction.

Medicinal plants and treatment regimes based on medicinal plants play a key role in the healthcare system of tribal communities of India. Reinvention and scientific validation of this valuable knowledge can lead to the development of novel drugs and valid treatment practices for the entire world. The tribes of Dangs district in the Indian state, Gujarat, which is one of the rich biodiversity areas, use the plants from the Dangs forest for the treatment of various diseases, and they have very good knowledge about the location, distribution, propagation, and medicinal values of the plants (Pandey et al. 2005). There are about 234 ethnobotanically important plant species that represent 63 families, and they used these in the treatment of nearly 32 different types of diseases. The tribes of Chhattisgarh state in India mainly depend on the forest resources for food and medicine. Ethnic communities of the state use these plants as a source of ethno-veterinary drugs (Kala 2009).

Herbal preparations were widely available to take care of pregnant woman during ancient times itself. This was more prevalent in South India. Ayurvedic texts contain valuable information regarding these types of herbal preparations for mother care. Another sector of Indian dominance in herbal remedies is against snakebite. Nearly, 198 plant species were reported from India used against snakebite by various tribal communities (Dey and De 2012). Some of the examples are *Casearia graveolens* Dalz. used by the tribal community of Nasik district, Maharashtra (Patil and Patil 2005), and *Isonandra lanceolata* Wight used by tribal community of Tamil Nadu (Ayyanar and Ignacimuthu 2005). The Northeast sector of India is one of the best biodiversity hot spots which cover the Himalayan area also. This area harbors rarest of the rare herbs, which are overexploited for commercial production. Northeast tribal communities like Mikir, Monpas, and Reangs that inhabit in forest regions heavily depend on folklore medicines. Documents of northeastern Indian states' ethnobotany give the information of 37 antifertility plants used by the tribal communities of Assam (Kalita et al. 2011). The country's traditional knowledge wealth is perishing due to lack of proper documentation and due to overexploitation and environmental degradation. Many of the important medicinal herbs are at the verge of extinction. The need of the hour is a thorough understanding of the materials and practices involved in ethnomedicine.

18.2 Biodiversity of Medicinal Plants

The whole world depends on the variety of life on earth for its survival. India is one among the 12 mega-biodiversity nations with 3 out of the 36 biodiversity hot spots present in India. India supports the life of nearly 40,000 plants, which is nearly 12% of the overall plant wealth of the universe. The Indian flora includes 17,500 flowering plants with 3000 of them reported and documented for their medicinal use (Sharma et al. 1997).

Indian medicinal plants were used in three different and distinct sectors. Nearly, 3000 species are used in folk medicine by the ethnic communities, which are yet to get properly documented. Modern medicine utilizes the active principle isolated from nearly 700 species for treating various diseases. Over 1500 species are used in classical system of medicine like Ayurveda, Siddha, and Unani (Prakash 2002). Medicinal plants can be divided into the following three categories: (i) plants used in ethnomedicine include undocumented folk medicine used by the ethnic communities (3000 species); (ii) plants used in modern medicine, the active principle isolated from the plants, are used to treat various ailments (700 species); and (iii) plants used in the classical system of medicine like Ayurveda, Unani, and Siddha (over 2500 species) (Prakash 2002). Ancient Vedic treaties of India too had recorded the medicinal value of a variety of plant species. The Rigveda contains the reports of 67 medicinal plants, Yajurveda 81 species, and Atharvaveda 290 species, and Charaka Samhita and Sushruta Samhita explained the properties of 1100 and 1270 species, respectively (Joy et al. 1998) (Fig. 18.1).

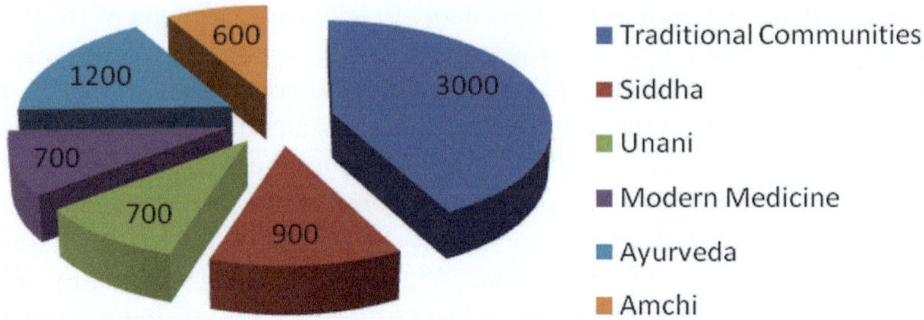

Fig. 18.1 Medicinal plants used in Indian system of medicine

The Indian biodiversity is largely accumulated in the Western Ghats, Northeast, Himalaya, Eastern Ghats, and the Andaman and Nicobar Islands (Ramesh and Pascal 1997). Dense forest areas in the country provide shelter to a great majority of the medicinal plants and spices. The thick forests in India serve as the primary source of large number of medicinal plants for manufacture of drugs and perfumery products. Endemism is the peculiarity of Indian biodiversity where about 33% of the recorded plant species are unique.

In the Indian medicinal plant treasure, nearly two-thirds is located in the Western Ghats forests situated in the chain of mountains spread over the Indian states of Gujarat, Maharashtra, Karnataka, Kerala, Goa, and Tamil Nadu (Suresh et al. 2006). Western Ghats consists of 5800 plant species, out of which 3500 are flowering plants and with great majority of them being endemic. In addition to the land of economically and medicinally important plants like pepper, cardamom, ginger, mango, jackfruit, variety of millets, and rice, the Western Ghats harbors medicinally important ferns and other lower group of plants. Traditionally important pteridophytes like *Acrostichum aureum* Linn. and *Marattia fraxinea* Sw. which are distributed across the mountains of Western Ghats are widely used for the treatment of variety of ailments. Eastern Ghats consists of 2600 plant species belonging to the category of angiosperms, gymnosperms, and pteridophytes. Out of these, *Acorus calamus*, *Aegle marmelos*, *Costus speciosus*, *Gloriosa superba*, *Gymnema sylvestre*, *Mucuna pruriens*, *Plumbago indica*, *Rauvolfia serpentina*, and *Piper nigrum* are reported in literature for their medicinal applications.

As given in Table 18.1, out of these 80 species, 51 have a global RL status as these are endemic (or nearly so) to the state/region for which the assessment was undertaken. Thirteen species have been assigned critically endangered (CR) status, 24 species are endangered (EN), 33 species are vulnerable (VU), and 10 species are near threatened (NT). Eighteen of these red-listed medicinal plant species have been recorded in high volume trade in the national level trade study.

For instance, Andaman and Nicobar Islands (situated 1200 km away from the mainland) are another hot spot of medicinal plant biodiversity from where more than 2500 angiosperm species were reported with 52 of them with potential medicinal property. *Calotropis gigantea*, *Cynodon dactylon*, *Centella asiatica*, *Jatropha gossypiifolia*, and *Tamarindus indica* are major multipurpose medicinal plants used

Table 18.1 Medicinal plant species of conservation concern identified for Tamil Nadu

S. no.	Botanical name of the plant	Habit	Status in Tamil Nadu	RL status (global)	Recorded in trade	Trade 100 MT/yr dry wt.
1.	*Adhatoda beddomei* C.B. Clarke (Gaertn.)	S	CR	CR(G)	Td	
2.	*Coscinium fenestratum* Coleb.	C	CR	–	Td	HT
3.	*Cycas circinalis* L.	T	CR	–	Td	
4.	*Eulophia cullenii* (Wight) Blume	H	CR	CR(G)		
5.	*Helminthostachys zeylanica* (L.) Hook.	H	CR	–		
6.	*Decalepis arayalpathra* J. Joseph and V. Chandras.	S	CR	CR(G)		
7.	*Paphiopedilum druryi* (Bedd.) Pfitz	H	CR	CR(G)		
8.	*Piper barberi* Gamble	C	CR	CR(G)		
9.	*Pterocarpus santalinus* L.f.	T	CR	CR(G)	Td	HT
10.	*Shorea tumbaggaia* Roxb.	T	CR	CR(G)		
11.	*Utleria salicifolia* Bedd.	S	CR	CR(G)		
12.	*Valeriana leschenaultii* DC.	H	CR	CR(G)		
13.	*Vateria macrocarpa* B.L. Gupta	T	CR	CR(G)		
14.	*Ampelocissus indica* (L.) Planch.	C	EN	–		
15.	*Cayratia pedata* (Lam.) Juss. ex Gagnep. var. *glabra*	C	EN	EN(G)	Td	
16.	*Cinnamomum wightii* Meisn.	T	EN	EN(G)	Td	
17.	*Decalepis hamiltonii* Wight and Arn.	C	EN	EN(G)	Td	HT
18.	*Dipterocarpus indicus* Bedd.	T	EN	EN(G)		
19.	*Drosera peltata* J.E.Sm. ex Willd.	H	EN	–		
20.	*Dysoxylum malabaricum* Bedd. ex Hiern	T	EN	EN(G)	Td	
21.	*Garcinia travancorica* Bedd.	T	EN	EN(G)		
22.	*Gymnema khandalense* Santapau	C	EN	EN(G)		
23.	*Gymnema montanum* (Roxb.) Hook.f.	C	EN	EN(G)		
24.	*Humboldtia vahliana* Wight	T	EN	EN(G)		
25.	*Hydnocarpus macrocarpa* (Bedd.) Warb.	T	EN	EN(G)		
26.	*Nervilia aragoana* Gaud.	H	EN	–	Td	

(continued)

Table 18.1 (continued)

S. no.	Botanical name of the plant	Habit	Status in Tamil Nadu	RL status (global)	Recorded in trade	Trade 100 MT/yr dry wt.
27.	*Nilgirianthus ciliatus* (Nees) Bremek.	S	EN	EN(G)	Td	HT
28.	*Persea macrantha* (Nees) Kosterm.	T	EN	–		
29.	*Piper longum* L.	H	EN	–	Td	HT
30.	*Plectranthus nilgherricus* Benth.	H	EN	EN(G)		
31.	*Rauvolfia serpentina* (L.) Benth. ex Kurz	H	EN	–	Td	HT
32.	*Santalum album* L.	T	EN	–	Td	HT
33.	*Semecarpus travancorica* Bedd.	T	EN	EN(G)		
34.	*Strychnos aenea* A.W.Hill	C	EN	EN(G)		
35.	*Swertia lawii* (Wight ex C.B. Clarke) Burkill	H	EN	EN(G)		
36.	*Syzygium travancoricum* Gamble	T	EN	EN(G)		
37.	*Trichopus zeylanicus* Gaertn. subsp. *travancoricus* (Bedd.) Burkill	H	EN	EN(G)		
38.	*Acorus calamus* L.	H	VU	–	Td	HT
39.	*Aegle marmelos* (L.) Corr.	T	VU	–	Td	HT
40.	*Amorphophallus commutatus* (Schott) Engl.	H	VU	VU(G)		
41.	*Amorphophallus paeoniifolius* (Dennst.) Nicolson	H	VU	–	Td	
42.	*Ampelocissus araneosa* (Dalz. and Gibson) Planch.	C	VU	VU(G)		
43.	*Artocarpus hirsutus* Lam.	T	VU	VU(G)		
44.	*Calophyllum apetalum* Willd.	T	VU	VU(G)	Td	
45.	*Canarium strictum* Roxb.	T	VU	–	Td	
46.	*Cinnamomum macrocarpum* Hook.f.	T	VU	VU(G)	Td	
47.	*Cinnamomum sulphuratum* Nees	T	VU	VU(G)	Td	HT
48.	*Curcuma pseudomontana* Graham	H	VU	VU(G)		
49.	*Diospyros candolleana* Wight	T	VU	VU(G)		
50.	*Diospyros paniculata* Dalz.	T	VU	VU(G)		
51.	*Embelia ribes* Burm.f	C	VU	–	Td	Td

(continued)

Table 18.1 (continued)

S. no.	Botanical name of the plant	Habit	Status in Tamil Nadu	RL status (global)	Recorded in trade	Trade 100 MT/yr dry wt.
52.	*Embelia tsjeriam-cottam* (Roem. and Schult.) A. DC	S	VU	–	Td	HT
53.	*Garcinia morella* (Gaertn.) Desr	T	VU	–	Td	
54.	*Gardenia gummifera* L.f.	T	VU	VU(G)	Td	
55.	*Glycosmis macrocarpa* Wight	S	VU	VU(G)		
56.	*Heracleum candolleanum* (Wight and Arn.) Gamble	H	VU	VU(G)	Td	
57.	*Hydnocarpus alpina* Wight	T	VU	VU(G)		
58.	*Hydnocarpus pentandra* (Buch.-Ham.) Oken	T	VU	VU(G)	Td	
59.	*Kingiodendron pinnatum* (Roxb. ex DC.) Harms	T	VU	VU(G)		
60.	*Michelia champaca* L.	T	VU	–	Td	
61.	*Michelia nilagirica* Zenk.	T	VU	VU(G)		
62.	*Myristica malabarica* Lam.	T	VU	VU(G)	Td	
63.	*Nothapodytes nimmoniana* (Graham) Mabber.	T	VU	–	Td	
64.	*Ochreinauclea missionis* (Wall. ex G. Don) Ridsdale	T	VU	VU(G)		
65.	*Piper mullesua* Buch.-Ham. ex D. Don	C	VU	–		
66.	*Pueraria tuberosa* (Roxb. ex Willd.) DC.	C	VU	–	Td	
67.	*Salacia oblonga* Wall. ex Wight and Arn.	C	VU	VU(G)	Td	
68.	*Swertia corymbosa* (Griseb.) Wight ex C.B. Clarke	H	VU	VU(G)	Td	
69.	*Tragia bicolor* Miq.	C	VU	VU(G)		
70.	*Vateria indica* L.	T	VU	VU(G)	Td	HT
71.	*Adenia hondala* (Gaertn.) W.J.de Wilde	C	NT	–	Td	
72.	*Celastrus paniculatus* Willd.	C	NT	–	Td	HT
73.	*Garcinia gummi-gutta* (L.) Robson	T	NT	NT(G)	Td	
74.	*Holostemma ada-kodien* Schult.	C	NT	–	Td	HT
75.	*Knema attenuata* (Hook.f. and Thoms.) Warb.	T	NT	NT(G)		
76.	*Operculina turpethum* (L.) Silva Manso	C	NT	–	Td	HT

(continued)

Table 18.1 (continued)

S. no.	Botanical name of the plant	Habit	Status in Tamil Nadu	RL status (global)	Recorded in trade	Trade 100 MT/yr dry wt.
77.	*Merremia turpethum* (L.) Shah and Bhat					
78.	*Piper nigrum* L.	C	NT	–	Td	
79.	*Pseudarthria viscida* (L.) Wight and Arn.	H	NT	–	Td	HT
80.	*Rhaphidophora pertusa* (Roxb.) Schott	C	NT	–	Td	
81.	*Symplocos racemosa* Roxb.	T	NT	–	Td	HT

Source: http://envis.frlht.org-ENVIS Centre on Conservation of Medicinal Plants, FRLHT, Bangalore (22 September 2010)

EX extinct, *CR* critically endangered, *EN* endangered, *VU* vulnerable, *NT* near threatened, *G* global, *H* herb, *S* shrub, *T* tree, *C* climber, *Td* recorded in trade, *HT* high volume traded

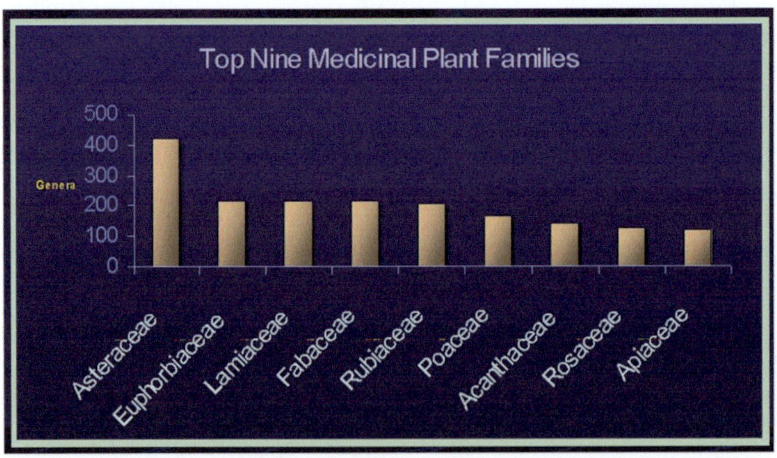

Fig. 18.2 Important medicinal plant families in India (**66**, Ravikumar K)

by the tribal communities of the island (Joy et al. 2001; Suresh et al. 2006). Northeast region of India accounts for 50% of total plant species, and many of the plant species are endemic to this region which include *Artocarpus lakoocha*, *Magnolia griffithii*, and *Sapindus mukorossi*. A few species like *Acorus calamus* and *Rauvolfia serpentina* are at the risk of extinction (Dhar et al. 2000). The rich Himalayan medicinal plant biodiversity owes it to the unique climatic conditions and topography of the region where most of the available plants are habitat specific. The Himalayas are reported to have 8000 angiosperms, 44 gymnosperms, 600 pteridophytes, and 1737 bryophytes along with 1159 lichens (Singh and Hajara 1996).

Medicinal plants identified and documented from different parts of India come under 386 families and 2200 genera. The top nine medicinally important plant families are shown in Fig. 18.2.

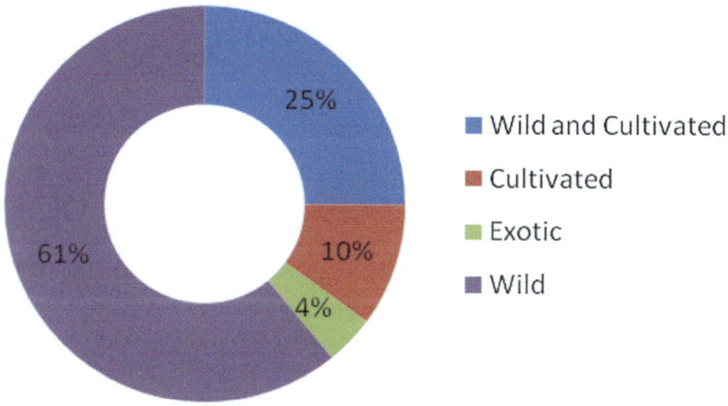

Fig. 18.3 Sources of medicinal plants traded in India (Suresh et al. 2006)

In India, around 8000 plant species are used for various purposes, and only 10% of total plant species are in active trade. Eighteen percent of traded species are from Himalayan and Northeast regions of India. Four percent and 0.5% are from Western Ghats and desert zones, respectively. Nearly, 61% of the species occurs in wild (Suresh et al. 2006). Figure 18.3 shows the source of medicinal plants traded in India.

The International Union for Conservation of Nature (IUCN) reports that medicinally important and other wild plants are at the verge of extinction, and the extinction rate is between 1,000 and 10,000 times (IUCN). Many of the traditionally valuable medicinal plants are going to extinct due to overexploitation. Increased human activity, habitat loss, and marketing of medicinal plants by biopharmaceutical industries are main reasons for the biodiversity crisis. Many of the medicinal plants endemic to India are going to extinct and immediate conservation action is essential.

18.3 Threat to Biodiversity

When the generations keep a belief that natural resources are a finite one, they exploit the resources in a reckless manner, and this results in the rapid diminution of the same. Along with other resources, bioresources are also disappearing at a rapid pace. It is high time to start the conservation strategies aided with high technology to preserve the remaining resources for the future generations. Indiscriminate and irresponsible exploitation of the natural resources by mankind has been cited as the major cause for this critical situation. Many hardwoods and medicinal plants are overexploited and their populations exhausted (Namboodiri 1994).

The earth has rich and diverse array of living organisms, and India maintains its status as one of the 12 mega-biodiversity countries of the world despite the emerging threats on its biodiversity treasures. The country is the second largest medicinal plant exporter, but this in turn affects the diversity due to reckless overharvesting. In order to preserve the serene biodiversity centers, a medicinal plant conservation action has been established in some of the Indian states like Kerala, Tamil Nadu, Karnataka, Himachal Pradesh, Maharashtra, Sikkim, etc.

Generally, finding their use in treatment or prevention of specific ailments, medicinal herbs were playing a very favorable and potential role in traditional rural healthcare. Later they were adopted to a great extent by modern medicine with detection of active substances present in selected herbs. Despite their importance, medicinal plants are rarely handled within a controlled regulated zone. Most are still exploited either little or no regard to the future. Generally every developing nation is using plants in their medicinal preparation, but now as a result of rising population and of aging, many medically important plant species are becoming insufficient. Most of them are at the verge of extinction (Chantia 2003).

The major threats to the diminishing Indian medicinal plant biodiversity are low annual rainfall due to climate change, sand mining from river basins and subsequent silting, and environmental impact of major human-centered developments, including commercial farming practices aided with intense application of chemical fertilizers and pesticides/insecticides. The change in the pattern of availability of the required insects for pollination of these precious plants is a result of intense agricultural activities. It is extremely essential to conserve important medicinal plants species by the restoration of its natural habitats like shrub lands and woodlands (Singh 2005).

IUCN defines conservation as a positive, embracing preservation, maintenance, sustainable utilization, restoration, and enhancement of the natural environment (IUCN 1980). Conservation and sustainable management presents vital role for all countries, especially India and its Western Ghats. Northeastern Indian states are hilly area widely rich in biodiversity. Different hills provide the distribution of specific medicinal plants. With increasing populations, increased demands for crude drugs, and lack of a comprehensive knowledge base, the continued availability of plant material from the wild cannot be scientifically or practically assured (Aryal 1993).

The demand for raw materials by the Indian Ayurvedic industry sector and that for international market is largely met by the forests of the Indian state, Himachal Pradesh, as high as 80 % of all Ayurvedic drugs, 46% of all Unani drugs, and 33% of all allopathic drugs developed. The Western Ghats, one of the hot spot areas of mega-diversity, forms the second major source of medicinal plants. In situ conservation, conserving species in their natural habitat, and ex situ conservation, cultivation of species in controlled conditions, are the corrective measure for conservation.

18.4 *Ex Situ* Conservation

The Indian *ex situ* conservation scenario develops many innovative methodologies to increase the quality and yield many of the cultivated medicinal plants. Standard protocols were developed for the propagation of almost all the cultivated medicinal plant varieties. India's premier research agencies including the Council of Scientific and Industrial Research (CSIR), Department of Science and Technology, Department of Biotechnology, and Indian Council of Agricultural Research and the various research institutes coming under these agencies including the National Bureau of Plant Genetic Resources, National Botanical Research Institute, and Central Institute of Medicinal and Aromatic Plants along with their state level counterparts like the Jawaharlal Nehru Tropical Botanical Garden and Research Institute (Kerala) and National Research Centre for Medicinal and Aromatic Plants (Gujarat) specialize in plant improvement, domestication, and strategies for implementation of modern biology in medicinal plants. Besides this, two other research efforts the central government initiated a five-year program implemented by the Ministry of Agriculture to accelerate research and development of medicinal plants. With the support of 16 state agricultural universities, state horticulture, and agriculture departments, regional laboratories and the International Crop Research Institute for the Semi-Arid Tropics (ICRISAT) also engaged in this field. On identification of the potential of this billion dollar business, private entrepreneurs also started diverting their attention to this sector. All those big pharma companies who were basically interested only in the exploitation of the rich medicinal resources started investing for conservation. They created large farms of medicinal plants for the manufacture of their own herbal products, and the export of raw materials shifted from harvested wild plants to cultivated species. Companies are actively involved in the development of cultivation method of medicinal plants of importance to them, with the direct participation and partnership of local farmers and tribal women (Lambert et al. 1996). The state of Kerala in India is known for its tradition in Ayurveda and other herbal treatments. This tradition has invited lot of attention to this small state for investment in the traditional healthcare sector. As a result, along with traditional Ayurvedic hospitals including the Kottakkal Arya Vaidya Sala and Vaidyaratnam Oushadhasala, several other small and big players came into limelight. To cater to the needs of the Ayurveda pharmacies, several commercial manufacturers also came in to picture with huge investment. There is an enhanced annual growth rate in the sector, and at the moment India is having 900 companies involved in the manufacture of Ayurvedic medicines. This has resulted in the overharvesting of the precious medicinal herbs resulting in mass disappearance. Realizing this situation, there is an enhanced interest in both at the government level and at the industry level to preserve the remaining diversity. With soft loans from government and banks, self-help groups of women and individuals are now cultivating medicinal plants, and the industry is buying back the produce from farms. Utilization of tissue culture techniques for the mass propagation is also utilized for the supply of quality seedlings. In Kerala, government institutes like the Jawaharlal Nehru Tropical Botanical Garden and Research Institute (JNTBGRI), Malabar Botanical Garden and Institute for Plant Sciences (MBGIPS), and Kerala Forest Research Institute (KFRI) are taking lead role in this aspect. The M.S. Swaminathan Botanical Garden located at the biodiversity-rich Western Ghats region of Wayanad is also an important center for conservation. The increased awareness among the public, entrepreneurs, and the policy makers is helping in the preservation and sustainable utilization of the remaining medicinal plant diversity. In addition to this, at national level, the Tropical Forest Research Institute (TFRI, Jabalpur) has established a herbal garden with a collection of 461 species, including rare and endangered species. The Forest Research Institute at Dehradun has collected significant medicinal and aromatic plants of the Himalayan region. The Department of Indian System of Medicine and Homeopathy and Ministry of Health and Family Welfare have promoted the establishment of small herbal gardens in educational institutes. The Central Council of Research in Ayurveda and Siddha situated in Pune maintains a Medical Plant Garden and houses 320 species of medical and aromatic plants, including a number of endangered species (Ilah et al. 2007).

18.5 *In Situ* Conservation

To retain the quality characteristics of the naturally available plants, *in situ* conservation is highly relevant. Various five-year plans of the country gave enough emphasis on *in situ* conservation resulting in the establishment of 13 biosphere reserves, 65 national parks, 426 sanctuaries, 316 preservation plots, and more than 7200 sacred/religious groves. These networks cover more than 5% of geographical area of the nation, and these are coordinated by the Ministry of Environment and Forests (Venkatesh et al. 2010). The country has also framed stringent wildlife laws to protect its forests and associated resources. A key organization involved in mobilizing a movement for medicinal plants conservation policy is Foundation for Revitalization of Local Health Traditions (FRLHT) founded by Danish International Development Agency (DANIDA). They have established coordinated networks of *in situ* research areas in the Indian states of Tamil Nadu, Karnataka, and Kerala, which harbor the precious Western Ghats. These medicinal plant conservation areas are approximately 200 hectares in size and are located within notified forest reserves or wildlife sanctuaries (FRLHT 1995).

18.6 Knowledge Base of Traditional Medical Practice

"Traditional medicine and medical practice is based on the knowledge, skills and practices based on the theories, beliefs and experiences of different cultures, used in the upholding of health and in the prevention, diagnosis, or treatment of physical and mental illness" (WHO). It ranges from managing simple nutritional deficiency symptoms to treatment of chronic illness. Traditional systems of medicine such as Ayurveda, Siddha, and Unani are deep rooted in the Indian society and well accepted as a treatment regime without side effects. About 60% of world populations are using traditional medicine for primary healthcare, and many of the practices and active molecules are now getting adapted by modern medicine. Folk healers in remote places use local flora for treating and preventing ailments and are generally considered as healthcare resource in rural places difficult to get modern healthcare services (Frawley 2003). The eastern Himalayan region along with Western Ghats and Andaman and Nicobar Islands gives shelter to about 45,000 plant species, and out of this more than 6000 species are reported to be used by traditional healers. In rural India, 70% of the population depended on traditional system of medicine. Ayurvedic medicine, which is native to Indian Subcontinent, has got worldwide acceptance. A systematic approach through experimental evidence and clinical validation of efficacy is required for a plant identified for traditional medicine, as is done in modern medicine (Farnsworth 1998).

Ayurveda utilizes medicinal plants and their components including roots, leaves, fruits, bark, and seeds for the preparation of various medicinal formulations, and this practice started for about 3000 years (Sharma and Dash 1998). Charaka Samhita and Sushruta Samhita are the two classical authentic Sanskrit texts for Ayurveda. In Ayurvedic system, there are combinations of medicinal plants for the effective management of many diseases. Some of such well-known combinations are triphala, trikatu, trimad, trijat, ashtavarga, dasamula, and dasapushpam.

Triphala is one such combination, and as the name implies, triphala contains three fruits: Amalaki (*Emblica officinalis*), Bibhitaki (*Terminalia bellerica*), and Haritaki (*Terminalia chebula*). Triphala is an important rasayana drug having balancing rejuvenating effect on human body, and it is reported to have various medicinal properties such as antimicrobial, antiaging, antimutagenic, anticancer, anti-stress, hepatoprotective, and cleanser of colon (Sharma and Dash 1998).

Another combination ashtavarga is prepared from a group of eight medicinal plants, which are threatened medicinal herbs of Northwest Himalaya. *Malaxis acuminta*, *Microstylis muscifera*, *Polygonatum verticillatum*, *P. cirrhifolium*, *Roscoea procera*, *Lilium polyphyllum*, *Habenaria intermedia*, and *H. edgeworthii* are the eight medicinal plants used in ashtavarga. These herbs are considered to be energy boosters to restore health, to strengthen the immune system, and to have good regeneration capacity. They are used for healing fractures, for promoting body fat, for treating seminal weakness, fever, diabetic condition, and abnormal thirst (Balakrishna et al. 2012).

Dasapushpam is another important combination of medicinal herbs widely used in Ayurvedic system of medicine. Dasapushpam literally means "ten flowers" with many curative properties. They are therapeutically active against various diseases like fever, dysentery, hemorrhage, and constipation. Dasapushpam herbs are usually found in wild habitats of Kerala state (Mini et al. 2010). The ten plants include *Cardiospermum halicacabum*, *Vernonia cinerea*, *Biophytum sensitivum*, *Emilia sonchifolia*, *Cynodon dactylon*, *Curculigo orchioides*, *Evolvulus alsinoides*, *Eclipta alba*, *Aerva lanata*, and *Ipomoea obscura*.

Dasamula is a combination of ten medicinal plants in which roots are used in preparation of Ayurvedic formulations. Each of these plants shows incredible medicinal property. Dasamula is a major compound formulation of Ayurveda and has been supported with promising result in different ailments. Dasamula plants have become vulnerable in Karnataka and Andhra Pradesh and endangered in Kerala (Samatha et al. 2014). Roots of *Aegle marmelos*, *Desmodium gangeticum*, *Gmelina arborea*, *Oroxylum indicum*, *Premna integrifolia*, *Solanum anguivi*, *Solanum virginianum*, *Stereospermum colais*, *Uraria picta*, and *Tribulus terrestris* plants are used in Dasamula preparation.

18.7 Conservation Priorities

Demand for medicinal plants is increasing in both developed and developing countries. Traditional medicine is based on medicinal plants and medicinal plant-derived compounds. About 15,000 species of medicinal plants are globally threatened. The extinction of these plants results in serious problems for people's health and livelihoods. Sustainable management of medicinal plants is important not only because of their value as conservation strategies of the potential drug but also the reliance on medicinal plants for healthcare. Interest and support for the conservation and development of medicinal plants are increasing in all parts of the world (Cunningam 1993). We are yet to frame conservation strategies for even the endangered species. There is very little material of them in the Gene Banks. Stringent regulation is required on overexploitation and export to ensure their availability for the future generation.

Effective conservation strategies for medicinal plants are essential considering many factors including the presence of valuable bioactive compounds, which can be converted to potential pharmacologically important molecules (Okigbo et al. 2008). According to WHO estimate, about 80% of population in developing countries relies on traditional medicine (Bhandary and Chandrashekar 2002; Rao et al. 2003). Recently, a red book issued by IUCN noted that 91% of all species are threatened with extinction principally due to environmental degradation. To control the situation, various measures like commercial cultivation, habitat conservation, setting up of natural resources, implementation of laws for restricting export of plants, etc. have to be taken.

It is a challenging task for conserving our medicinal plant resources as they occur in wide range of habitats, but conservation of these important groups of plants can be achieved through an integrated approach with societal participation where we will have to depend on both *ex situ* and *in situ* conservation strategies. However, *in situ* conservation is more advantageous because it allows natural selection, in which the risk of organizing conservation steps in a wild habitat lead us to put more efforts in *ex situ* strategies. Medicinal plant resources can be conserved through methods like seed banks, in vitro storage methods, pollen banks, and DNA banks as well as the maintenance of living collections. This is particularly important for threatened medicinal plants where protection in the wild may not be realistic in the short term or where ex situ work cannot be directly linked to *in situ*. The resources provided by these collections are important for *in situ* recovery programs (Hawkins 2008). Consistent information on the threat status of medicinal plants at a global level is not yet available, and so there is a need for prioritization of plant species which needs to be conserved. This can be done on the basis of information collected on value of a particular species in terms of healthcare, livelihood support, and their role in the ecosystem. India leads in studies on conservation priorities of medicinal plants. Conservation strategies must be on a local-to-local basis as the ecosystems are also diverse. Medicinal plants can be symbolically very important to people. In a country like India where we people consider most of the medicinal plants with special religious, nationalistic, or ideological esteem, these plants are getting better attention for its preservations. The best example is the biodiversity-rich sacred grooves and ponds in the country. Because so many species of plants are medicinal, medicinal plant conservation in some ways means plant conservation as a whole.

Studies have also been conducted by NGOs and research organizations to evaluate the status of sacred groves. The CPR Environmental Education Centre, Chennai, is one of such an autonomous centers of excellence of the Ministry of Environment and Forests. They have published books on the Sacred Trees of Tamil Nadu and Ecological Traditions of Tamil Nadu. A UNESCO study on the sacred groves of India provides a comprehensive picture of these groves along with their status. Some of the large groves and community associated with them have been presented as case studies. However, a systematic nationwide survey of sacred groves has not been undertaken to account for the status of these groves in terms of the conservation of biological diversity and the traditional particular of the community. The most important of all is the legal status and ownership of the grove. The changing pattern of lifestyles and religious beliefs are also responsible for the deteriorating conditions of some of the groves (Government of India, Planning Commission Report 2000).

18.8 Implications in Modern Medicine

Plants have long been used as a source of therapeutic agents. Medicinal plants play an important role in modern medicine. Many of the newly discovered drugs are derived from the medicinal plant sources. The use of plants in medicine ranges from crude preparations or extracts to refined extracts and single molecular species. There are also several plant extracts or phytomedicine in clinical trial for the treatment of various diseases. Human civilization was depending on medicinal plants thousands of years ago. Initially, these medicines are used in the form of crude drugs like tinctures, tea, poultices, powders, and other herbal formulations (Samuelsson 2004). The use of medicinal plants as medicine has evolved into the isolation of compounds, beginning with the isolation of morphine from opium in the early nineteenth century (Samuelsson 2004). Drug discovery from medicinal plants led to the isolation of early drugs like cocaine, codeine, digitoxin, and quinine (Newman et al. 2000). Plants are the most perfect natural sources for drug development. Present drug discovery from medicinal plants has mainly relied on biological activity-guided isolation methods (Fabricant and Farnsworth 2001; Lahlou 2012). Isolation and characterization of pharmacologically active compounds from medicinal plants continue till today.

The correlation between the ethnomedical usage of medicinal plants and modern medicines discovered from those plants has been studied by Fabricant and Farnsworth (2001). Based on their analysis, 88 single chemical entities isolated from 72 medicinal plants have been introduced into modern therapy (Fabricant and Farnsworth 2001). Medicinal plants are important for pharmacological research and drug development. Plant constituents are directly used as therapeutic agents. They are also used as starting material for the synthesis of drugs or as model for pharmacologically active compounds. Some of these plant-derived compounds like atropine (anticholinergic), codeine (cough suppressant), colchicine (antigout), ephedrine (bronchodilator), morphine (analgesic), pilocarpine (parasympathomimetic), and physostigmine (cholinesterase inhibitor) are widely used in prescription drugs. Drugs discovered from medicinal plants serve as a solution for various pharmacological targets including cancer, HIV, Alzheimer's, malaria, and pain. Numerous compounds from tropical rainforest plant species with potential anticancer activity have been identified (Marcy and Douglas 2005).

Drug discovery from medicinal plants played an important role in the treatment of cancer. Most new clinical applications of plant secondary metabolites and their derivatives are targeted for the treatment of cancer. Anticancer agents from plants that are currently in use can be classified into four main classes of compounds – vinca alkaloid, epipodophyllotoxins, taxanes, and camptothecins – and have been used clinically for over 40 years (van Der Heijden et al. 2004).

Plants play a major role in the introduction of new therapeutic agents and have received much attention as sources of biologically active substances. The global demand for herbal medicine is not only large but growing. Plants have contributed hugely to Western medicine, through providing ingredients for drugs or having played central roles in drug discovery. Some drugs, having botanical origins, are still extracted directly from plants, and others are made through transformation of chemicals found within them, while yet others are today synthesized from inorganic materials but have their historical origins in research into the active compounds found in plants. There are undoubtedly many more secrets still hidden in the world of plants.

Plant materials provide various secondary metabolites, which facilitate certain therapeutic applications. There are many plants having some compounds common among them. Also, there may be many compounds in a single plant. Such a scenario would help substitutions in case certain plant materials turn to be unavailable. It would promote both intensive and extensive search for authenticity and novel applications or applicators (Preethidan et al. 2013). This would make such searches exciting to help provide valuable leads for other systems of medicine. This is the fact that, even now, the secondary metabolites of plants are highly significant for all systems of medicine (Aparna et al. 2012). Herbal decoctions are used as drugs for millennia and described in classical texts of Ayurveda (Valiathan 2007). It is found used in almost all native and folk practices. However, their use in fermented form is more prevalent in Ayurvedic practice. In most of the cases, the use of the plant decoctions or the fermented decoctions is decided by the doctor (Valiathan 2007). But it is based on the condition of the patient, since the fermentation of the plant parts or the fermentation of the plant part decoctions would make the compounds present in the decoctions absorbed better and faster (Chandra et al. 2011). And, in most cases, the fermentative protocol would make decoctions forming easily absorbable and higher active compounds in large quantity (Chandra et al. 2012). Such changes due to fermentation are more with the fermentation of secondary metabolites of plant, and they become the drugs, whereas the primary metabolites provide energy for biomass production to perform fermentation.

Berberine is a major isoquinoline alkaloid in herbs such as goldenseal, berberis, and *Coptis chinensis* and has been used for about 1000 years to treat gastrointestinal diarrhea (Lahiri and Dutta 1967; Amin et al. 1969). Presently, berberine is widely used in oriental countries for diabetes, hypercholesterolemia, and other medical conditions (Vuddanda et al. 2010). Goldenseal, a main source of berberine (Abourashed and Khan 2001), is ranked as the sixth most commonly used herbal supplement for children in America (Barnes et al. 2008; Guo et al. 2011). Biotransformation of berberine by fermentation leads to its partial or complete demethylation, producing hydroxyl derivatives, as revealed by Fourier transform infrared spectroscopy and nuclear magnetic resonance and electrospray ionization mass spectrometric analyses. Surface plasmon resonance and enzyme kinetic studies showed that biotransformed derivatives of berberine had a higher inhibitory potential than berberine toward phospholipase A_2. X-ray crystal structures demonstrated that biotransformed berberine binds to PLA_2 in an entirely different, inverted orientation with respect to the binding of berberine (Chandra et al. 2015).

18.9 Allopathic and Ayurvedic Approaches to Leukoderma

Leukoderma, also known as vitiligo, is a skin disorder that affects nearly 2% of the world population (Zhang et al. 2004). Allopathy, the system of medicine in widespread practice today, offers palliative measures, but no cure. By contrast, the Charaka Samhita, a 2000-year-old Indian Ayurvedic text (Pillai 1979), provides a description of leukoderma with recommendations for treatment. In consideration of no known medical cure, allopathic treatment of leukoderma offers two approaches to the management of this disorder: protection of depigmented patches and repigmentation therapies. Due to the absence of melanocytes in depigmented skin, the body is unable to provide adequate UV light shielding from the harmful rays of the sun; thus protection of depigmented patches is of paramount importance. To minimize exposure to UV-concentrated light, protective measures are encouraged, such as the application of sunscreen 15–30 SPF or the use of protective clothing. In addition to protecting leukodermic patches, many cosmetic products are available that act to camouflage the affected skin. The second approach to the management of leukoderma focuses on repigmentation therapies. Common therapies are of three types: the topical application of potent corticosteroids, psoralen photochemotherapy (PUVA), and surgical therapies. Of these interventions PUVA is considered the most effective treatment available in the United States; however, the treatments are time-consuming, side effects can be severe, and complete repigmentation only occurs in 15–20%. Interestingly, psoralens used with photochemotherapy are obtained from the ancient Ayurvedic herb known as *Psoralea corylifolia* (locally called as Vakuchi) (Pathak and Fitzpatrick 1992).

Another herb, *Vernonia anthelmintica* (also called as Purple fleabane), is also reported to be an effective remedy for vitiligo. The powdered seeds are taken with black pepper or black sesame seeds in equal parts, daily in the morning, with warm water, just after perspiring. It is indicated that if the above methods is followed for one year, resolution of vitiligo will occur (Nadkarni 2002). In Western medicine, reduction of symptoms alone is all too often the result. However, leukoderma stubbornly refuses to be cured by allopathic methods. On the other hand, many cases of leukoderma have been resolved with Ayurveda. The key to successfully treating disease is a comprehensive approach. Ayurveda offers this (http://www.ayurved-acollege.com/articles/ students/leucoderma).

18.10 Status of Medicinal Plants in India

A report published by Food and Agriculture Organization of the United Nations (FAO 2002) on the impact of cultivation and gathering of medicinal plants on biodiversity states that species most susceptible to overharvest are habitat specific, slow growing, and destructively harvested for their bark, roots, or the whole plant. These species suffer most from harvesting, and many of them have been seriously depleted, for example, *Prunus africana* in West Africa, *Warburgia salutaris* in southern Africa, and *Saussurea costus* in the Himalayas. For threatened medicinal plant species, cultivation is a conservation option because the constant drain of material from their populations is much higher than the annual sustained yield. If the demand for these species can be met from cultivated sources, the pressure on the wild populations will be relieved. In these cases, the need for strict conservation of remaining populations, improved security of germplasm ex situ, and investment in selection and improvement programs is extremely urgent as the example of Jaborandi (*Pilocarpus jaborandi*) in Brazil shows (Pinheiro 1997).

18.11 Importance of Insect Pest Management in Conservation

One of the major threats to medicinal plants, which are under cultivation, is the infestation by insects. These infestations, in most of the cases, reduce either the medicinal quality of the plant or the production of the plant compound that is responsible for that property. Sathe et al. (2014) reported that sucking insect pests like lacewing *Cochlochila bullita* (Stal.), whiteflies *Aleurodicus dispersus* Russell, *Dialeurodes* sp., and aphids *Macrosiphum* sp. suck the cell sap, secrete honeydew-like substance, create sooty molds, and affect photosynthesis and yield of Tulsi *Ocimum sanctum* L. (Lamiaceae), which in turn affects their medicinal properties. Eldredge and Baird (1995) recorded for the first time the orange mint moth, *Pyrausta orphisalis*, in a peppermint (*Mentha piperita*) field in Idaho. Mareggiani and Bachur (1998) noted the occurrence of *Kaltenbachiella pallida* on mint (*Mentha* spp.) in Argentina. Chopra (1933) reported two pests, viz., *Agrotis flammatra* Schf. and larvae of tenebrionid, *Gonocephalum* sp. on *Atropa acuminata* Royle ex Lindley. While surveying the insect pests of medicinal plants at Jammu and Kashmir, Mathur and Srivastava (1964) recorded *Epilachna vigintioctopunctata* Fab. feeding on *Datura stramonium* Linn., *D. metel* Linn., *D. inoxia* Mill., *Solanum aviculare* Forst., *S. surrathensis* Burn, *S. nigrum* Linn., and *Withania somnifera* Dunal. Ali (1980) reported Bombay locust, *Patanga succincta* Linn., damaging *Datura stramonium* Linn., *Ricinus communis* Linn., *Achyranthes aspera* Linn., and *Abutilon indicum* (Linn.). Some of the insect pests recorded on medicinal plants are given in Table 18.2. Adoption of

Table 18.2 Insect pests recorded during survey at different locations in Himachal Pradesh

S. no.	Host plant	Insects recorded	Intensity[a]	Plant part damaged	Locations
1.	*Withania somnifera* (Dunal.) (Ashwagandha)	*Henosepilachna vigintioctopunctata* (Fabr.)	Medium	Leaves	Baldook (Hamirpur)
		Nezara viridula (Linn.)	Low	Leaves and inflorescence	
		Dysdercus cingulatus (Fabr.)	Low	Leaves and inflorescence	
		Pentatomid bug (unidentified)	Low[b]	Developing seeds	
		Pentatomid bug (unidentified)	Medium	Leaves and inflorescence	Shilly, Solan
		Heliothis armigera Hübner	Low	Leaves and seeds	Berthin (Bilaspur)
		Drosicha mangiferae (Green)	Medium	Apical twig	Jogindernagar
2.	*Arctium lappa* (C.B. Clarke) (Kuth)	*Thysanoplusia orichalcea* (Fabr.)	Low	Leaves	Panthaghati, Shimla
3.	*Angelica glauca* Edgew (Chura)	*Thysanoplusia orichalcea* (Fabr.)	Low	Leaves	
4.	*Aegle marmelos* (Correa Ex. Roxb.) Bael.	*Papilio* sp.	Medium	Leaves	
5.	*Digitalis lanata* (Ehrh.) (Foxglove)	Scale insects	Medium	Leaves	
6.	*Sauserra costus* (C.B. Clarke) (Kuth)	*Thysanoplusia orichalcea* (Fabr.)	Low	Leaves	Panthaghati, Shimla
		Alcidodes crinalifer (Marshall)	Low	Leaves	Shilly, Solan
		Condica conducta (Walker)	Low	Leaves	
		C. albigutta (Wileman)	Low	Leaves	
		T. orichalcea	Low	Leaves	Kullu
7.	*Picrorhiza kurooa* (Karu)	*Thysanoplusia orichalcea* (Fabr.)	Low	Leaves	Shilly, Solan
8.	*Rauwolfia serpentina* (Sarpagandha)	*Drosicha mangiferae* (Green) mealybugs (unidentified)	Medium	Apical twig	Jogindernagar
9.	*Celastrus paniculatus* (Wild) (Peng Black Oil Tree)	Corid bugs (unidentified)	Medium	Leaves and developing seeds	
10.	*Bacopa monnieri* (Linn) (Brahmi)	*Spodoptera litura* (Fabr.)	Low	Leaves	

Source: Sharma et al. (2014)

[a]Low, up to 10% leaf infestation; medium, 11–30% leaf infestation

[b]Based on visual observations

various control strategies has been followed for the improvement of the productivity and quality of medicinal plants (Verma 2006). But before adoption of any pest control strategies, assessment of population and estimation of the damage caused are necessary, in order to determine their pest status.

18.12 Conclusions

Increase in demand for medicinal products based on natural source is the key to reports on endangered plant species. Overexploitation and refusal of conservation of plant species have led to extinction of most of the medicinally important species. At this juncture, increasing abundance of insect pests could also damage the medicinal plants that are cultivated today. Focus on cultivation leading to conservation as well as insect pest management and collecting data from the local people may help us in maintaining these precious resources for our upcoming generations. Hence, monitoring and further *in situ* conservation will be needed to save these plant species from further deterioration.

References

Abasta SP (1986) The useful plants of India. Council of scientific and industrial research, New Delhi, pp 1–918

Abourashed EA, Khan IA (2001) High performance liquid chromatography determination of hydrastine and berberine in dietary supplements containing goldenseal. J Pharm Sci 90:817–822

Ali S (1980) Life history of the Bombay locust *Patanga succincta* L. (Orthoptera: Acrididae). Bangladesh J Zool 81:49–52

Aparna V, Dileep KV, Mandal PK, Karthe P, Sadasivan C, Haridas M (2012) Anti-inflammatory property of n-hexadecanoic acid: structural evidence and Kinetic assessment. Chem Biol Drug Des 80:434–439

Amin AH, Subbaiah TV, Abbasi KM (1969) Berberine sulfate: antimicrobial activity, bioassay and mode of action. Can J Microbiol 15:1067–1076

Aryal M (1993) Diverted wealth: the trade in Himalayan herbs. Himal 6(1):52–58

Ayyanar M, Ignacimuthu S (2005) Medicinal plants used by the tribals of Tirunelveli hills, Tamil Nadu to treat poisonous bites and skin diseases. Indian J Tradit Knowl 4(3):229–236

Balakrishna A, Srivasthava A, Mishra RK, Patel SP, Vasishta RK, Sing A (2012) Ashtavarga plants – threatened Medicinal herbs of North West Himalaya. Int J Med Aromat Plants 2(4):661–675

Barnes PM, Bloom B, Nahin RL (2008) Complementary and alternative medicine are among adults and children: United States. Natl Health Stat Rep 10(12):1–23

Bhandary MJ, Chandrashekar KR (2002) Glimpses of ethnic herbal medicine of coastal Karnataka. Ethnobotany 14:1–12

Chandra DN, Abhilash J, Prasanth GK, Sabu A, Sadasivan C, Haridas M (2012) Inverted binding due to a minor structural change in berberine enhances its phospholipase A(2) inhibitory effect. Int J Biol Macromol 50:578–585

Chandra DN, Prasanth GK, Singh N, Kumar S, Jithesh O, Sadasivan C, Sharma S, Singh TP, Haridas M (2011) Identification of a novel and potent inhibitor of phospholipase A2 in a medicinal plant: crystal structure at 1.93 Å and surface plasmon resonance analysis of phospholipase A2 complexed with berberine. Biochim Biophys Acta 1814:657–663

Chandra DN, Preethidan DS, Sabu A, Haridas M (2015) Traditional fermentation of Ayurvedic medicine yields higher pro-inflammatory enzyme inhibition compared to wine-model product. Front Life Sci. doi:10.1080/21553769.2015.1005245
Chantia A (2003) Traditional knowledge of ethno medicine in Jaunsarbawar, Dehradun district. Indian J Tradit Knowl 2(4):397–399
Chopra RN (1933) Indigenous drugs of India. The Art Press, Calcutta, p 655
Cunningam AB (1993) African medicinal plants – setting priorities at the interface between conservation and primary health care, Working paper 1, WWF/UNESCO/Kew people and plants initiative. UNESCO, Paris
Dey A, De JN (2012) Traditional use of plants against snakebite in Indian subcontinent: a review of the recent literature. Afr J Tradit Complement Altern Med 9(1):153–174
Dhar U, Rawal RS, Upreti J (2000) Setting Prioritization for conservation of medicinal plants: a case study in the Indian Himalaya. Biol Conserv 95:57–65
Eldredge EP, Baird CR (1995) First record of orange mint moth (Lepidoptera: Pyralidae) on commercial peppermint in Idaho. J Entomol Soc Br Columbia 92:107–109
Electronic textbook of dermatology—botanical dermatology, Phytophoto-dermatitis, www.telemedicine.org/botanica/bot5.htm. 01/06: p 1
Fabricant DS, Farnsworth NR (2001) The value of plants used in traditional medicine for drug discovery. Environ Health Perspect 109:69–75
FAO (2002) Impact of cultivation and gathering of medicinal plants on biodiversity: global trends and issues by Uwe Schippmann, Danna J. Leaman and A. B. Cunningham. Published in Biodiversity and the Ecosystem Approach in Agriculture, Forestry and Fisheries, Rome
Farnsworth NR (1998) Screening plants for new medicines. In: Wilson EO (ed) Biodiversity. National Academy Press, Washington, DC, pp 83–97
Fisseha M (2007) An ethnobotanical study of medicinal plants in Wonago Woreda, SNNPR, ethiopia. M.Sc thesis, Addis Ababa University, Ethiopia
Frawley D (2003) Ayurvedic healing, a comprehensive guide, 2nd edn. Motilal Banarsidass, New Delhi, p 124
FRLHT (1995) Securing involvement of local communities for conservation of In situ. Medicinal Plants Conservation Areas (MPCA's): a case study in unpublished UNESCO project Proposal
Guo Y, Li F, Ma X, Cheng X, Zhou H, Klaassen CD (2011) CYP2D plays a major role in berberine metabolism in liver of mice and humans. Xenobiotica 41(11):996–1005
Hawkins B (2008) Plants for life: medicinal plant conservation and botanic gardens. Botanic Gardens Conservation International, Richmond
Ilahi I, Rahim F, Jabeen M (2007) Enhanced clonal propagation and alkaloid biosynthesis in cultures of Rauwolfia. Pak J Plant Sci 13(1):45–56
IUCN- https://www.iucn.org/what/biodiversity/
IUCN (1980) World conservation strategy. IUCN, Gland
Joy PP, Thomas J, Mathew S, Skaria BP (1998) Medicinal plants. Aromatic and Medicinal Plants Research Station, Kerala Agriculture University, Ernakulam, pp 193–194
Joy PP, Thomas J, Mathew S, Skaria BP (2001) Medicinal plants. In: Bose TK, Kabir J, Das P, Joy PP (eds) Tropical horticulture, vol 2. Naya Prokash, Calcutta, pp 449–632
Kala CP (2009) Aboriginal uses and management of ethnobotanical species in deciduous forests of Chhattisgarh state in India. J Ethnobiol Ethnomed 5:20
Kalita JC, Chakrabarty A, Tanti B (2011) Assessment of antifertility activity of some traditionally used plants by different ethnic communities in three district of Assam, India. J Herb Med Toxicol 5(2):65–67
Lahiri SC, Dutta NK (1967) Berberine and chloramphenicol in the treatment of cholera and severe diarrhoea. J Indian Med Assoc 48:1
Lahlou M (2012) De la Plante au Médicament: Une Passerelle entre Tradition et Science. Édition LAWNE, Maroc. ISBN 978–9954–31-103-5, p 272
Lambert JJ, Srivastava J, Vietmeyer N (1996) Medicinal plants: rescuing a traditional heritage. Agriculture and Natural Resources department, World Bank, Washington, DC

Marcy JB, Douglas KA (2005) Drug discovery from medicinal plants. Life Sci 78:431–441
Mareggiani G, Bachur C (1998) The mint woolly aphid *Kaltenbachiella pallida* Pemphiginae: Eriosomatini in Argentina. Revista de la Facultad de Agronomia Universidad de Buenos Aires 18(1/2):133–134
Mathur AC, Srivastava JB (1964) *Epilachna vigintioctopunctata* as a defoliator of some solanaceous medicinal plants. Indian J Entomol 26:246
Mini VN, Ida B, Seema D, Shital D, Riva DS, Astrida R (2010) Antimicrobial activity of ten common herbs, commonly known as 'Dashapushpam' from Kerala. Afr J Microbiol Res 4(22):2357–2362
Muth C, Ayyanar M, Raja N, Ignacimuthu S (2006) Medicinal plants used by traditional healers in kancheepuram district of Tamilnadu, India. J Ethnobiol Ethnomed 2:43
Nadkarni AK (2002) Indian Materia Medica. Popular Prakashan LTD, Reprinted, vol 1. p 1269
Namboodiri AN (1994) Scope of plant conservation and strategy for conserving threatened plants. In: Environmental problems and prospects. Himalaya Pub. Co, New Delhi
Newman DJ, Cragg GM, Snader KM (2000) The influence of natural products upon drug discovery. Nat Prod Rep 17(3):215–234
Okigbo RN, Eme UE, Ogbogu S (2008) Biodiversity and conservation of medicinal and aromatic plants in Africa. Biotechnol Mol Biol Rev 3(6):127–134
Pandey CN, Raval BR, Mali S, Harshad S (2005) Medicinal plants of Gujarat. Gujarat Ecology Education and Research Foundation, Gandhinagar, p 410
Pathak MA, Fitzpatrick TB (1992) The evolution of photochemotherapy with psoralens and UVA (PUVA): 2000 BC to 1992 AD. J Photochem Photobiol B:Biol 14:3–22
Patil MV, Patil DA (2005) Ethnomedicinal practices of Nasik district, Maharashtra. Indian J Tradit Knowl 4(3):287–290
Pillai K (1979) History of Siddha medicine, 1st edn. Government of Tamil Nadu, Manorama Press, Chennai, p 192
Pinheiro CUB (1997) Jaborandi (*Pilocarpus* sp., Rutaceae) A wild species and its rapid transformation into a crop. Econ Bot 51:49–58
Prakash V (2002) In: Samant SS, Dhar U, Palni LMS (eds) Himalayan medicinal plants potential and prospect. Gyanodaya Prakashan, Nainital
Preethidan DS, Arun G, Prathyusha S, Prasanth S, Sabu A, Sadasivan C, Haridas M (2013) Lipoxygenase inhibitory activity of some *Sida* species due to di (2-ethylhexyl) phthalate. Curr Sci 105:232–237
Ramesh BR, Pascal JP (1997) Atlas of endemics of Western Ghats (India) distribution of tree species in the ever-green and semi- evergreen forest. French Institute of Pondicherry, Puducherry
Rao CK, Geetha BL, Suresh G (2003) Red list of threatened vascular plant species in India. Director, Botanical Survey of India, ENVIS Centre for Floral Diversity, Botanical Survey of India, Howrah
Report of the Task Force on Conservation & Sustainable use of Medicinal Plants. Government of India, Planning Commission March – 2000
Samatha T, Akula S, Kuntamalla S, Nanna RS (2014) Effect of stem bark extracts of Oroxylum Indicum; an ethnomedicinal forest tree on silk production of Bombyx Mori. Int J Pharm Sci Res 5(2):568–571
Samuelsson G (2004) Drugs of natural origin: a textbook of pharmacognosy, 5th edn. Swedish Pharmaceutical Press, Stockholm
Sathe TV, Sathe NT, Ghodake D, Sathe A (2014) Sucking insect pests and medicinal value of Tulsi *Ocimum sanctum* L. (Lamiaceae). Indian J Appl Res 4(3):31–33
Sharma RK, Dash B (1998) Charaka Samhita, vol II. Chowkhamba Sanskrit Series Office, Varanasi
Sharma JR, Mudgal V, Hajra PK (1997) Floristic Diversity and Conservation Strategies in India. Botanical Survey of India, Ministry of Environment and Forests, Kolkata, p 5
Sharma PC, Kumar A, Mehta PK, Singh R (2014) Survey studies on insect-pests associated with important medicinal plants in Himachal Pradesh. Indian J Sci Res Technol 2(4):2–7
Singh PB (1993) Medicinal plants of ayurvedic importance from Mandi district of Himachal Pradesh. Bull Medico-Ethnobot Res 14(3–4):126–136

Singh S (2005) Medicinal plants? A natural gift to mankind. Agric Today 3(3):58–60
Singh DK, Hajara PK (1996) In: Gujral GS, Sharma V (eds) Changing perspectives of biodiversity status in the Himalaya. British Council Division, New Delhi, pp 23–38
Suresh J, Hemalatha P, Selvi R (2006) Biodiversity and conservation of medicinal plants. In: John William S (ed) Biodiversity: life to our mother earth. Loyola College, Chennai, pp 231–256
Valiathan MS (2007) Legacy of Susruta. Orient Long Man, Chennai
Van Der Heijden R, Jacobs DI, Snoeijer W, Hallard D, Verpoorte R (2004) The Catharanthus alkaloids: pharmacognosy and biotechnology. Curr Med Chem 11(5):607–628
Venkatesh K, Hari R, Sushrutha CK (2010) Flora of concern endangered medicinal plants – part I *Vateria indica* Linn, Dipterocarpaceae – a review. Int J Res Ayurveda Pharm 1(1):1–7
Verma JS (2006) Insect pest problem in medicinal plants- a review. Agric Rev 27:130–136
Vuddanda PR, Chakraborty S, Singh S (2010) Berberine: a potential phytochemical with multispectrum therapeutic activities. Expert Opin Investig Drugs 19:1297–1307
World Health Organization (WHO) (2014) Traditional medicine. http://www.who.int/topics/traditional_medicine/en/
Zhang XJ, Liu JB, Gui JP, Li M, Xiong QG, Wu HB, Li JX, Yang S, Wang HY, Gao M, Yang J, Yang Q (2004) Characteristics of genetic epidemiology and genetic model for vitiligo. J Am Acad Dermatol 51(3):383–390

MIX
Papier aus verantwortungsvollen Quellen
Paper from responsible sources
FSC® C105338

If you have any concerns about our products,
you can contact us on
ProductSafety@springernature.com

In case Publisher is established outside the EU,
the EU authorized representative is:
Springer Nature Customer Service Center GmbH
Europaplatz 3, 69115 Heidelberg, Germany

Printed by Libri Plureos GmbH
in Hamburg, Germany